Illustrated Notebook

PRINCIPLES OF ANATOMY AND PHYSIOLOGY

Ninth Edition

Gerard Tortora
Bergen Community College

Sandra Reynolds Grabowski
Purdue University

John Wiley & Sons, Inc.
New York Chichester Weinheim Brisbane Singapore Toronto

To order books or for customer service call 1-800-CALL-WILEY (225-5945).

To order books, please call 1 (800) 225-5945.

ISBN 0-471-37468-7

Printed in the United States of America

10 9 8 7 6 5 4 3 2 1

Printed and bound by Courier Westford, Inc.

NOTES

Figure 1.1 Levels of structural organization in the human body (page 3).

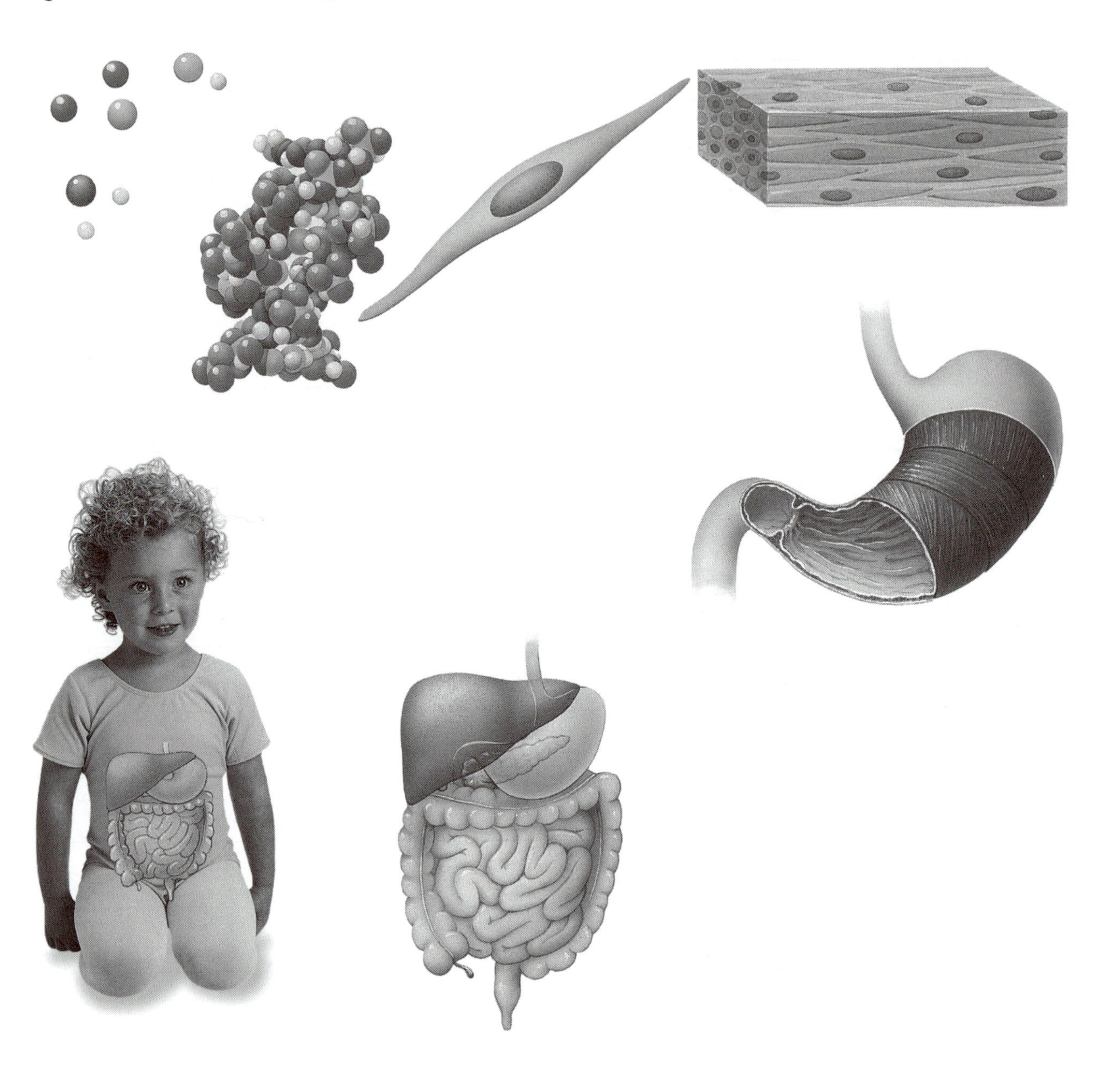

NOTES

Figure 1.2 Operation of a feedback system (page 8).

Figure 1.3 Homeostatic regulation of blood pressure by a negative feedback system (page 8).

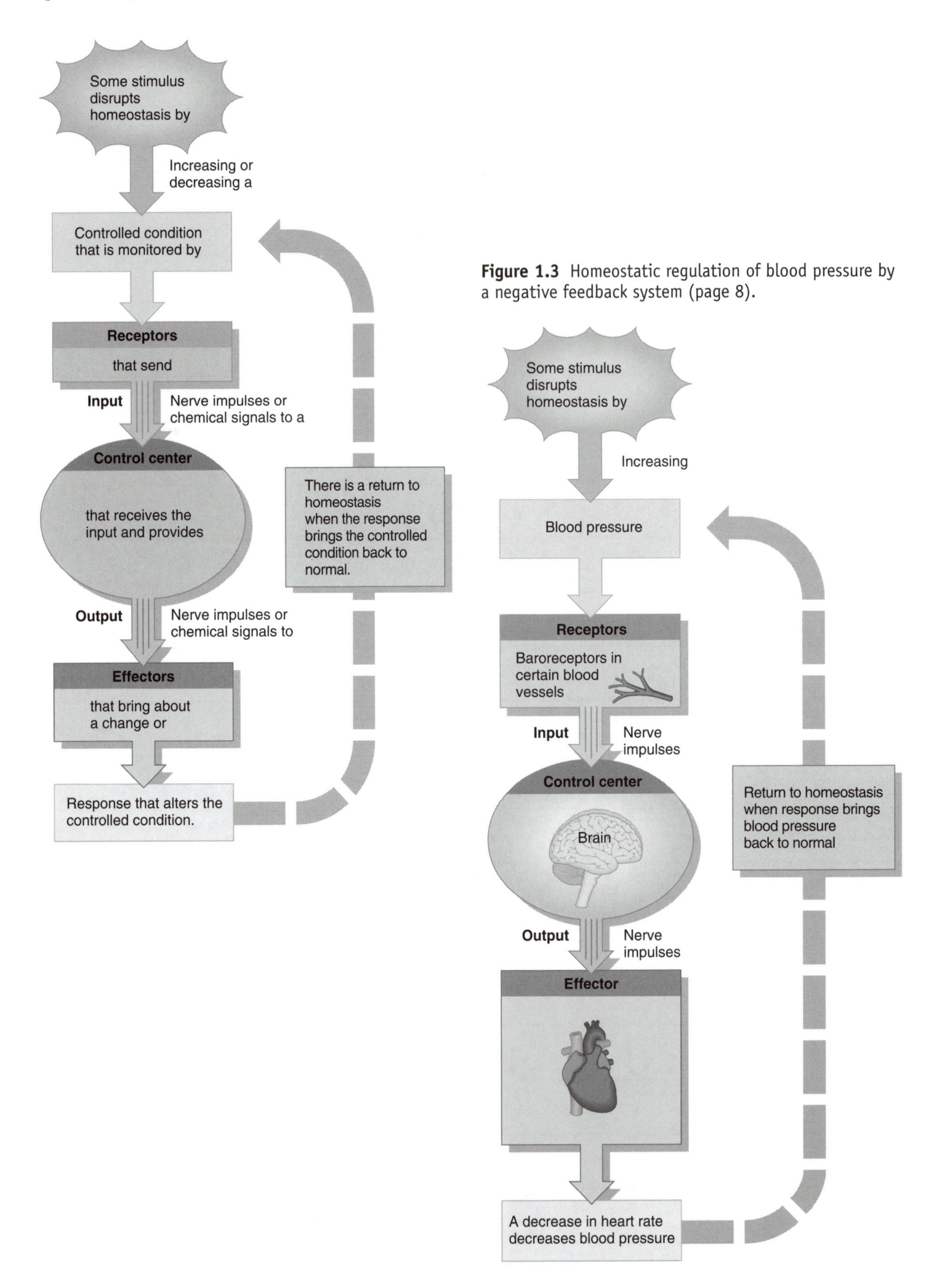

Some stimulus disrupts homeostasis by
Increasing or decreasing a
Controlled condition that is monitored by
Receptors
that send
Input
Nerve impulses or chemical signals to a
Control center
that receives the input and provides
Output
Nerve impulses or chemical signals to
Effectors
that bring about a change or
Response that alters the controlled condition.
There is a return to homeostasis when the response brings the controlled condition back to normal.
Some stimulus disrupts homeostasis by
Increasing
Blood pressure
Receptors
Baroreceptors in certain blood vessels
Input
Nerve impulses
Control center
Brain
Output
Nerve impulses
Effector
A decrease in heart rate decreases blood pressure
Return to homeostasis when response brings blood pressure back to normal

NOTES

Figure 1.4 Positive feedback control of labor contractions during birth of a baby (page 9).

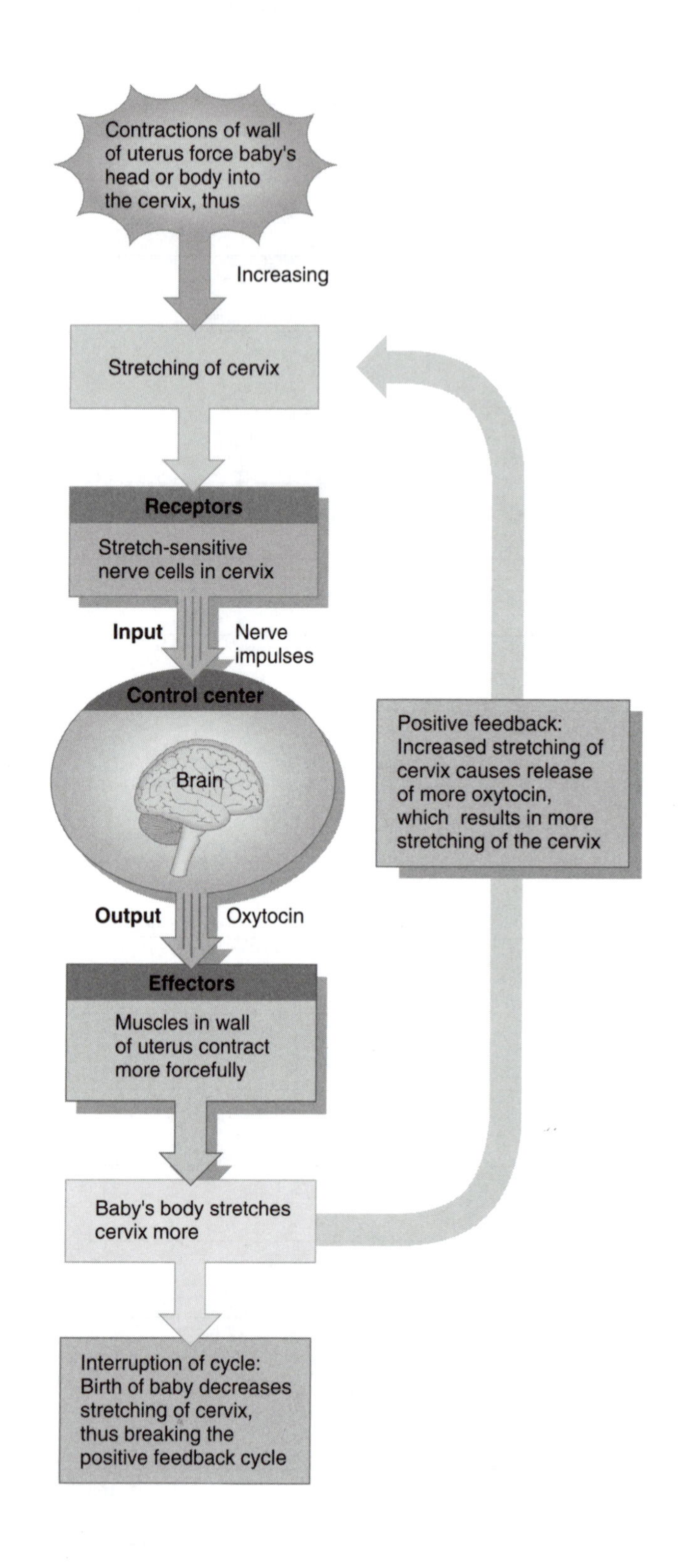

Contractions of wall of uterus force baby's head or body into the cervix, thus
Increasing
Stretching of cervix
Receptors
Stretch-sensitive nerve cells in cervix
Input
Nerve impulses
Control center
Brain
Output
Oxytocin
Effectors
Muscles in wall of uterus contract more forcefully
Baby's body stretches cervix more
Interruption of cycle: Birth of baby decreases stretching of cervix, thus breaking the positive feedback cycle
Positive feedback: Increased stretching of cervix causes release of more oxytocin, which results in more stretching of the cervix

NOTES

Figure 1.5 The anatomical position (page 11).

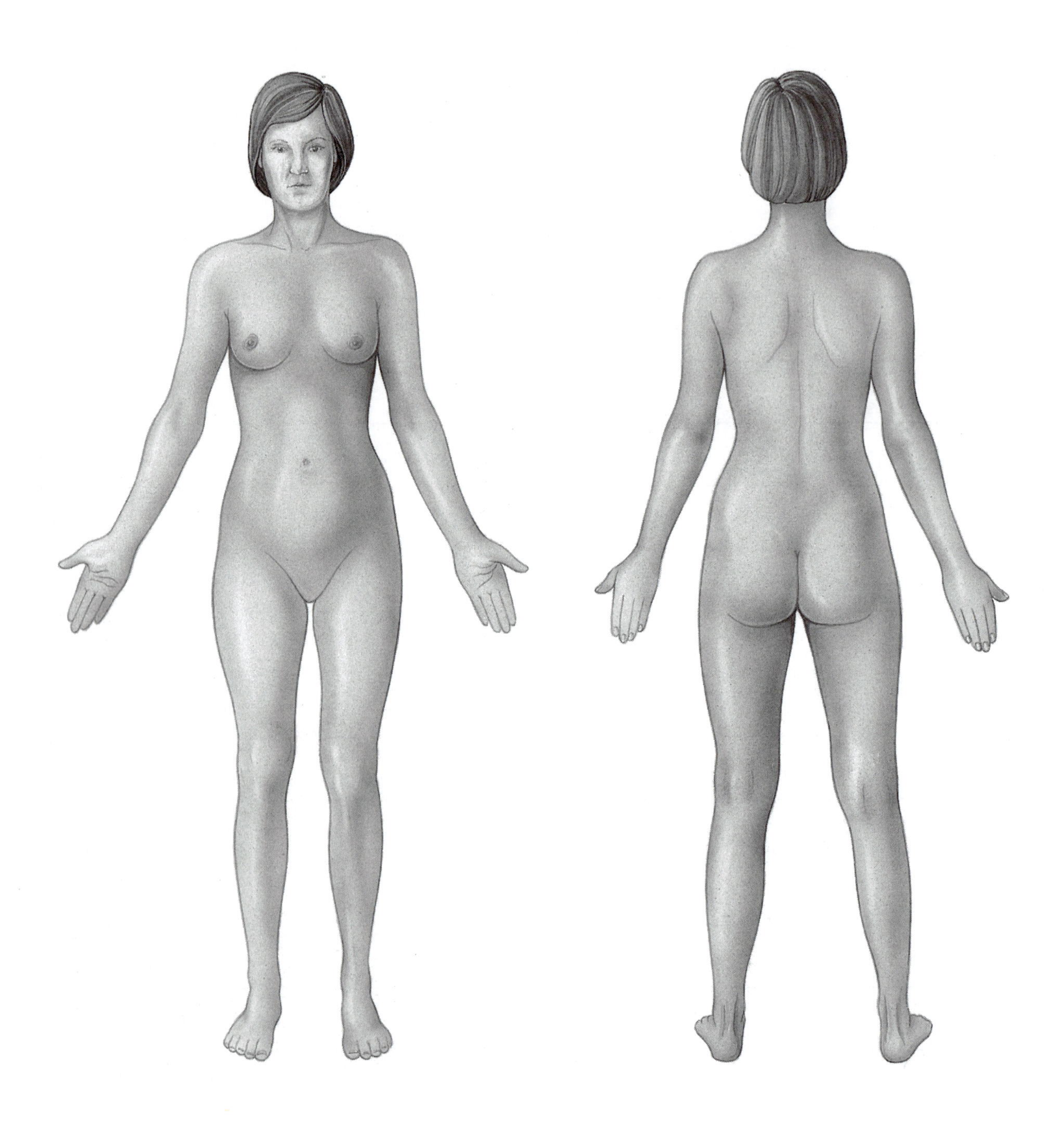

1

NOTES

Figure 1.6 Planes of the human body (page 12).

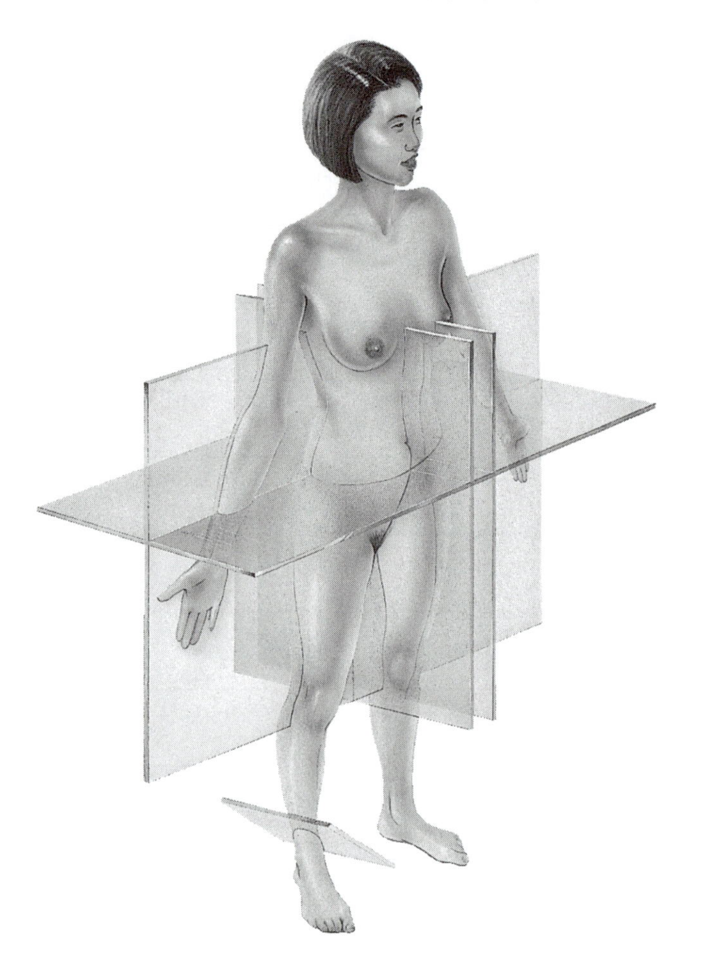

Figure 1.8 Planes and sections through different parts of the brain (page 13).

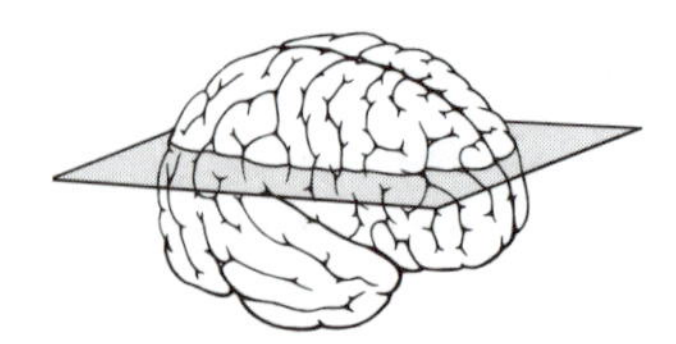

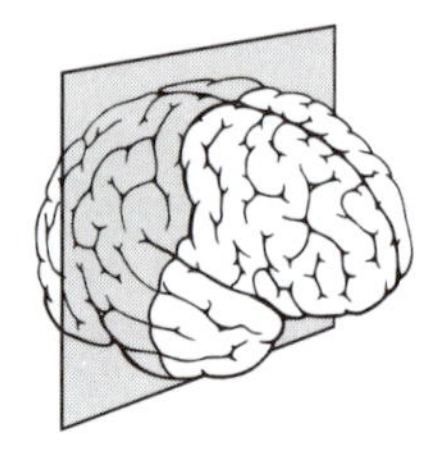

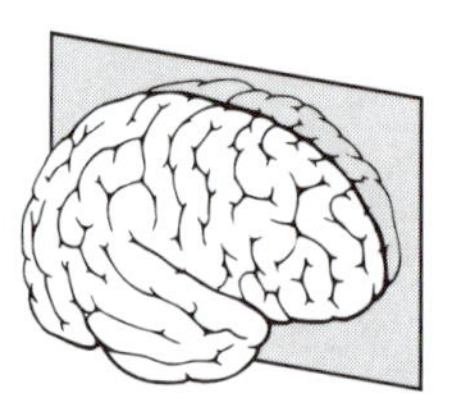

1

NOTES

Figure 1.8 Body cavities (page 13).

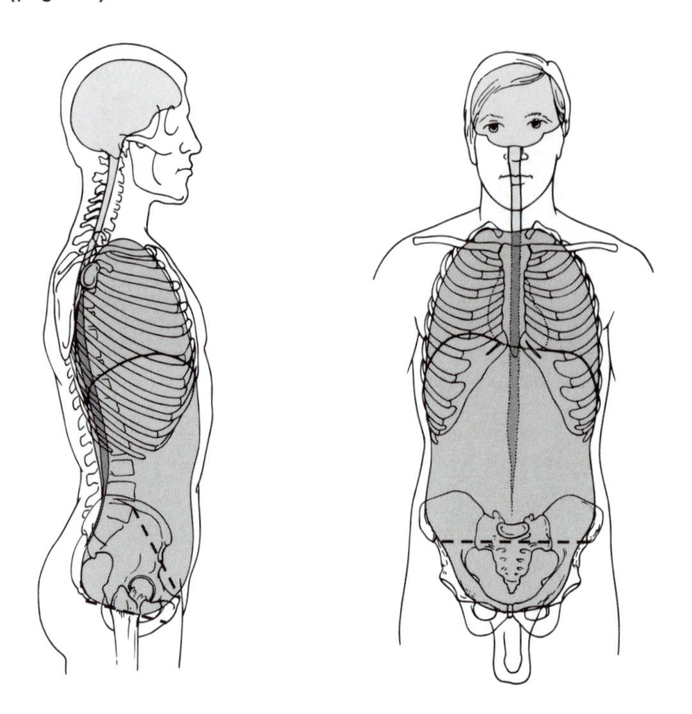

Figure 1.9 Directional terms (page 15).

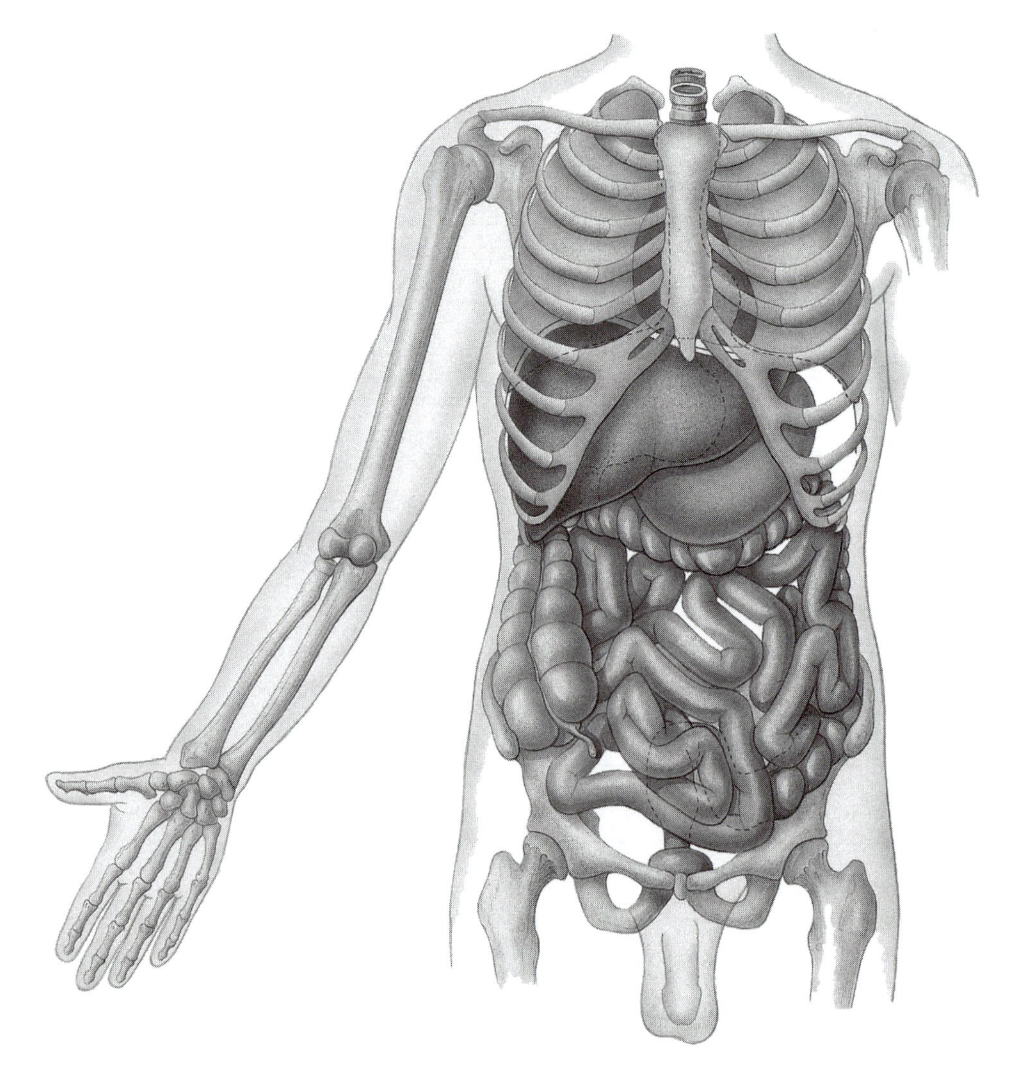

NOTES

Figure 1.10 The thoracic cavity (page 16).

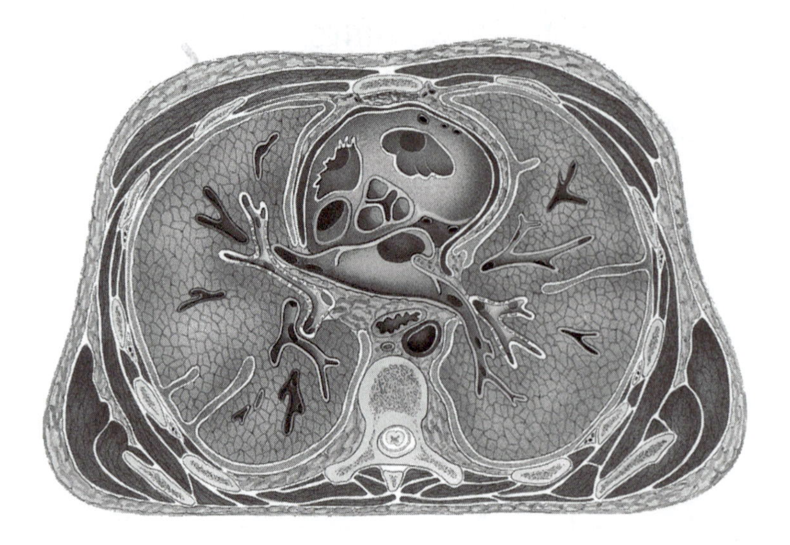

Figure 1.11 The abdominopelvic cavity (page 18).

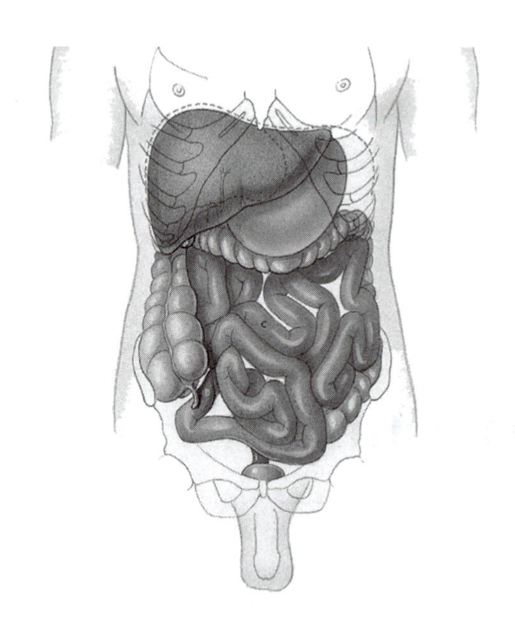

NOTES

Figure 1.12 Regions and quadrants of the abdominopelvic cavity (page 19).

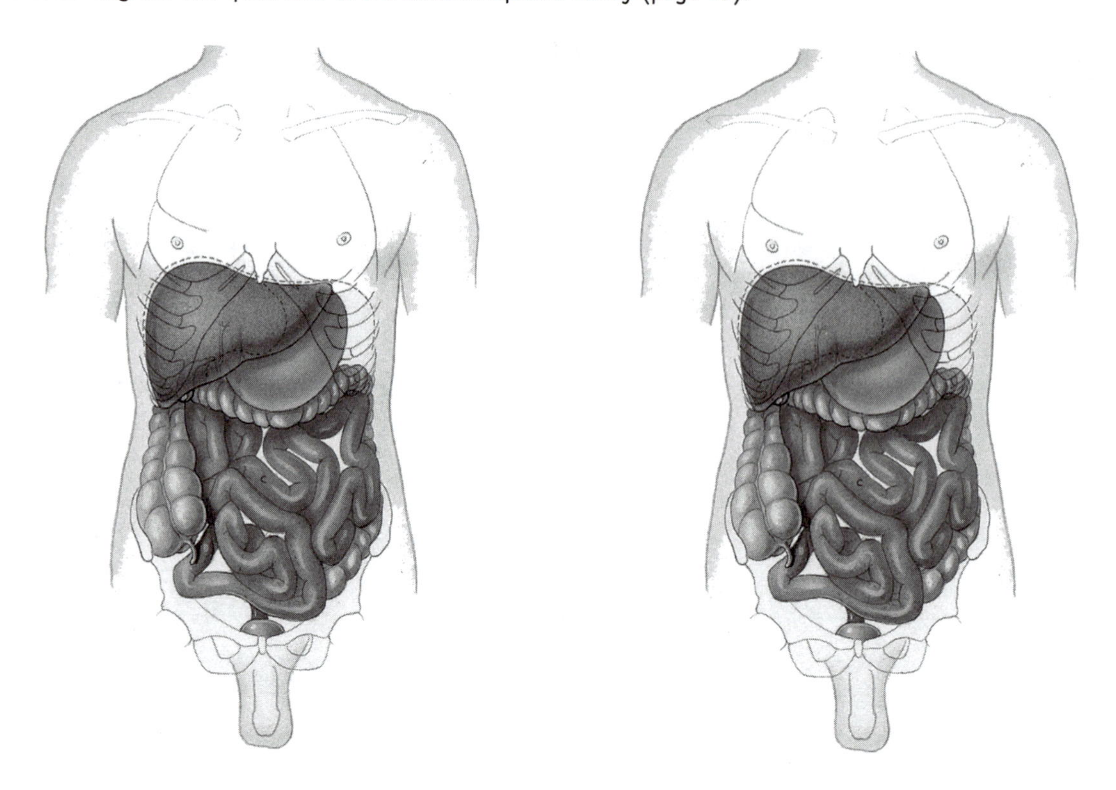

NOTES

Figure 2.1 Two representations of the structure of an atom (page 28).

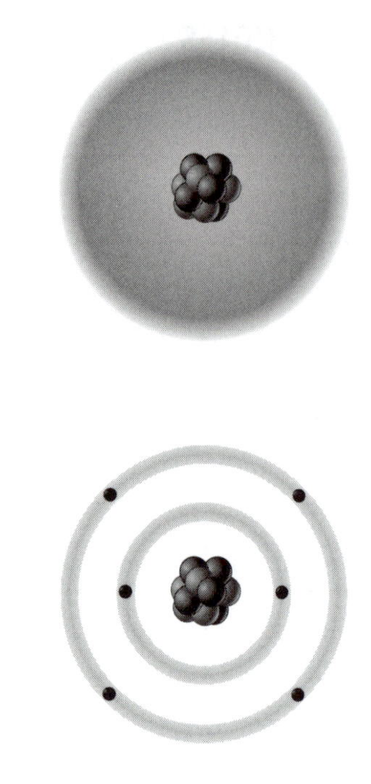

Figure 2.2 Atomic structures of several stable atoms (page 29).

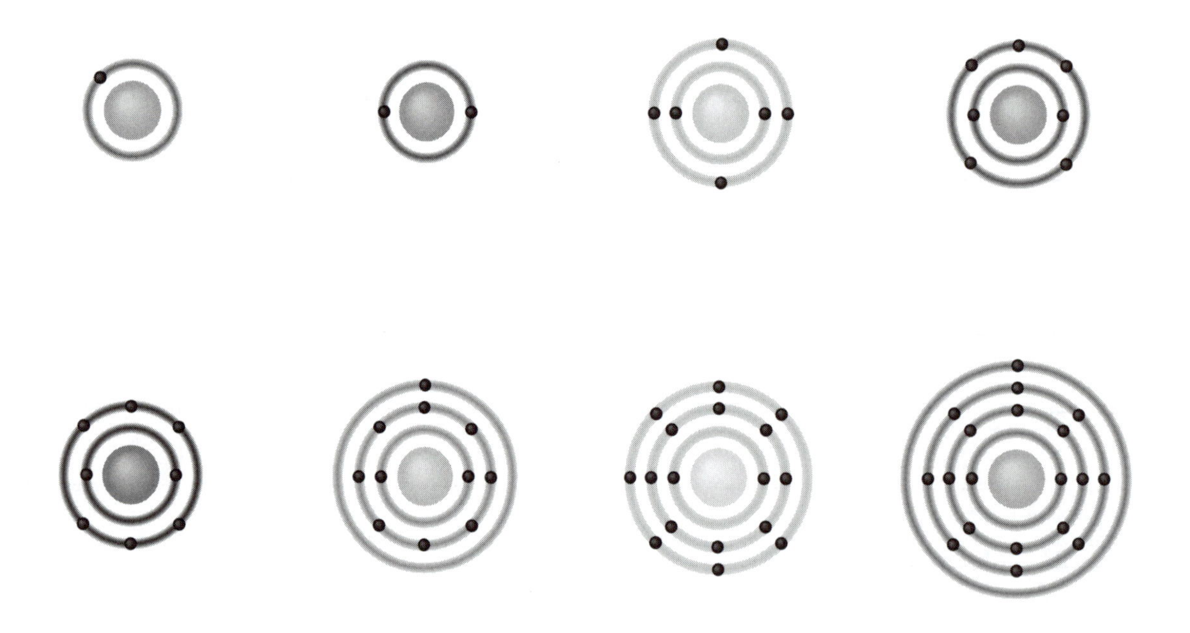

NOTES

2

Figure 2.3 Atomic structures of an oxygen molecule and a superoxide free radical (page 30).

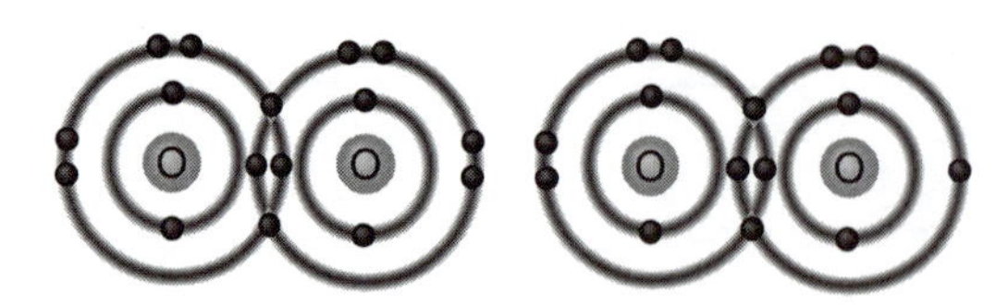

O
O
O
O

Figure 2.4 Ions and ionic bond formation (page 31).

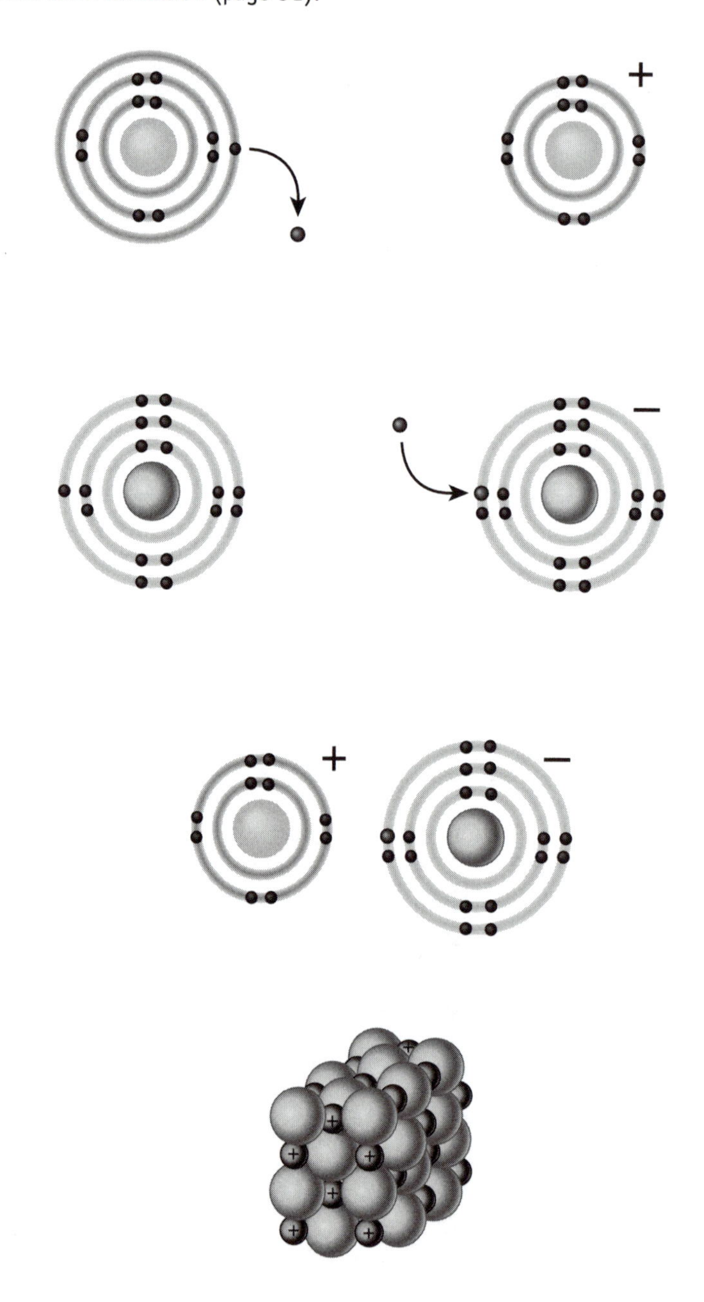

+
–
+
–

NOTES

2

Figure 2.5 Covalent bond formation (page 32).

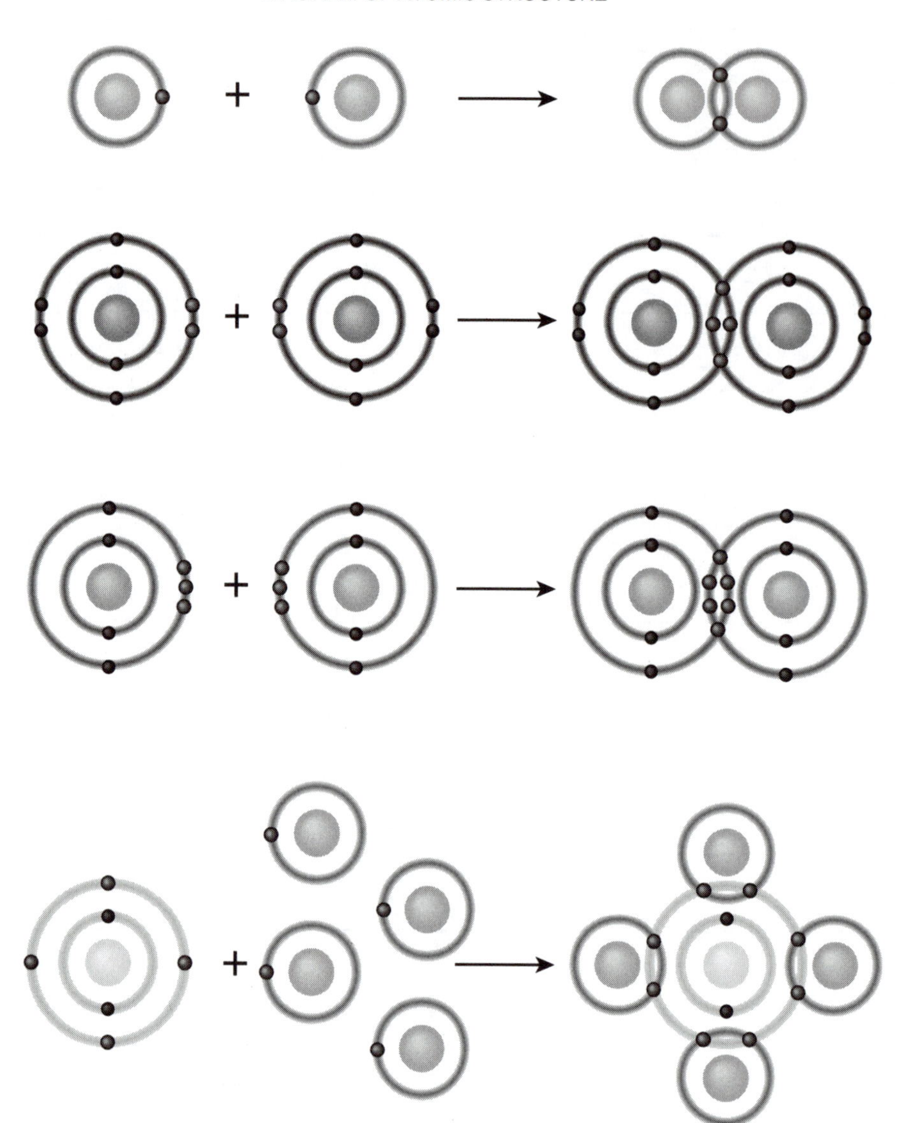
DIAGRAM OF ATOMIC STRUCTURE

STRUCTURAL FORMULA

MOLECULAR FORMULA

Figure 2.6 Polar covalent bonds (page 33).

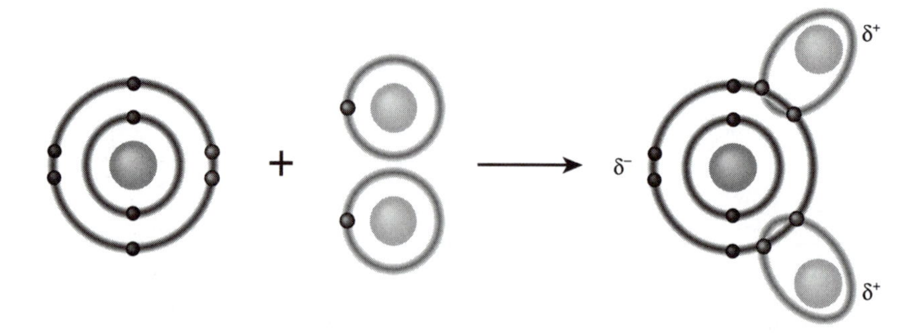
δ⁺
δ⁻
δ⁺

NOTES

Figure 2.7 Hydrogen bonding (page 33).

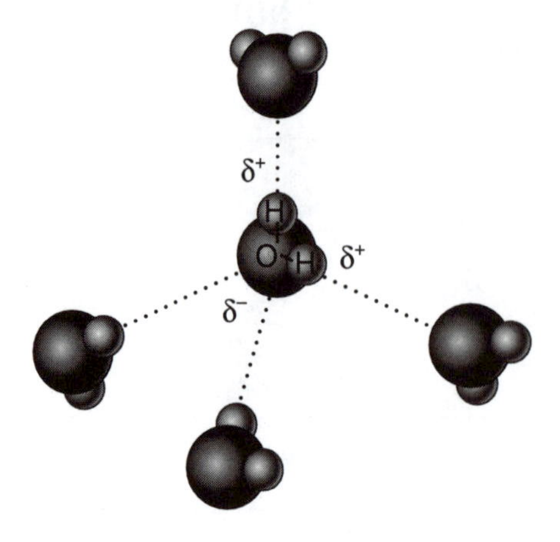
δ+
H
O
H
δ+
δ−

Figure 2.8 The chemical reaction (page 34).

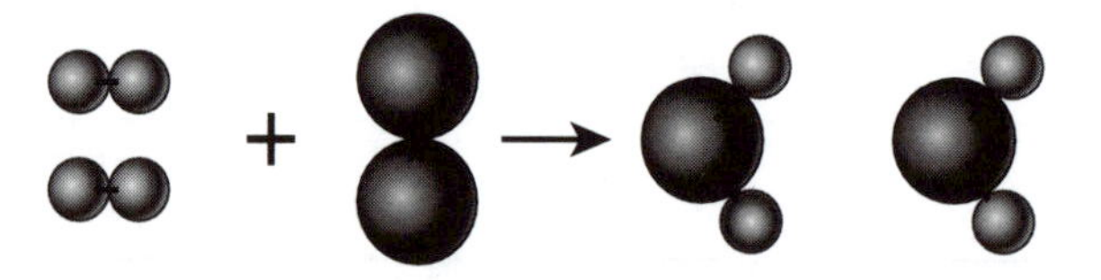

NOTES

2

Figure 2.9 Energy transfer during exergonic and endergonic reactions (page 35).

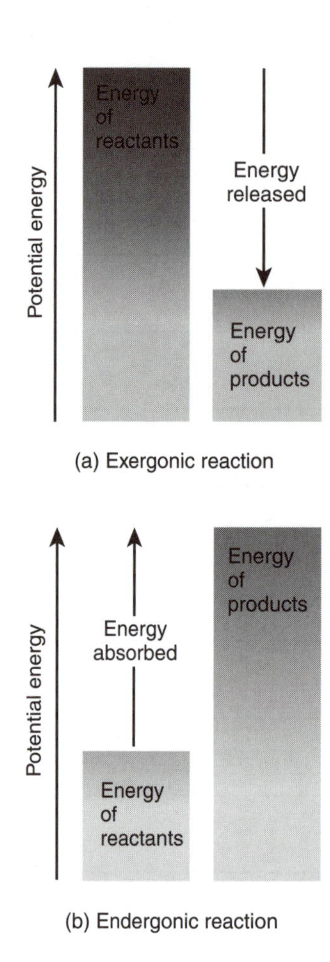
Potential energy
Energy of reactants
Energy released
Energy of products
(a) Exergonic reaction
Potential energy
Energy absorbed
Energy of reactants
Energy of products
(b) Endergonic reaction

Figure 2.10 Activation energy (page 35).

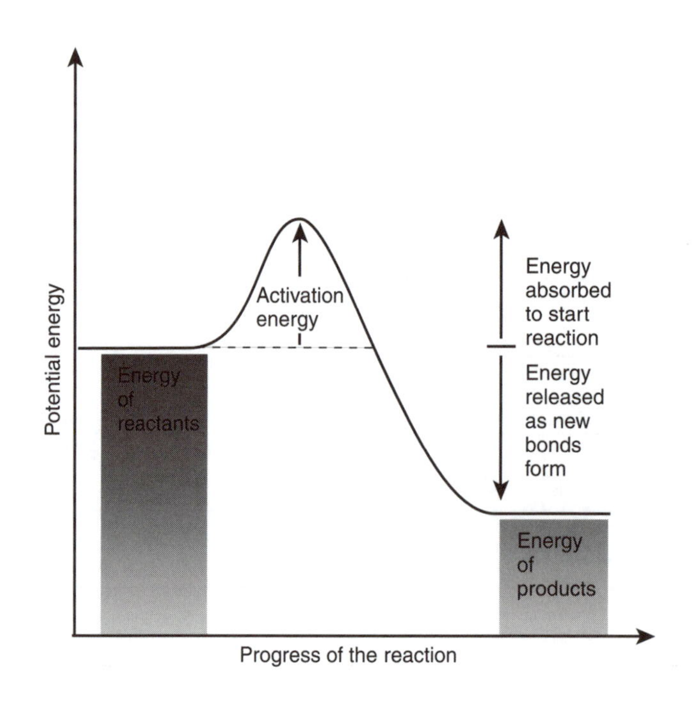
Potential energy
Activation energy
Energy absorbed to start reaction
Energy released as new bonds form
Energy of reactants
Energy of products
Progress of the reaction

NOTES

2

Figure 2.11 Energy needed for a chemical reaction (page 36).

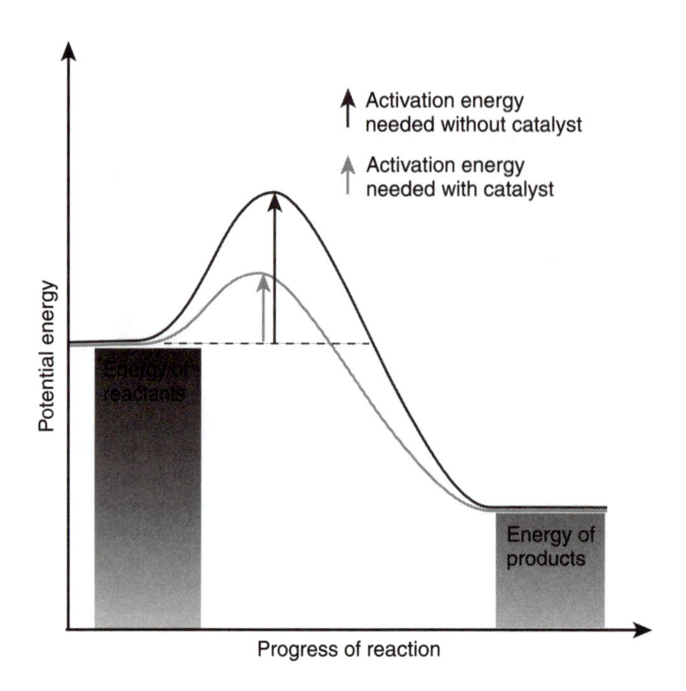
Activation energy needed without catalyst
Activation energy needed with catalyst
Potential energy
Energy of reactants
Energy of products
Progress of reaction

Figure 2.12 Dissociation of inorganic acids, bases, and salts (page 38).

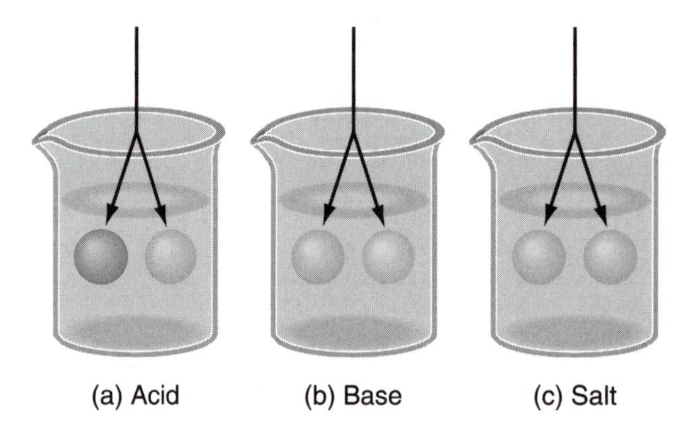
(a) Acid
(b) Base
(c) Salt

NOTES

Figure 2.13 How polar water molecules dissolve salts and polar substances (page 39).

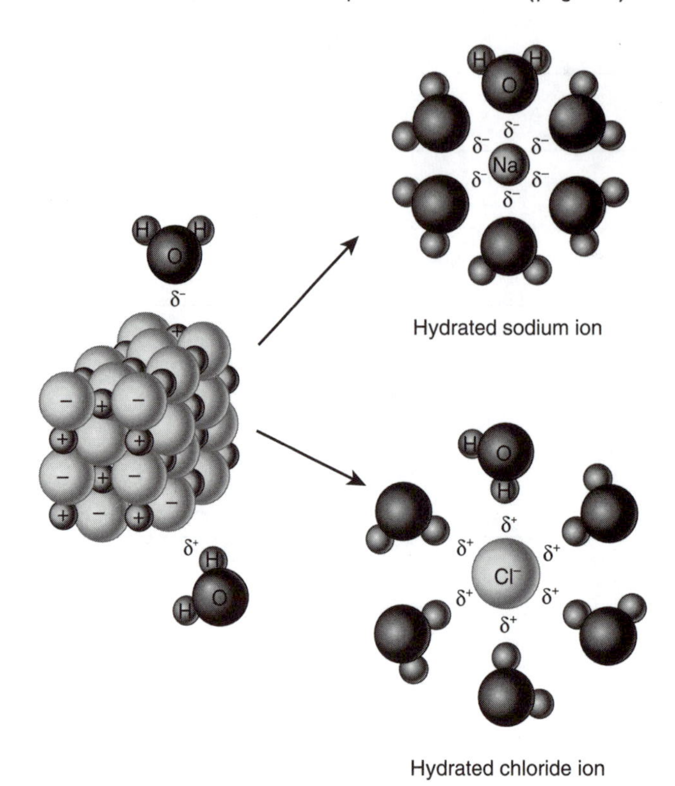

Hydrated sodium ion
Hydrated chloride ion

Adapted from Karen Timberlake, *Chemistry* 6e, F8.5, p255 (Menlo Park, CA; Addison Wesley Longman, 1999).

Figure 2.14 The pH scale (page 41).

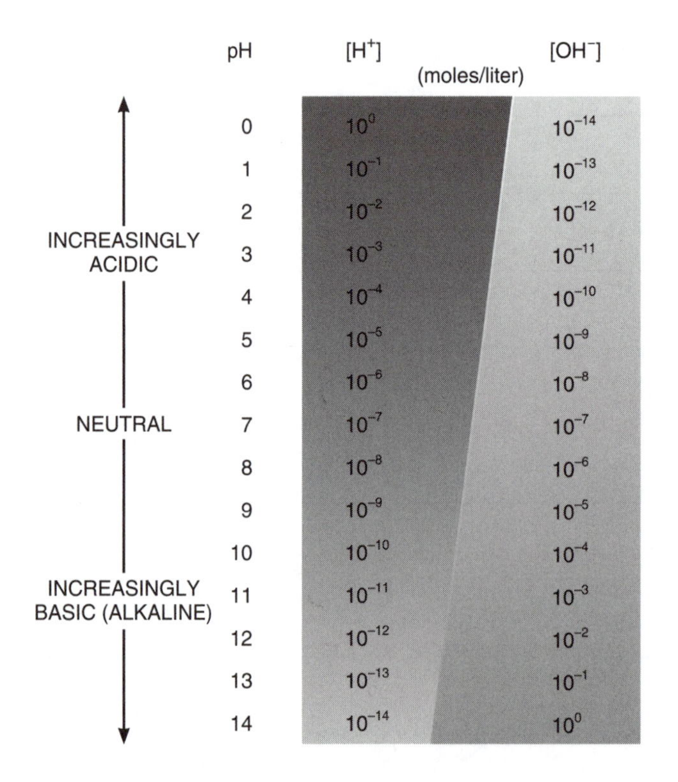

pH
[H$^+$]
(moles/liter)
[OH$^-$]
INCREASINGLY ACIDIC
NEUTRAL
INCREASINGLY BASIC (ALKALINE)
0 10^0 10^{-14}
1 10^{-1} 10^{-13}
2 10^{-2} 10^{-12}
3 10^{-3} 10^{-11}
4 10^{-4} 10^{-10}
5 10^{-5} 10^{-9}
6 10^{-6} 10^{-8}
7 10^{-7} 10^{-7}
8 10^{-8} 10^{-6}
9 10^{-9} 10^{-5}
10 10^{-10} 10^{-4}
11 10^{-11} 10^{-3}
12 10^{-12} 10^{-2}
13 10^{-13} 10^{-1}
14 10^{-14} 10^0

NOTES

2

Figure 2.15 Alternate ways to write the structural formula for glucose (page 43).

All atoms written out

Standard shorthand

Figure 2.16 The monosaccharides glucose and fructose and the disaccharide sucrose (page 44).

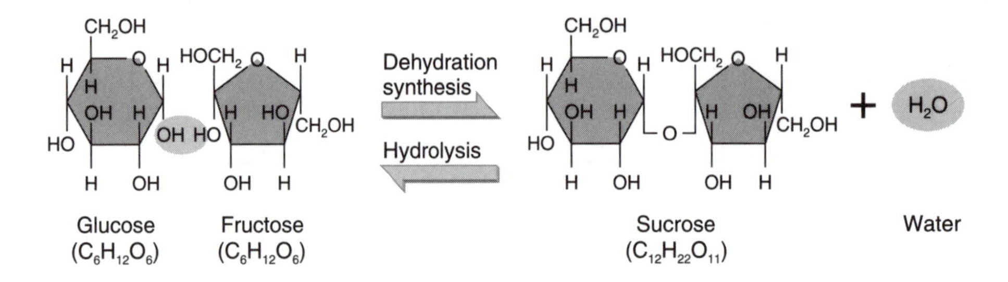

Dehydration synthesis
Hydrolysis
Glucose
($C_6H_{12}O_6$)
Fructose
($C_6H_{12}O_6$)
Sucrose
($C_{12}H_{22}O_{11}$)
H_2O
Water

Figure 2.17 The formation of a triglyceride (page 46).

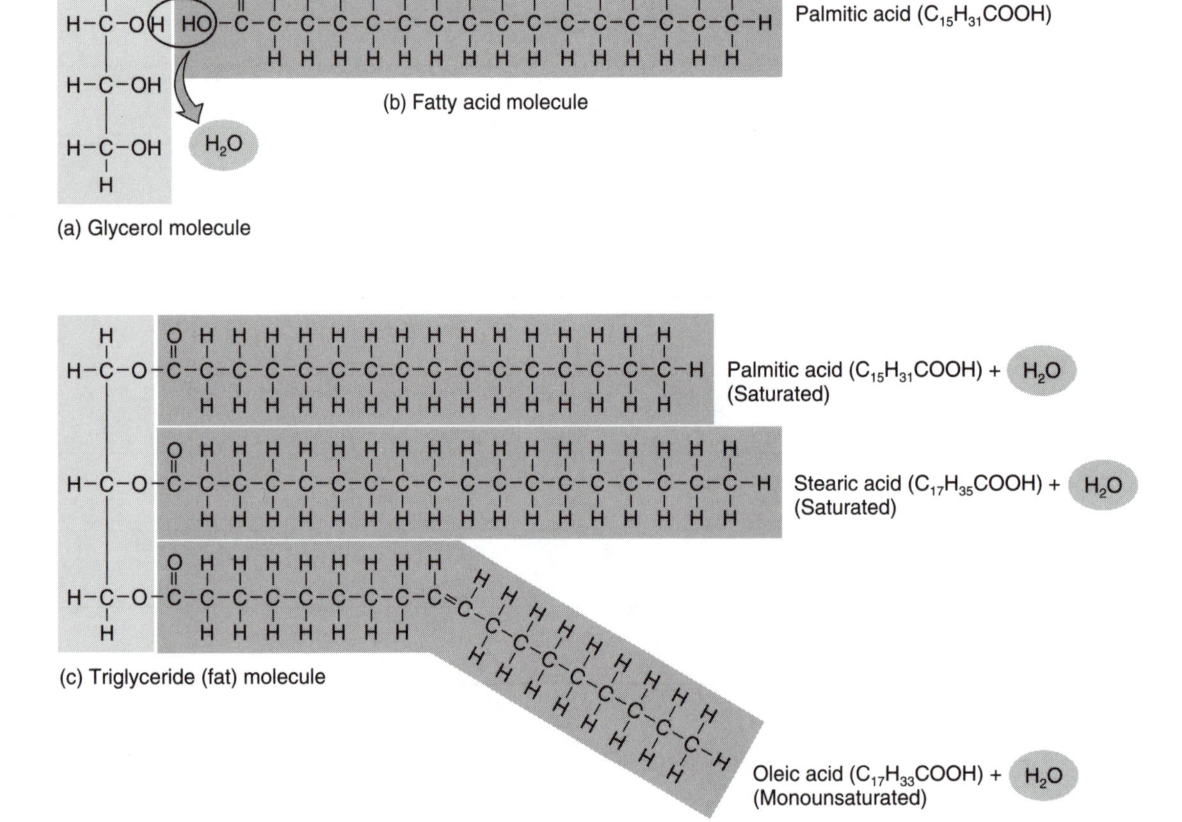

Palmitic acid ($C_{15}H_{31}COOH$)
(b) Fatty acid molecule
H_2O
(a) Glycerol molecule
Palmitic acid ($C_{15}H_{31}COOH$) + H_2O
(Saturated)
Stearic acid ($C_{17}H_{35}COOH$) + H_2O
(Saturated)
(c) Triglyceride (fat) molecule
Oleic acid ($C_{17}H_{33}COOH$) + H_2O
(Monounsaturated)

NOTES

2

Figure 2.18 Phospholipids (page 47).

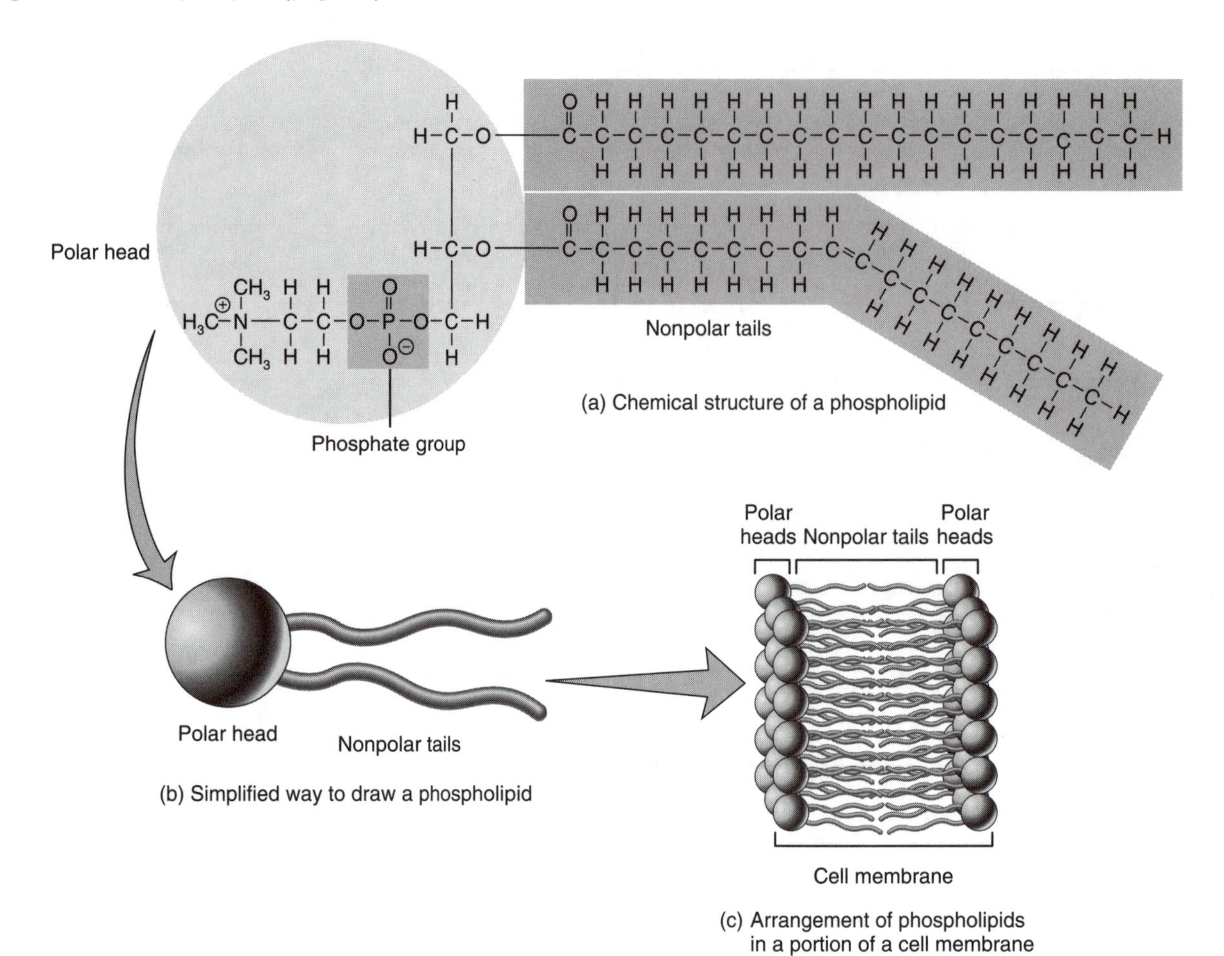

Polar head
Phosphate group
Nonpolar tails
(a) Chemical structure of a phospholipid
Polar head
Nonpolar tails
(b) Simplified way to draw a phospholipid
Polar heads
Nonpolar tails
Polar heads
Cell membrane
(c) Arrangement of phospholipids in a portion of a cell membrane

Figure 2.19 Steroids (page 48).

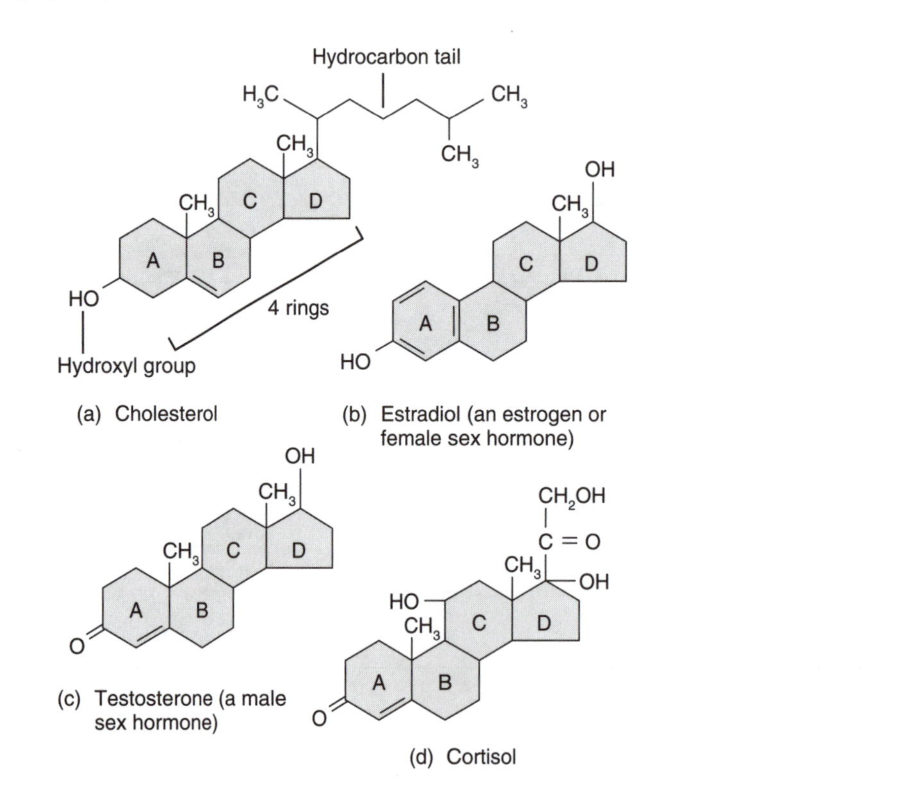

Hydrocarbon tail
4 rings
Hydroxyl group
(a) Cholesterol
(b) Estradiol (an estrogen or female sex hormone)
(c) Testosterone (a male sex hormone)
(d) Cortisol

NOTES

Figure 2.20 Amino acids (page 49).

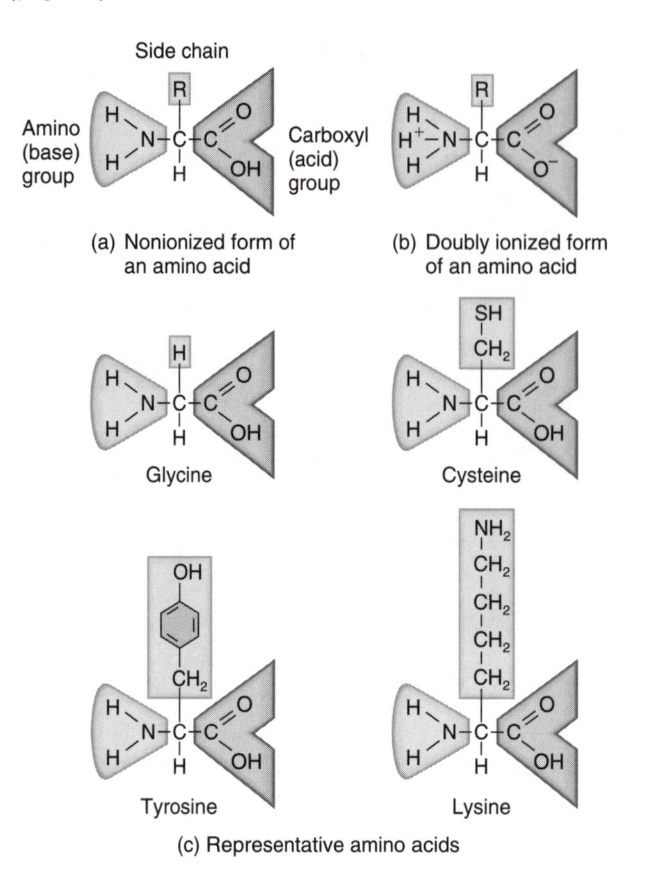

Side chain
Amino (base) group
Carboxyl (acid) group
(a) Nonionized form of an amino acid
(b) Doubly ionized form of an amino acid
Glycine
Cysteine
Tyrosine
Lysine
(c) Representative amino acids

Figure 2.21 Formation of a peptide bond (page 49).

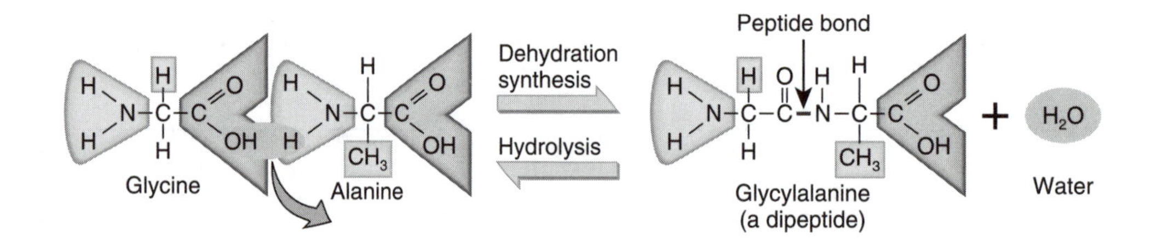

Glycine
Alanine
Dehydration synthesis
Hydrolysis
Peptide bond
Glycylalanine (a dipeptide)
Water

NOTES

2

Figure 2.22 Levels of structural organization in proteins (page 50).

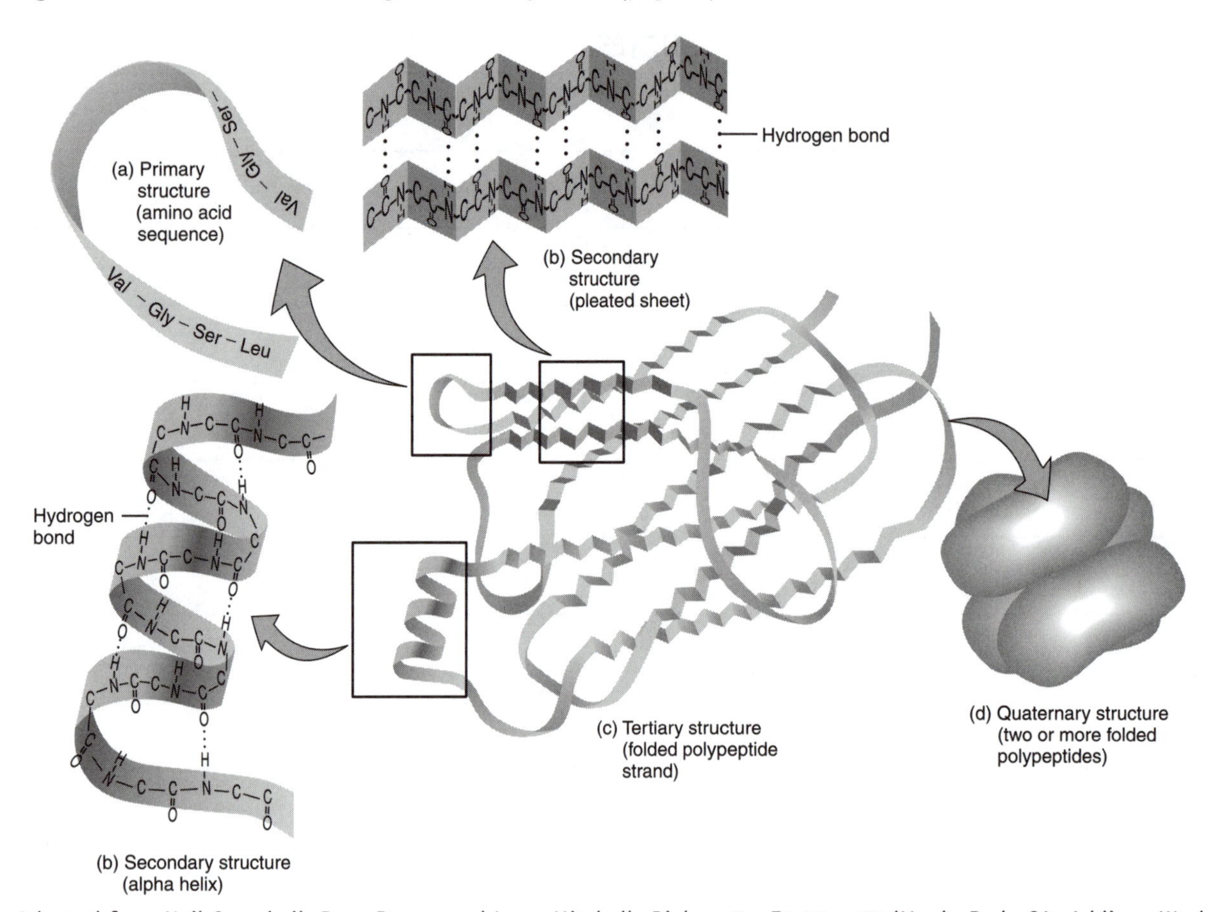

(a) Primary structure (amino acid sequence)
Val - Gly - Ser
Val - Gly - Ser - Leu
(b) Secondary structure (pleated sheet)
Hydrogen bond
Hydrogen bond
(c) Tertiary structure (folded polypeptide strand)
(d) Quaternary structure (two or more folded polypeptides)
(b) Secondary structure (alpha helix)

Adapted from Neil Campbell, Jane Reece, and Larry Mitchell, *Biology* 5e, F5.24, p75 (Menlo Park, CA; Addison Wesley Longman, 1999). ©1999 Addison Wesley Longman, Inc.

Figure 2.23 Types of bonds that stablilize the tertiary structure of proteins (page 51).

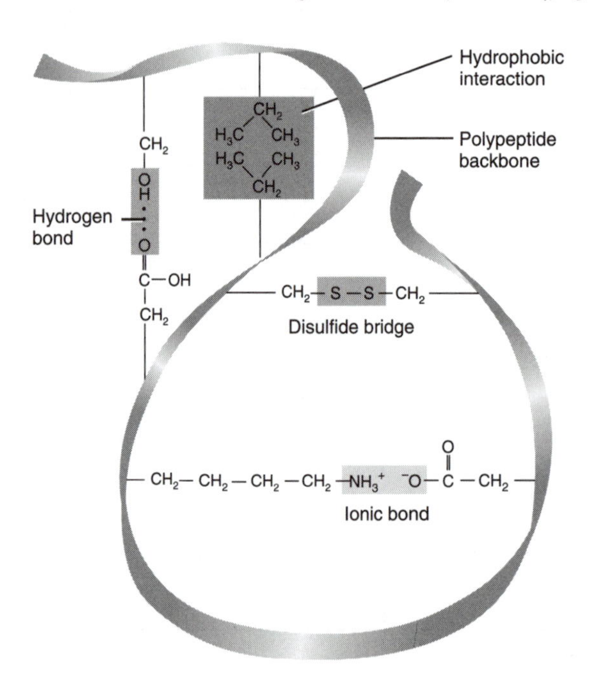

Hydrophobic interaction
Polypeptide backbone
Hydrogen bond
Disulfide bridge
Ionic bond

Adapted from Neil Campbell, Jane Reece, and Larry Mitchell, *Biology* 5e, F5.24, p75 (Menlo Park, CA; Addison Wesley Longman, 1999). ©1999 Addison Wesley Longman, Inc.

NOTES

2

Figure 2.24 How an enzyme works (page 52).

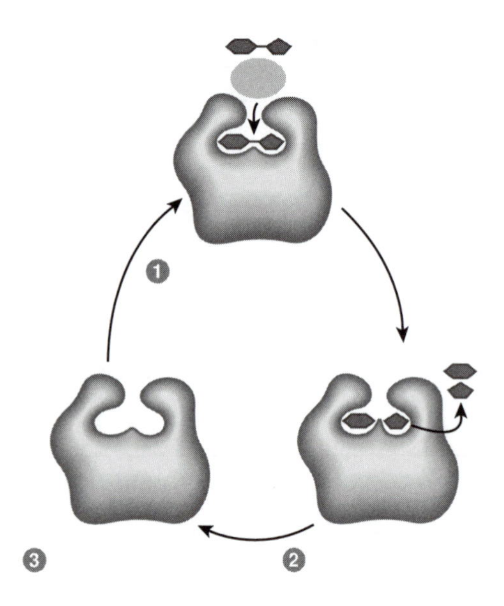

Figure 2.25 DNA molecule (page 53).

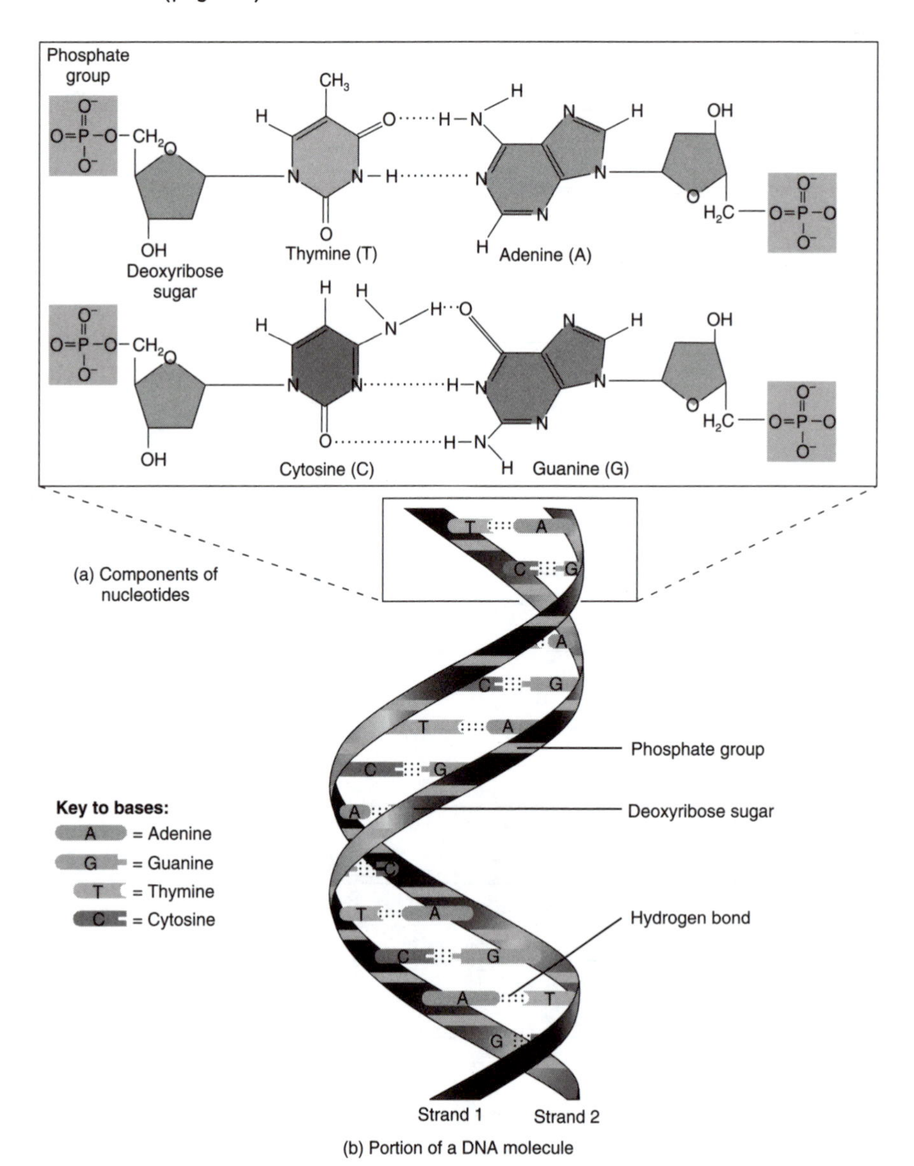

Phosphate group
Deoxyribose sugar
Thymine (T)
Adenine (A)
Cytosine (C)
Guanine (G)
(a) Components of nucleotides
Key to bases:
A = Adenine
G = Guanine
T = Thymine
C = Cytosine
Phosphate group
Deoxyribose sugar
Hydrogen bond
Strand 1
Strand 2
(b) Portion of a DNA molecule

NOTES

2

Figure 2.26 Structures of ATP and ADP (page 55).

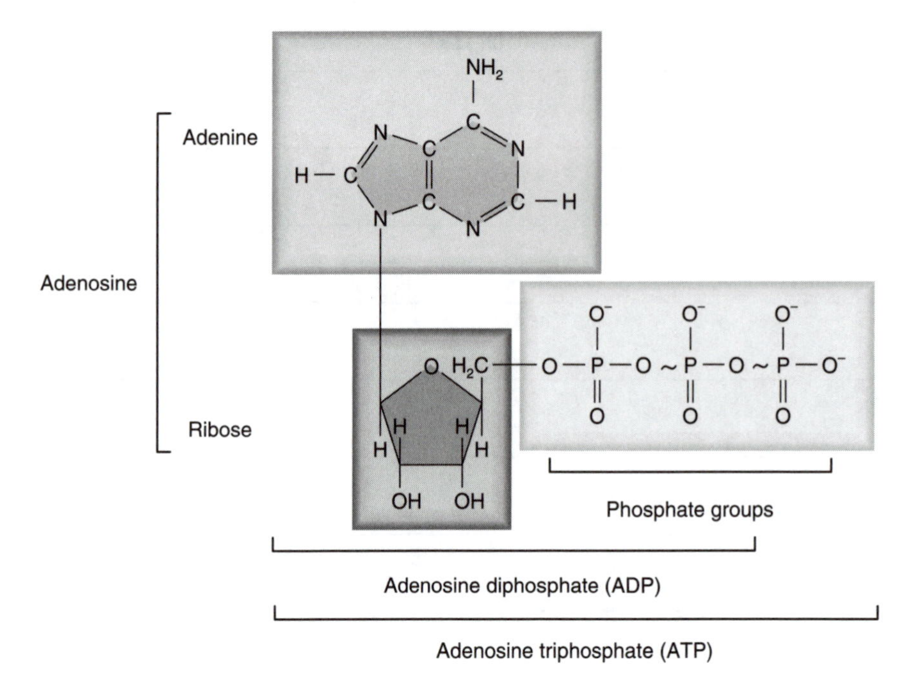

NH2
Adenine
Adenosine
Ribose
H2C
Phosphate groups
Adenosine diphosphate (ADP)
Adenosine triphosphate (ATP)

NOTES

2

Figure 3.1 Generalized view of a body cell (page 61).

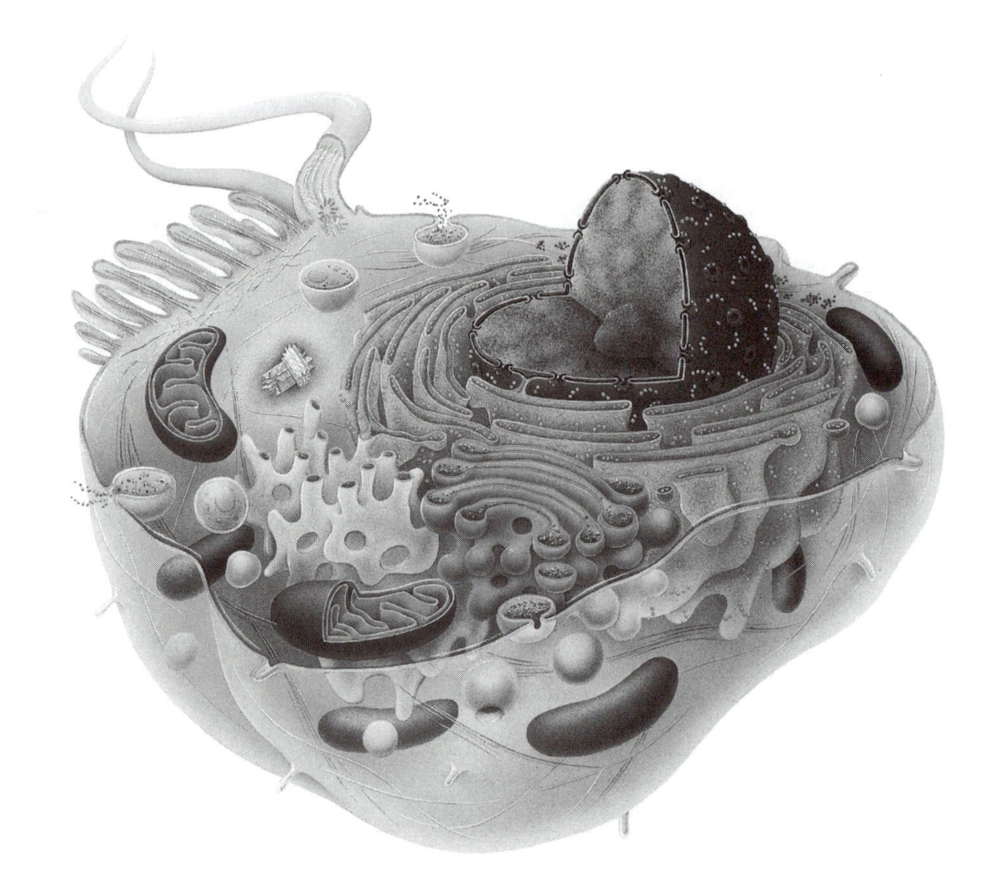

Figure 3.2 Plasma membrane (page 62).

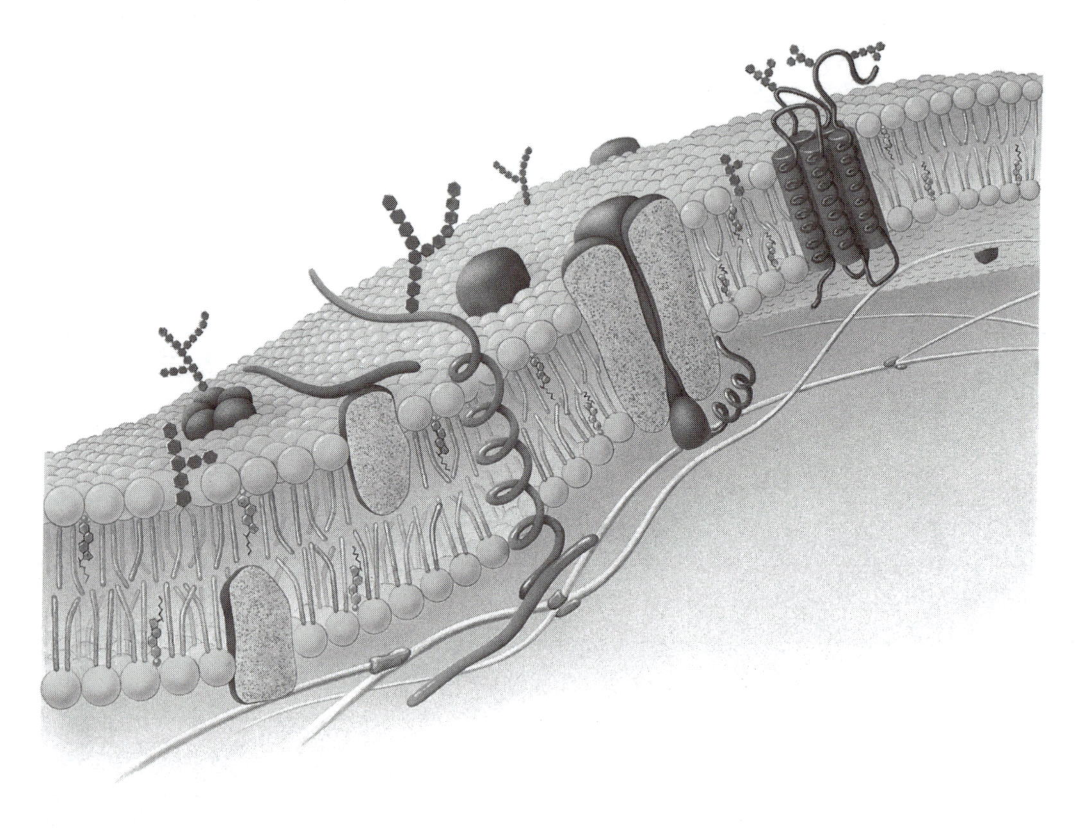

NOTES

3

Figure 3.3 Functions of membrane proteins (page 64).

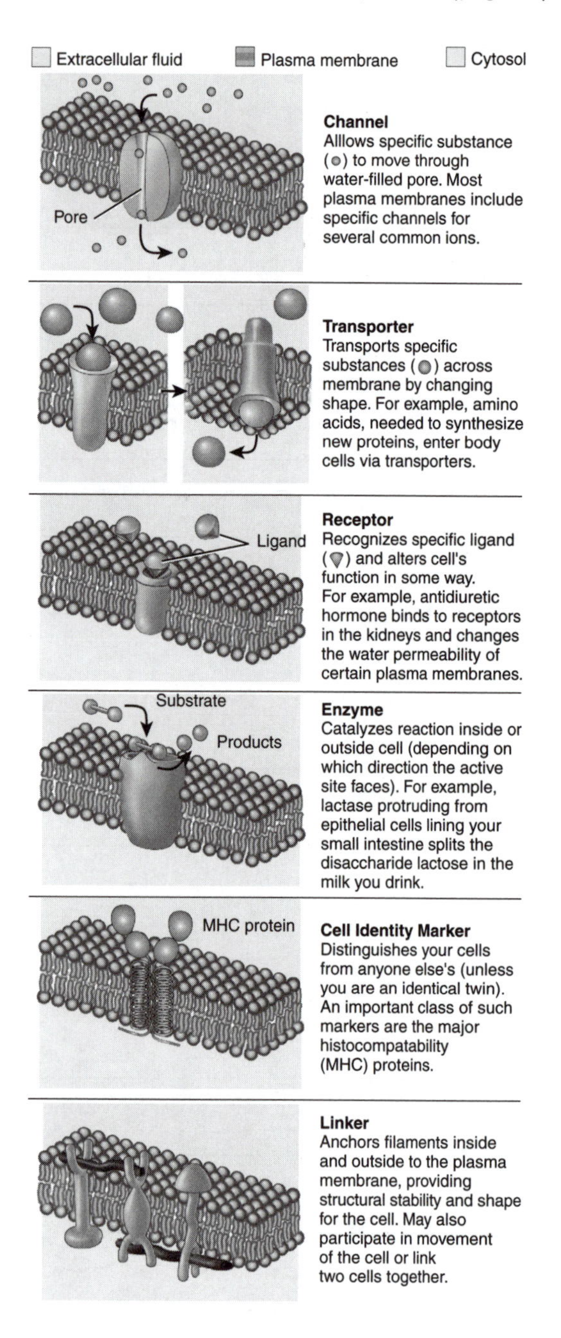

Extracellular fluid
Plasma membrane
Cytosol
Pore
Channel
Alllows specific substance (●) to move through water-filled pore. Most plasma membranes include specific channels for several common ions.
Transporter
Transports specific substances (●) across membrane by changing shape. For example, amino acids, needed to synthesize new proteins, enter body cells via transporters.
Ligand
Receptor
Recognizes specific ligand (▼) and alters cell's function in some way. For example, antidiuretic hormone binds to receptors in the kidneys and changes the water permeability of certain plasma membranes.
Substrate
Products
Enzyme
Catalyzes reaction inside or outside cell (depending on which direction the active site faces). For example, lactase protruding from epithelial cells lining your small intestine splits the disaccharide lactose in the milk you drink.
MHC protein
Cell Identity Marker
Distinguishes your cells from anyone else's (unless you are an identical twin). An important class of such markers are the major histocompatability (MHC) proteins.
Linker
Anchors filaments inside and outside to the plasma membrane, providing structural stability and shape for the cell. May also participate in movement of the cell or link two cells together.

Figure 3.4 Gradients across the plasma membrane (page 65).

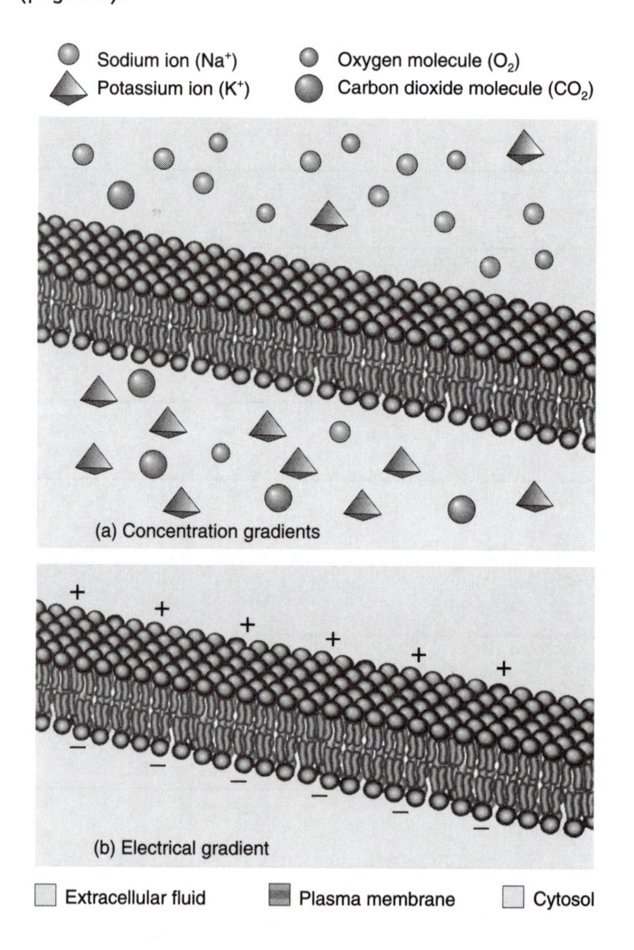

Sodium ion (Na^+)
Potassium ion (K^+)
Oxygen molecule (O_2)
Carbon dioxide molecule (CO_2)
(a) Concentration gradients
(b) Electrical gradient
Extracellular fluid
Plasma membrane
Cytosol

NOTES

3

Figure 3.5 Processes for transport of materials across the plasma membrane (page 66).

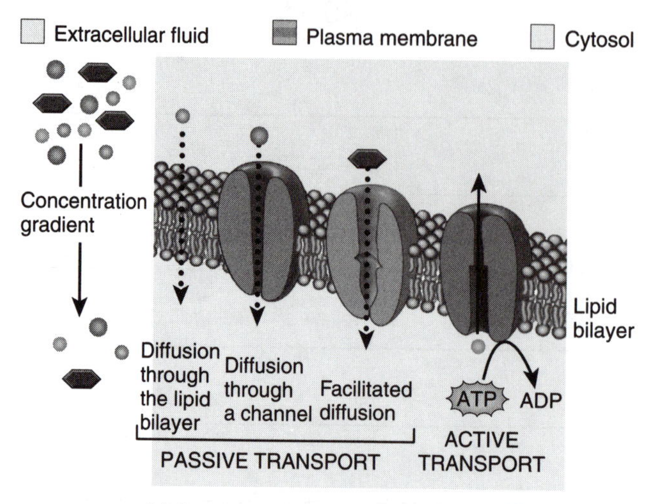
Extracellular fluid
Plasma membrane
Cytosol
Concentration gradient
Lipid bilayer
Diffusion through the lipid bilayer
Diffusion through a channel
Facilitated diffusion
ATP
ADP
PASSIVE TRANSPORT
ACTIVE TRANSPORT

(a) Passive and active transport

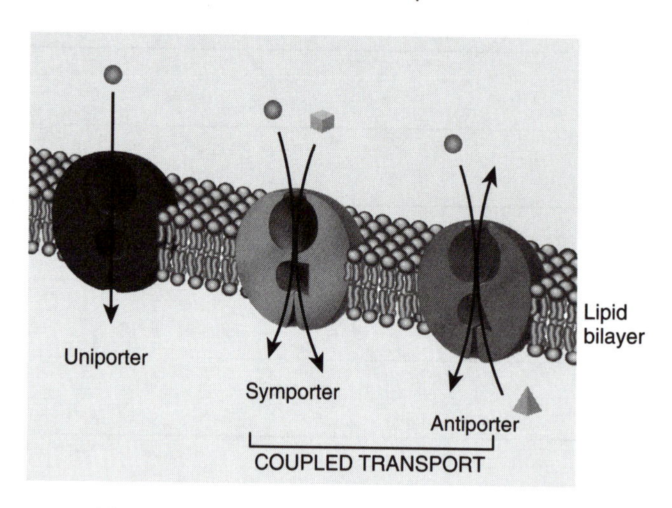
Lipid bilayer
Uniporter
Symporter
Antiporter
COUPLED TRANSPORT

(b) Types of transporters in mediated transport

NOTES

3

Figure 3.7 Principle of osmosis (page 68).

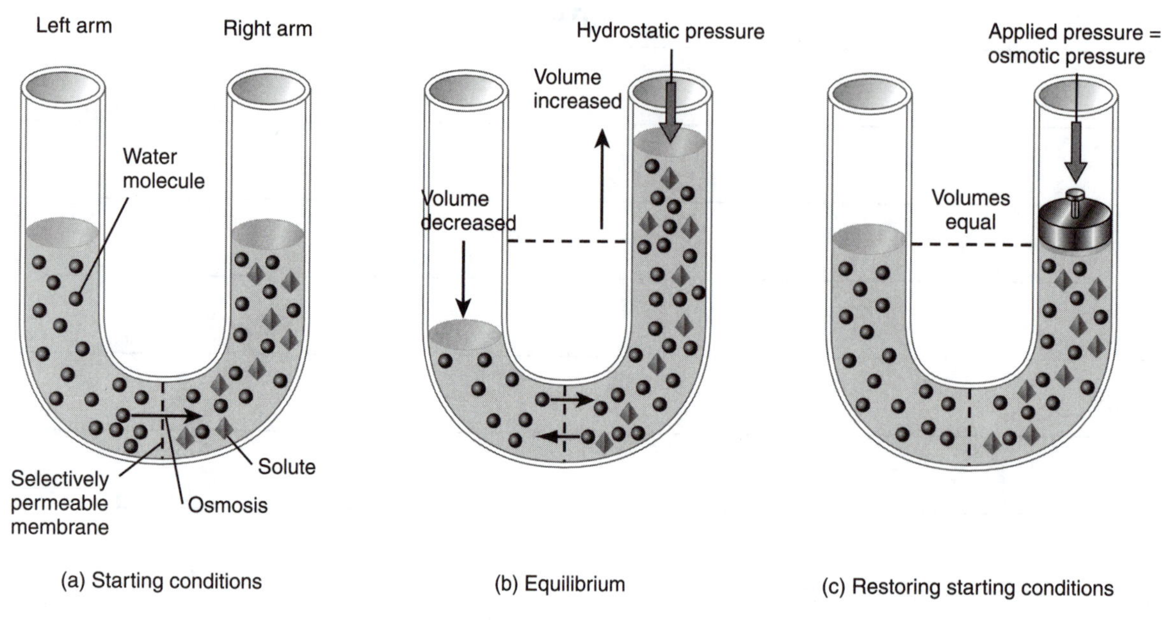

Left arm
Right arm
Water molecule
Selectively permeable membrane
Osmosis
Solute
Hydrostatic pressure
Volume increased
Volume decreased
Applied pressure = osmotic pressure
Volumes equal
(a) Starting conditions
(b) Equilibrium
(c) Restoring starting conditions

Adapted from Martini, *Fundamentals of Anatomy and Physiology* 4e, F3.7, p75 (Upper Saddle River, NJ: Prentice Hall, 1998).

Figure 3.8 Tonicity and its effects on red blood cells (page 69).

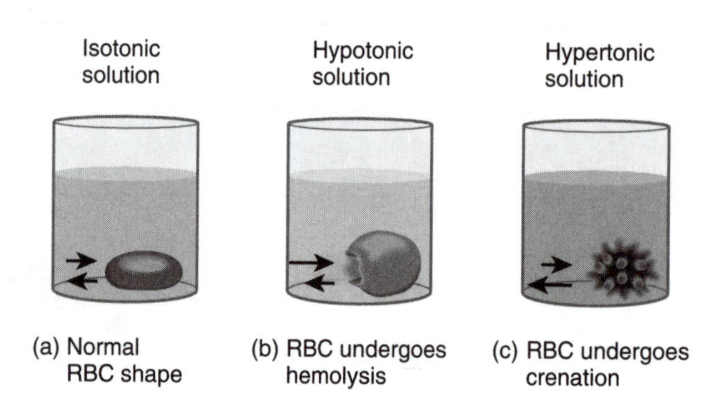

Isotonic solution
Hypotonic solution
Hypertonic solution
(a) Normal RBC shape
(b) RBC undergoes hemolysis
(c) RBC undergoes crenation

NOTES

3

Figure 3.9 Diffusion of K^+ through a gated membrane (page 70).

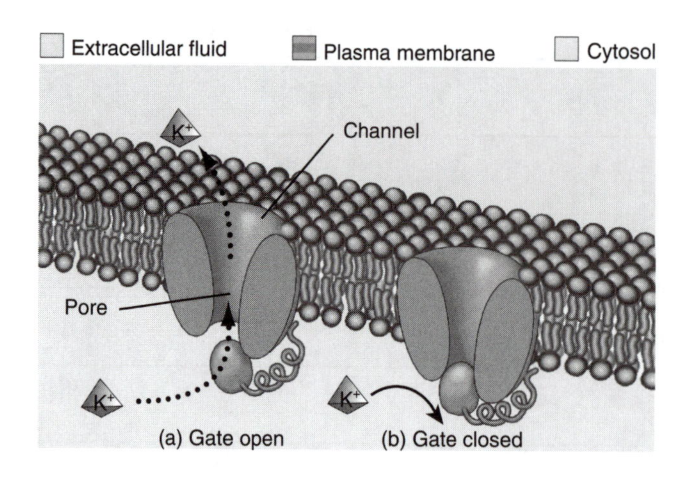
Extracellular fluid
Plasma membrane
Cytosol
K+
Channel
Pore
K+
K+
(a) Gate open
(b) Gate closed

Figure 3.10 Facilitated diffusion of glucose across a plasma membrane (page 70).

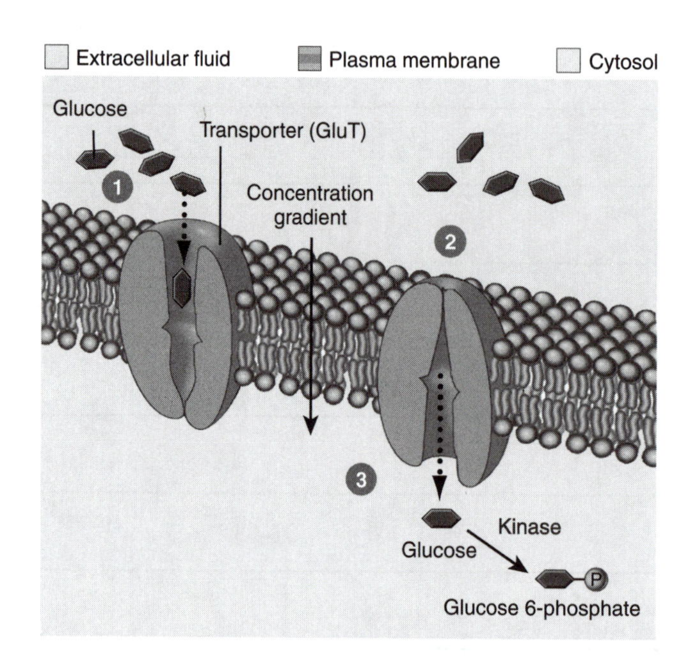
Extracellular fluid
Plasma membrane
Cytosol
Glucose
Transporter (GluT)
1
Concentration gradient
2
3
Kinase
Glucose
P
Glucose 6-phosphate

NOTES

3

Figure 3.11 The sodium pump (page 71).

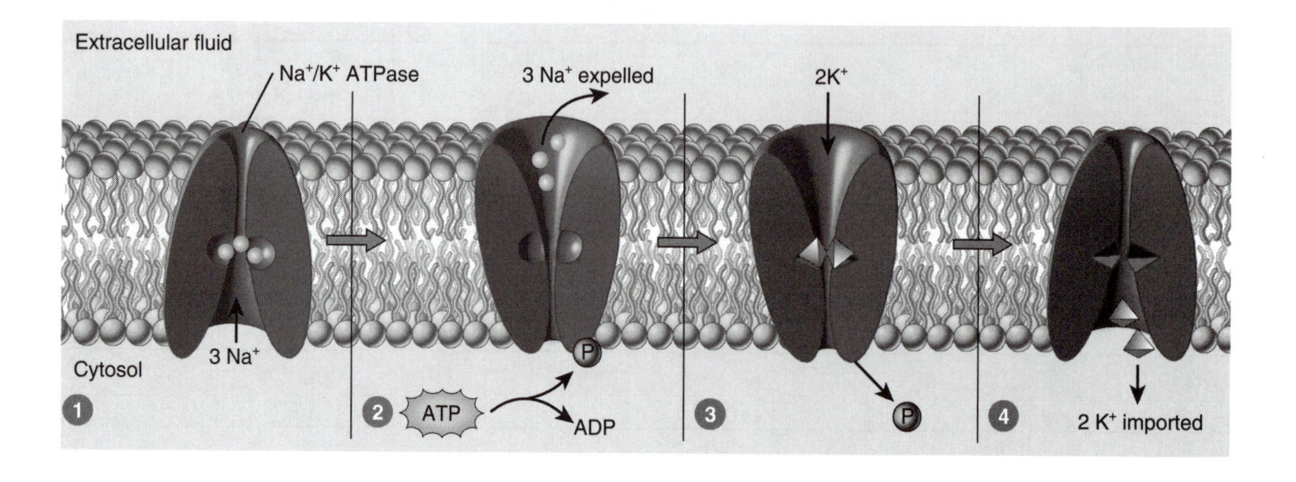

Extracellular fluid
Na⁺/K⁺ ATPase
3 Na⁺ expelled
2K⁺
Cytosol
3 Na⁺
P
1
2
ATP
ADP
3
P
4
2 K⁺ imported

Figure 3.12 Secondary active transport mechanisms (page 72).

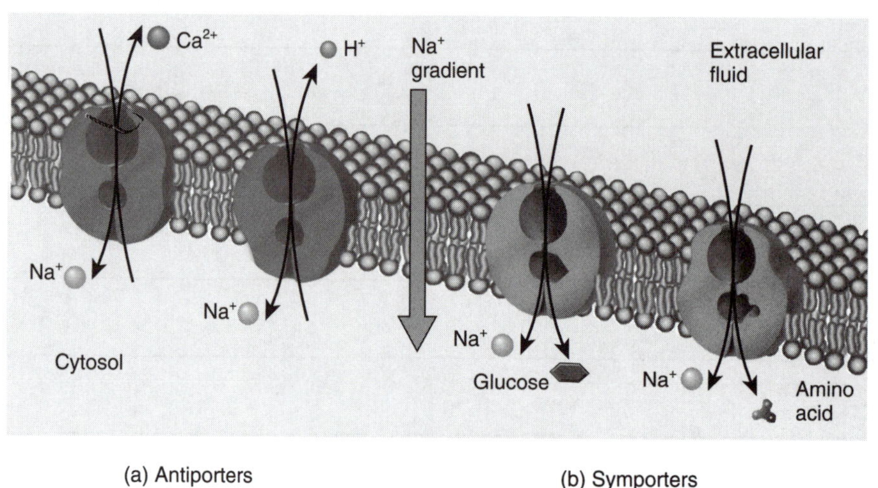

Ca²⁺
H⁺
Na⁺
gradient
Extracellular
fluid
Na⁺
Na⁺
Cytosol
Na⁺
Glucose
Na⁺
Amino
acid
(a) Antiporters
(b) Symporters

NOTES

3

Figure 3.13 Endocytosis: receptor-mediated endocytosis (page 73).

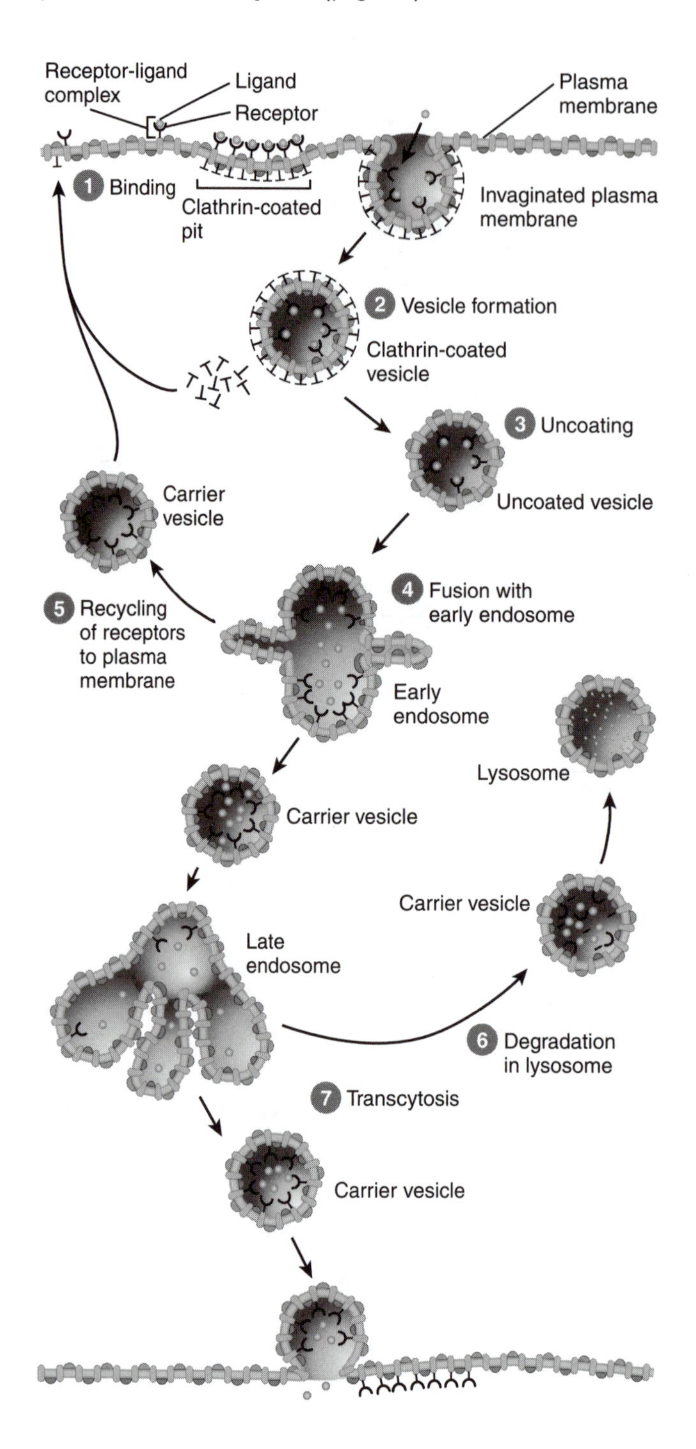

Receptor-ligand complex
Ligand
Receptor
Plasma membrane
1 Binding
Clathrin-coated pit
Invaginated plasma membrane
2 Vesicle formation
Clathrin-coated vesicle
3 Uncoating
Uncoated vesicle
Carrier vesicle
4 Fusion with early endosome
5 Recycling of receptors to plasma membrane
Early endosome
Lysosome
Carrier vesicle
Carrier vesicle
Late endosome
6 Degradation in lysosome
7 Transcytosis
Carrier vesicle

NOTES

Figure 3.14 Endocytosis: phagocytosis (page 74).

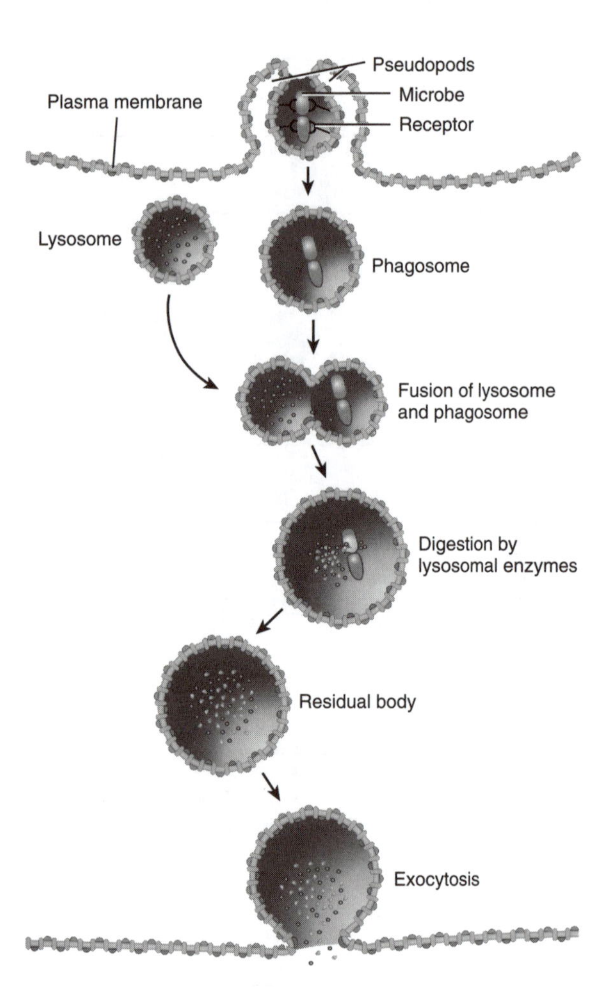

Pseudopods
Plasma membrane
Microbe
Receptor
Lysosome
Phagosome
Fusion of lysosome
and phagosome
Digestion by
lysosomal enzymes
Residual body
Exocytosis

Figure 3.15 Endocytosis: pinocytosis (page 75).

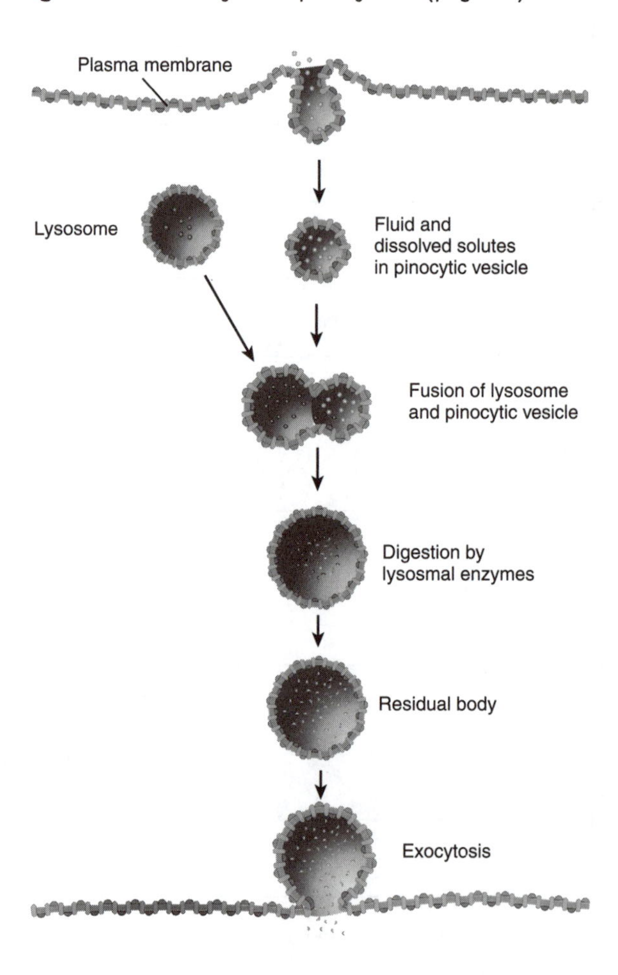

Plasma membrane
Lysosome
Fluid and
dissolved solutes
in pinocytic vesicle
Fusion of lysosome
and pinocytic vesicle
Digestion by
lysosmal enzymes
Residual body
Exocytosis

NOTES

3

Figure 3.16 Cytoskeleton (page 77).

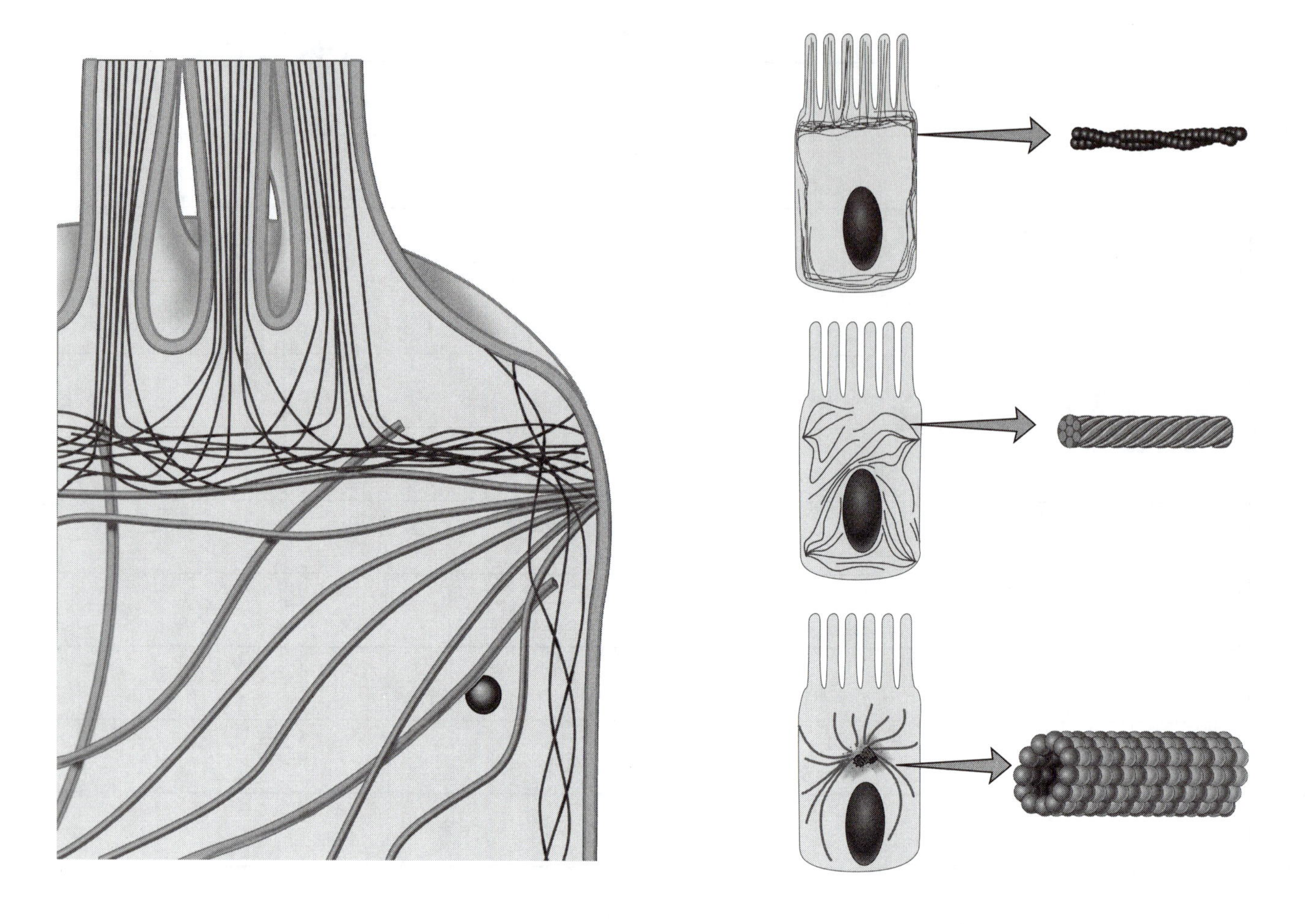

Figure 3.17 Centrosome (page 78).

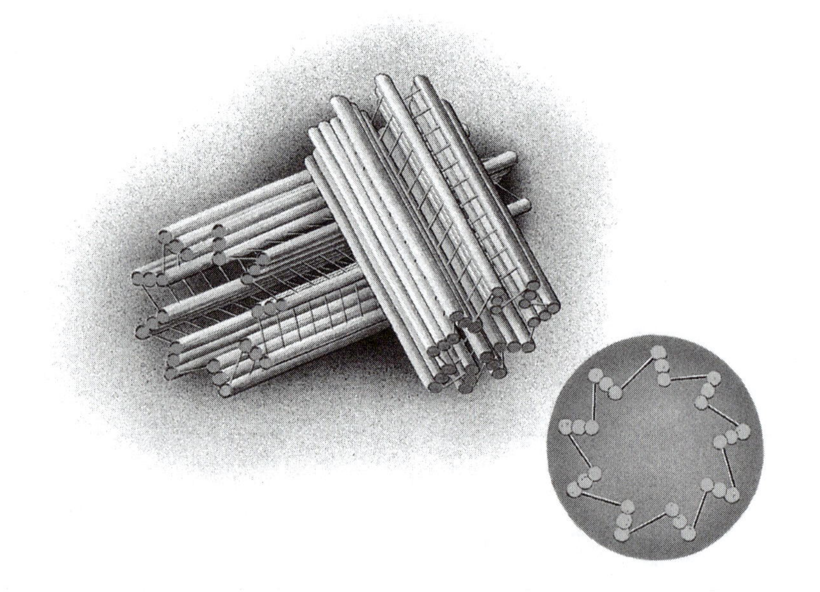

NOTES

Figure 3.18 Cilia and flagella (page 79).

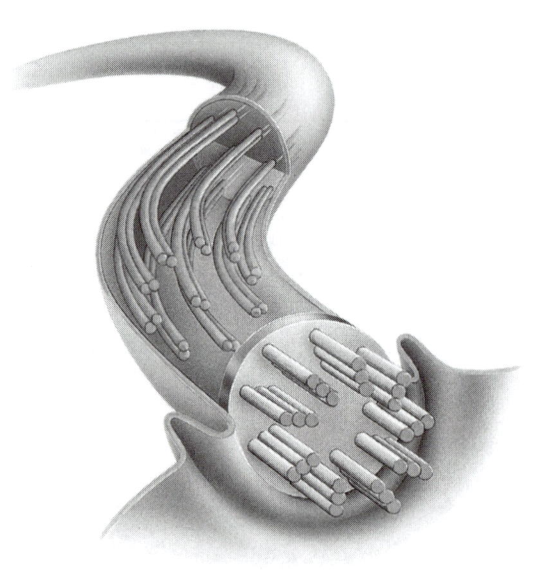

(a) 9 + 2 array of cilium or flagellum

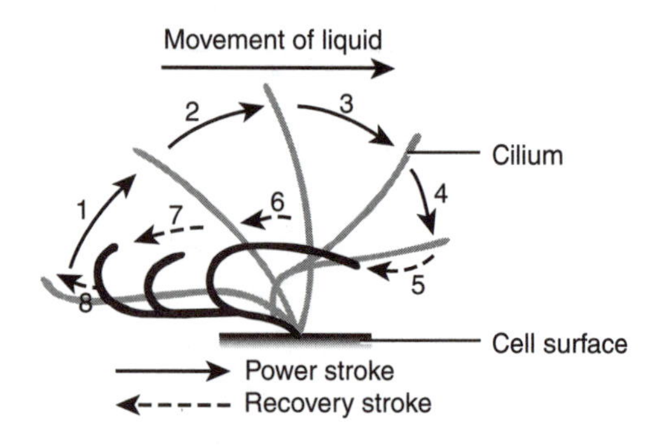

Movement of liquid
1
2
3
4
5
6
7
8
Cilium
Cell surface
Power stroke
Recovery stroke

(b) Ciliary movement

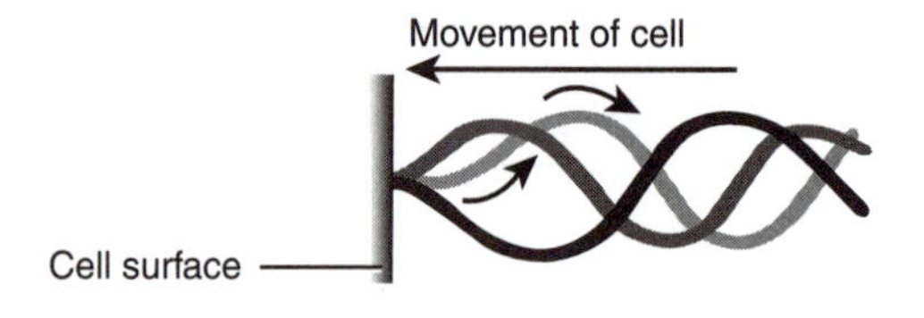

Movement of cell
Cell surface

(c) Flagellar movement

FUNCTIONS
1. Cilia move fluids along a cell's surface.
2. Flagella move an entire cell.

Figure 3.19 Ribosomes (page 80).

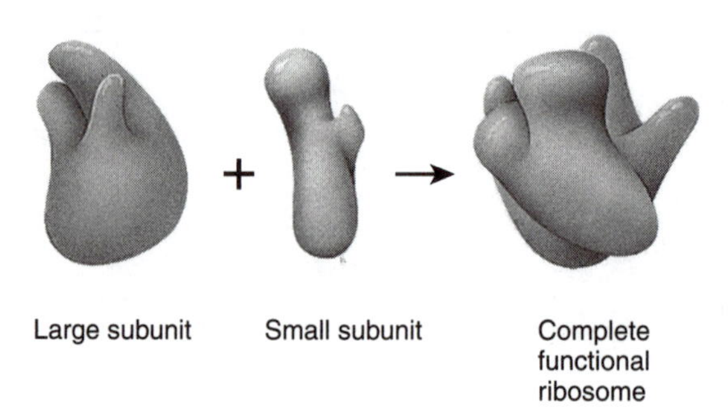

Large subunit
Small subunit
Complete functional ribosome

Details of ribosomal subunits

Figure 3.20 Endoplasmic reticulum (page 80).

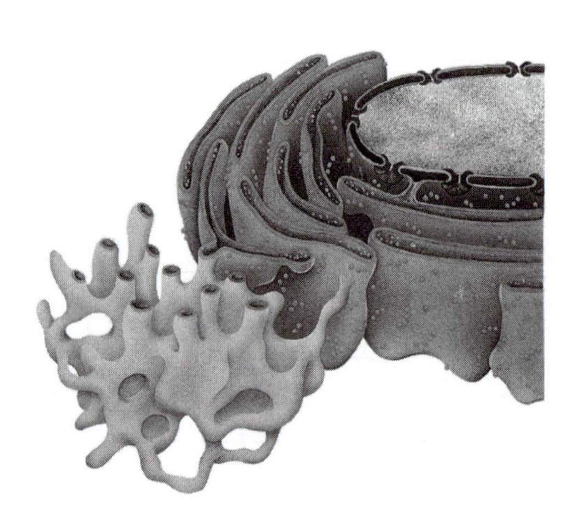

NOTES

3

Figure 3.21 Golgi complex (page 81).

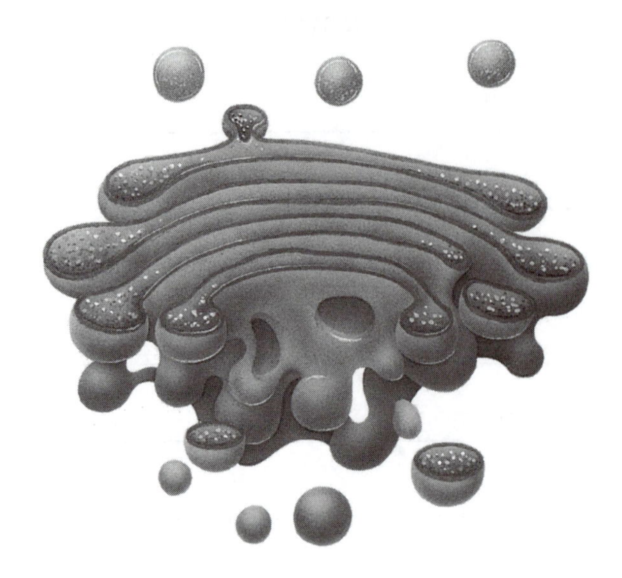

Figure 3.22 Packaging of synthesized proteins by the Golgi complex (page 82).

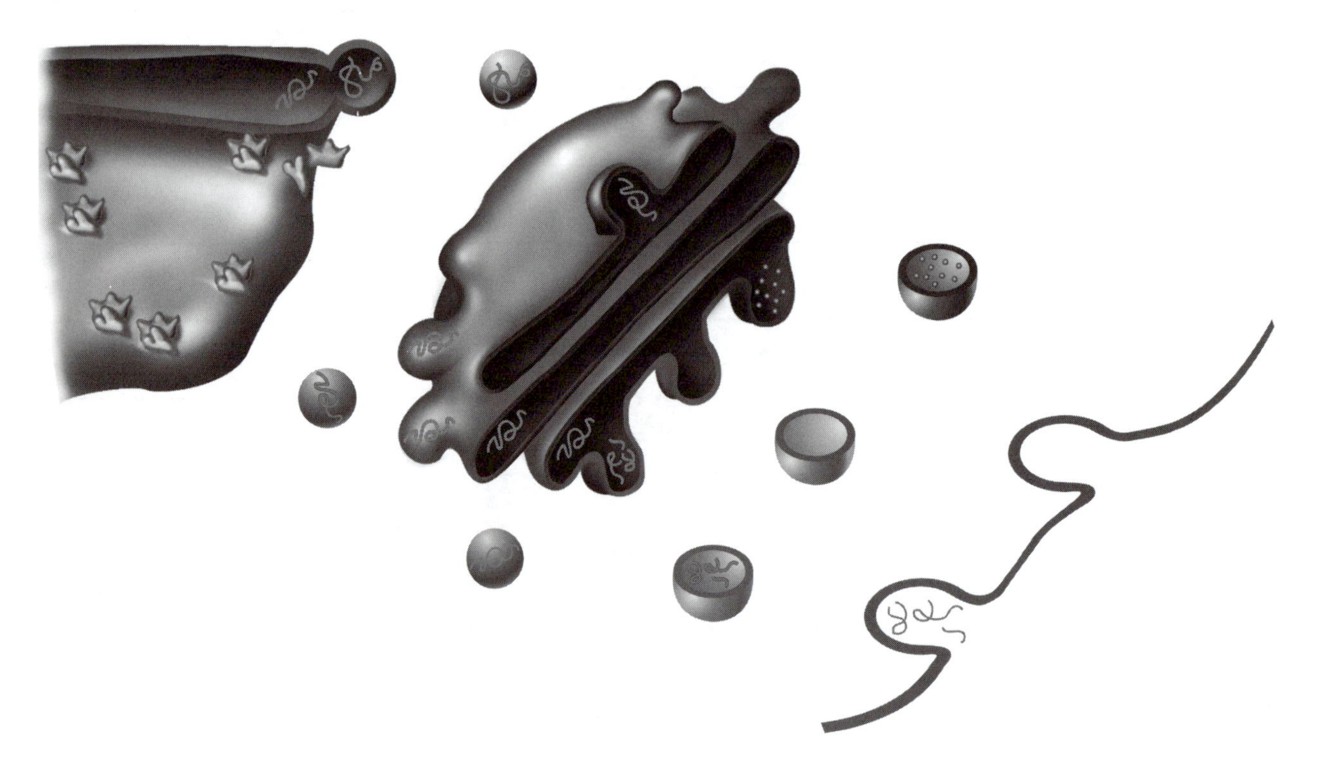

NOTES

Figure 3.22 Lysosomes (page 83).

Figure 3.23 Mitochondria (page 84).

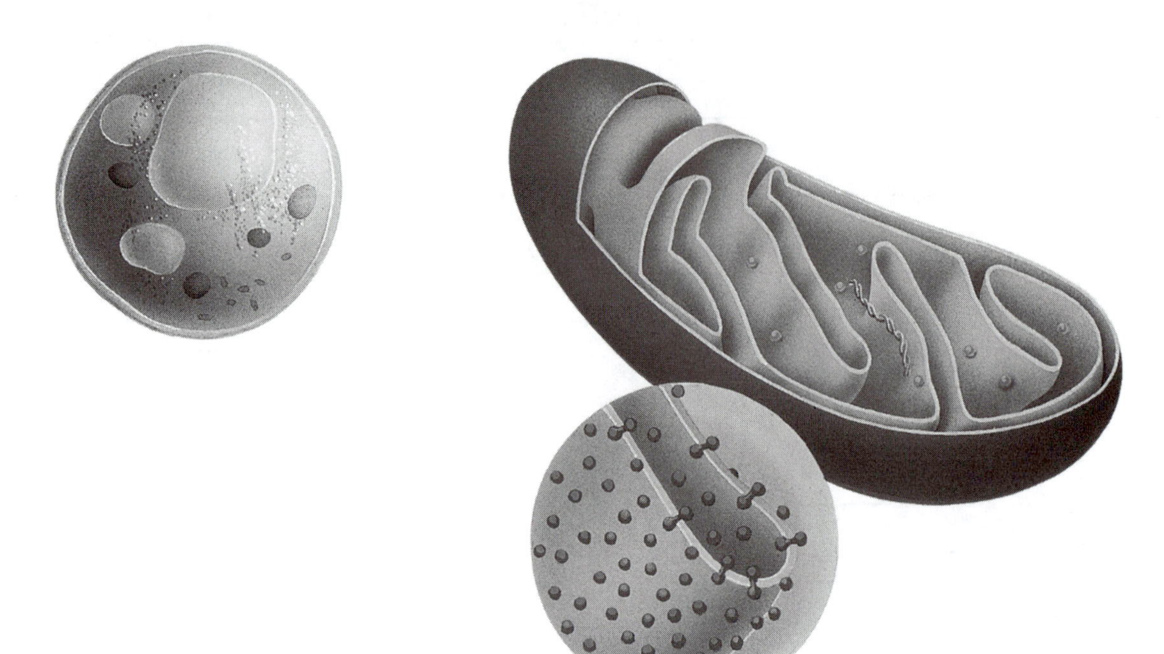

Figure 3.25 Nucleus (page 85).

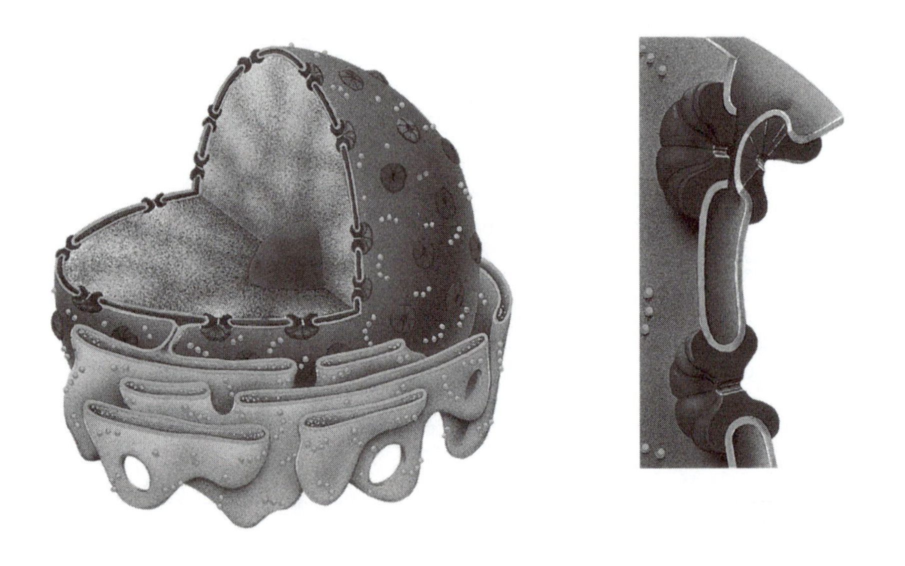

NOTES

3

Figure 3.26 Packing of DNA into a chromosome (page 86).

Figure 3.27 Overview of transcription and translation (page 88).

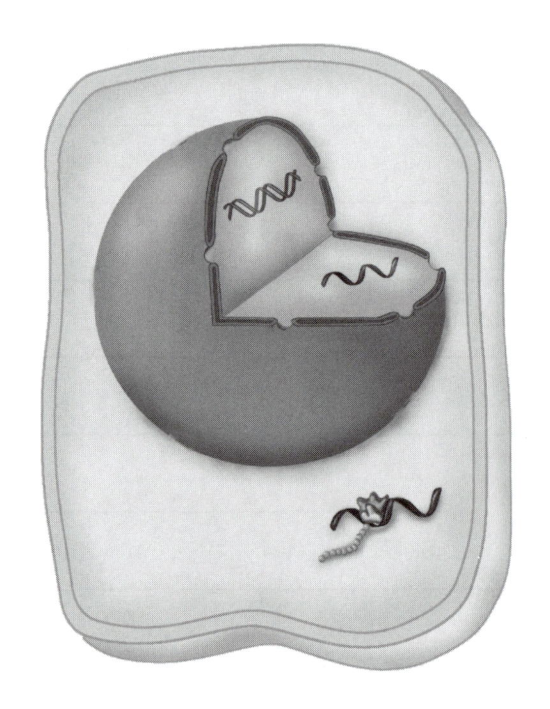

NOTES

Figure 3.28 Transcription (page 88).

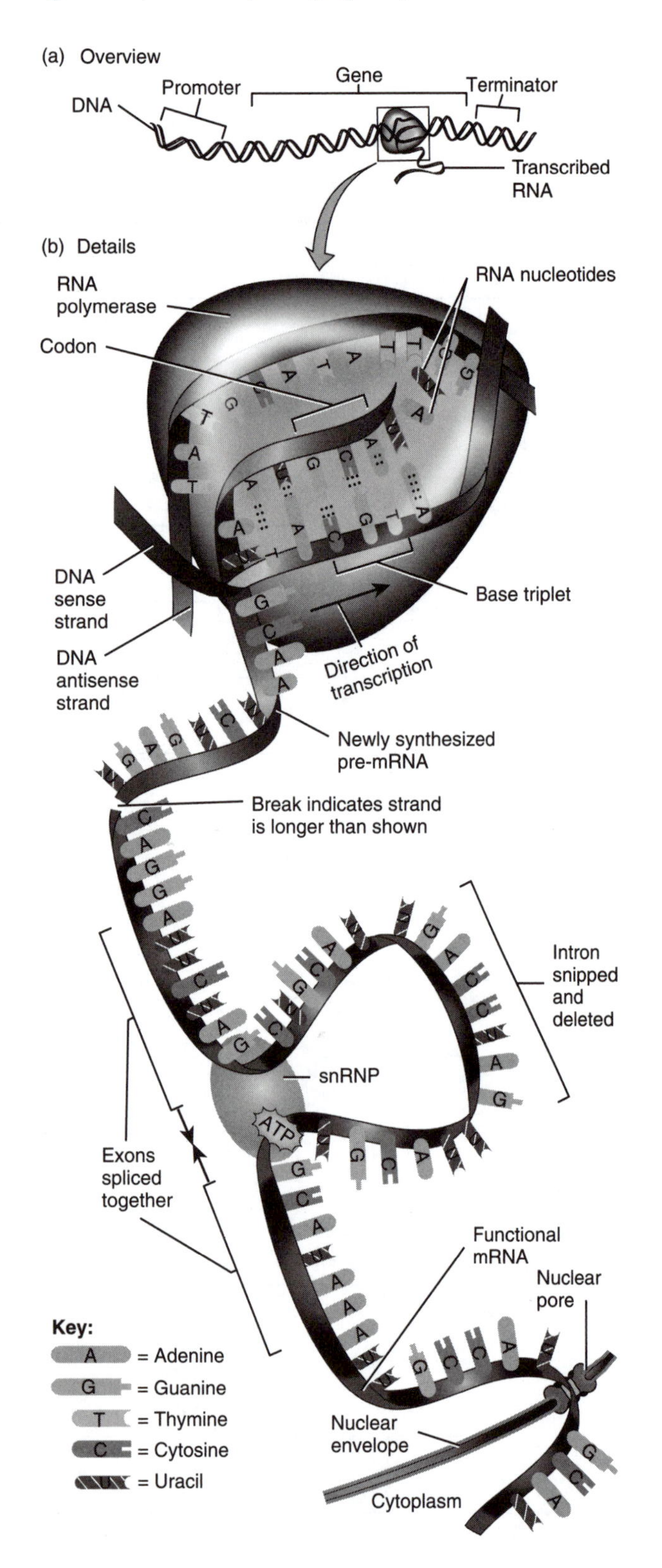
(a) Overview
Promoter
Gene
Terminator
DNA
Transcribed RNA
(b) Details
RNA polymerase
RNA nucleotides
Codon
DNA sense strand
DNA antisense strand
Base triplet
Direction of transcription
Newly synthesized pre-mRNA
Break indicates strand is longer than shown
Intron snipped and deleted
snRNP
ATP
Exons spliced together
Functional mRNA
Nuclear pore
Nuclear envelope
Cytoplasm
Key:
A = Adenine
G = Guanine
T = Thymine
C = Cytosine
= Uracil

Figure 3.29 Ribosomes and translation (page 89).

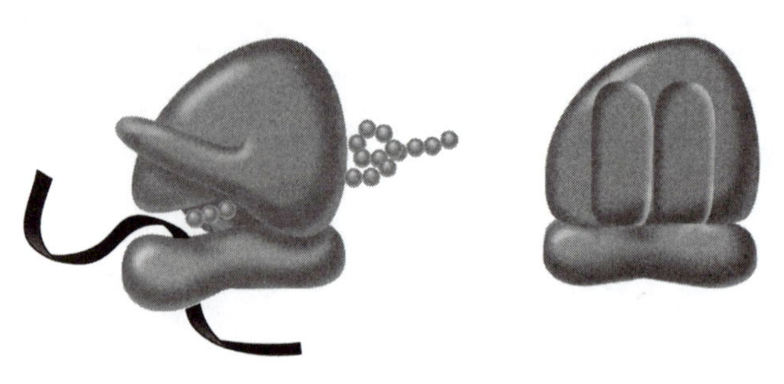

NOTES

3

Figure 3.30 Protein elongation and termination of protein synthesis during translation (page 90).

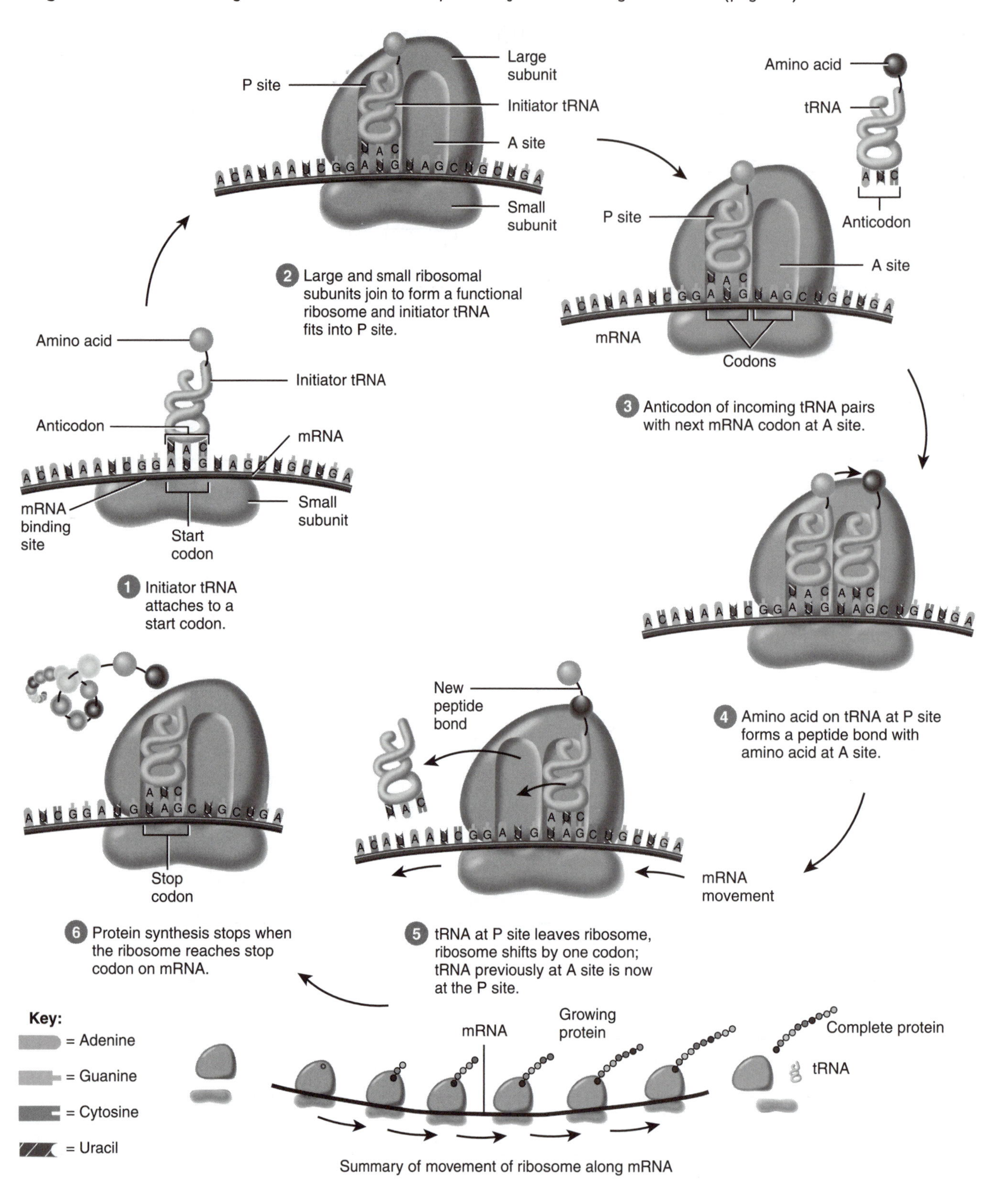

P site
Large subunit
Initiator tRNA
A site
Small subunit
2 Large and small ribosomal subunits join to form a functional ribosome and initiator tRNA fits into P site.
Amino acid
tRNA
Anticodon
P site
A site
mRNA
Codons
3 Anticodon of incoming tRNA pairs with next mRNA codon at A site.
Amino acid
Initiator tRNA
Anticodon
mRNA
mRNA binding site
Start codon
Small subunit
1 Initiator tRNA attaches to a start codon.
4 Amino acid on tRNA at P site forms a peptide bond with amino acid at A site.
New peptide bond
mRNA movement
Stop codon
6 Protein synthesis stops when the ribosome reaches stop codon on mRNA.
5 tRNA at P site leaves ribosome, ribosome shifts by one codon; tRNA previously at A site is now at the P site.
Key:
= Adenine
= Guanine
= Cytosine
= Uracil
mRNA
Growing protein
Complete protein
tRNA
Summary of movement of ribosome along mRNA

NOTES

Figure 3.31 The cell cycle (page 91).

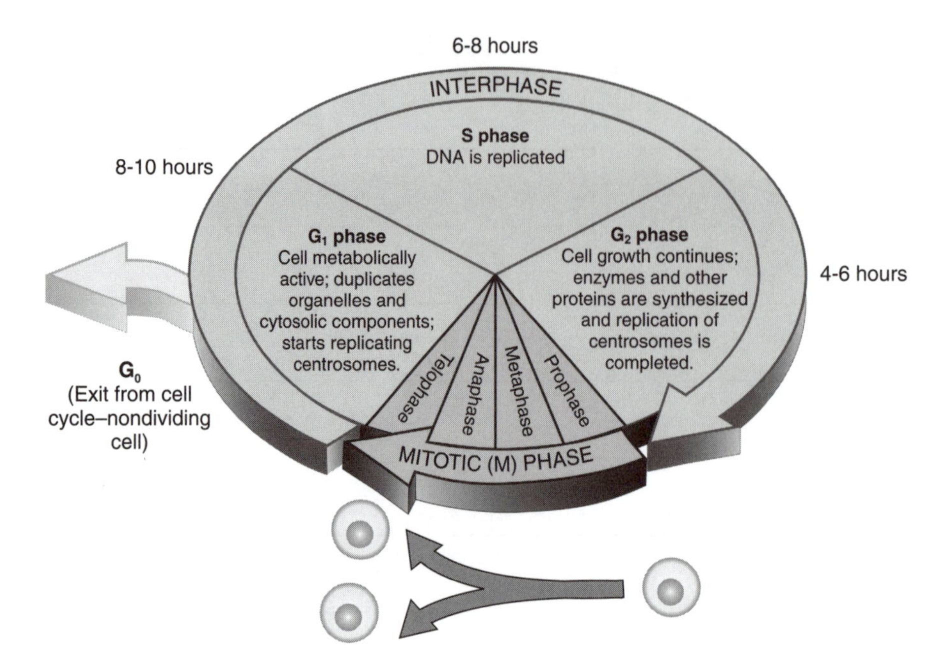

6-8 hours
INTERPHASE
S phase
DNA is replicated
8-10 hours
G_1 phase
Cell metabolically active; duplicates organelles and cytosolic components; starts replicating centrosomes.
G_2 phase
Cell growth continues; enzymes and other proteins are synthesized and replication of centrosomes is completed.
4-6 hours
G_0
(Exit from cell cycle–nondividing cell)
Telophase
Anaphase
Metaphase
Prophase
MITOTIC (M) PHASE

Figure 3.32 Replication of DNA (page 92).

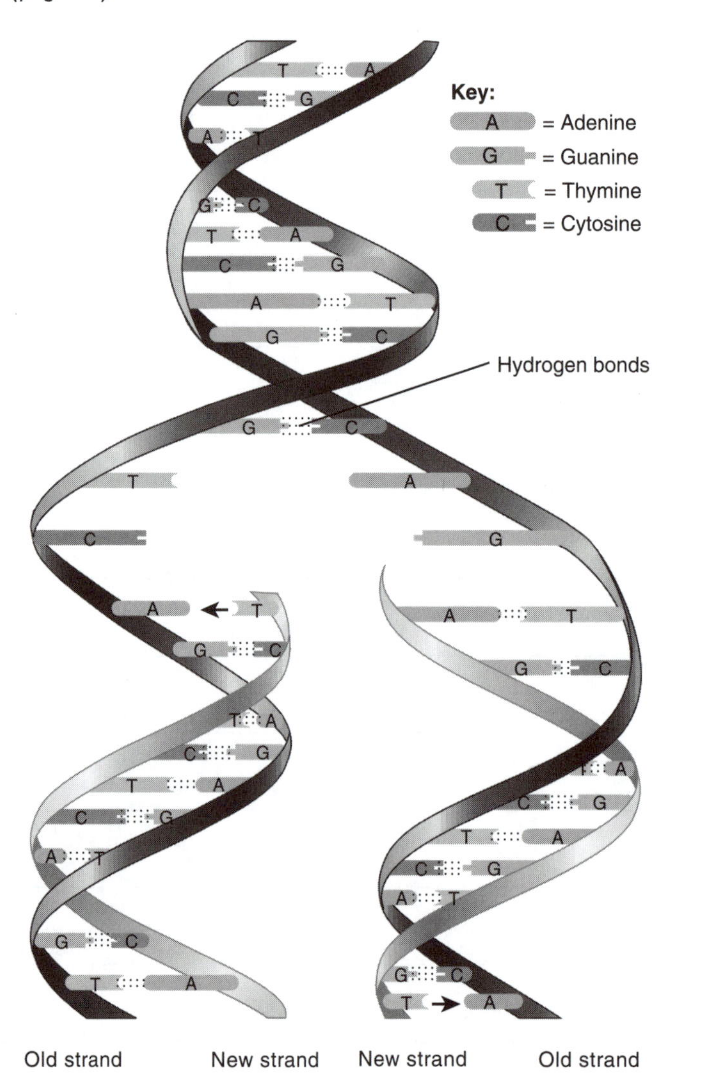

Key:
A = Adenine
G = Guanine
T = Thymine
C = Cytosine
Hydrogen bonds
Old strand
New strand
New strand
Old strand

NOTES

3

Figure 3.33 Cell division: mitosis and cytokinesis (page 93).

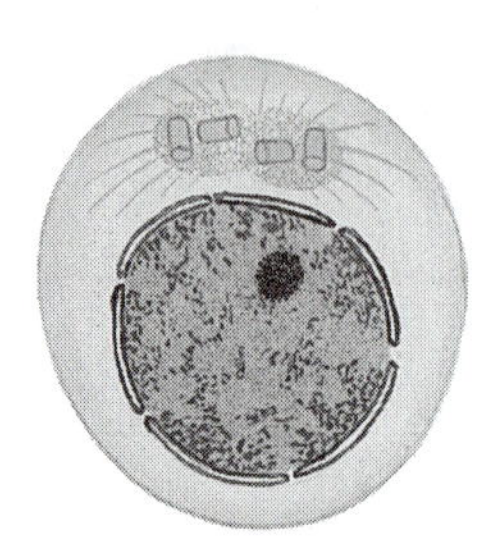

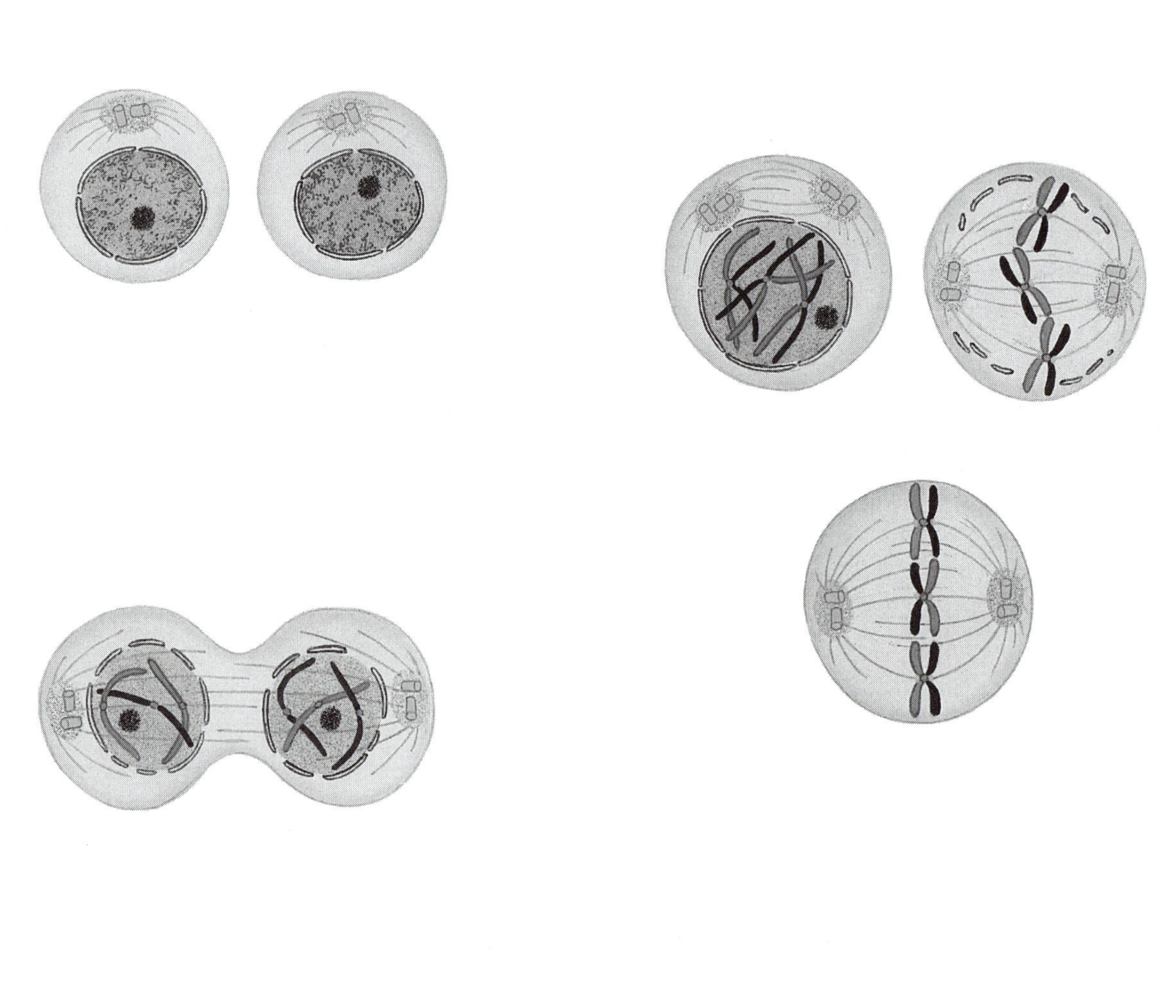

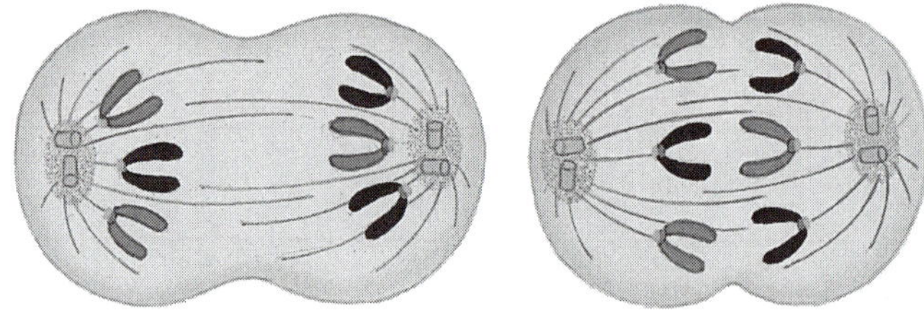

NOTES

3

Figure 3.34 Diverse shapes and sizes of human cells (page 96).

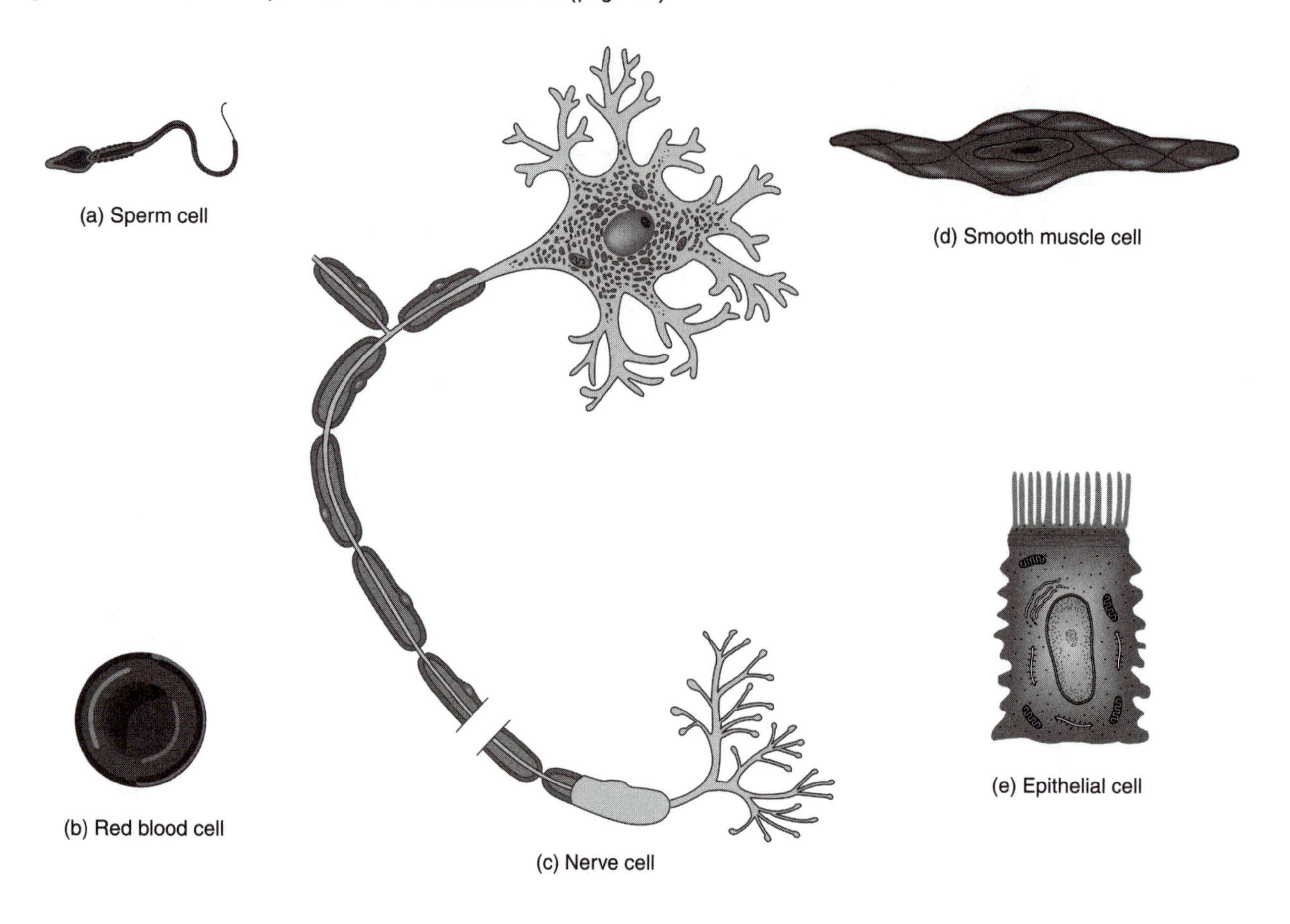
(a) Sperm cell
(b) Red blood cell
(c) Nerve cell
(d) Smooth muscle cell
(e) Epithelial cell

NOTES

3

Figure 4.1 Cell junctions (page 106).

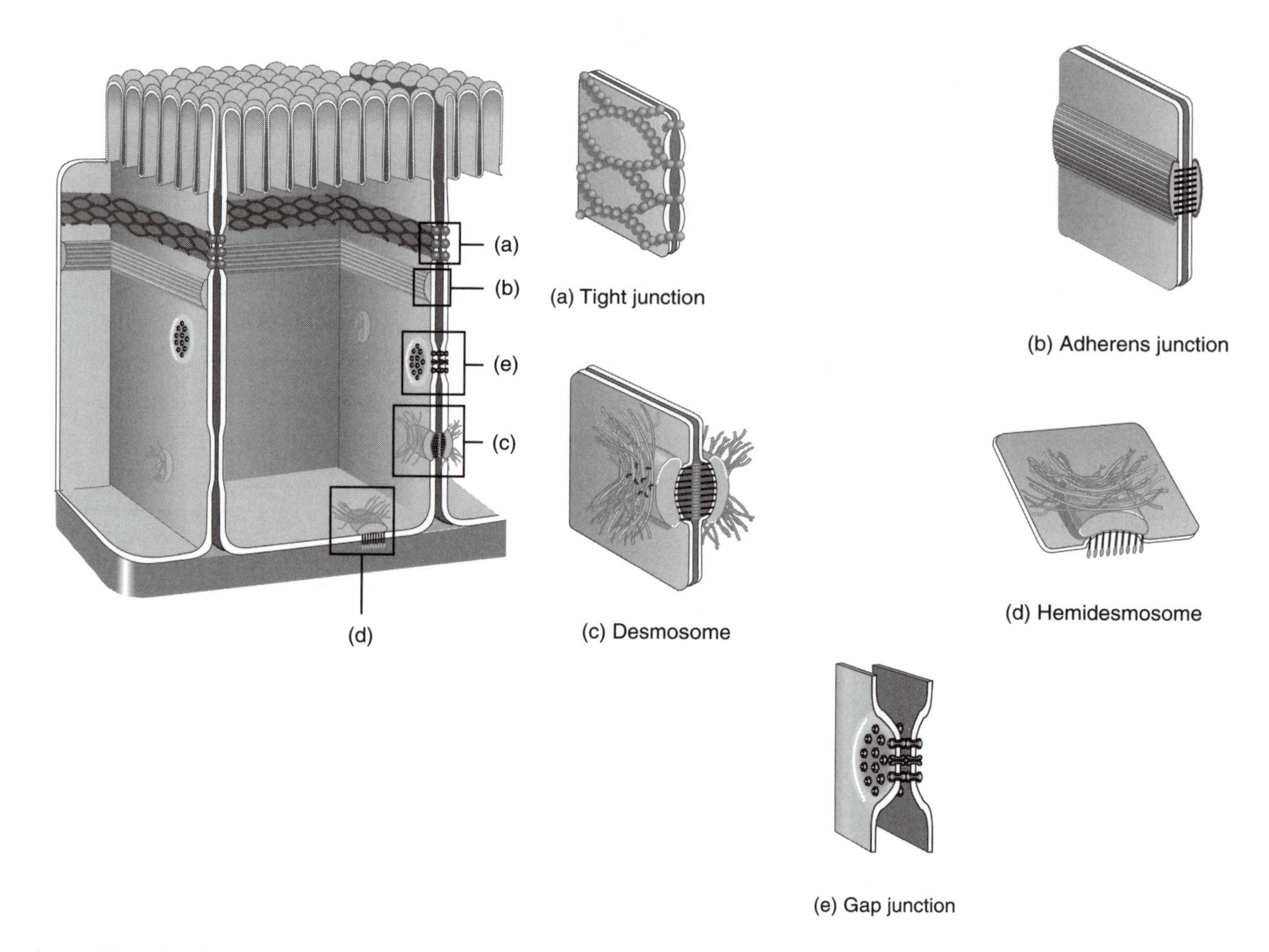
(a)
(b)
(e)
(c)
(d)
(a) Tight junction
(b) Adherens junction
(c) Desmosome
(d) Hemidesmosome
(e) Gap junction

Adapted from Lewis Kleinsmith and Valerie Kish, *Principles of Cell and Molecular Biology* 2e, F6.44, p237; F6.46, p238; F6.47, p239; F6.50, p241 (New York: HarperCollins, 1995). © 1995 HarperCollins College Publishers. By permission of Addison Wesley Longman.

Figure 4.2 Surfaces of epithelial cells and the structure and location of the basement membrane (page 107).

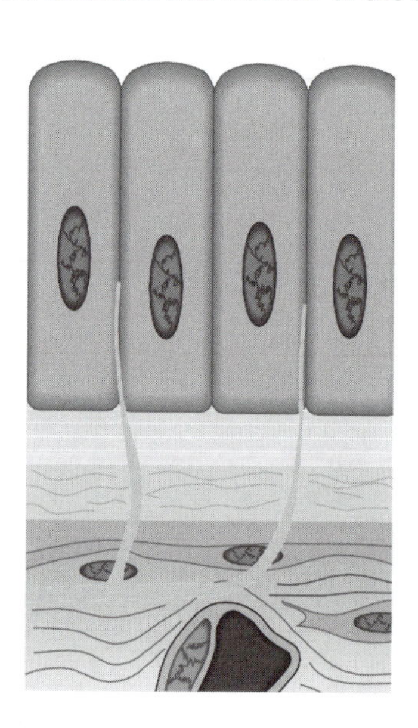

NOTES

4

Table 4.1a Simple squamous epithelium (page 108).

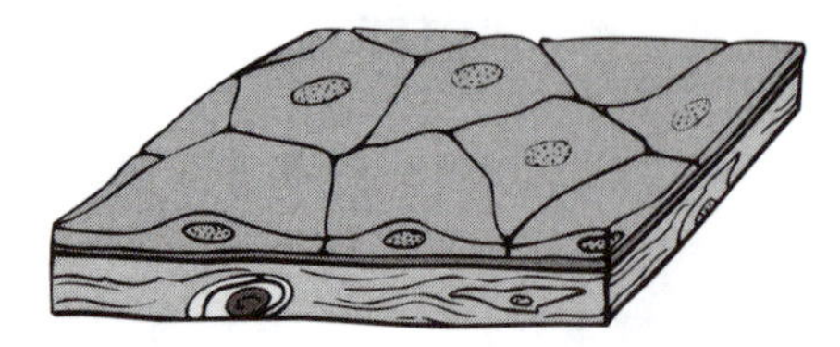

Table 4.1b A simple cuboidal epithelium (page 108).

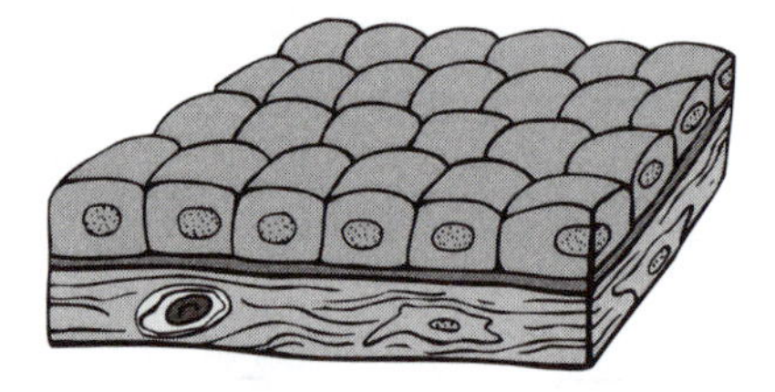

Table 4.1c Nonciliated simple columnar epithelium (page 109).

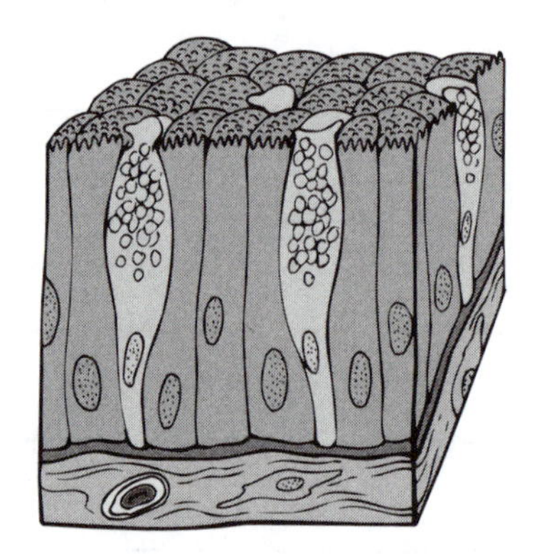

NOTES

4

Table 4.1d Ciliated simple columnar epithelium (page 109).

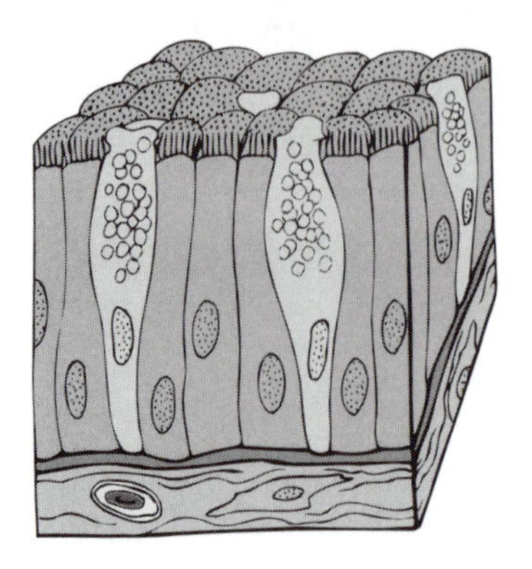

Table 4.1e Stratified squamous epithelium (page 110).

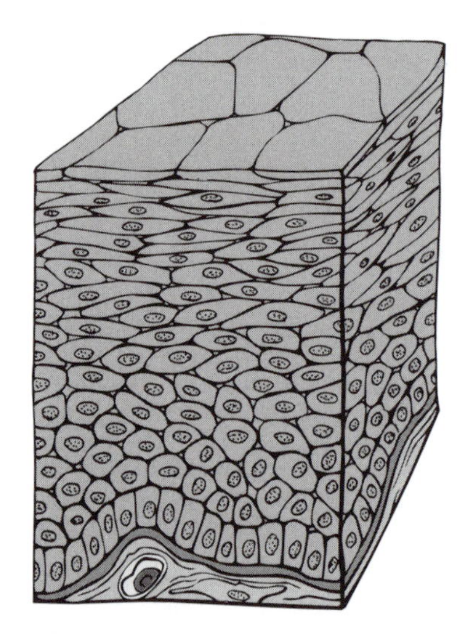

Table 4.1f Stratified cuboidal epithelium (pate 111).

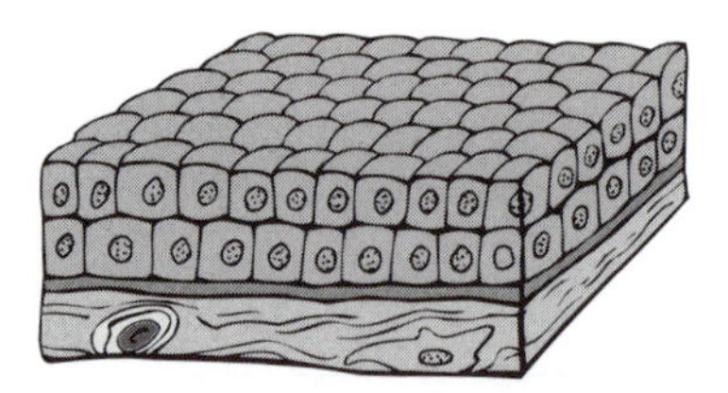

NOTES

4

Table 4.1g Stratified columnar epithelium (page 111).

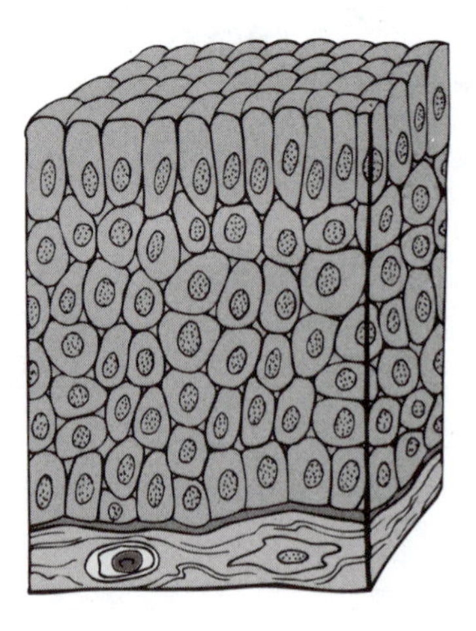

Table 4.1h Transitional epithelium (page 112).

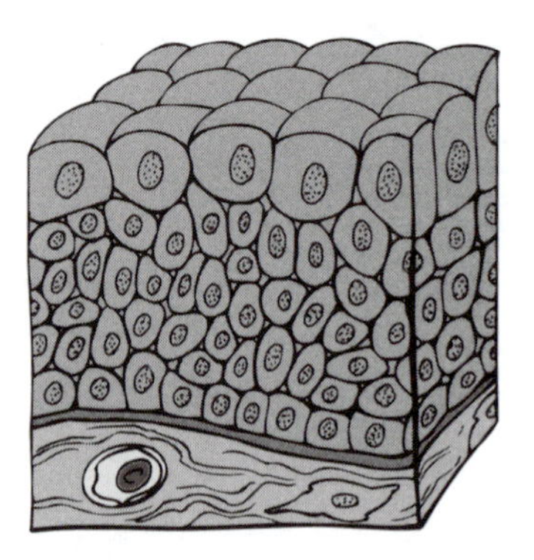

Table 4.1i Pseudostratified columnar epithelium (page 113).

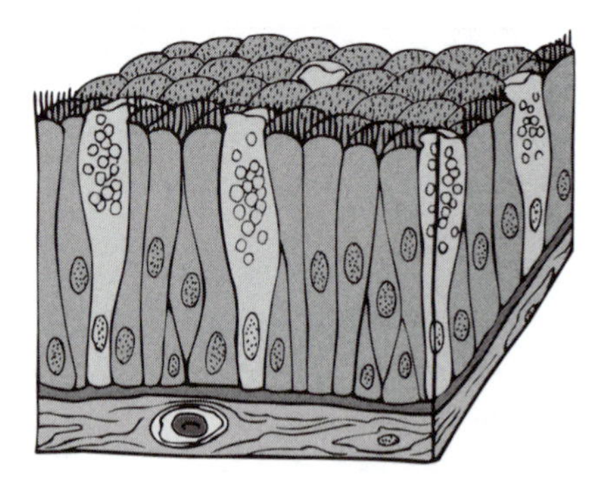

NOTES

Table 4.1j Endocrine glands (page 114).

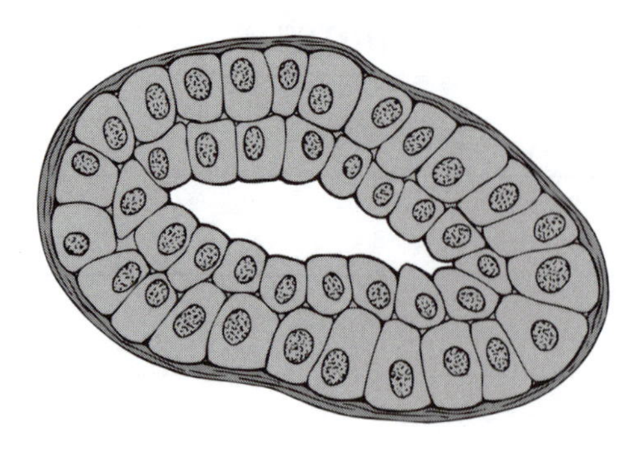

Table 4.1k Exocrine glands (page 114).

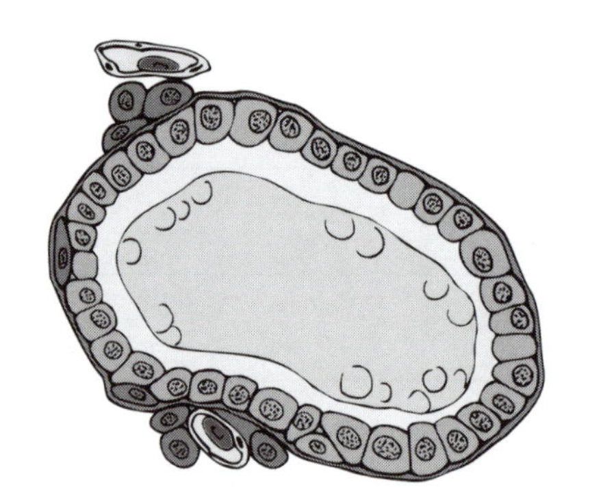

NOTES

Figure 4.3 Multicellular exocrine glands (page 116).

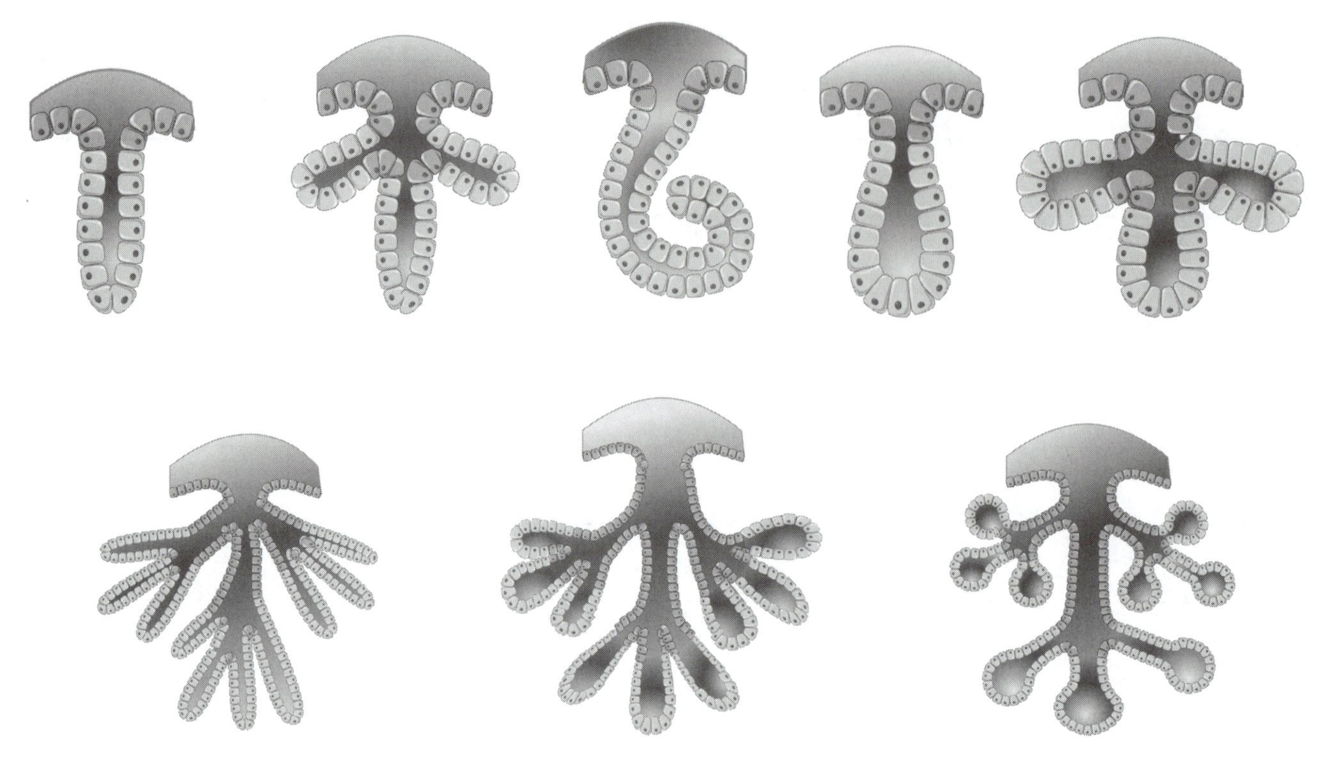

Figure 4.4 Functional classification of multicellular exocrine glands (page 117).

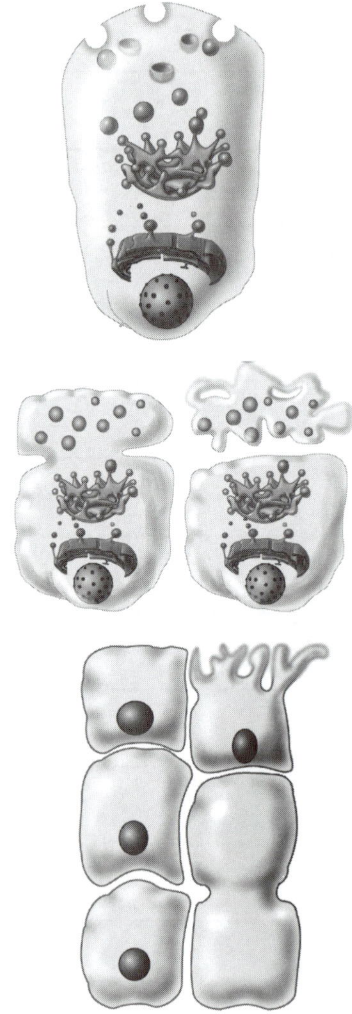

NOTES

4

Figure 4.5 Representative cells and fibers present in connective tissues (page 119).

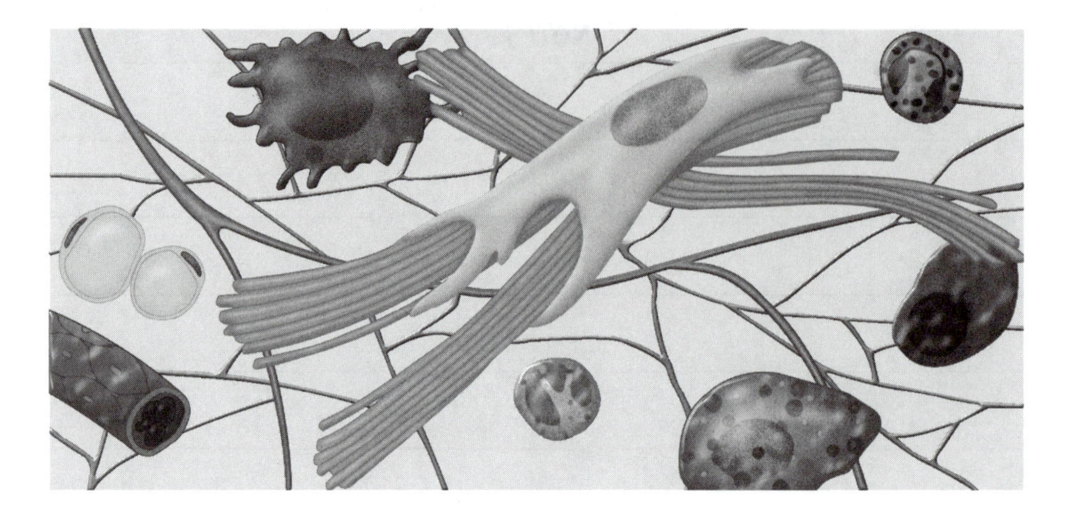

Table 4.2a Mesenchyme (page 121).

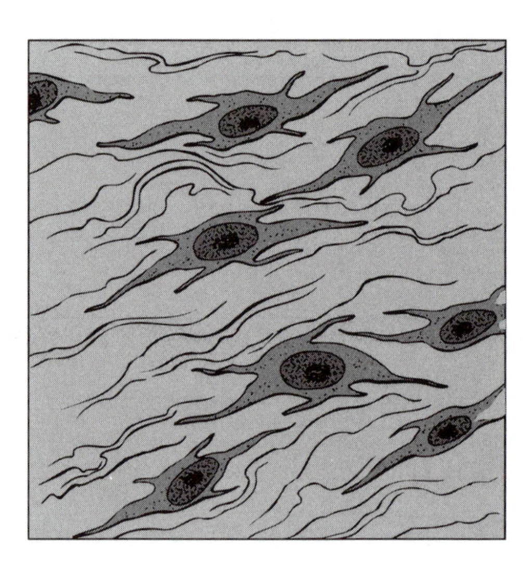

Table 4.2b Mucous connective tissue (page 121).

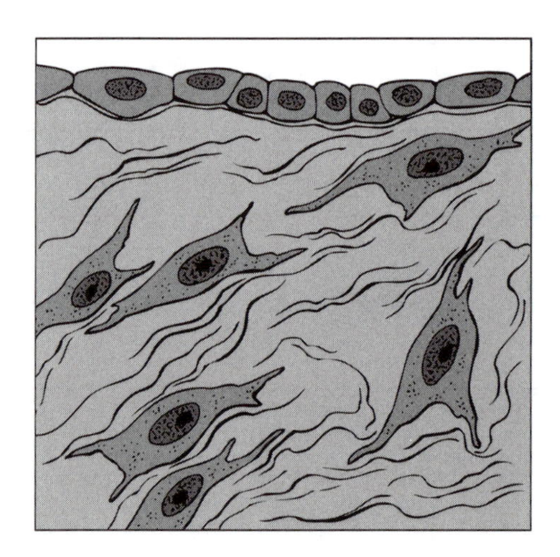

NOTES

Table 4.3a Areolar connective tissue (page 122).

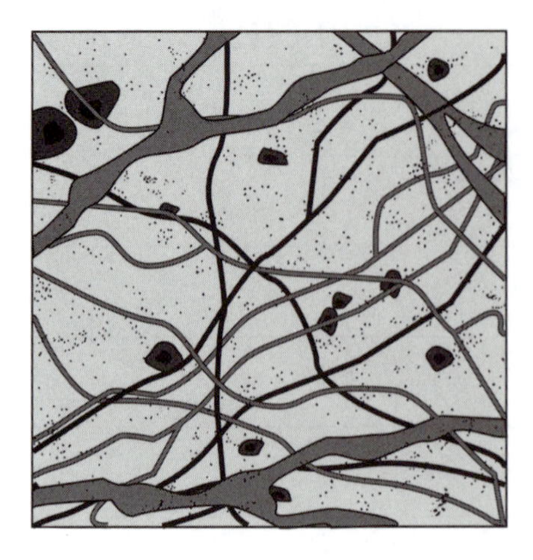

Table 4.3b Adipose tissue (page 122).

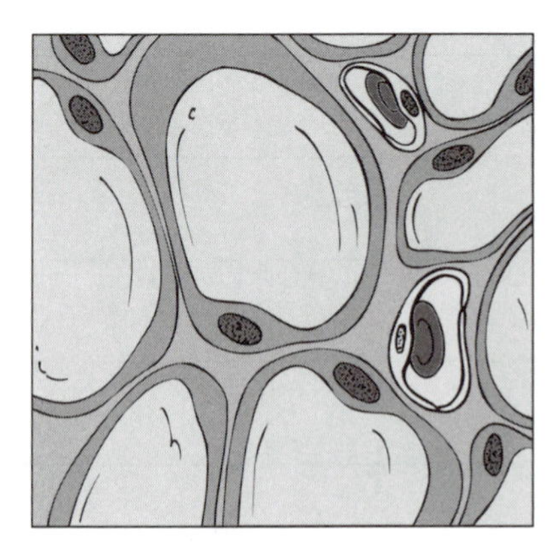

Table 4.3c Reticular connective tissue (page 123).

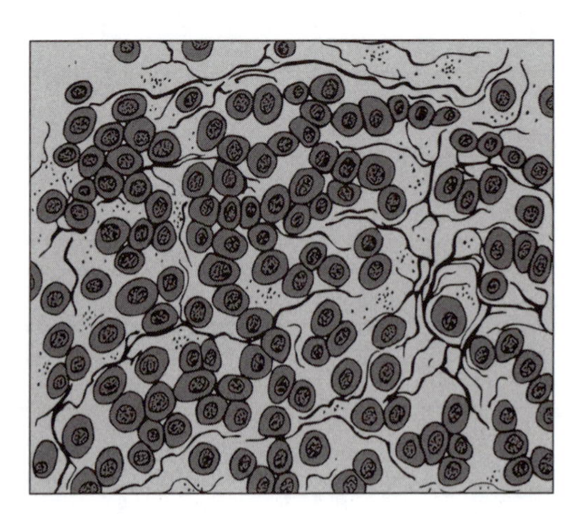

NOTES

4

Table 4.3d Dense regular connective tissue (page 123).

Table 4.3e Dense irregular connective tissue (page 124).

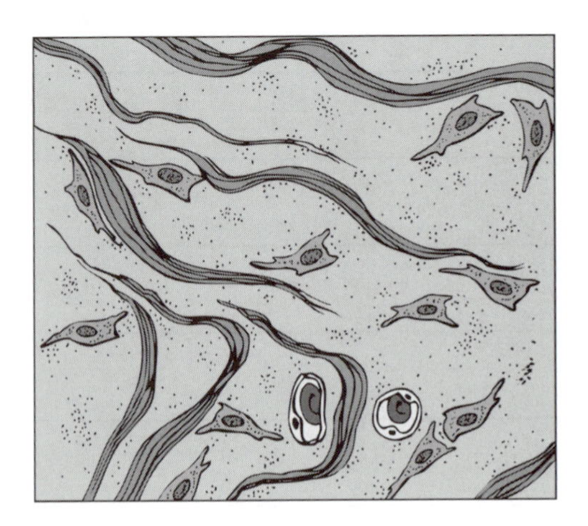

Table 4.3f Elastic connective tissue (page 124).

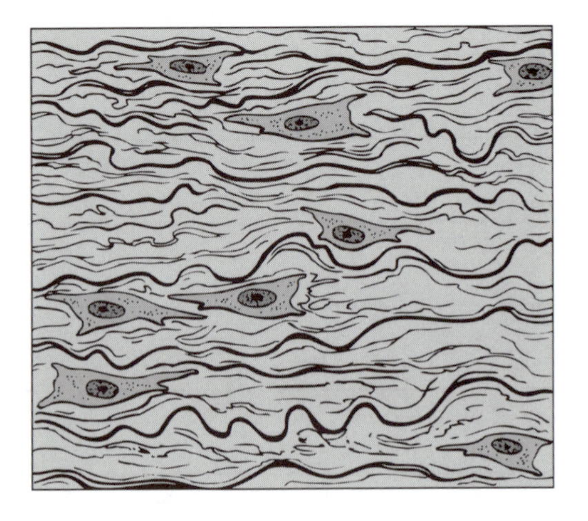

NOTES

Table 4.3g Hyaline cartilage (page 125).

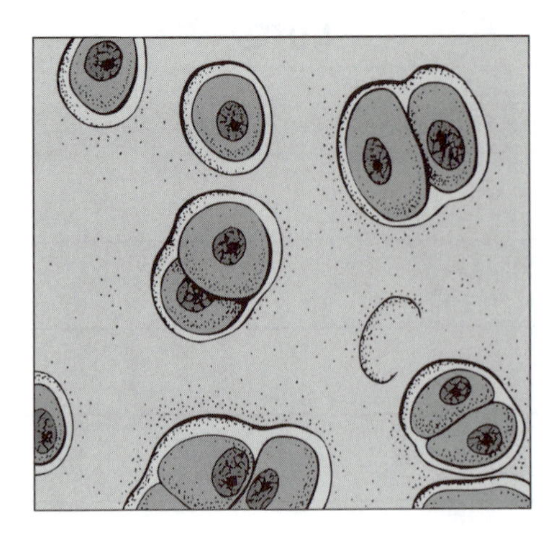

Table 4.3h Fibrocartilage (page 125).

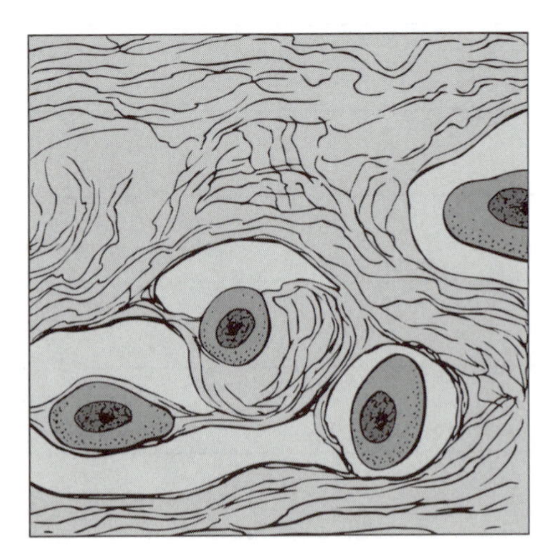

Table 4.3i Elastic cartilage (page 126).

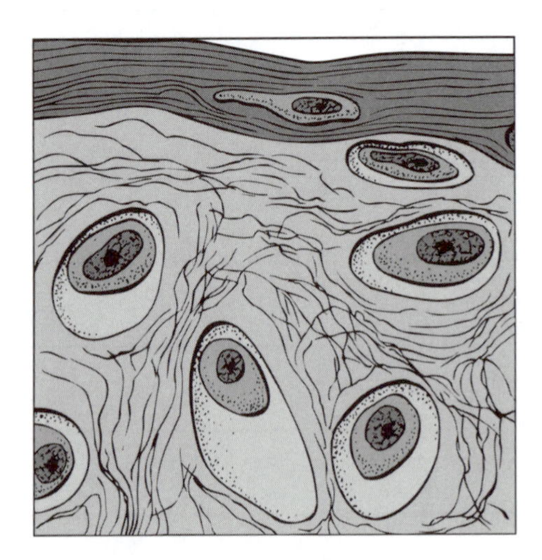

NOTES

4

Table 4.3j Compact bone (page 126).

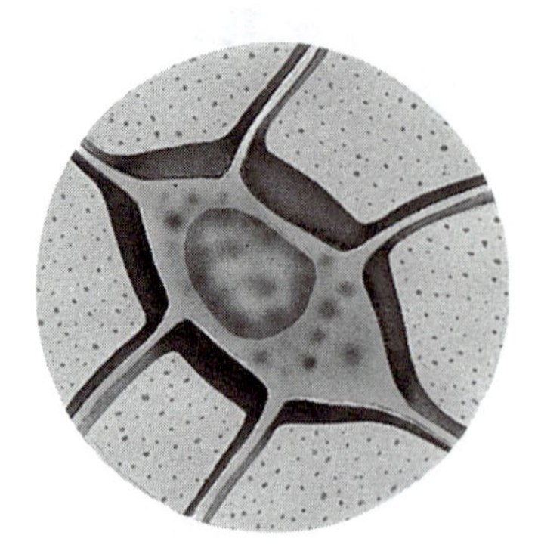

Table 4.3k Blood (page 127).

NOTES

4

Table 4.4a Skeletal muscle tissue (page 131).

Table 4.4b Cardiac muscle tissue (page 131).

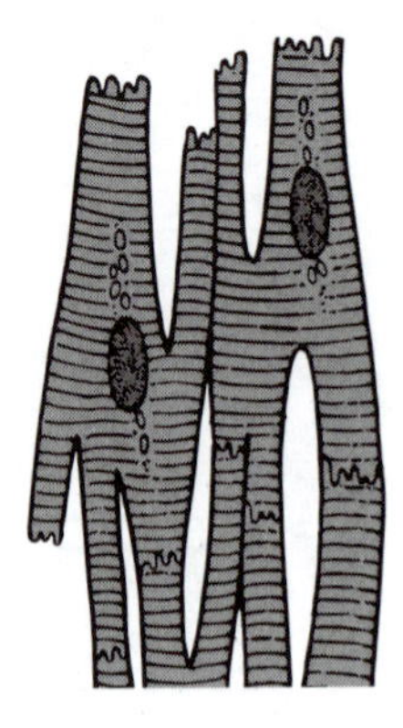

Table 4.4c Smooth muscle tissue (page 132).

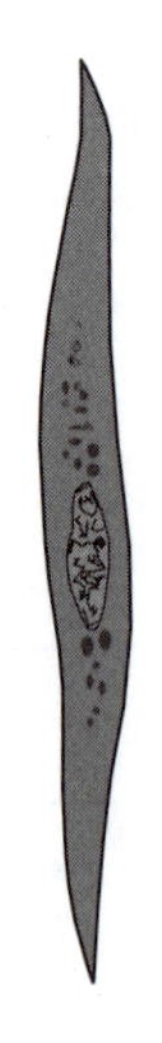

NOTES

Figure 5.1 Components of the integumentary system (page 141).

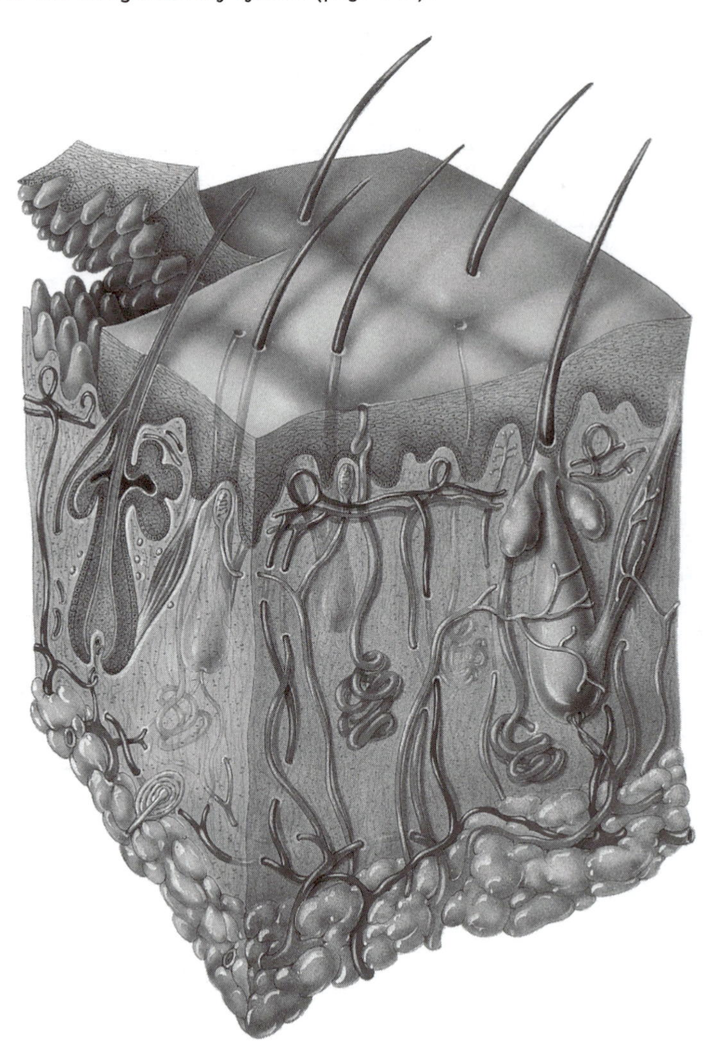

Figure 5.2 Types of cells in the epidermis (page 142).

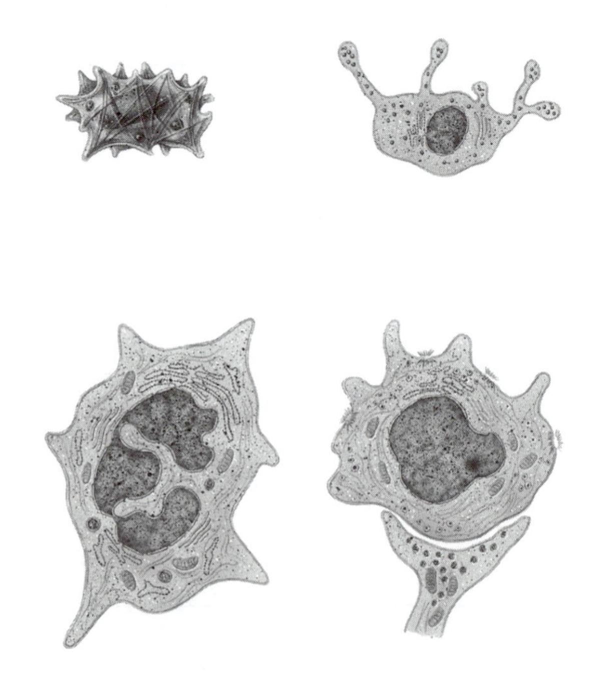

Adapted from Ira Telford and Charles Bridgeman, *Introduction to Functional Histology* 2e, p84, p261, p262 (New York: HarperCollins, 1995). ©1995 HarperCollins College Publishers. By permission of Addison Wesley Longman.

NOTES

Figure 5.3 Layers of the epidermis (page 143).

Figure 5.4 Hair (page 146).

NOTES

5

Figure 5.5 Nails (page 149).

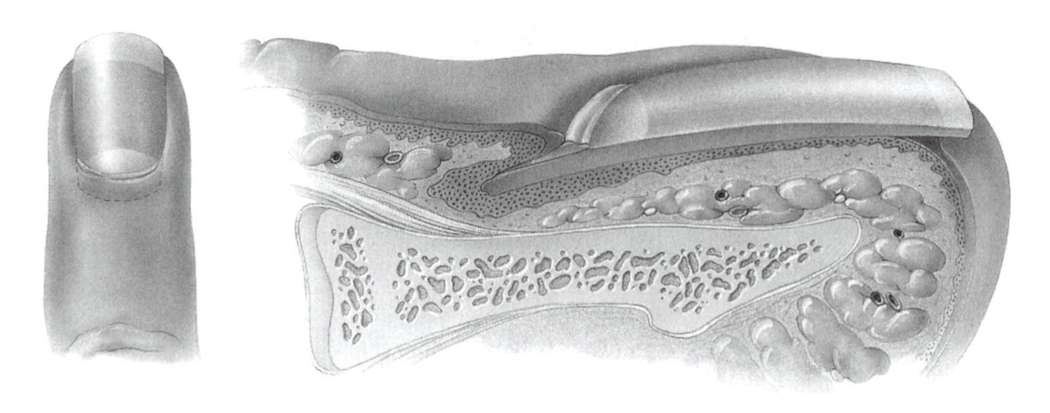

Figure 5.6 Epidermal wound healing (page 153).

NOTES

5

Figure 5.7 Deep wound healing (page 153).

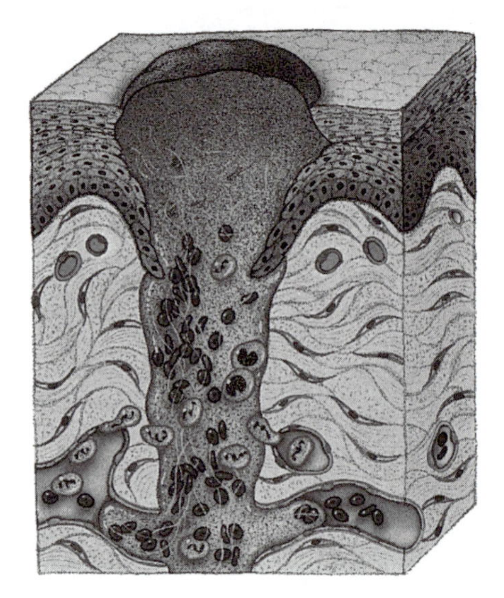

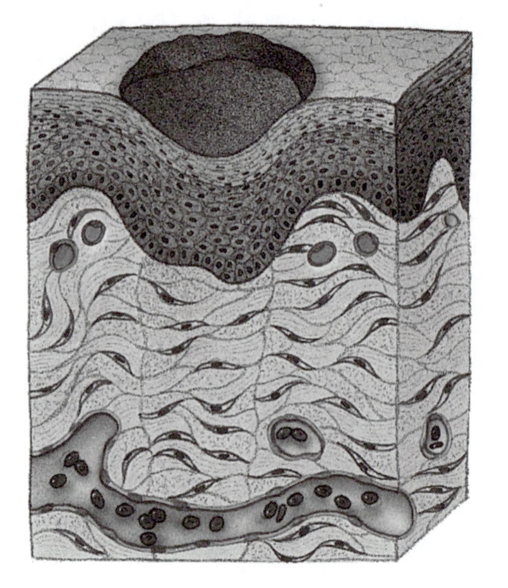

NOTES

5

Figure 6.1 Parts of a long bone (page 161).

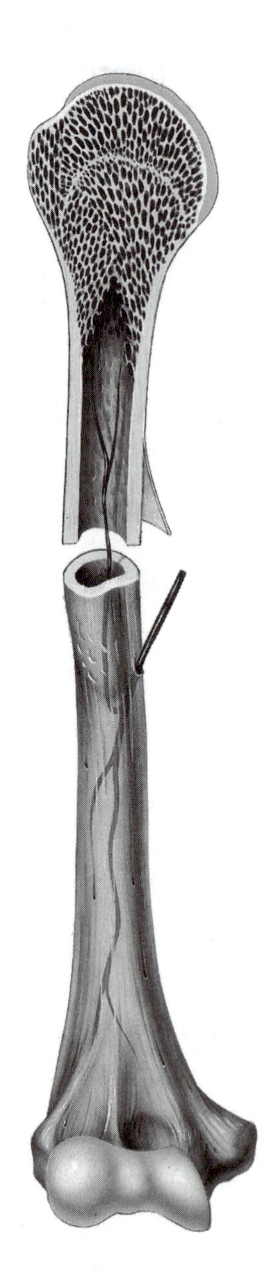

Figure 6.2 Types of cells in bone tissue (page 162).

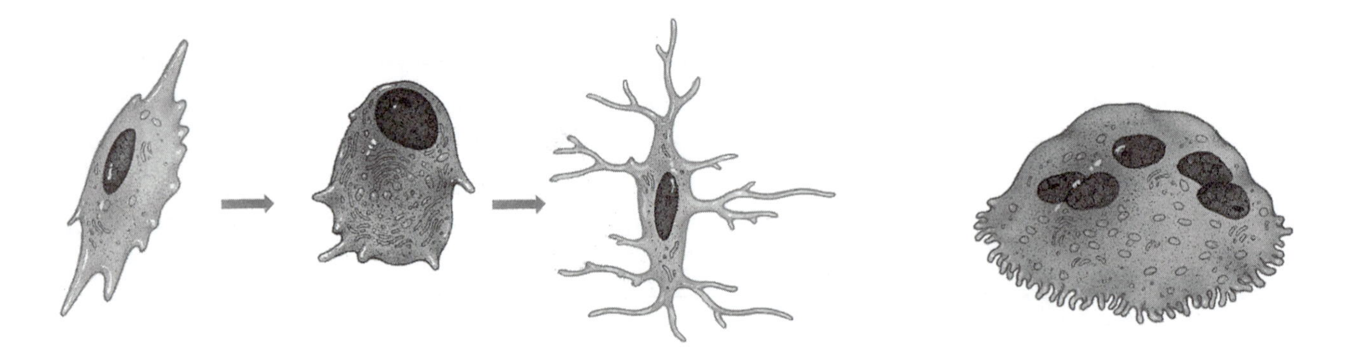

NOTES

6

Figure 6.3 Histology of compact and spongy bone (page 164).

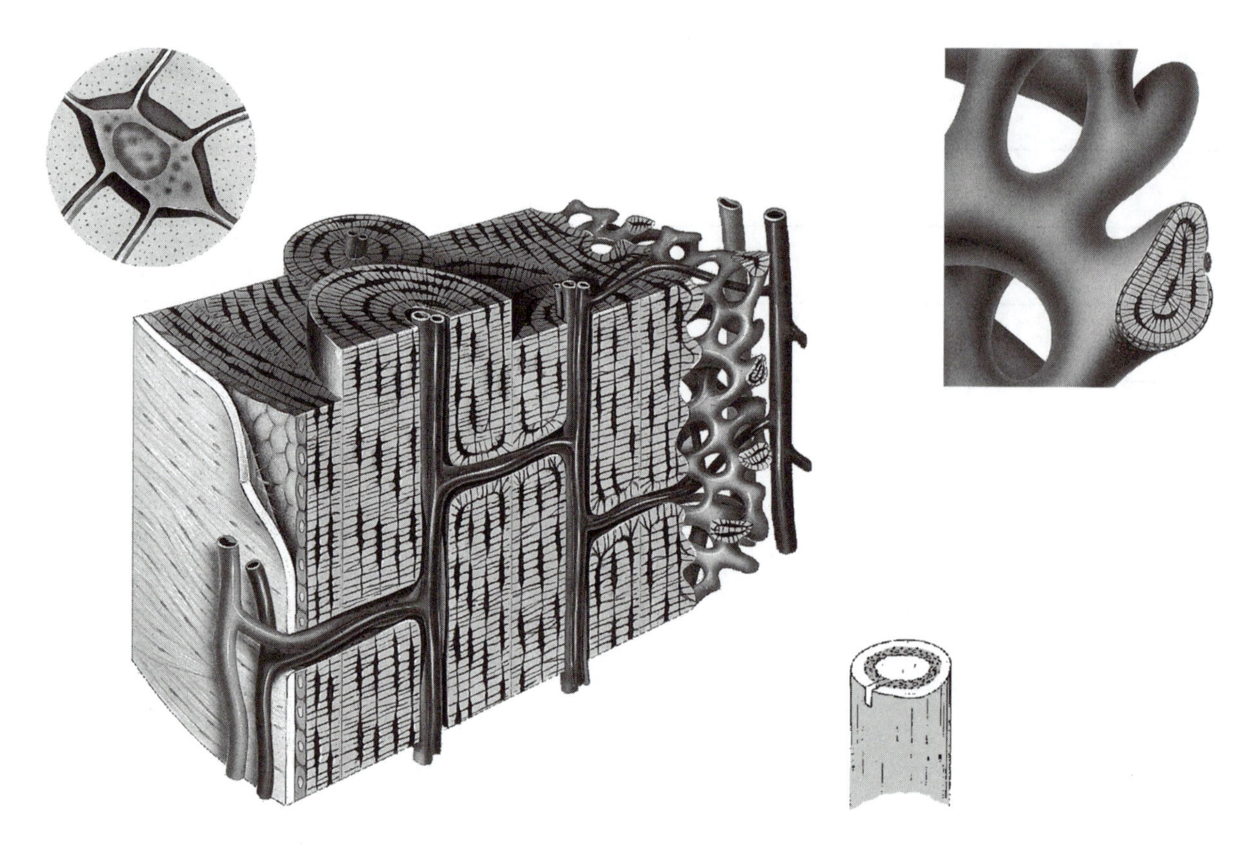

Figure 6.4 Blood supply of a mature long bone (page 166).

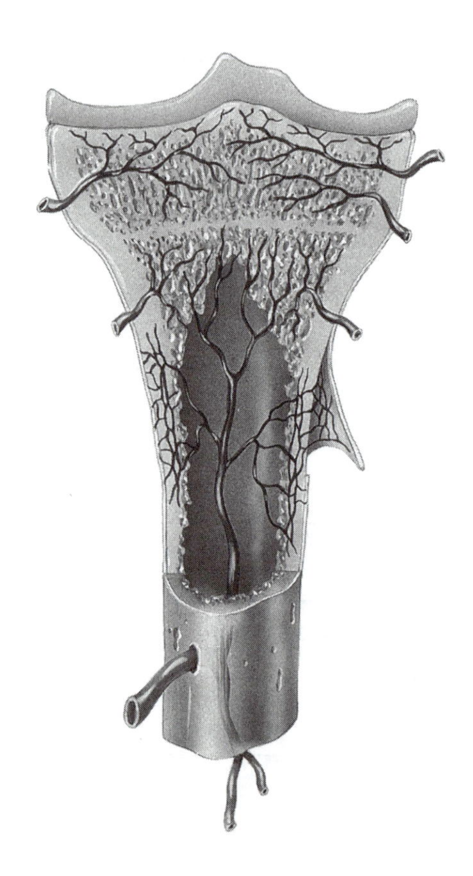

NOTES

Figure 6.5 Intramembranous ossification (page 167).

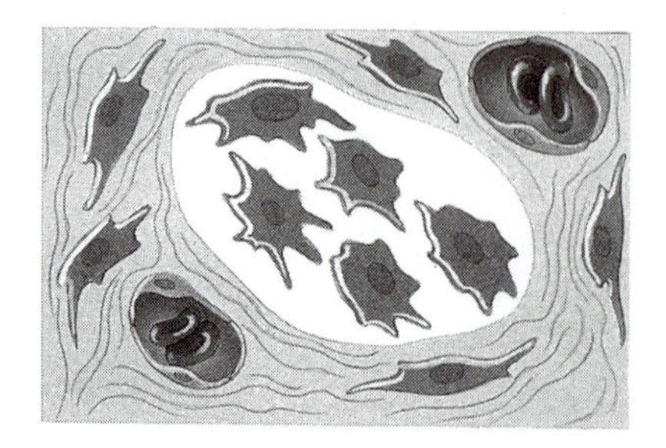

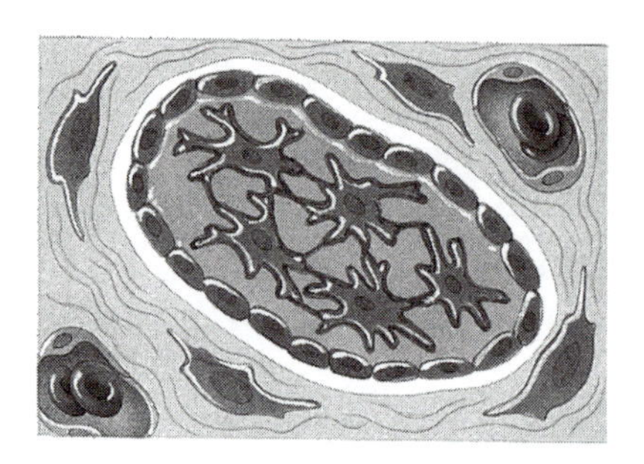

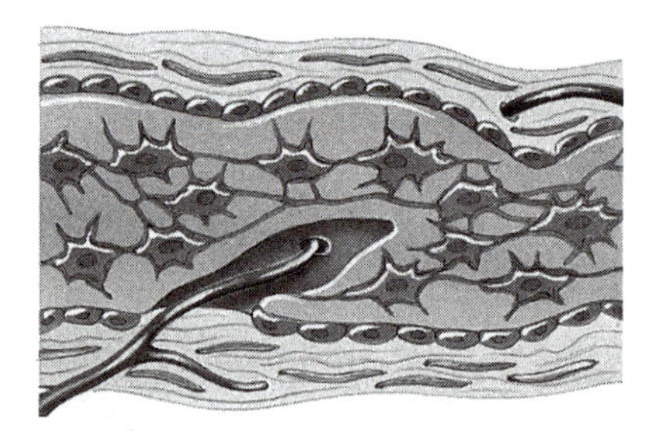

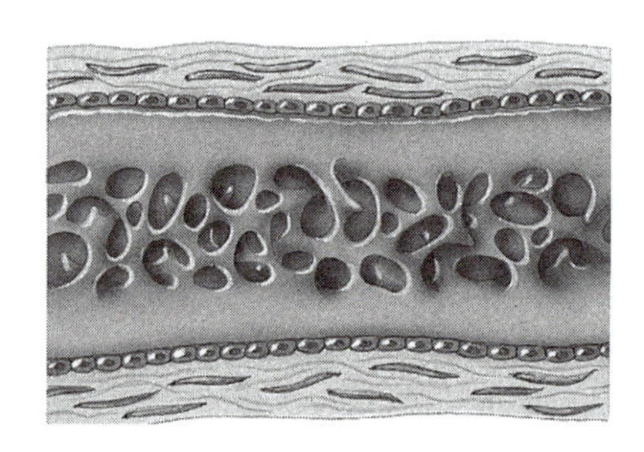

NOTES

6

Figure 6.6 Endochondral ossification (page 168).

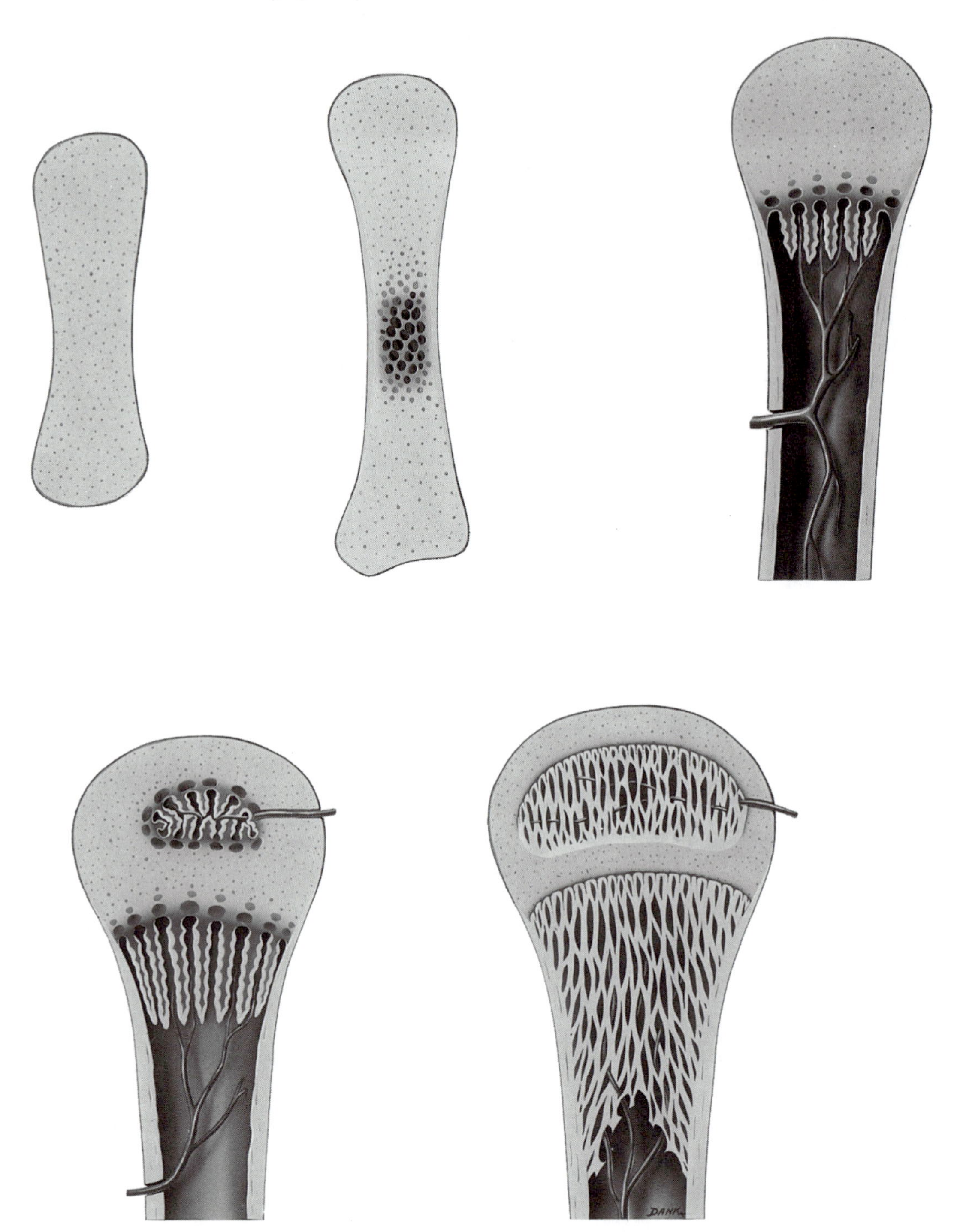

NOTES

6

Figure 6.8 Bone growth in diameter: appositional growth (page 171).

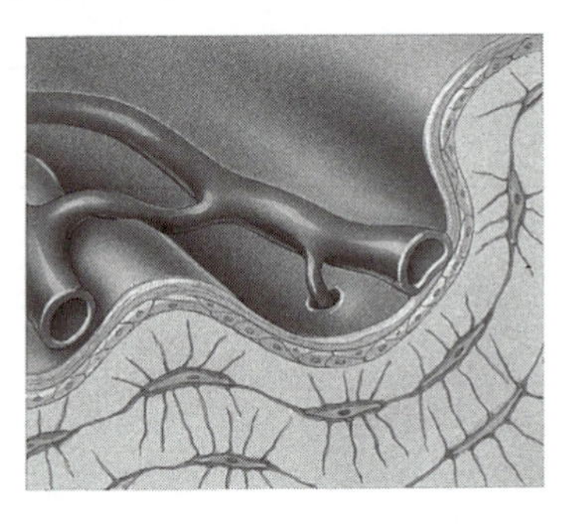

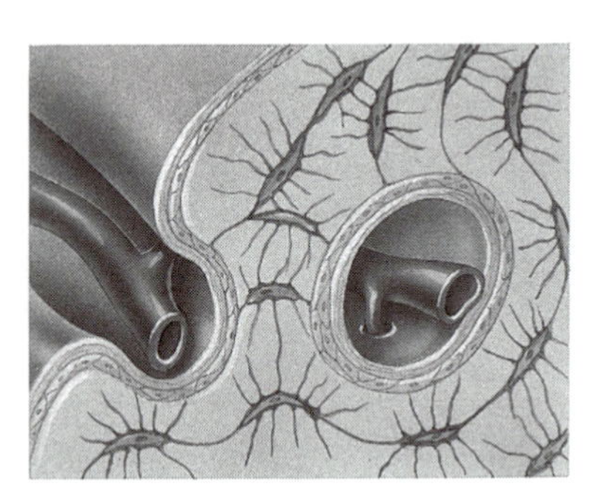

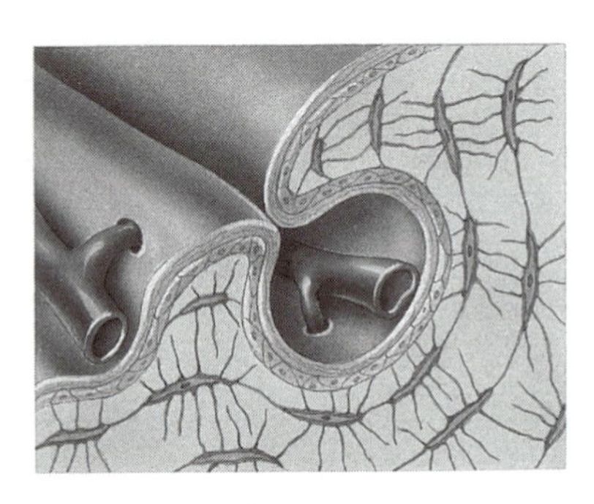

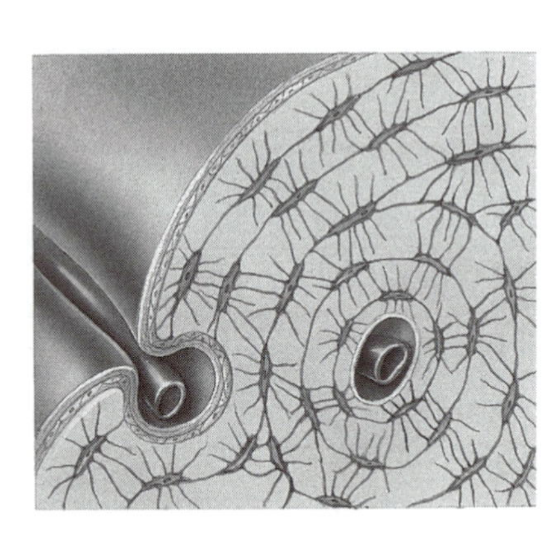

Figure 6.9 Types of bone fractures (page 173).

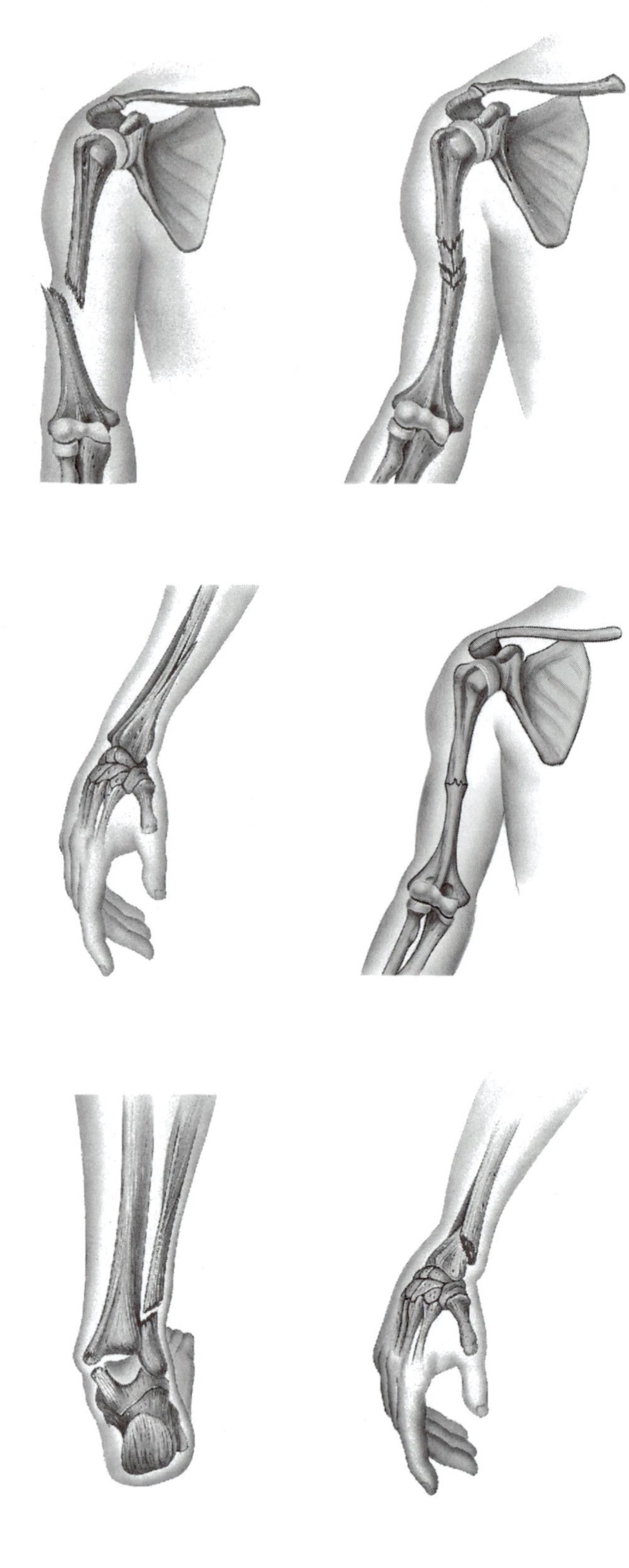

NOTES

6

Figure 6.10 Repair of a bone fracture (page 174).

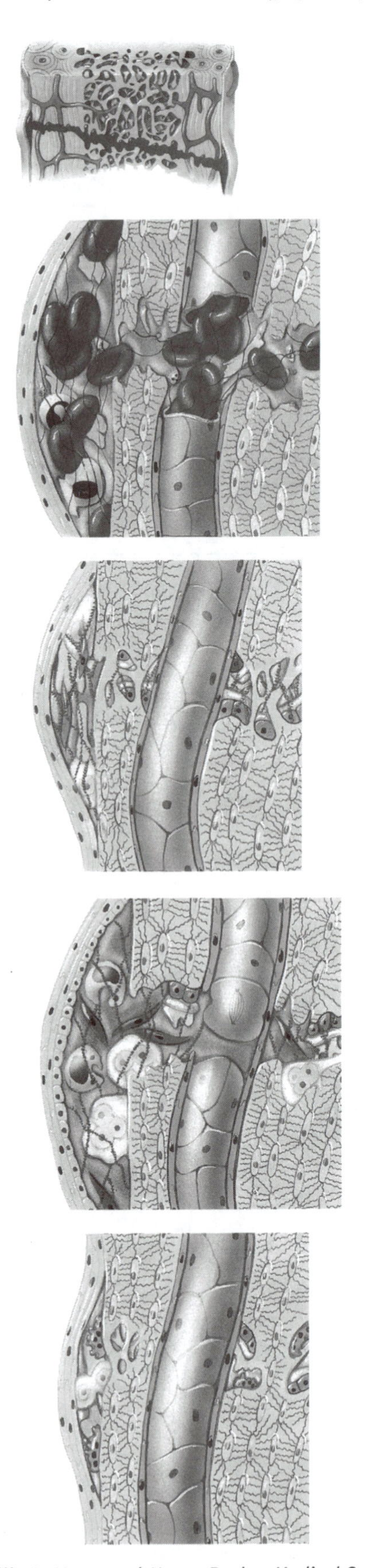

From Priscilla LeMone and Karen Burke, *Medical-Surgical Nursing,* p1560 (Menlo Park, CA: Benjamin/Cummings, 1996).

Figure 6.11 Regulation of blood calcium concentration (page 176).

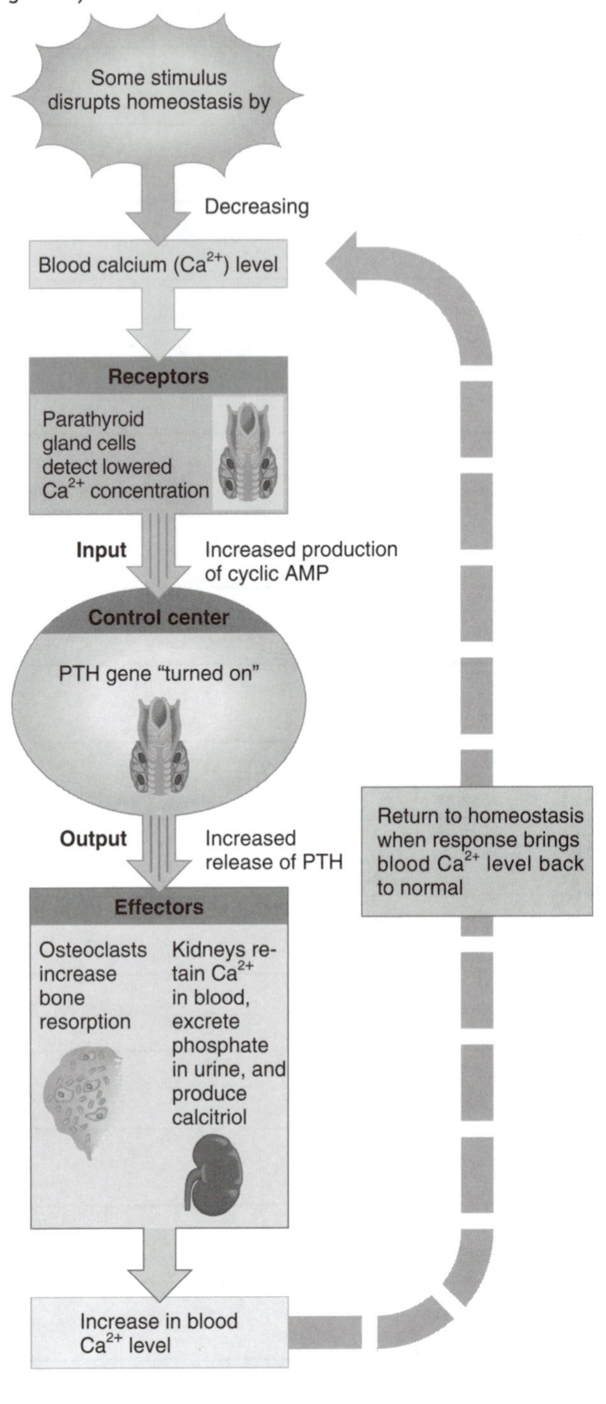
Some stimulus disrupts homeostasis by
Decreasing
Blood calcium (Ca^{2+}) level
Receptors
Parathyroid gland cells detect lowered Ca^{2+} concentration
Input
Increased production of cyclic AMP
Control center
PTH gene "turned on"
Output
Increased release of PTH
Effectors
Osteoclasts increase bone resorption
Kidneys retain Ca^{2+} in blood, excrete phosphate in urine, and produce calcitriol
Increase in blood Ca^{2+} level
Return to homeostasis when response brings blood Ca^{2+} level back to normal

NOTES

6

Figure 6.12 Features of a developing human embryo (page 177).

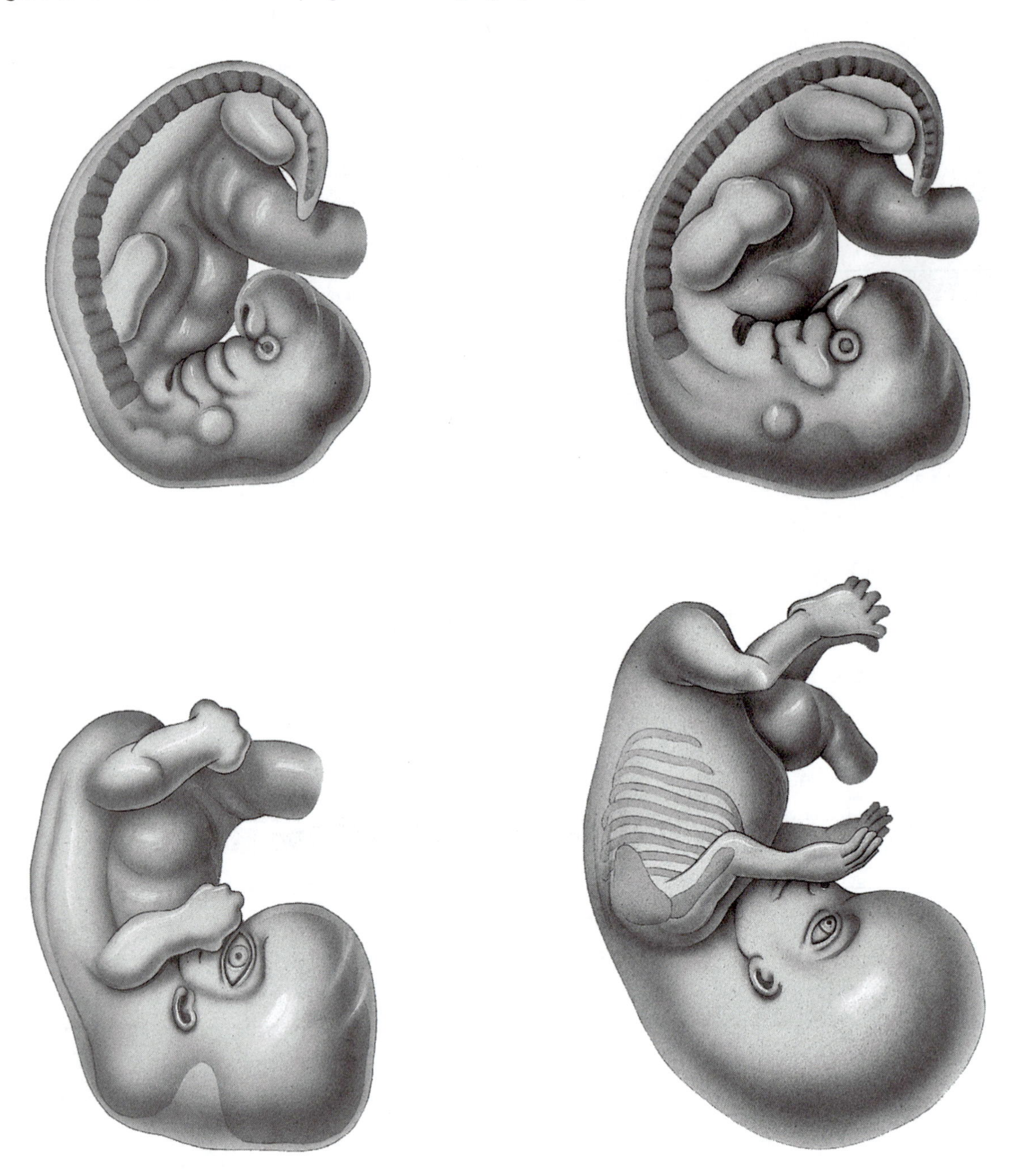

NOTES

6

Figure 7.1 Divisions of the skeletal system (page 184).

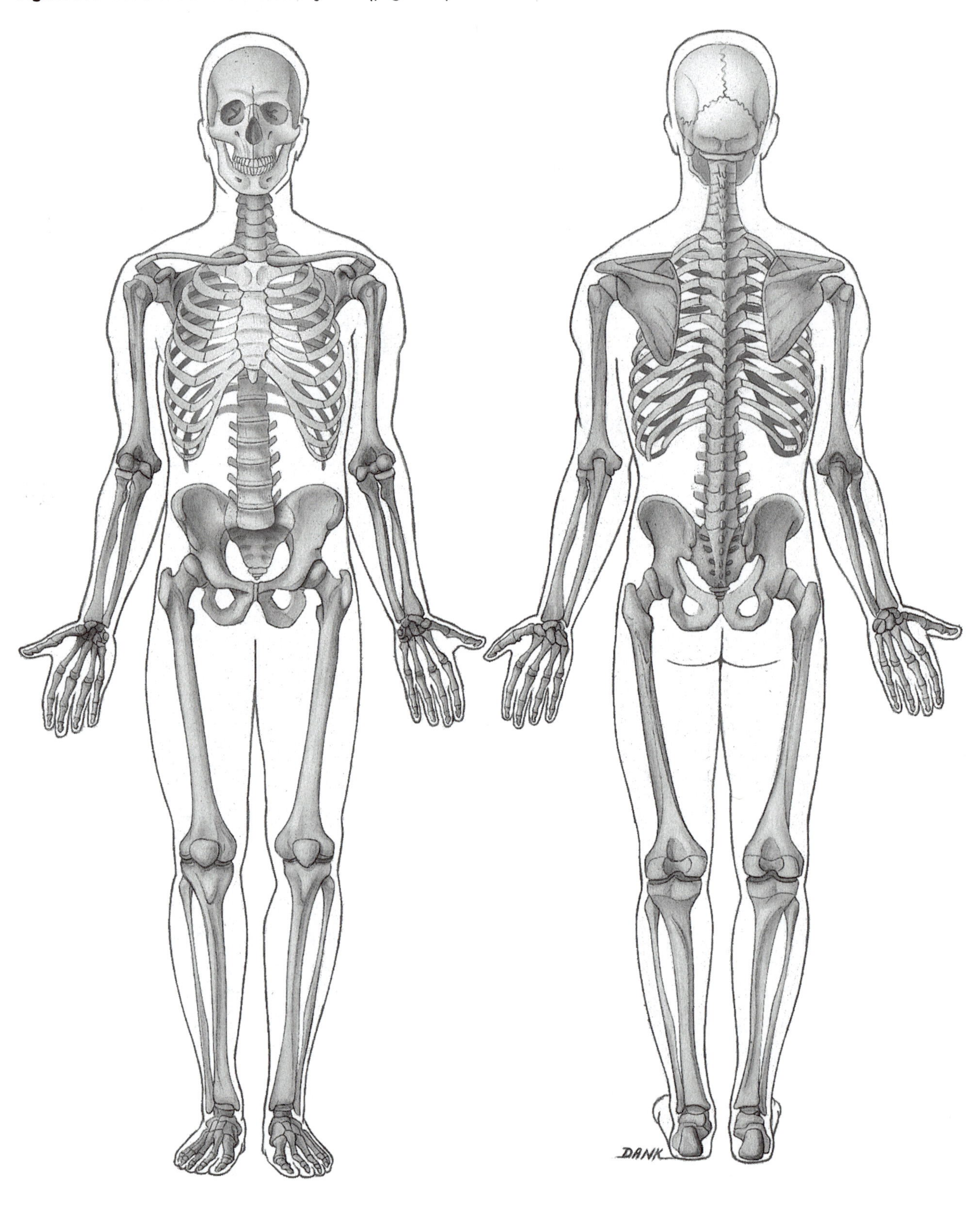
DANK

NOTES

7

Figure 7.2 Types of bones based on shape (page 186).

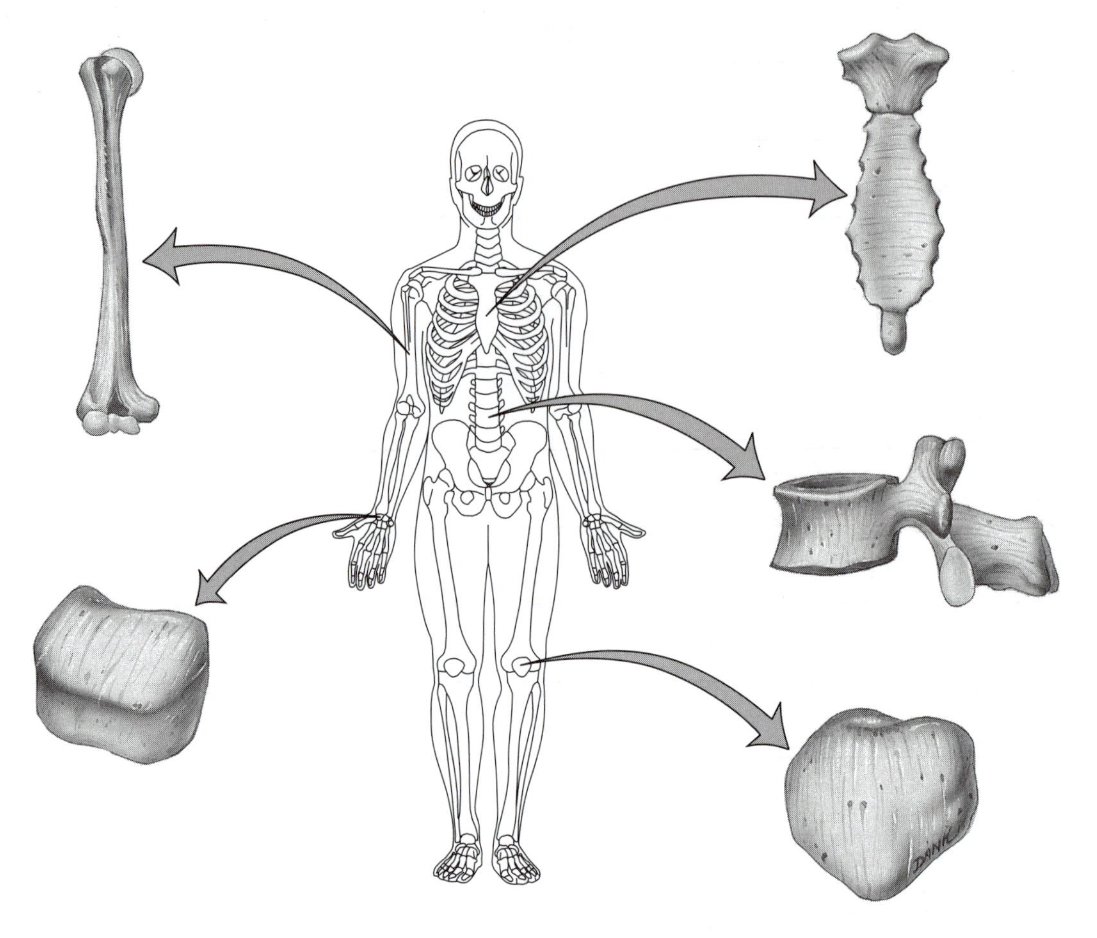

Figure 7.3 Skull, anterior view (page 188).

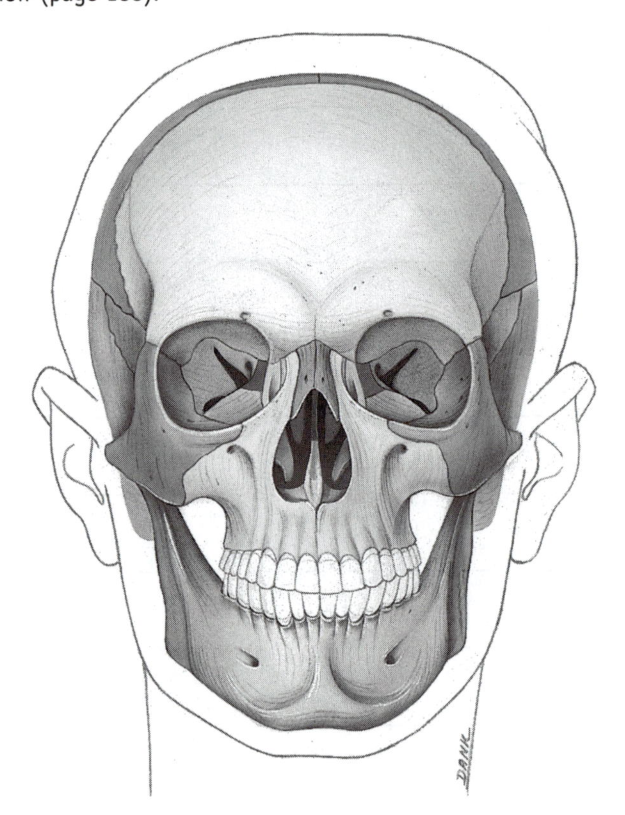

NOTES

7

Figure 7.4 Skull, right lateral view (page 189).

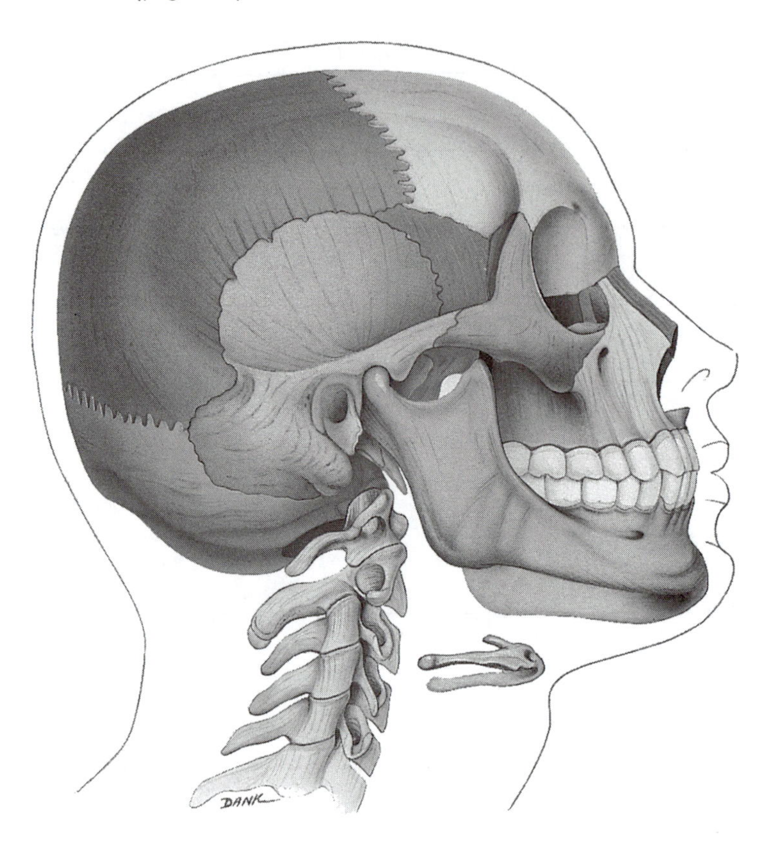

Figure 7.5 Skull, sagittal section (page 190).

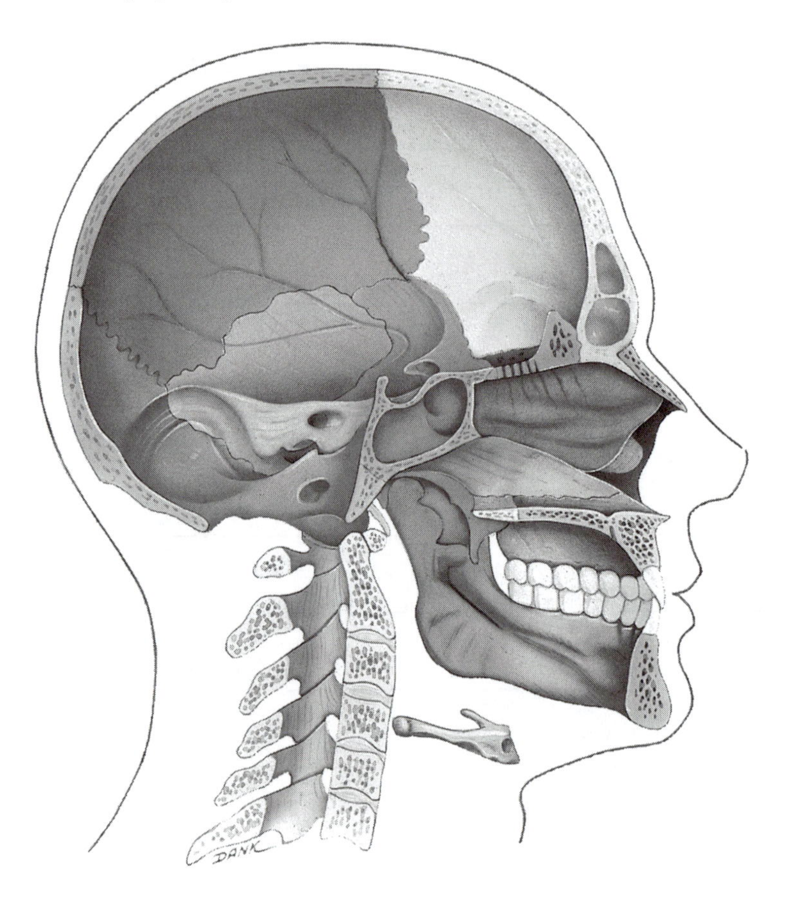

NOTES

7

Figure 7.6 Skull, posterior view (page 191).

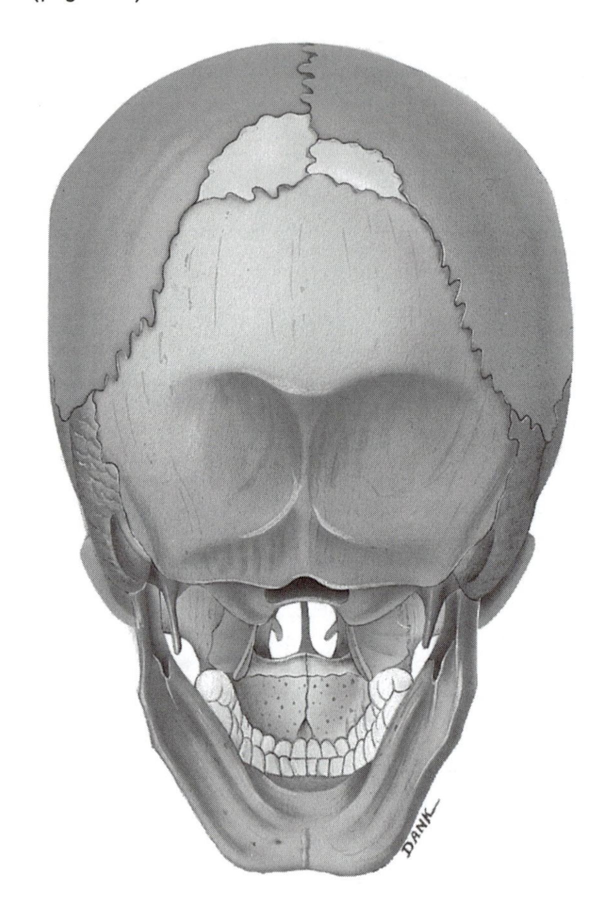
DANK

Figure 7.7 Skull, inferior view (page 192).

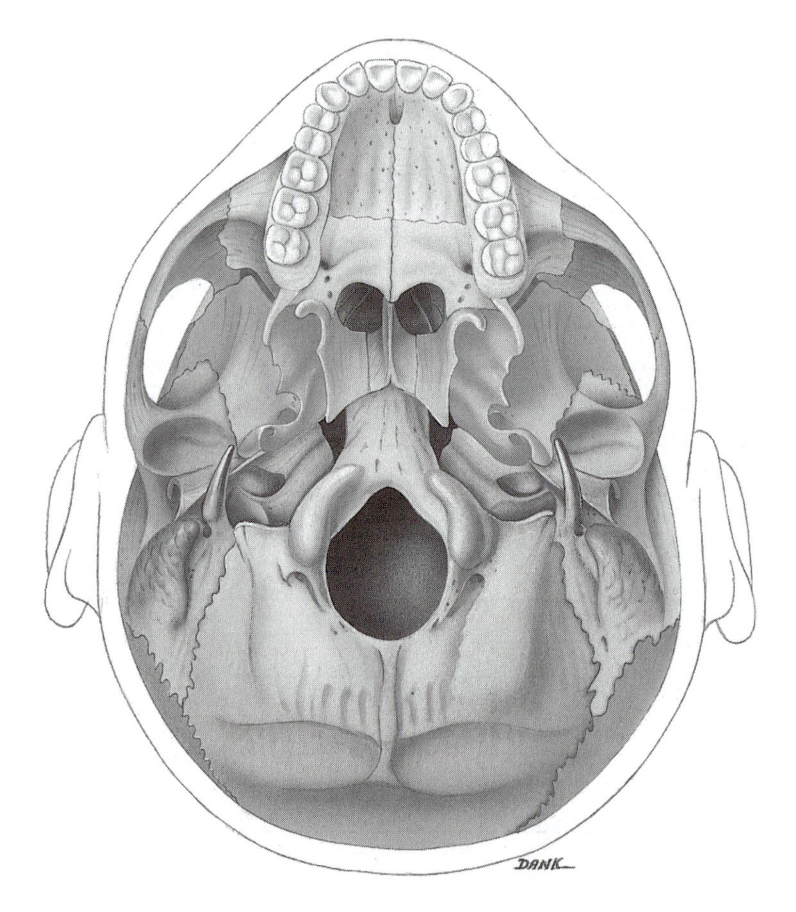
DANK

NOTES

7

Figure 7.8 Sphenoid bone (page 193).

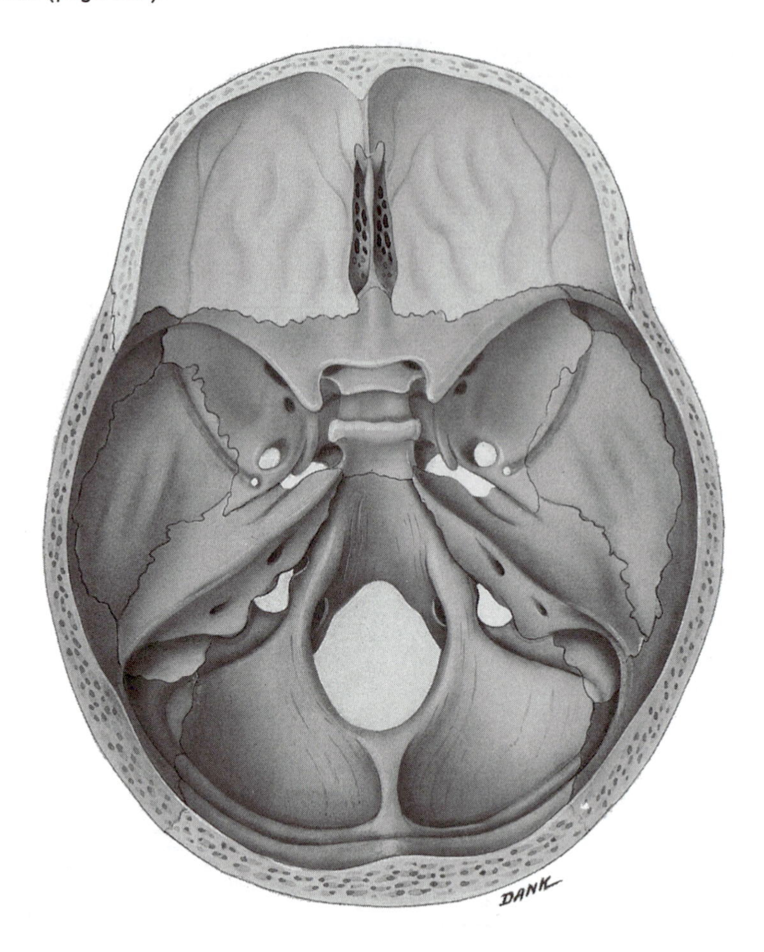
DANK

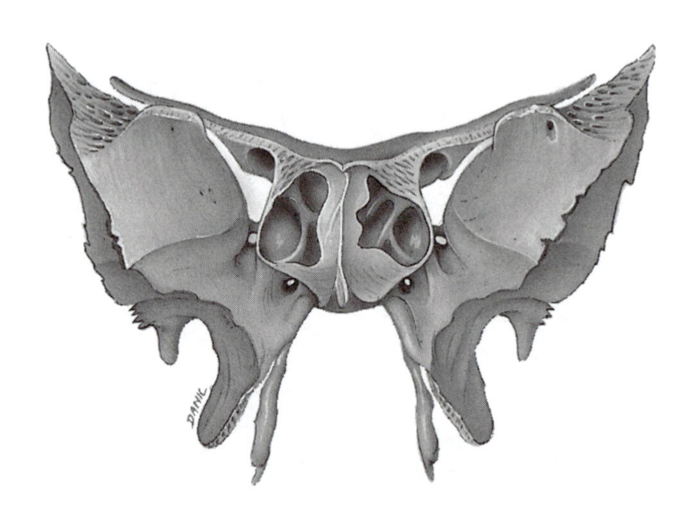
DANK

NOTES

7

Figure 7.9 Ethmoid bone (page 195).

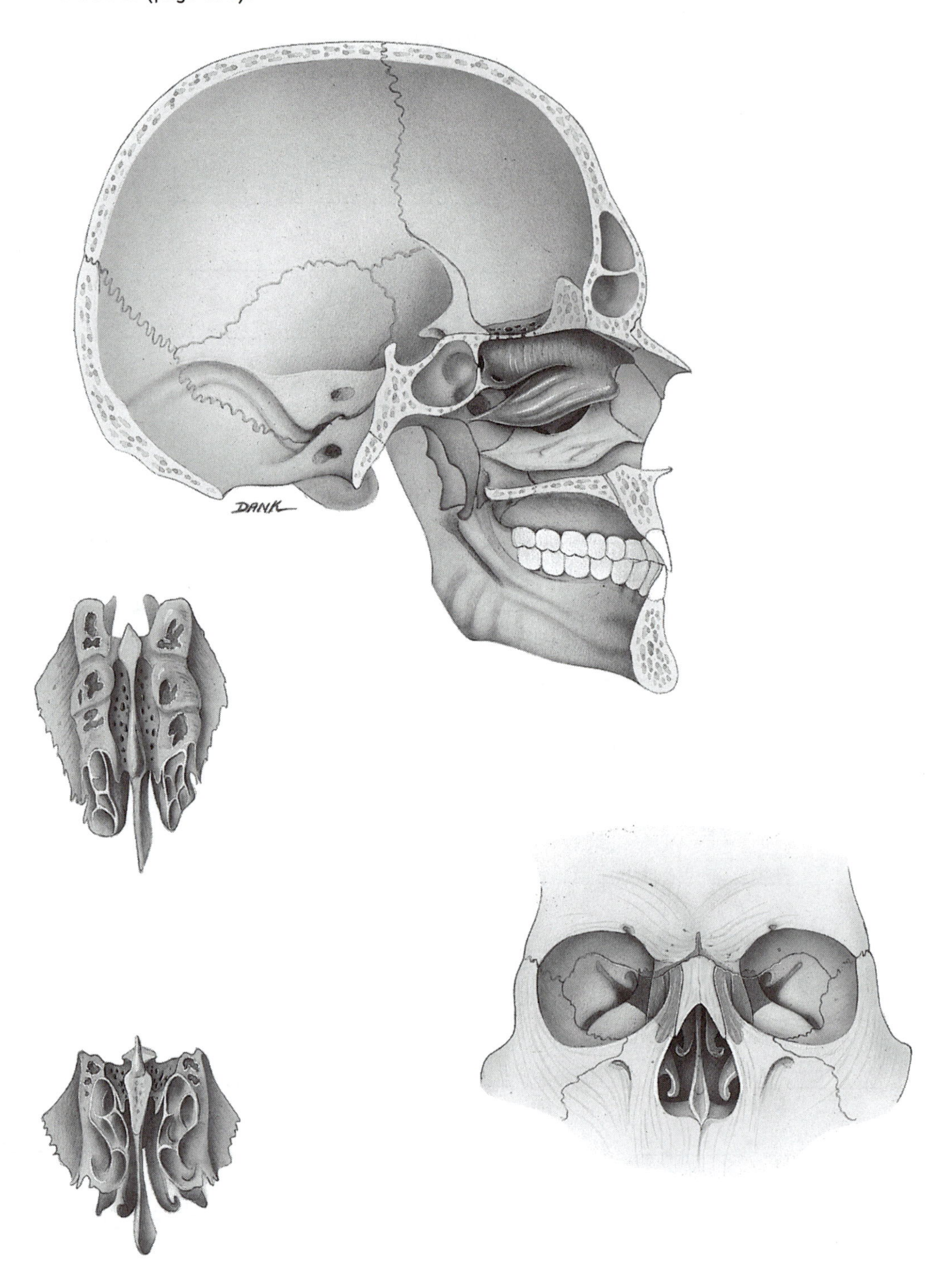
DANK

NOTES

7

Figure 7.10 Mandible (page 196).

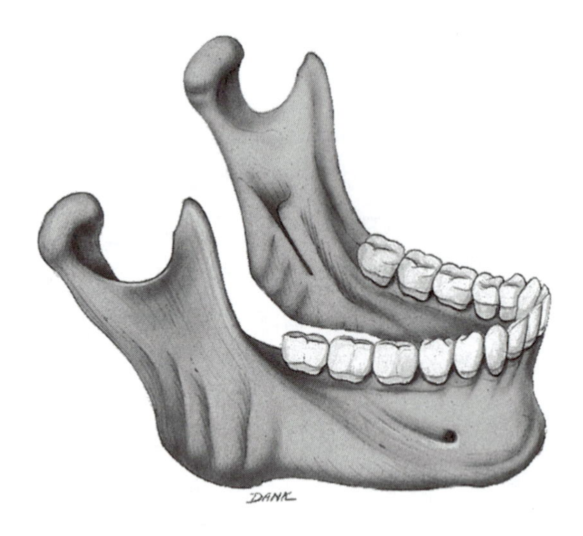
DANK

Figure 7.11 Paranasal sinuses (page 198).

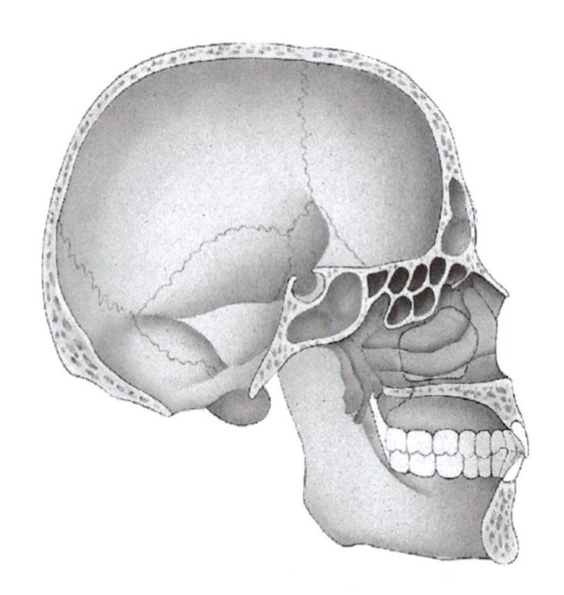

NOTES

7

Figure 7.12 Fontanels at birth (page 198).

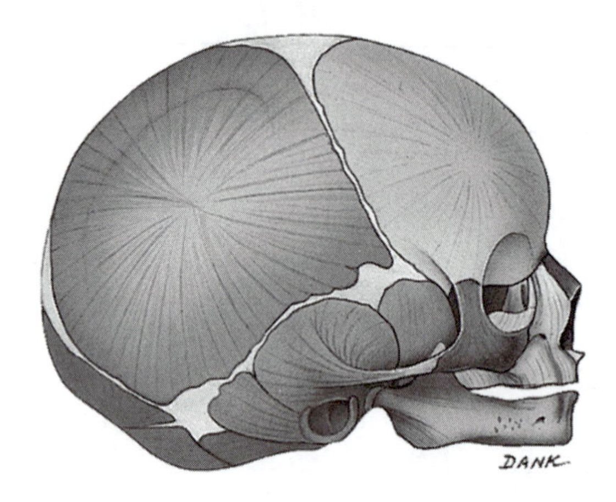
DANK

Figure 7.13 Details of the orbit (page 200).

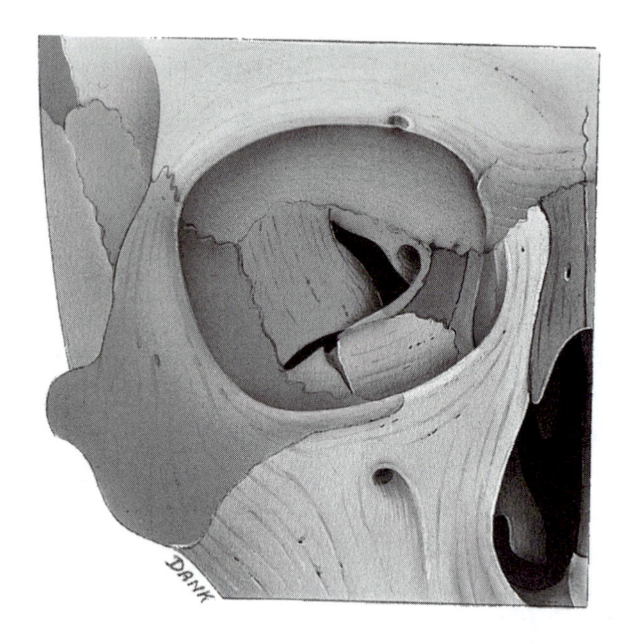
DANK

NOTES

Figure 7.14 Nasal septum (page 200).

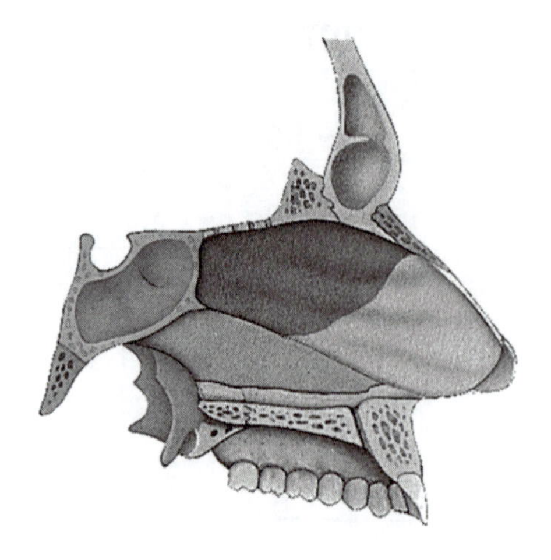

Figure 7.15 Hyoid bone (page 201).

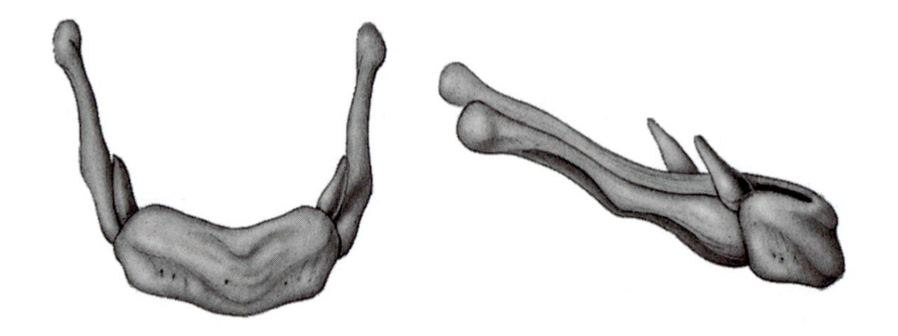

NOTES

7

Figure 7.16 Vertebral column (page 202).

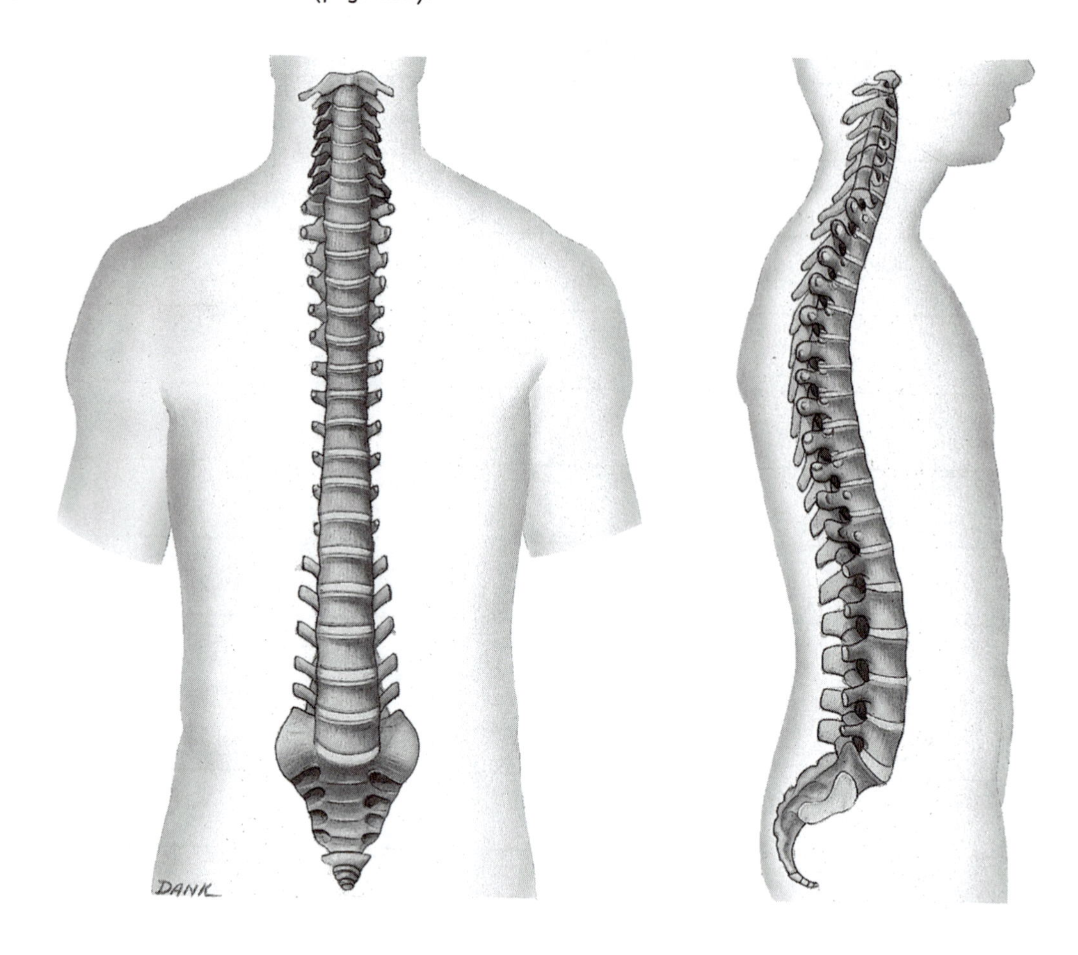

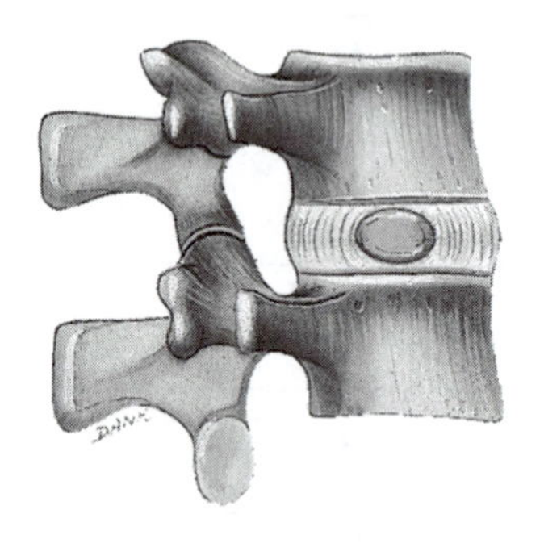

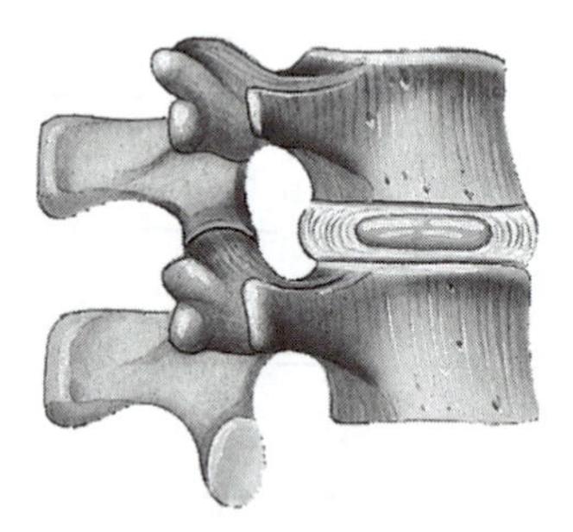

NOTES

7

Figure 7.17 Structure of a typical vertebra (page 204).

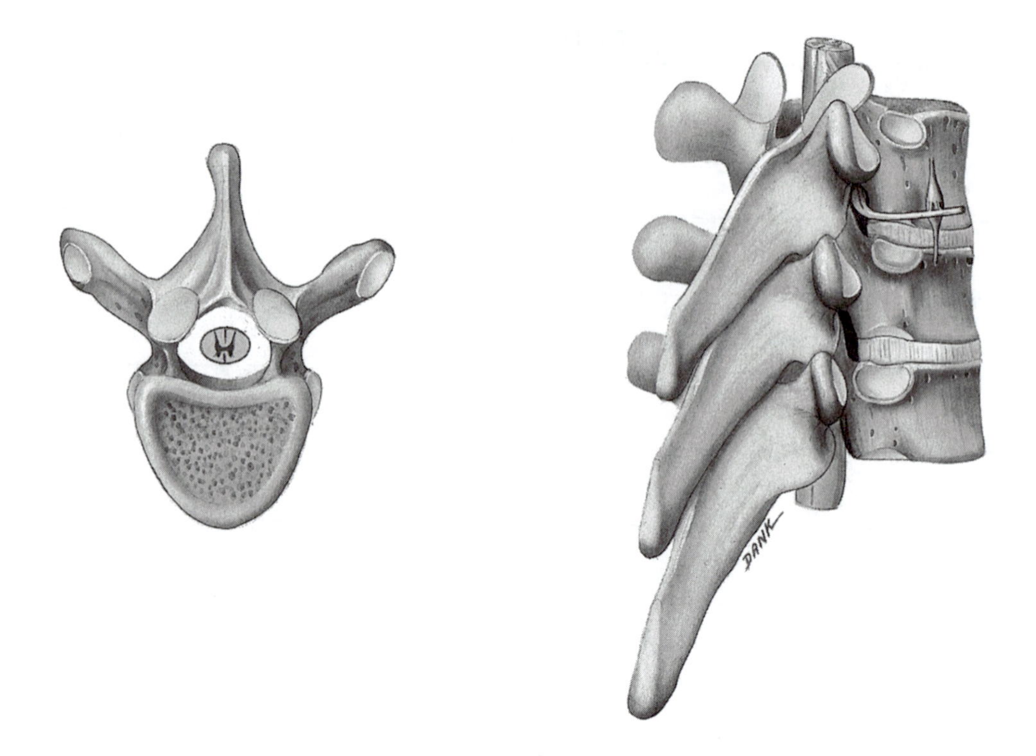

Figure 7.18a Cervical vertebrae: posterior view (page 205).

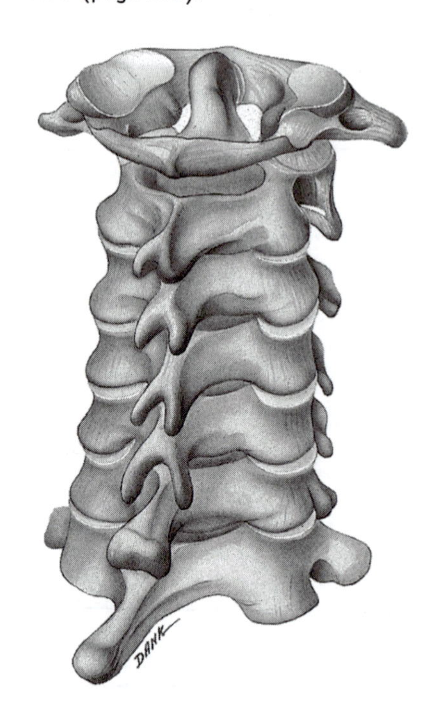

NOTES

7

Figure 7.18b–d Cervical vertebrae (page 206).

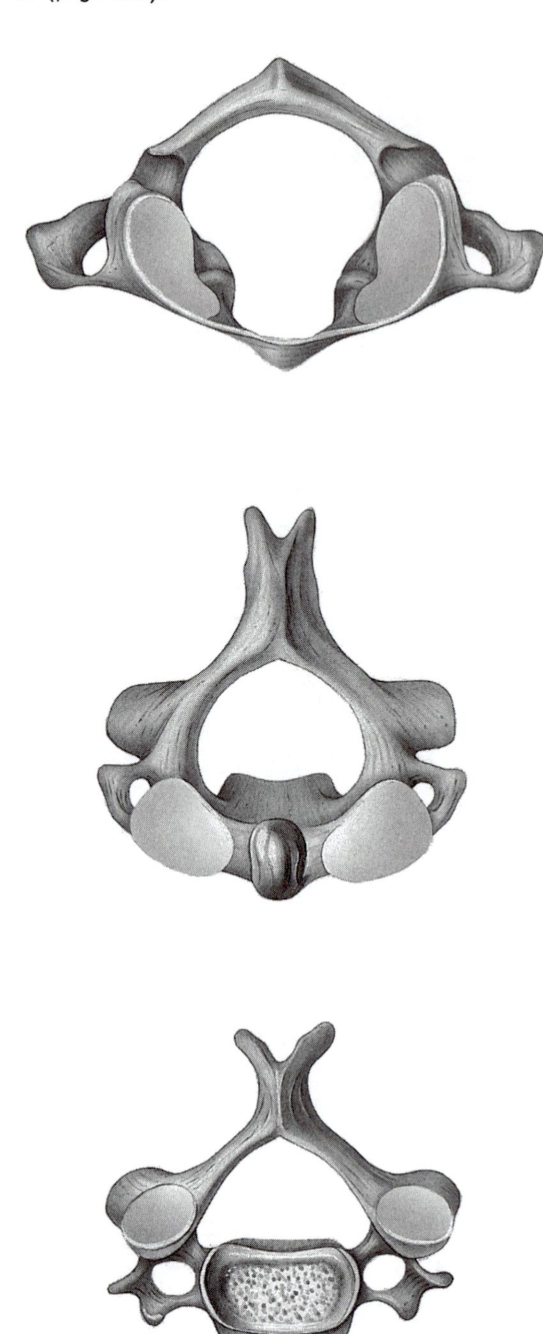
DANK

NOTES

7

Figure 7.19 Thoracic vertebrae (page 207).

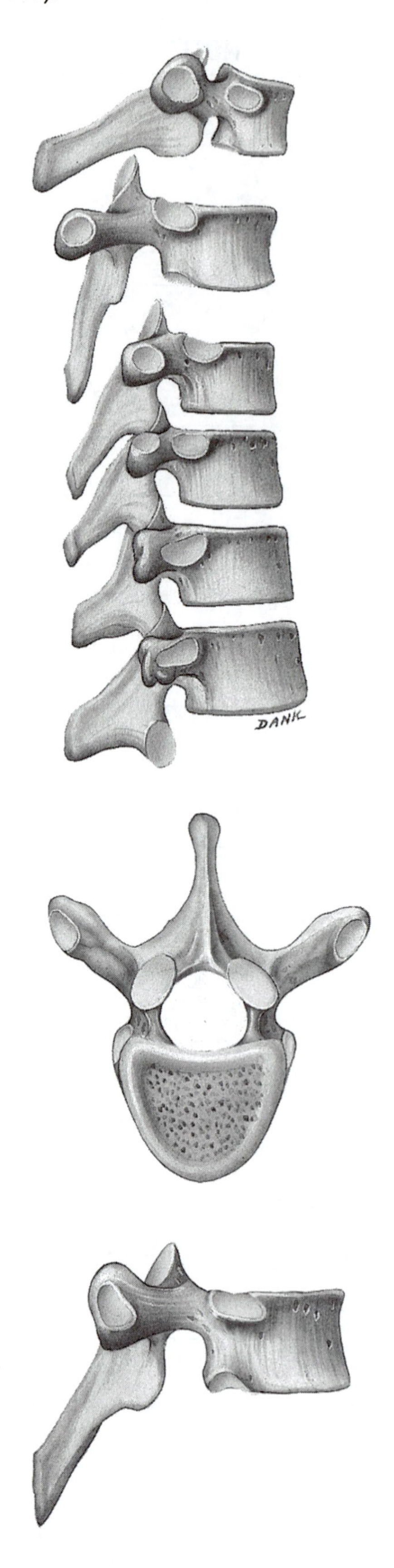
DANK

NOTES

7

Figure 7.20 Lumbar vertebrae (page 208).

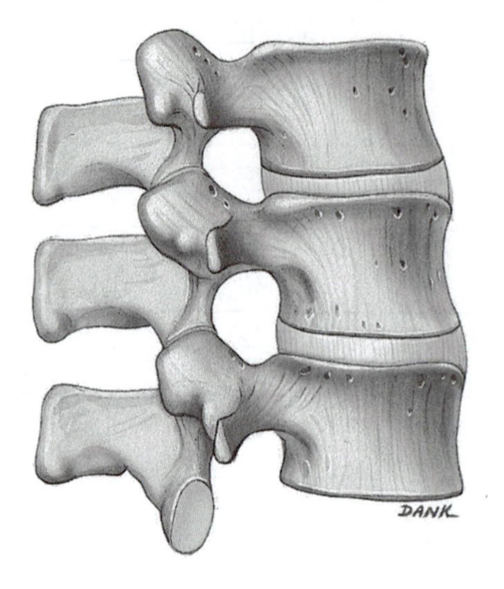
DANK

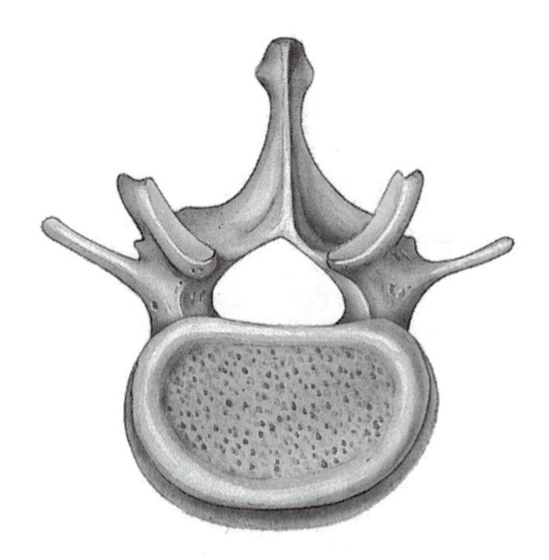

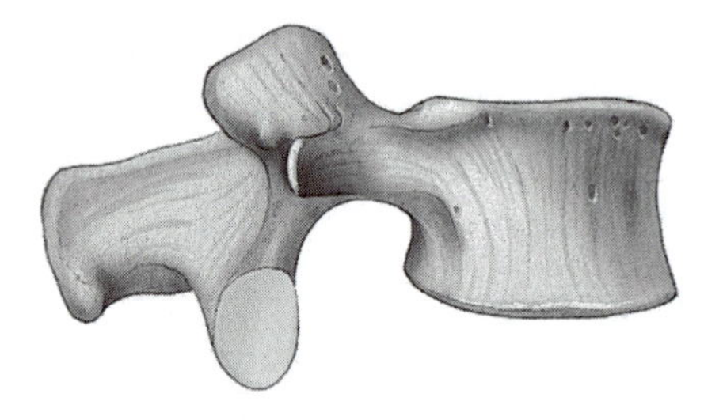

NOTES

7

Figure 7.21 Sacrum and coccyx (page 210).

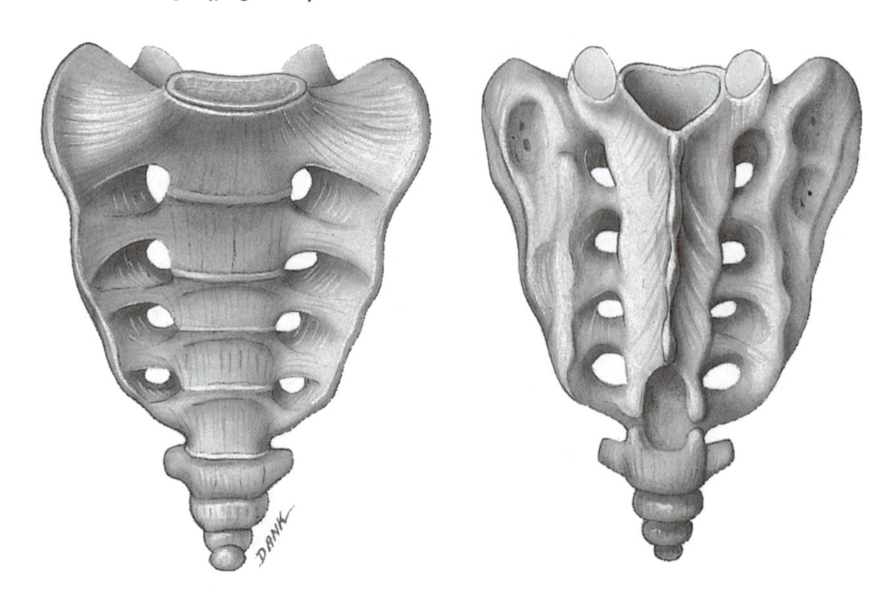
DANK

Figure 7.22 Skeleton of the thorax (page 211).

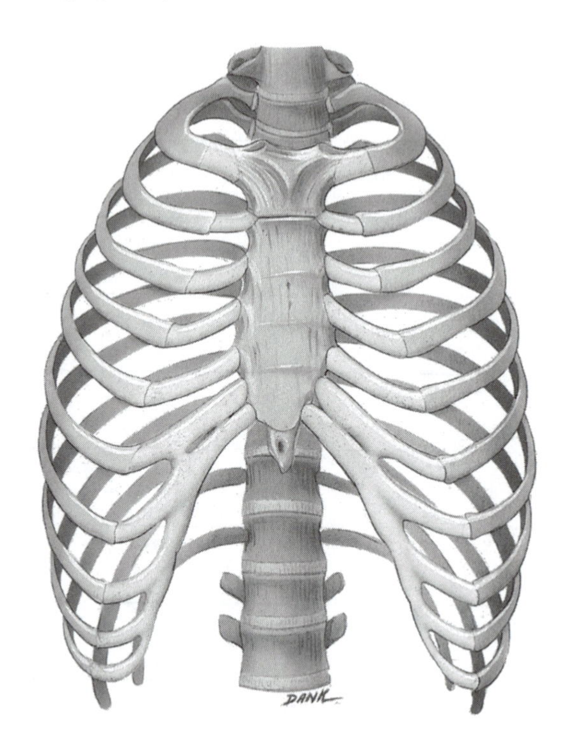
DANK

NOTES

7

Figure 7.23 The structure of ribs (page 212).

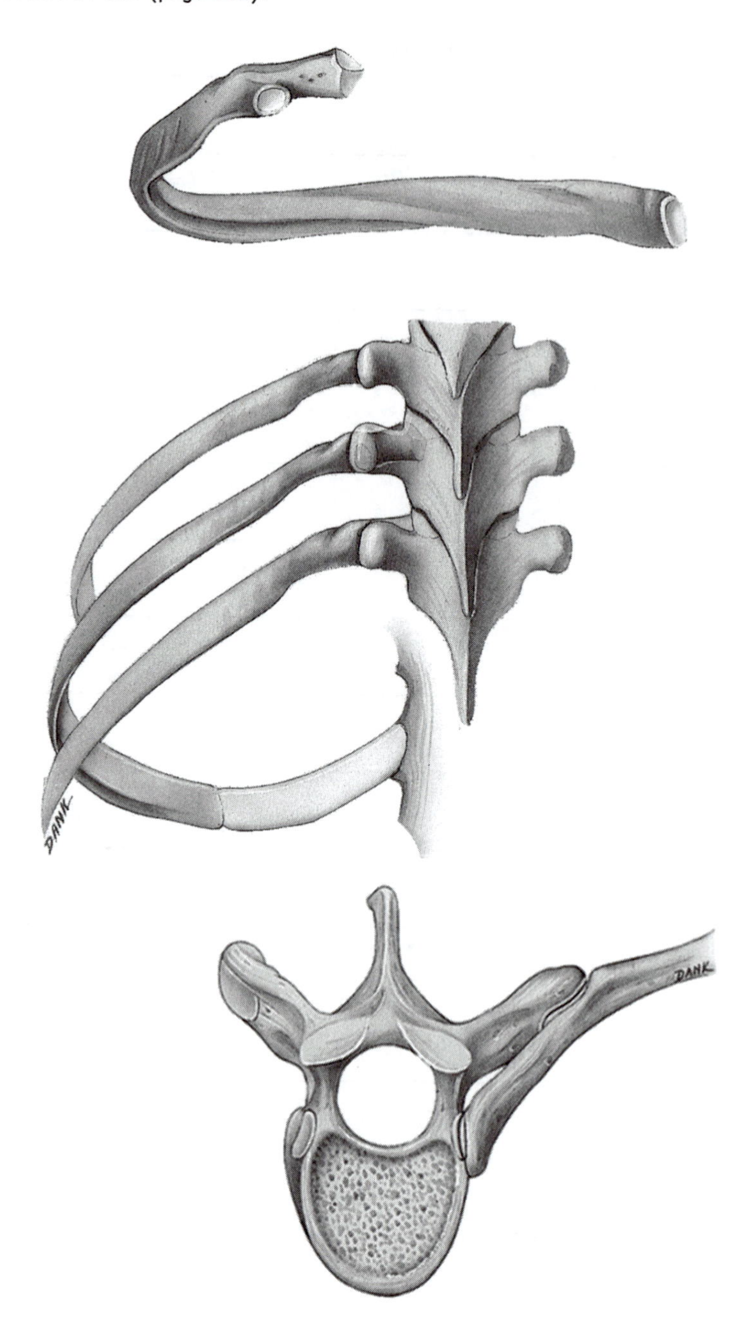

Figure 7.24 Herniated disc (page 213).

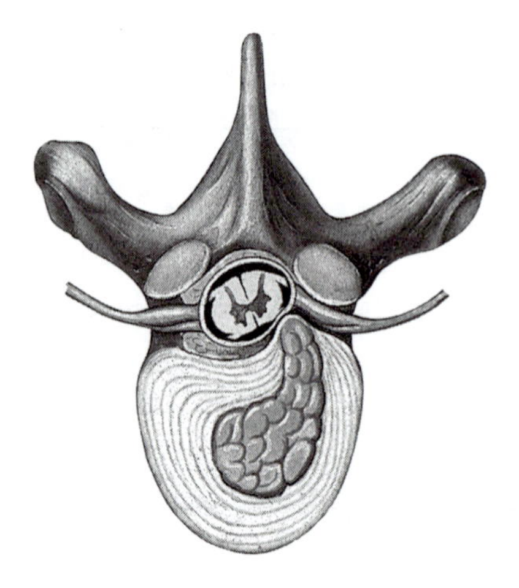

NOTES

7

Figure 8.1 Right pectoral girdle (page 219).

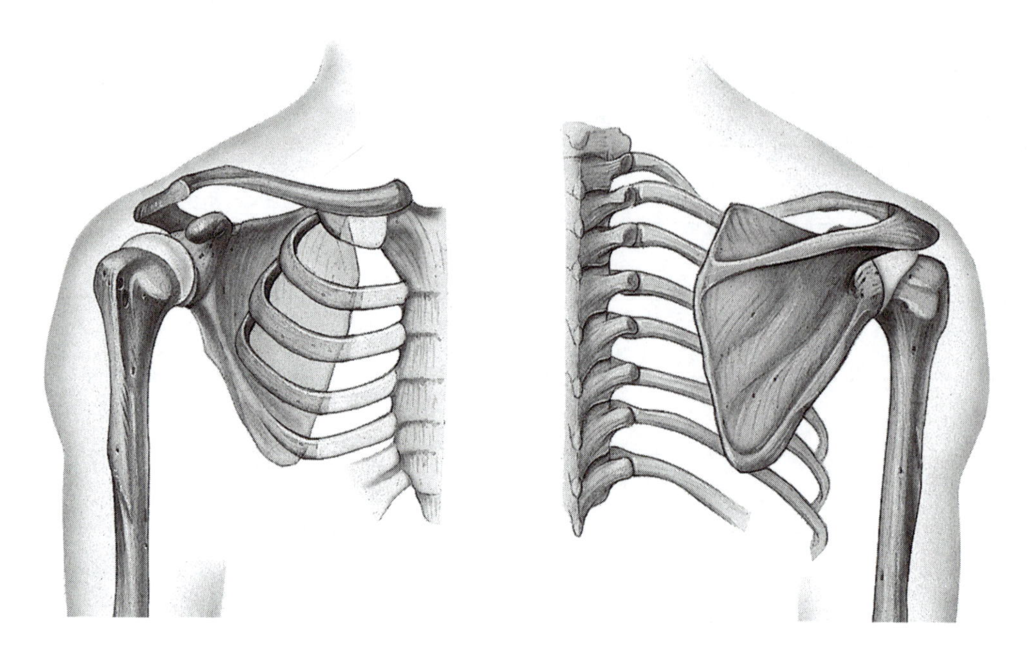

Figure 8.2 Right clavicle (page 219).

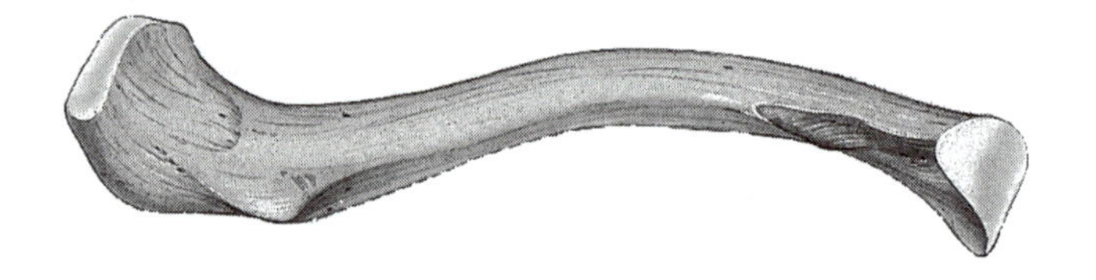

NOTES

8

Figure 8.3 Right scapula (page 220).

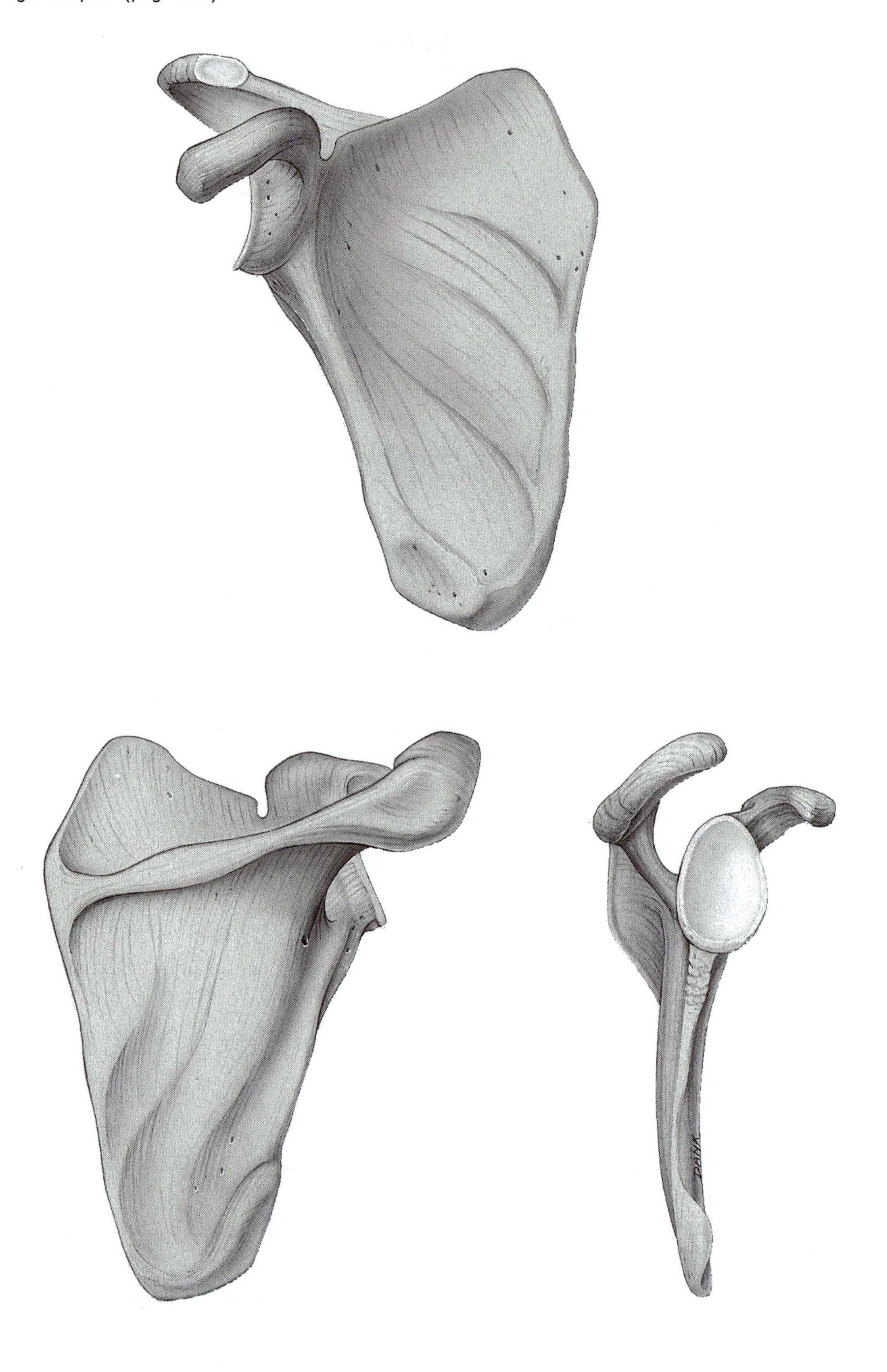

NOTES

Figure 8.4 Right upper limb (221).

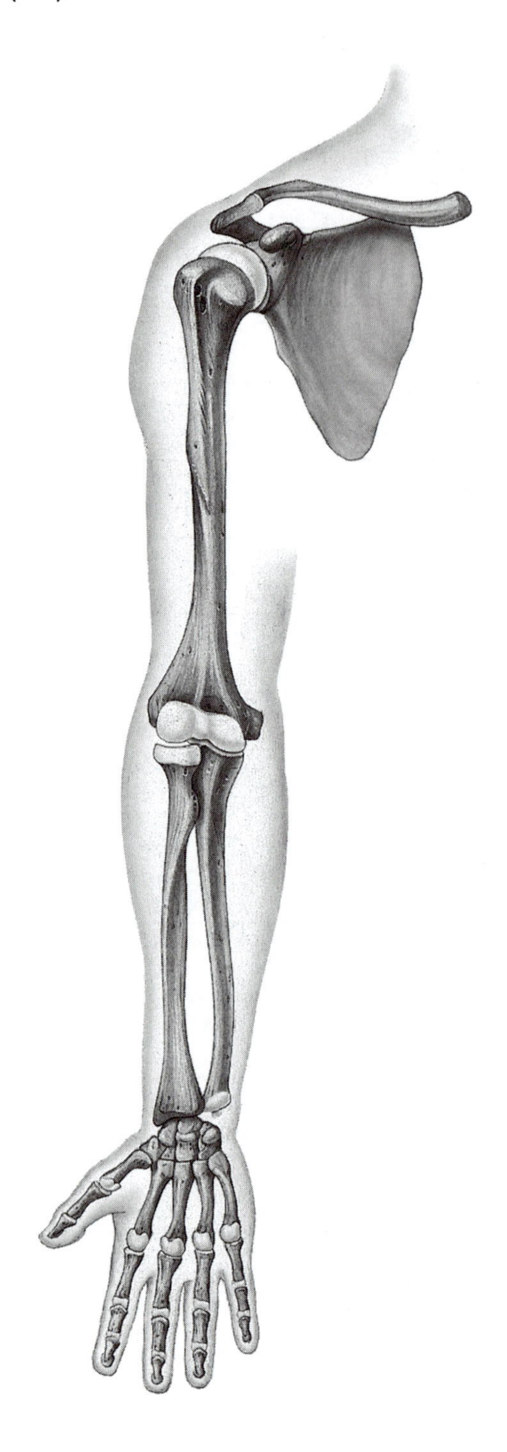

NOTES

8

Figure 8.5 Right humerus in relation to the scapula, ulna, and radius (page 222).

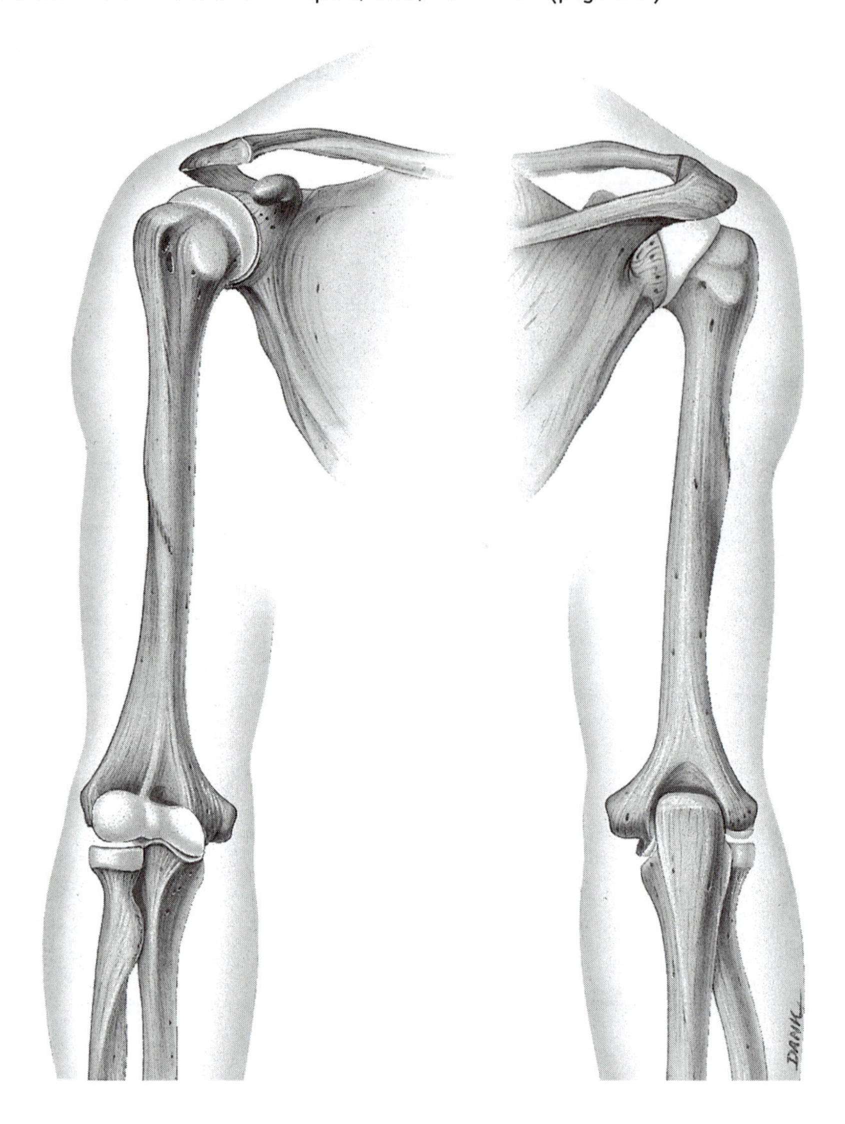

NOTES

8

Figure 8.6 Right ulna and radius in relation to the humerus and carpals (page 223).

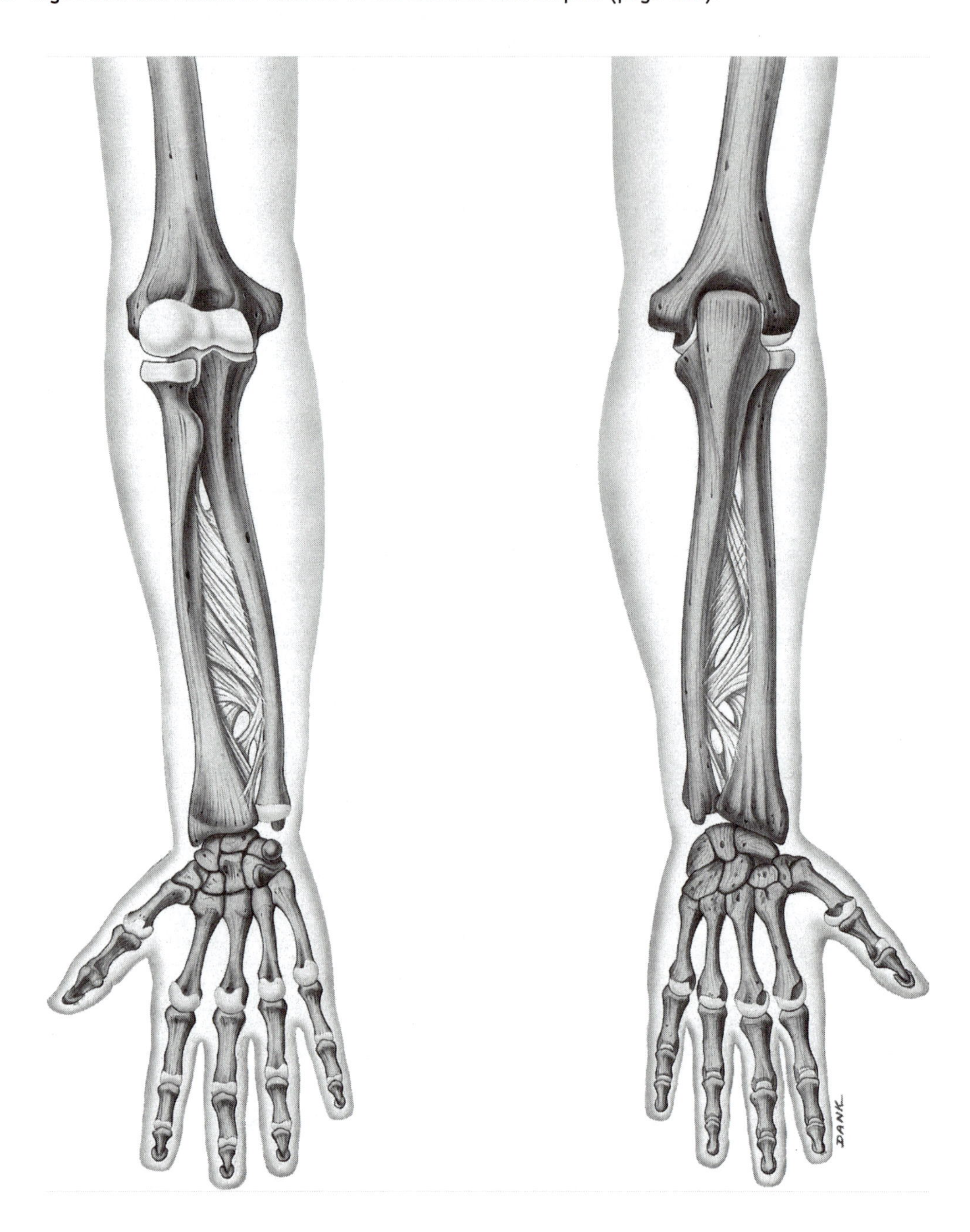
DANK

NOTES

8

Figure 8.7 Articulations formed by the ulna and radius (page 224).

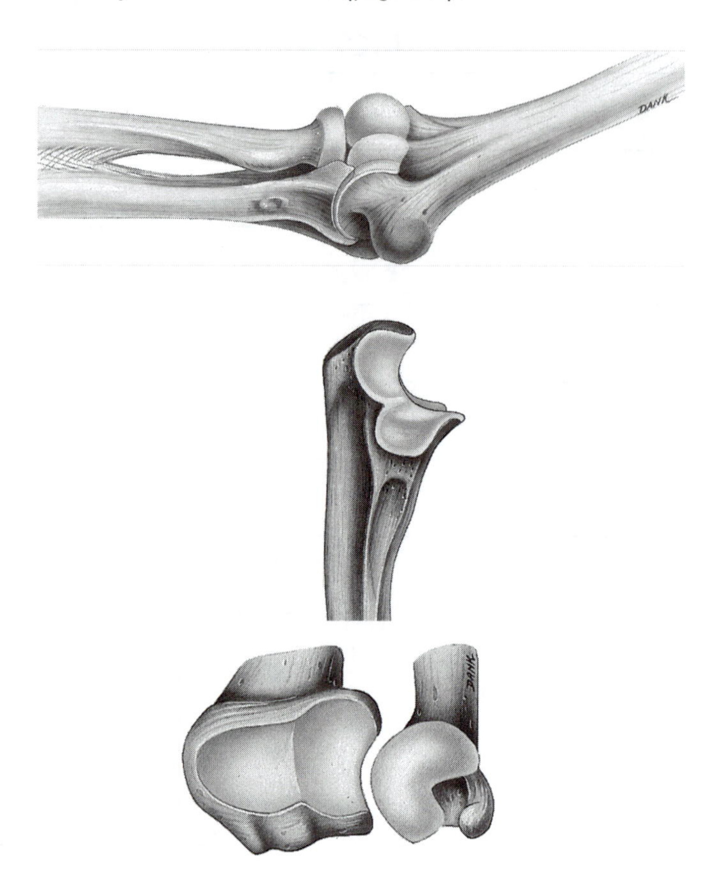

Figure 8.8 Right wrist and hand (page 226).

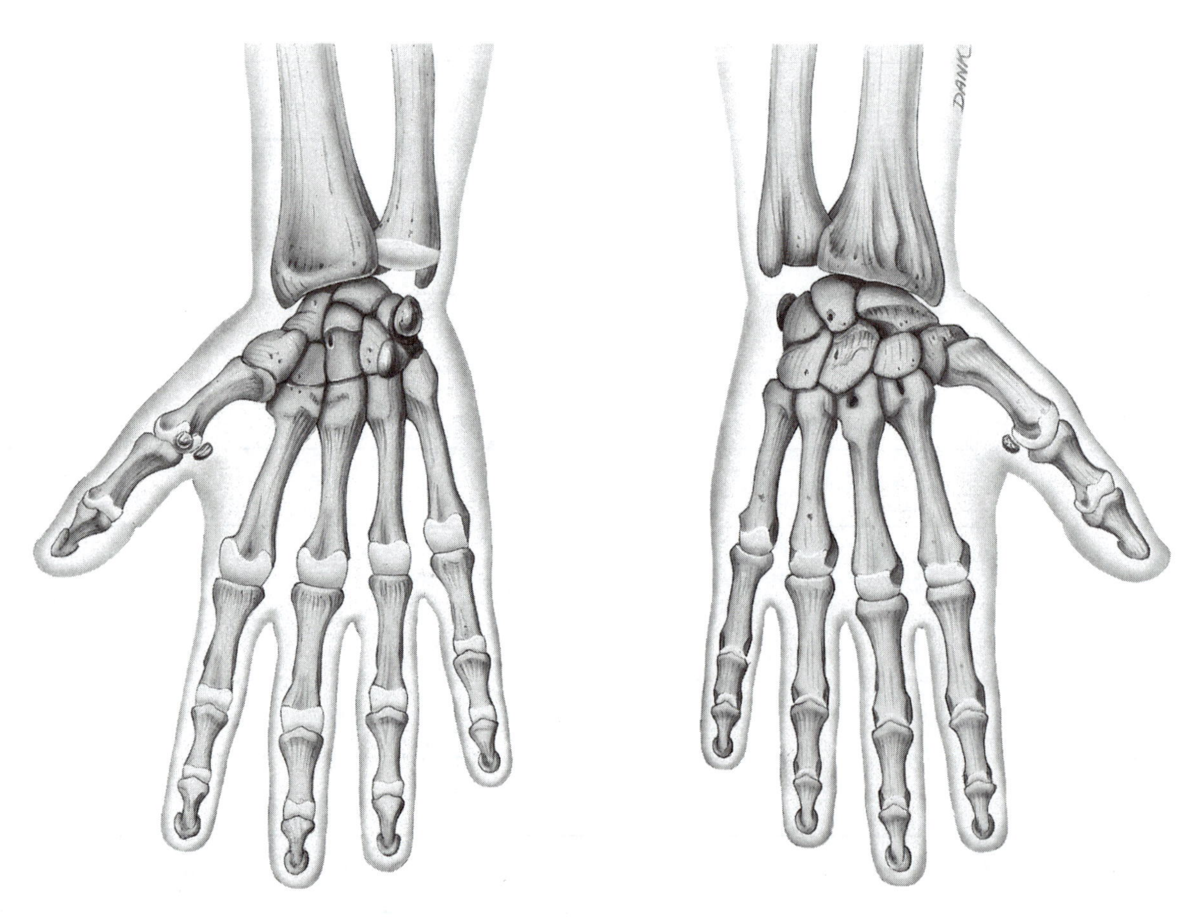

NOTES

8

Figure 8.9 Bony pelvis (page 227).

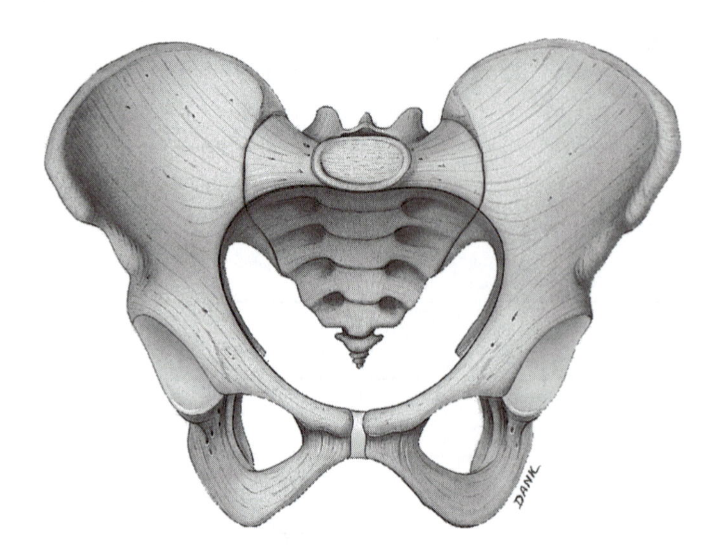
DANK

Figure 8.10 Right hip bone (page 228).

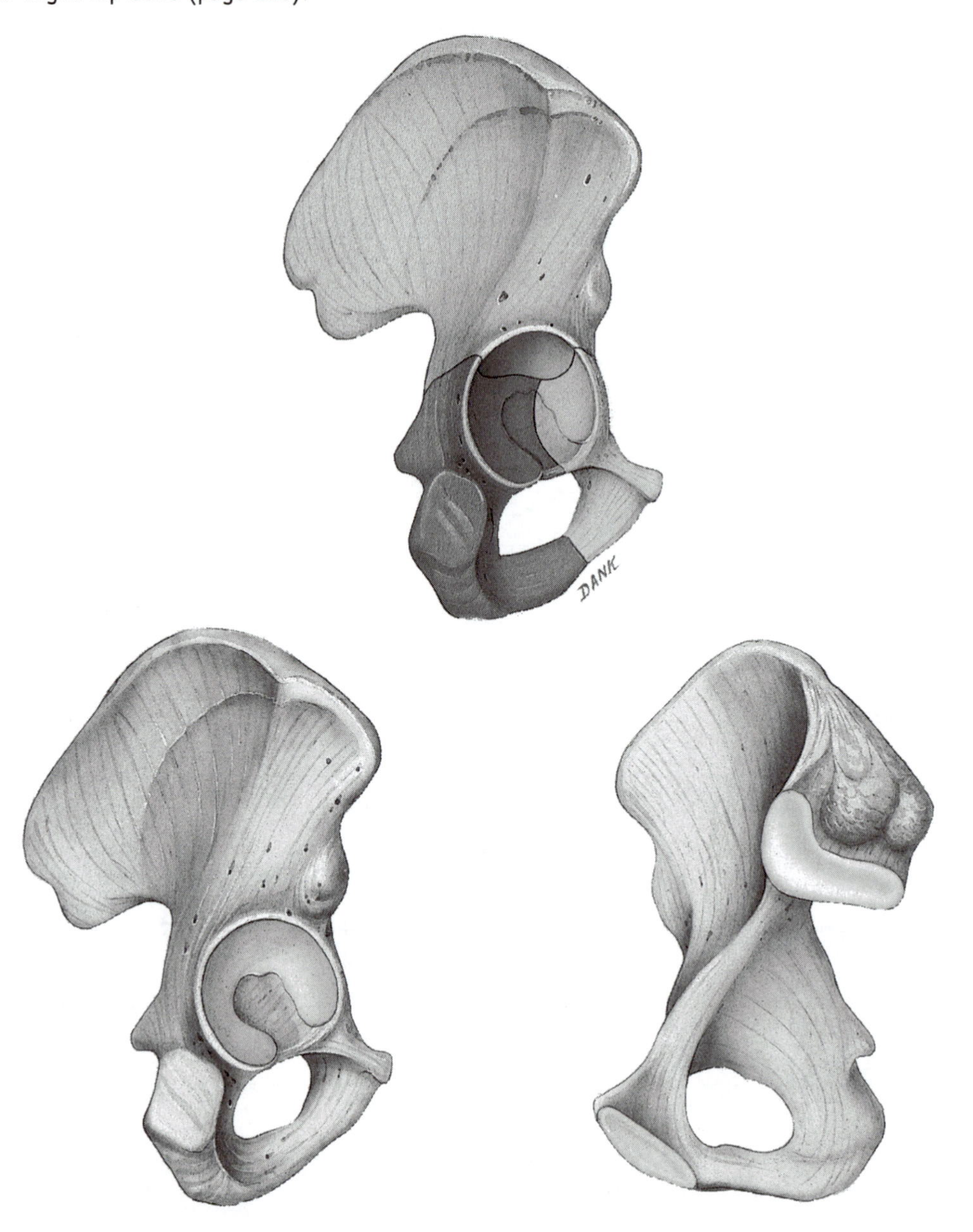
DANK

NOTES

8

Figure 8.11 True and false pelves (page 229).

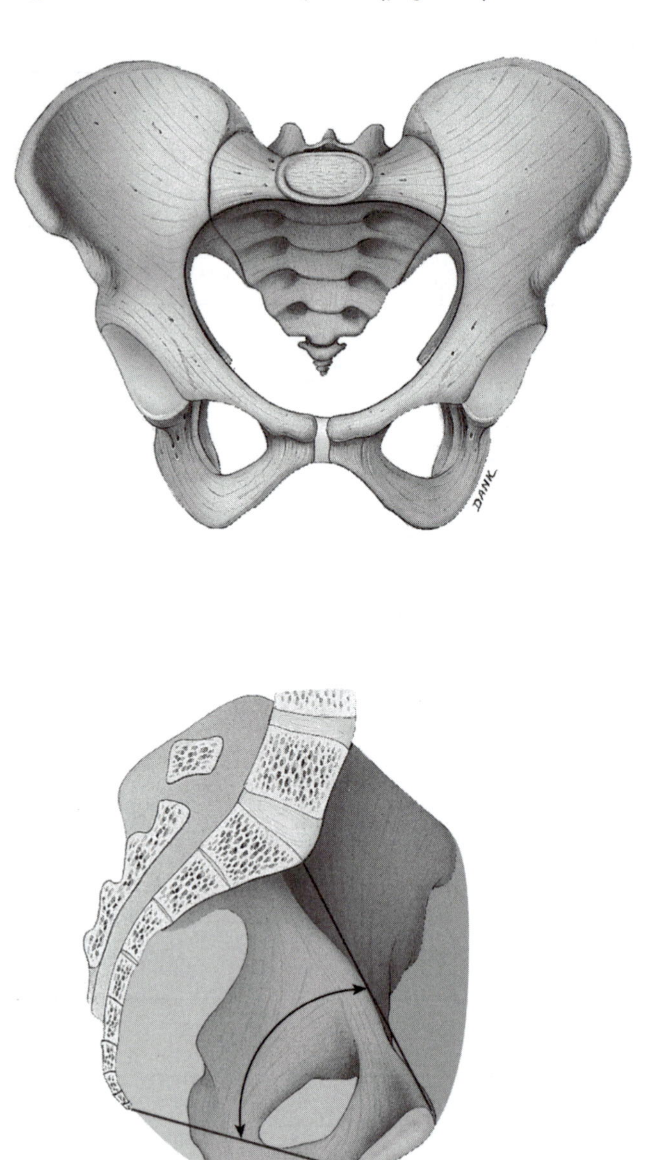

Figure 8.12 Right lower limb (page 232).

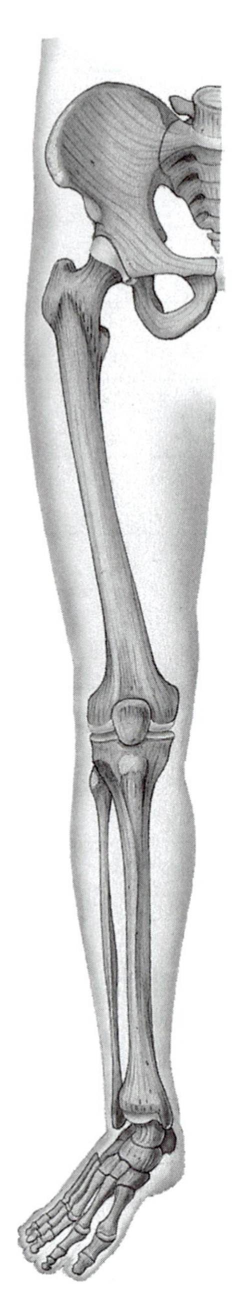

NOTES

Figure 8.13 Right femur (page 233).

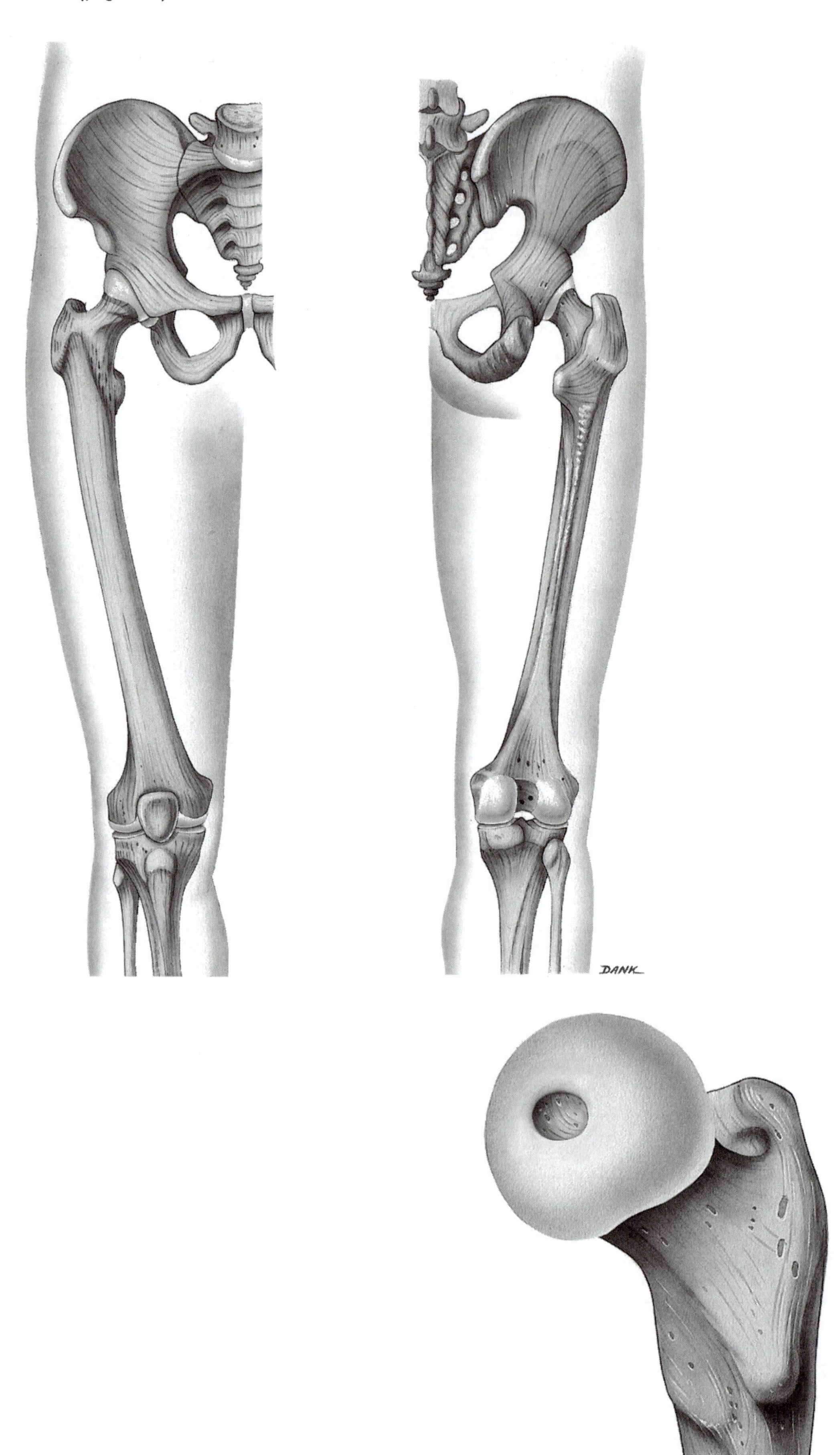
DANK

NOTES

8

Figure 8.14 Right patella (page 234).

Figure 8.15 Right tibia and fibula (page 235).

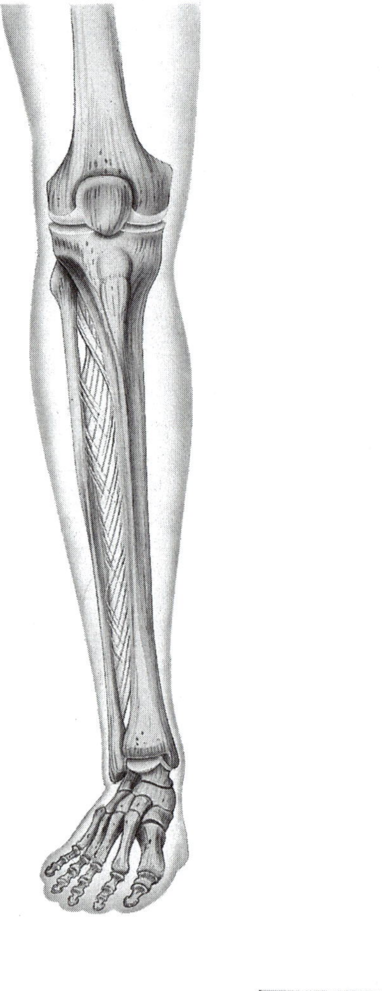

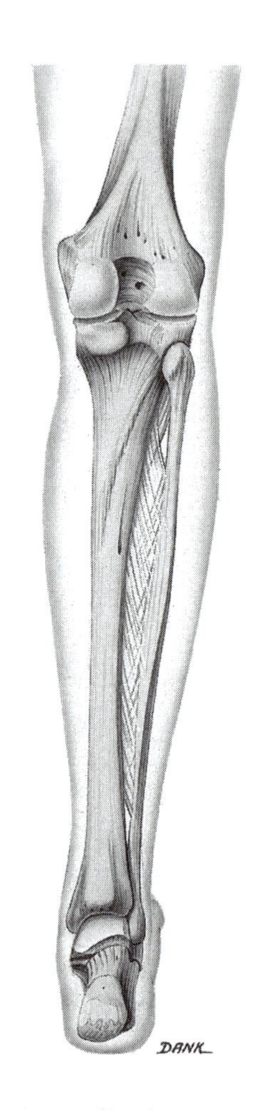

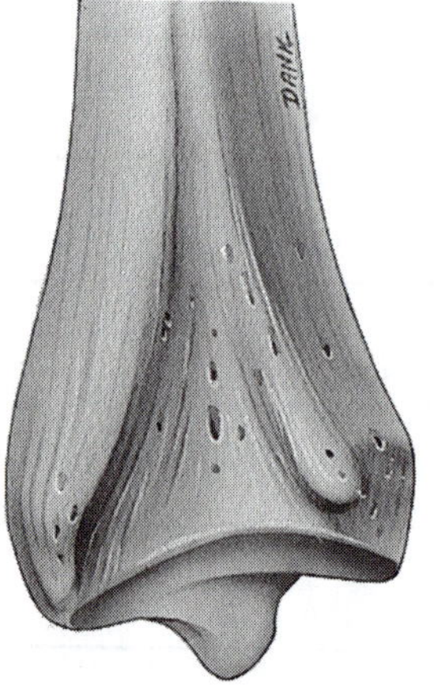

NOTES

8

Figure 8.16 Right foot (page 236).

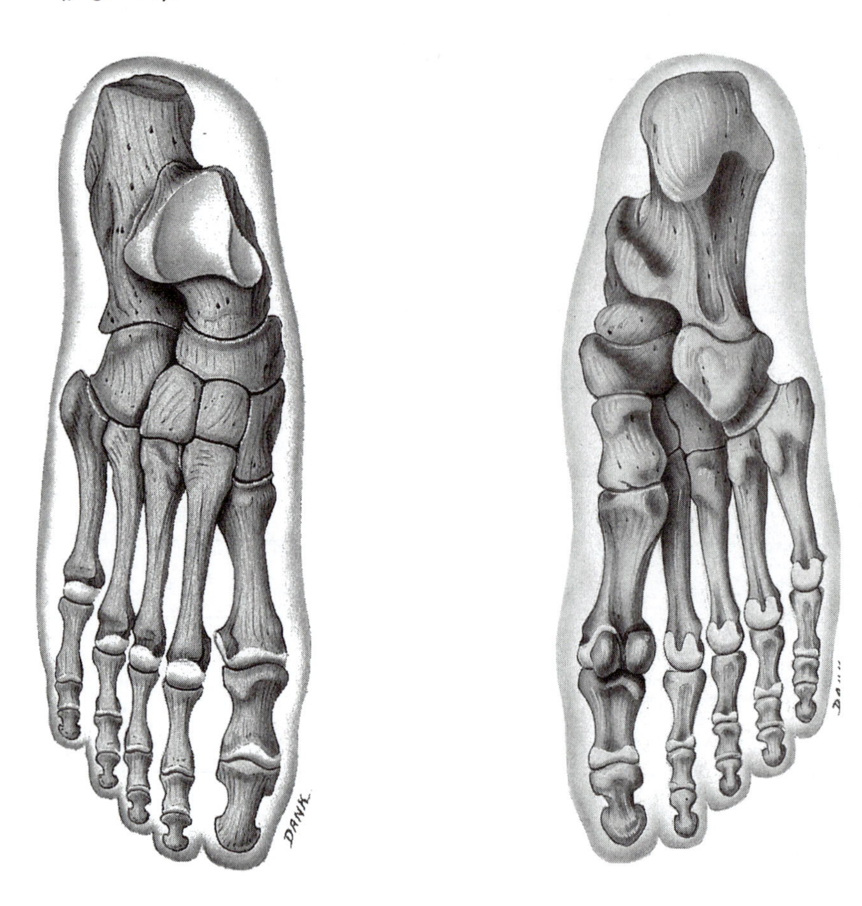
DANK

Figure 8.17 Arches of the right foot (page 237).

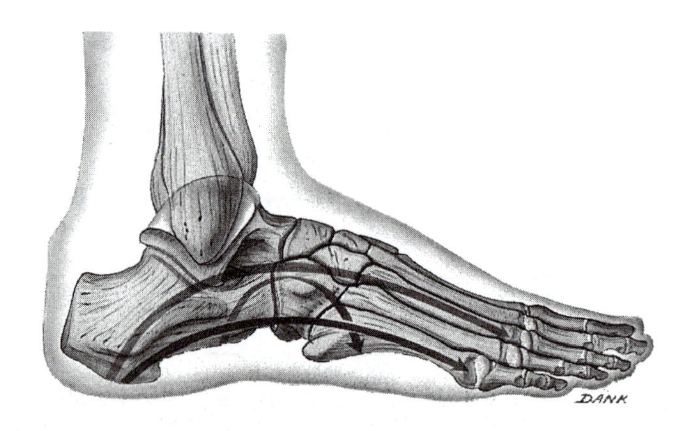
DANK

NOTES

8

Figure 9.1 Fibrous joints (page 242).

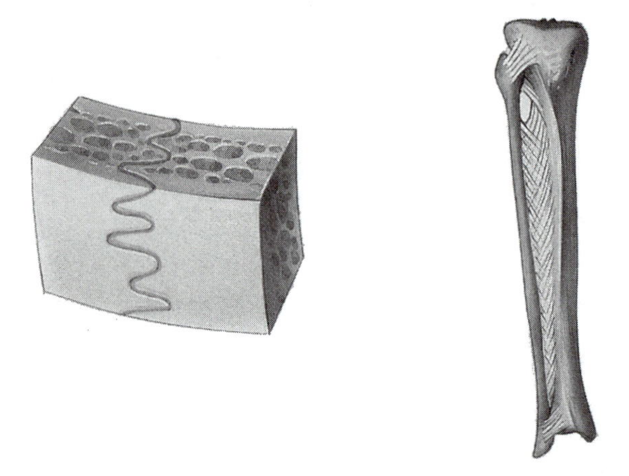

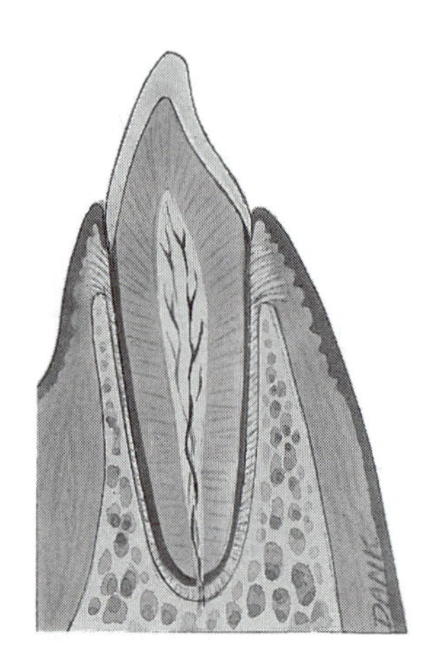

Figure 9.2 Cartilaginous joints (page 243).

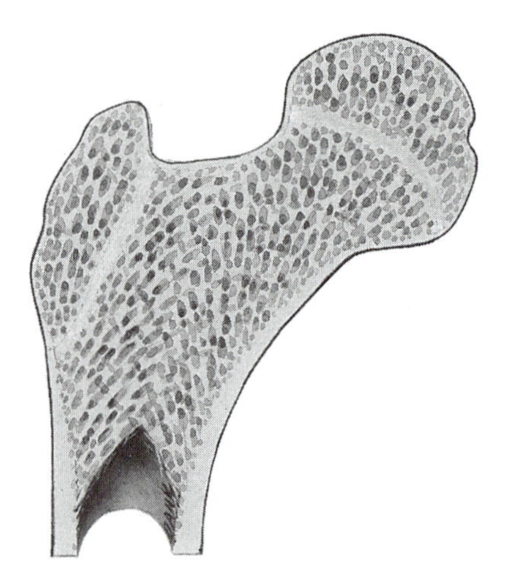

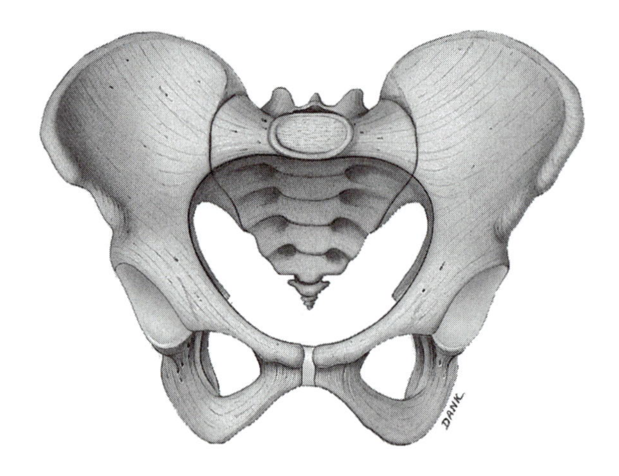

Figure 9.3 Structure of a typical synovial joint (page 244).

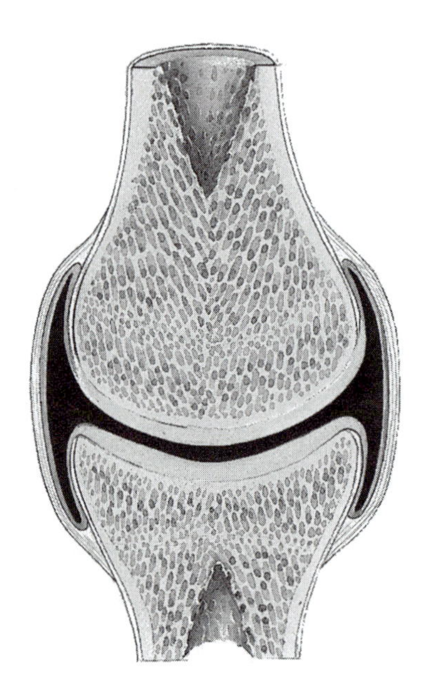

NOTES

9

Figure 9.4 Subtypes of synovial joints (page 246).

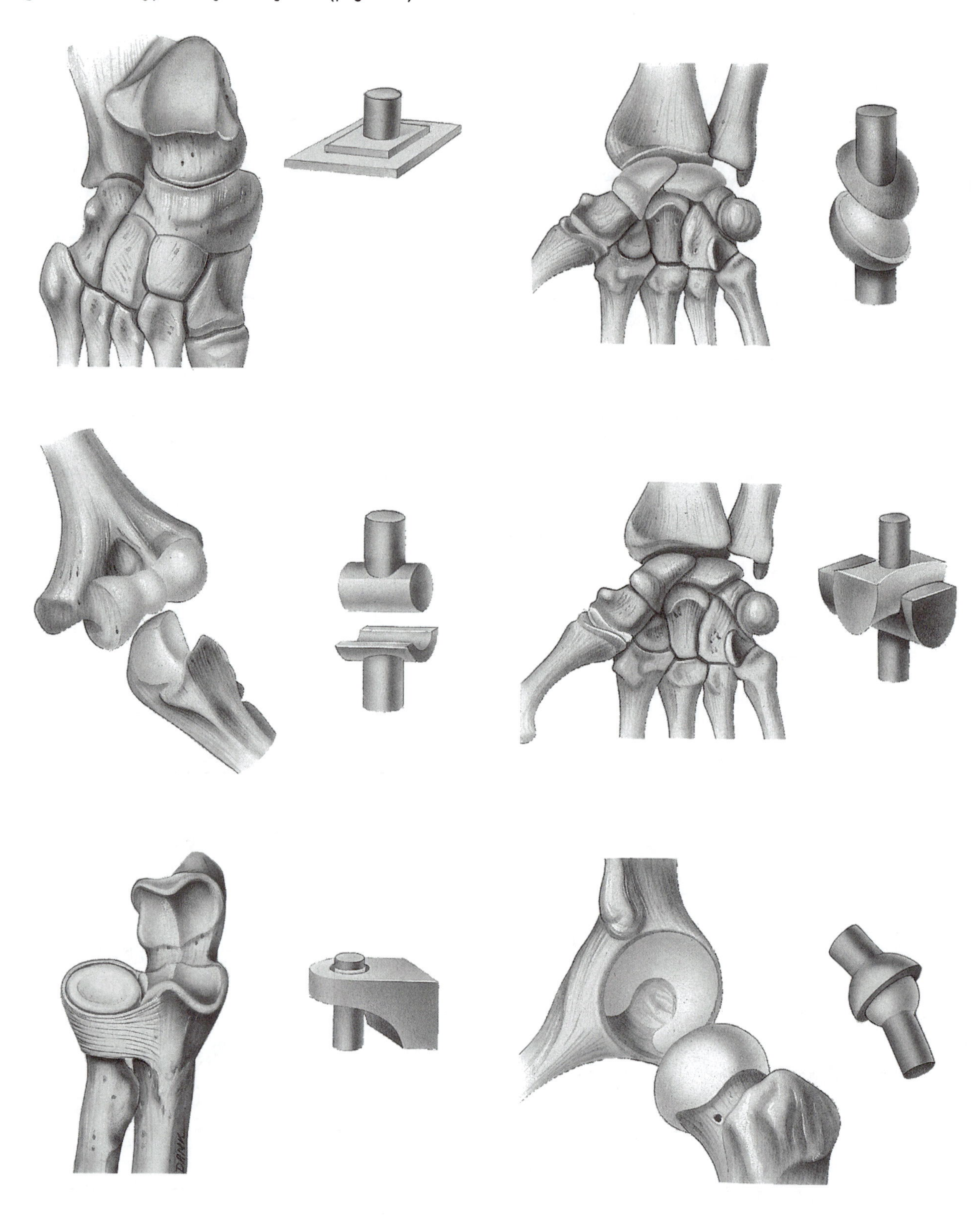

NOTES

Figure 9.11 Right shoulder joint (page 254).

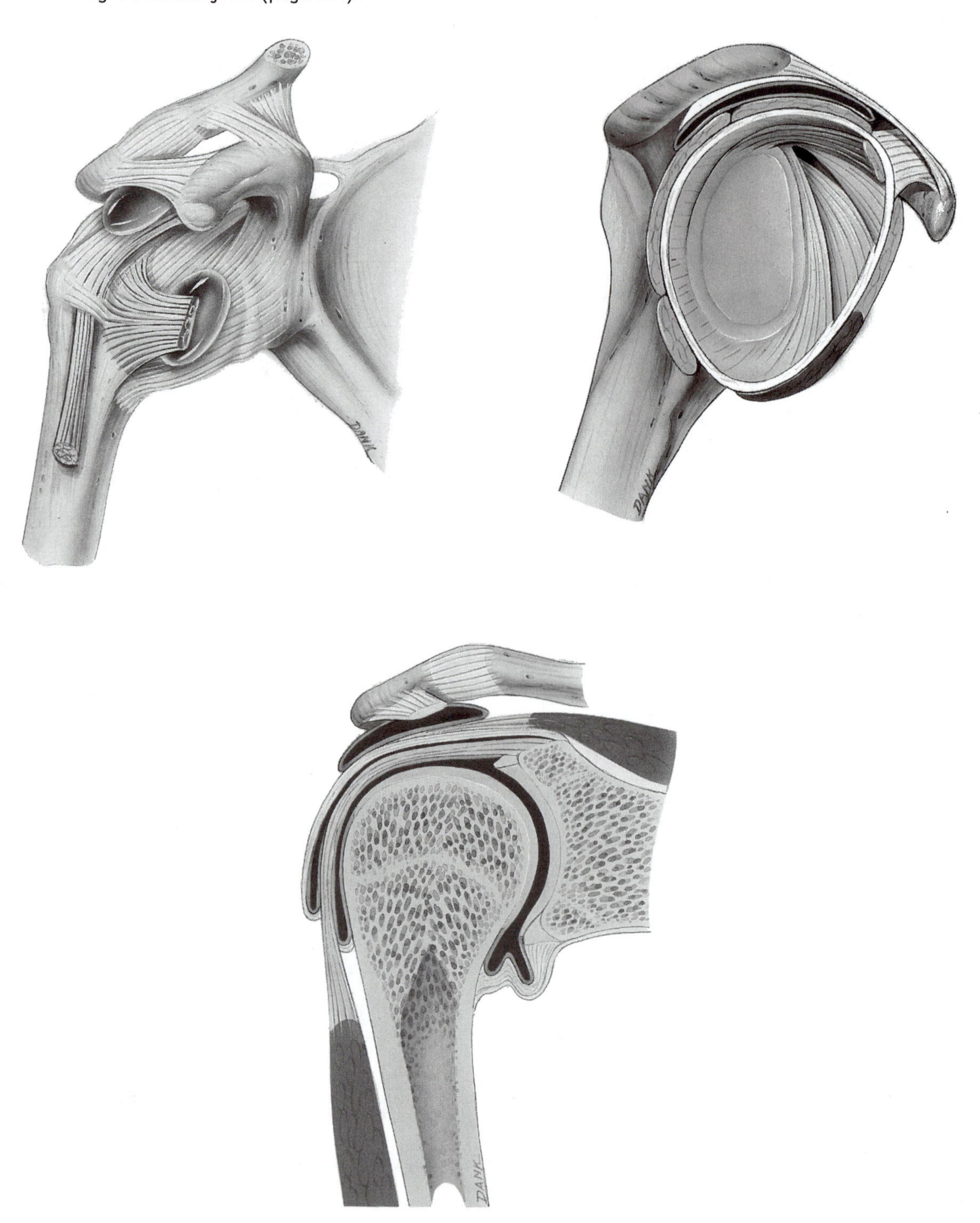

NOTES

9

Figure 9.12 Right elbow joint (page 256).

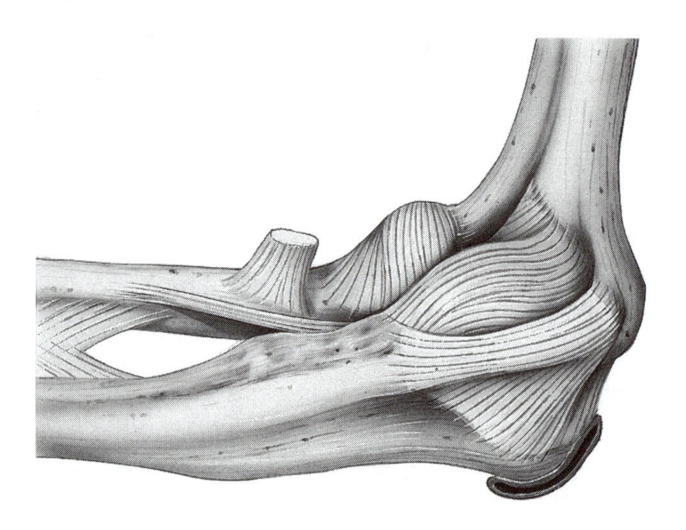

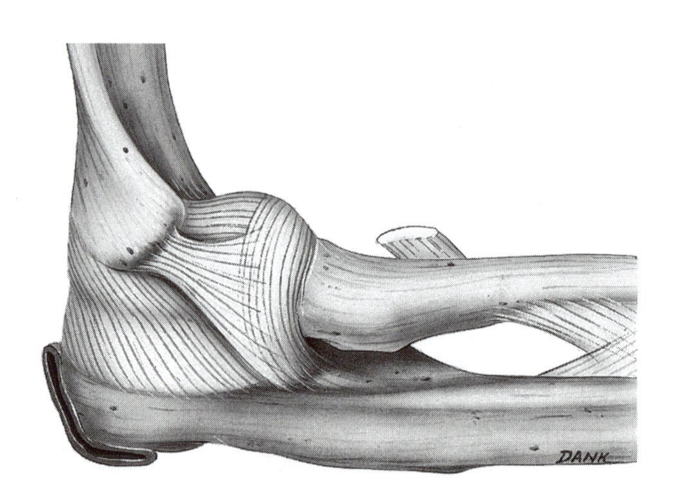
DANK

NOTES

9

Figure 9.13 Right hip joint (pae 257).

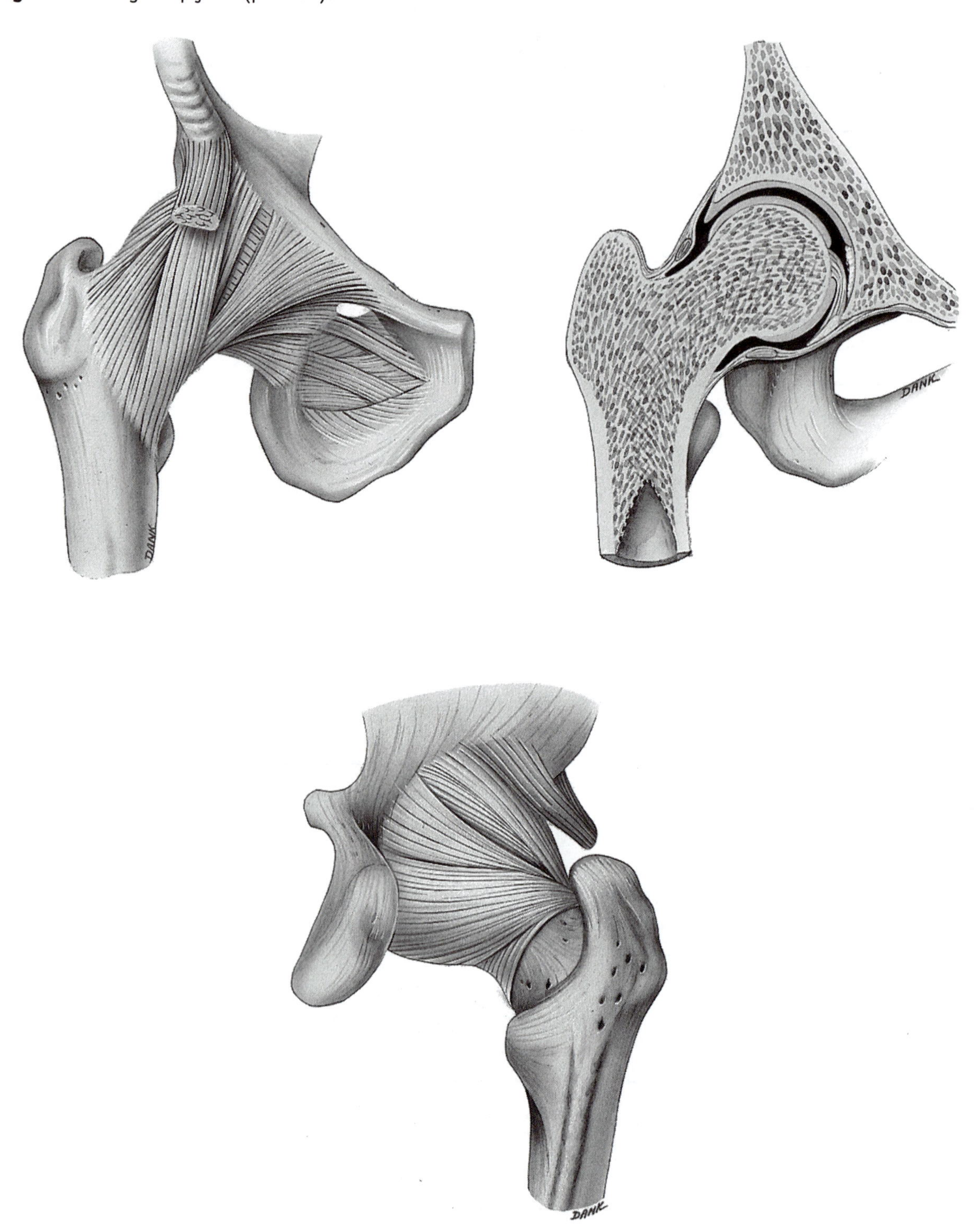
DANK
DANK
DANK

NOTES

9

Figure 9.14 Right knee joint (page 260).

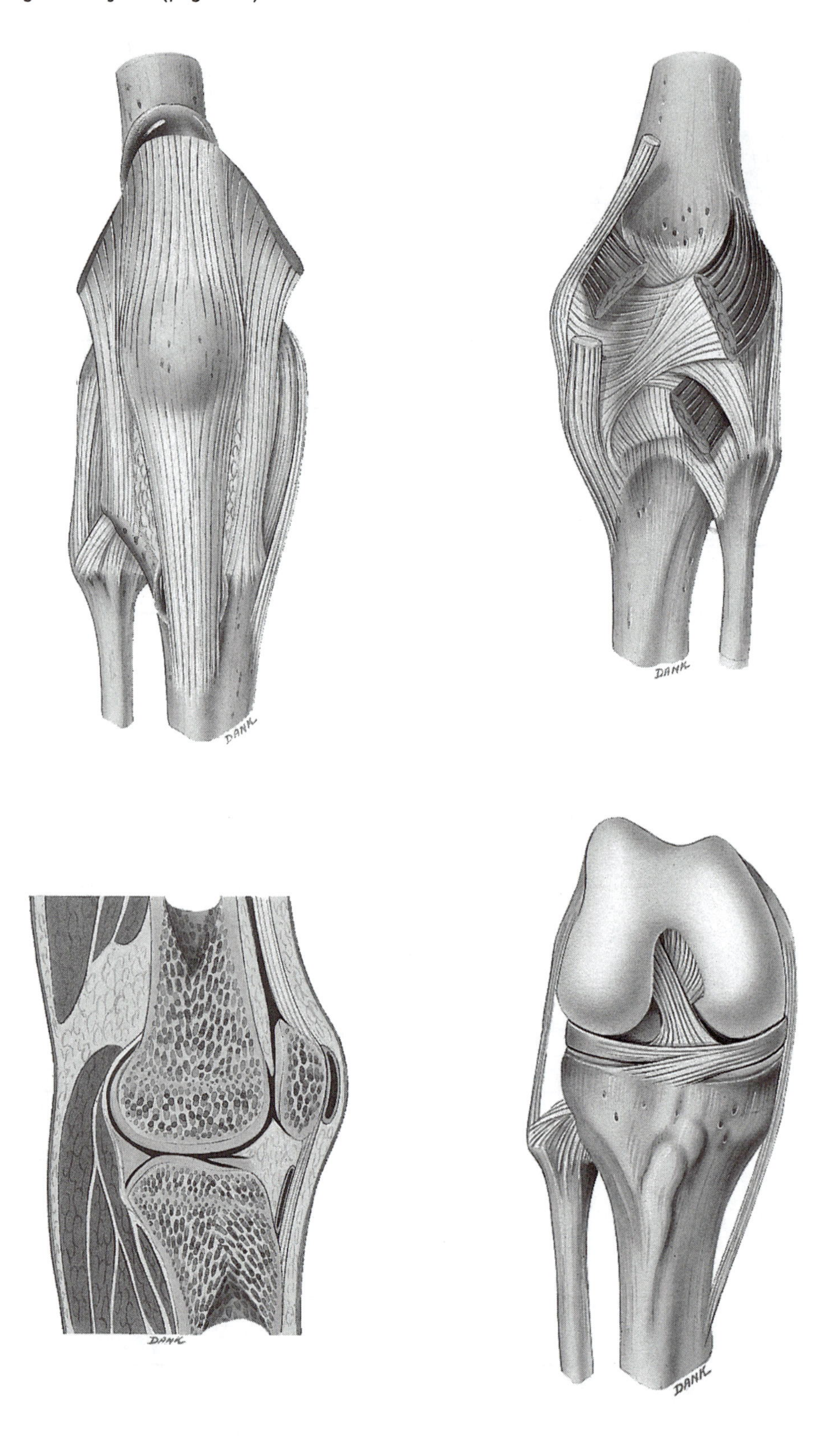

DANK
DANK
DANK
DANK

NOTES

9

Figure 10.1 Organization of skeletal muscle and its connective tissue coverings (page 270).

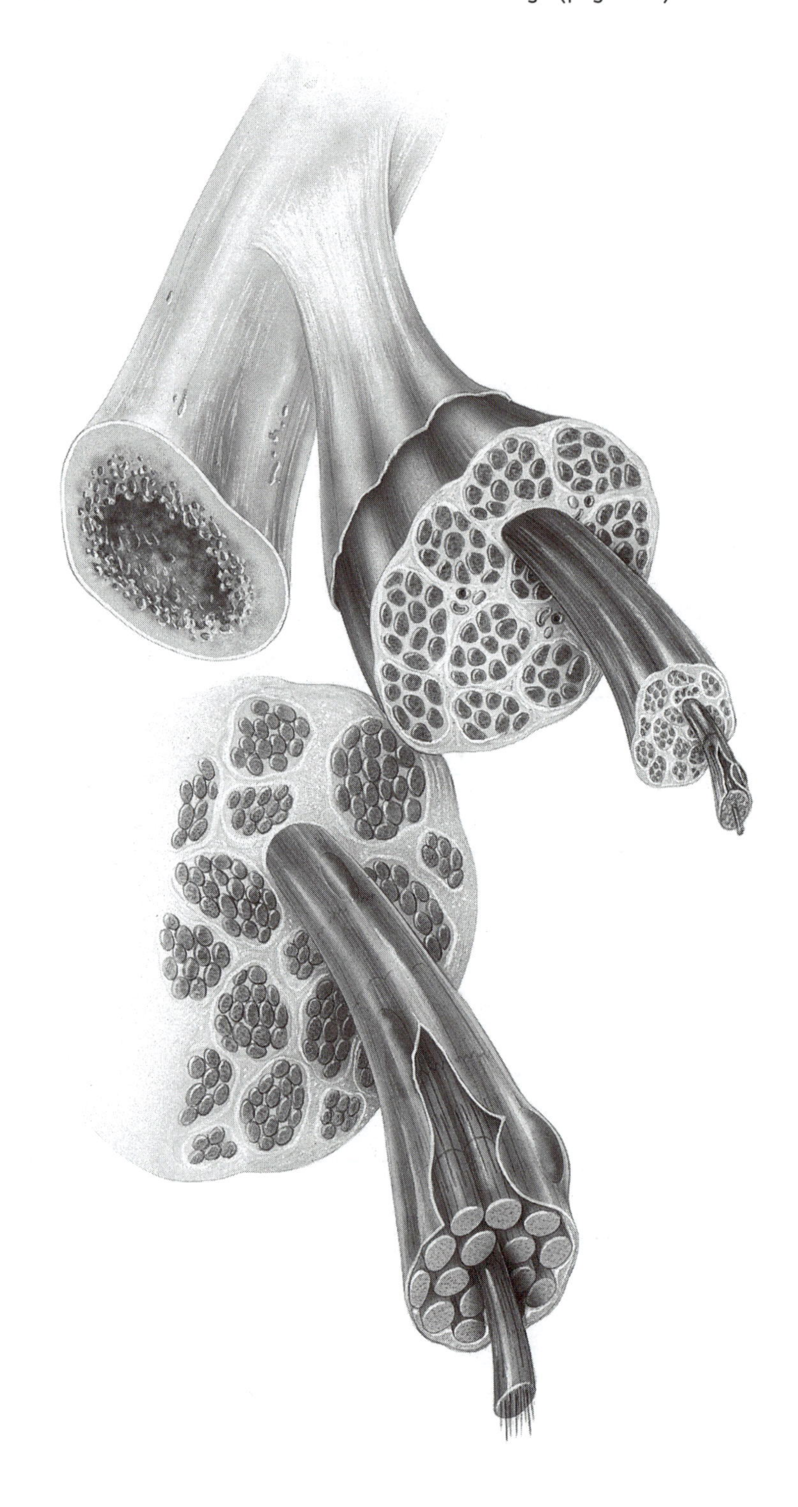

NOTES

10

Figure 10.3 Microscopic organization of skeletal muscle (page 272).

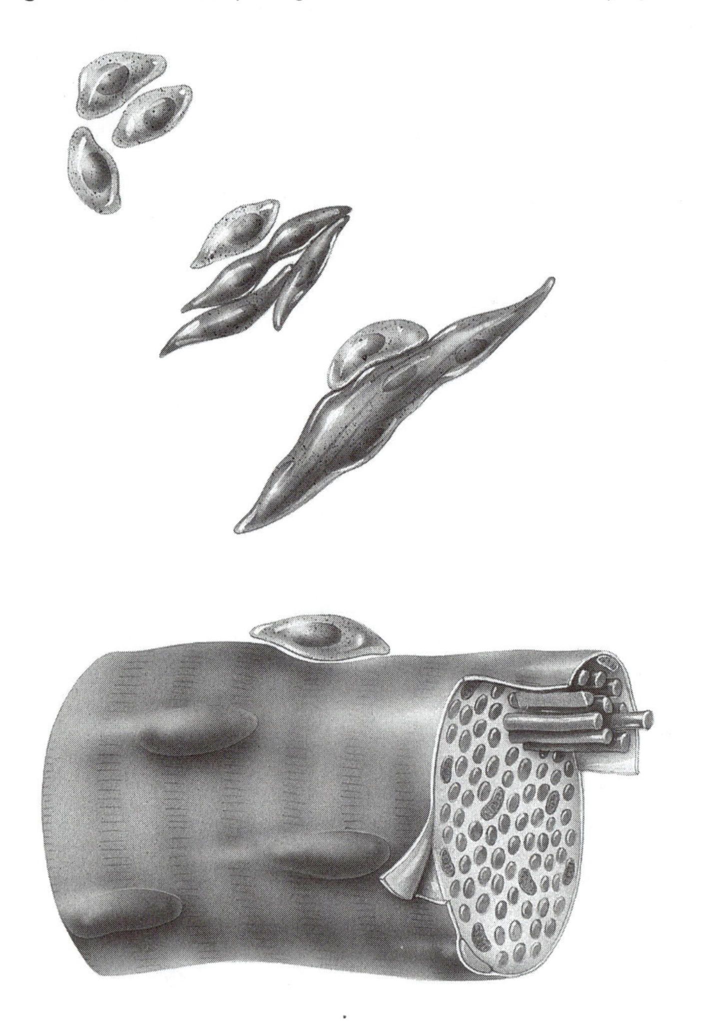

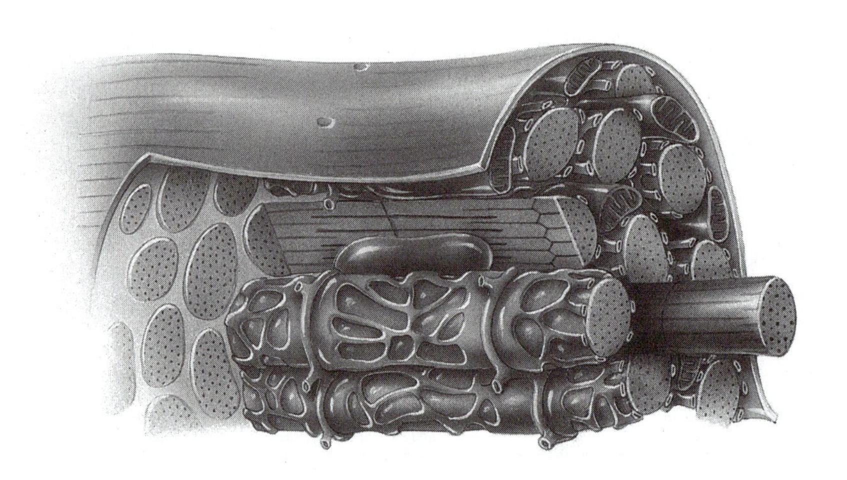

NOTES

10

Figure 10.4 The arrangement of filaments within a sarcomere (page 274).

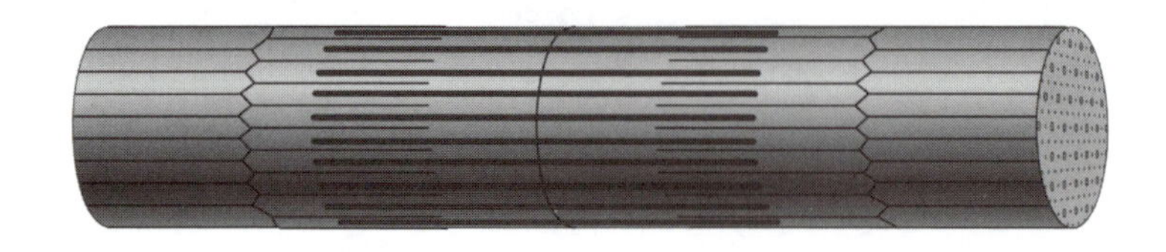

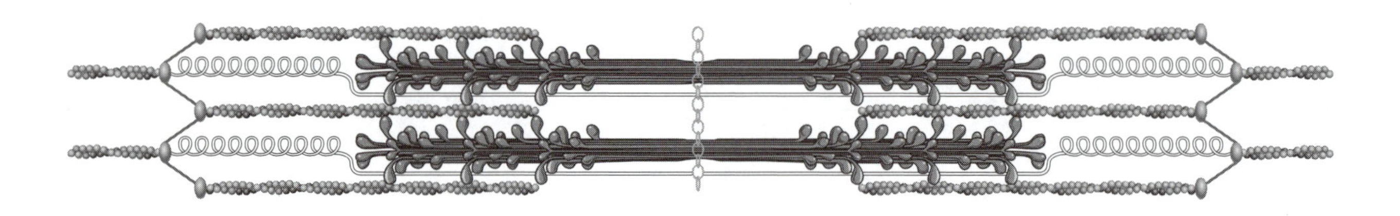

Figure 10.6 Structure of thick and thin filaments (page 275).

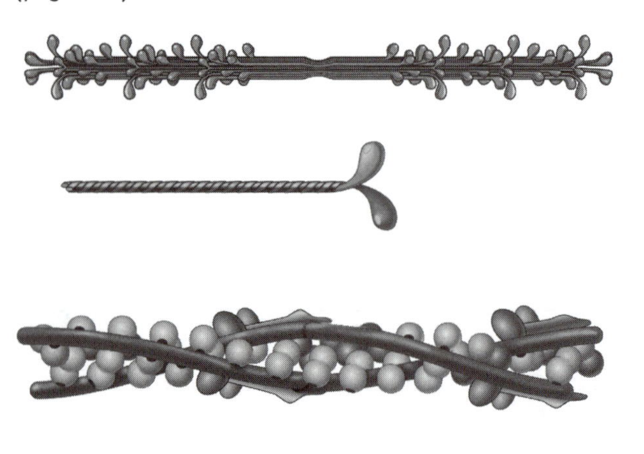

Figure 10.7 Sliding filament mechanism (page 276).

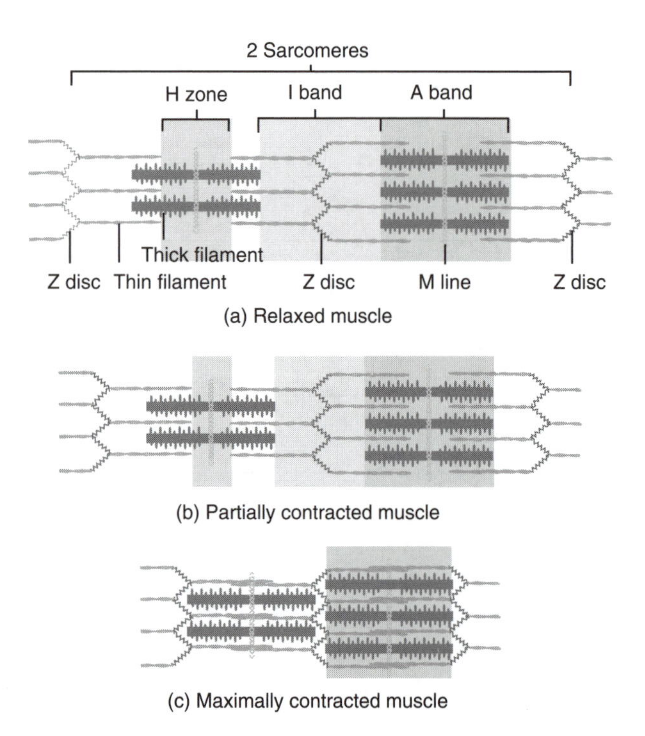
2 Sarcomeres
H zone
I band
A band
Thick filament
Z disc
Thin filament
Z disc
M line
Z disc
(a) Relaxed muscle
(b) Partially contracted muscle
(c) Maximally contracted muscle

NOTES

10

Figure 10.8 The contraction cycle (page 277).

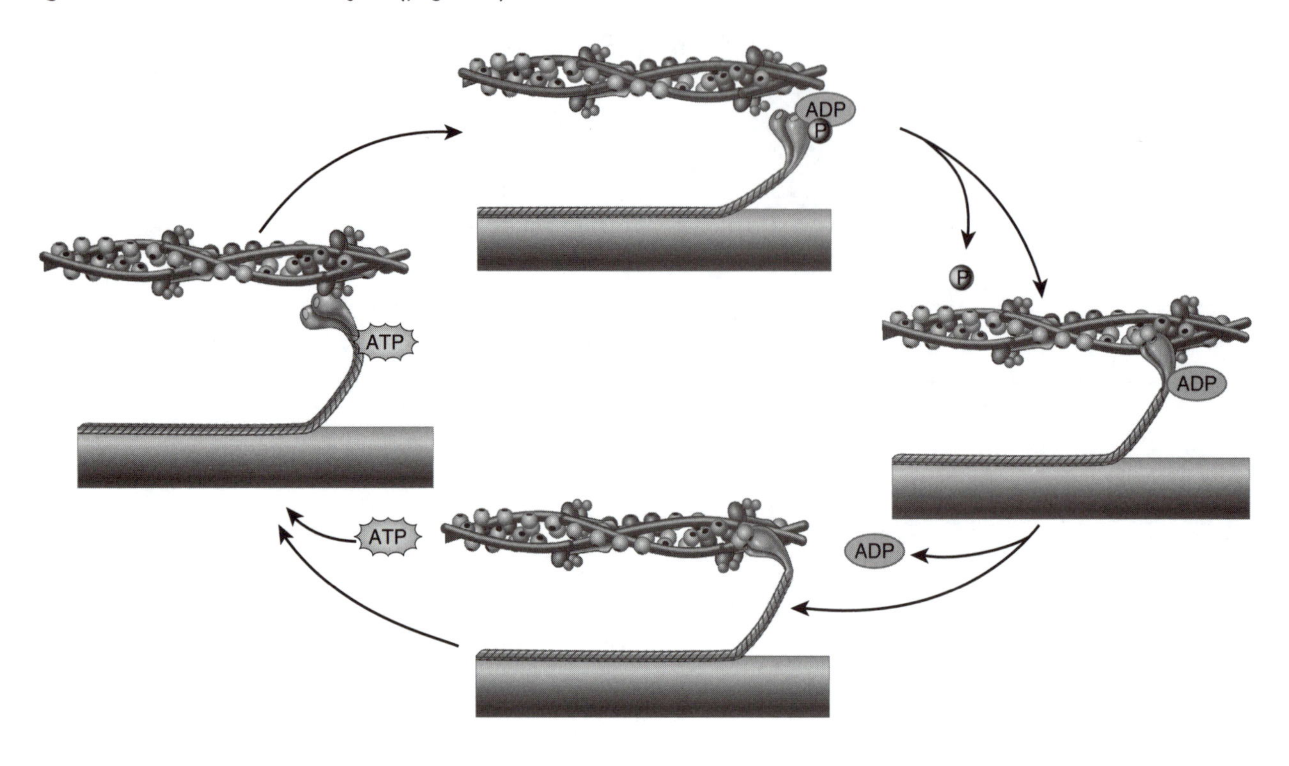

ADP
P
P
ATP
ADP
ATP
ADP

Figure 10.9 The role of Ca^{2+} in the regulation of contraction by troponin and tropomyosin (page 279).

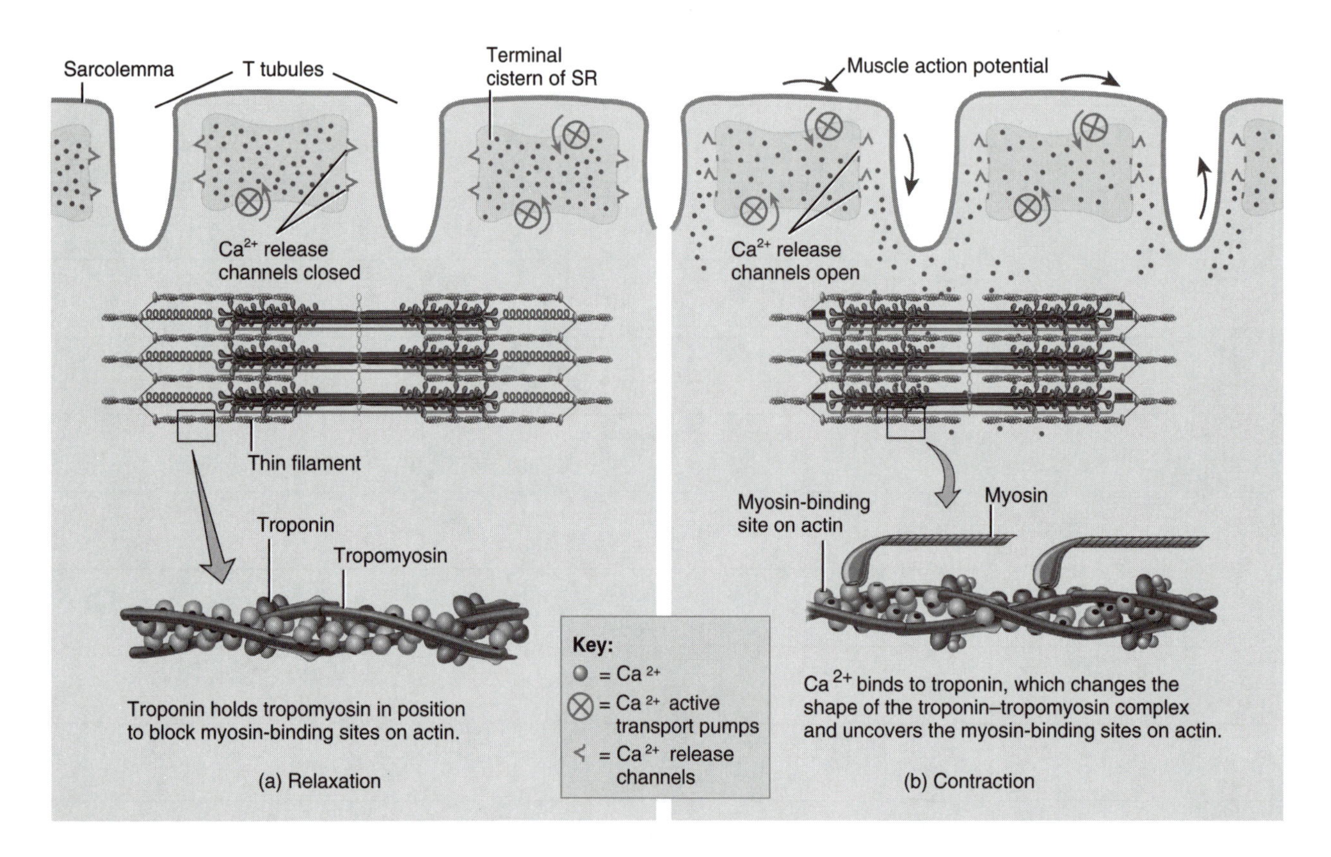

Sarcolemma
T tubules
Terminal cistern of SR
Muscle action potential
Ca2+ release channels closed
Ca2+ release channels open
Thin filament
Troponin
Tropomyosin
Myosin-binding site on actin
Myosin
Troponin holds tropomyosin in position to block myosin-binding sites on actin.
(a) Relaxation
Key:
= Ca 2+
= Ca 2+ active transport pumps
= Ca 2+ release channels
Ca 2+ binds to troponin, which changes the shape of the troponin–tropomyosin complex and uncovers the myosin-binding sites on actin.
(b) Contraction

NOTES

Figure 10.10 Length–tension relationship in a skeletal muscle fiber (page 279).

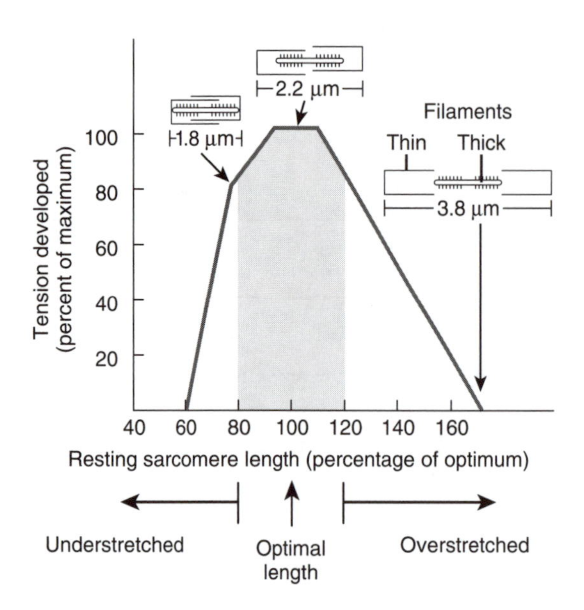

2.2 μm
1.8 μm
Filaments
Thin
Thick
3.8 μm
100
80
60
40
20
Tension developed
(percent of maximum)
40
60
80
100
120
140
160
Resting sarcomere length (percentage of optimum)
Understretched
Optimal
length
Overstretched

Figure 10.11 Structure of the neuromuscular junction (page 281).

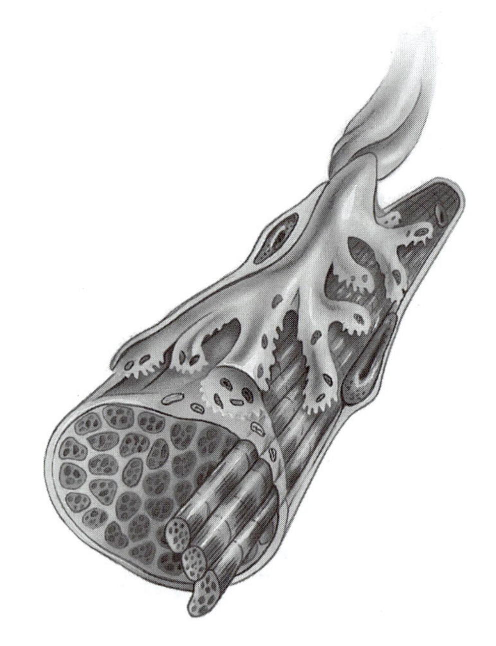

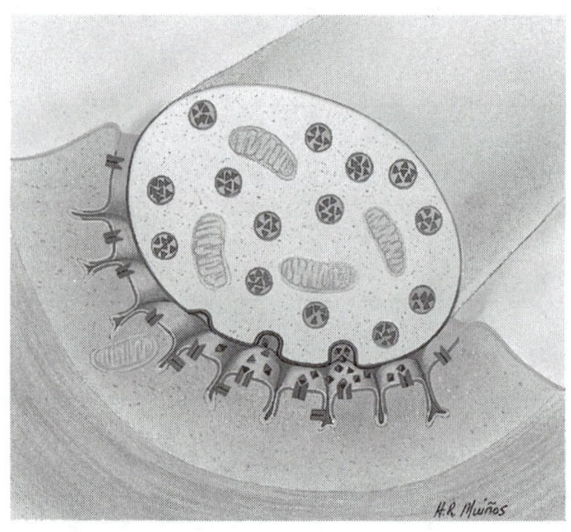

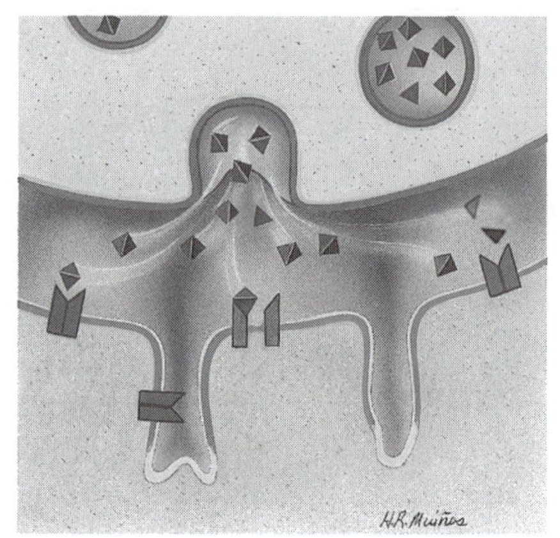

NOTES

10

Figure 10.12 Summary of the events of contraction and relaxation (page 282).

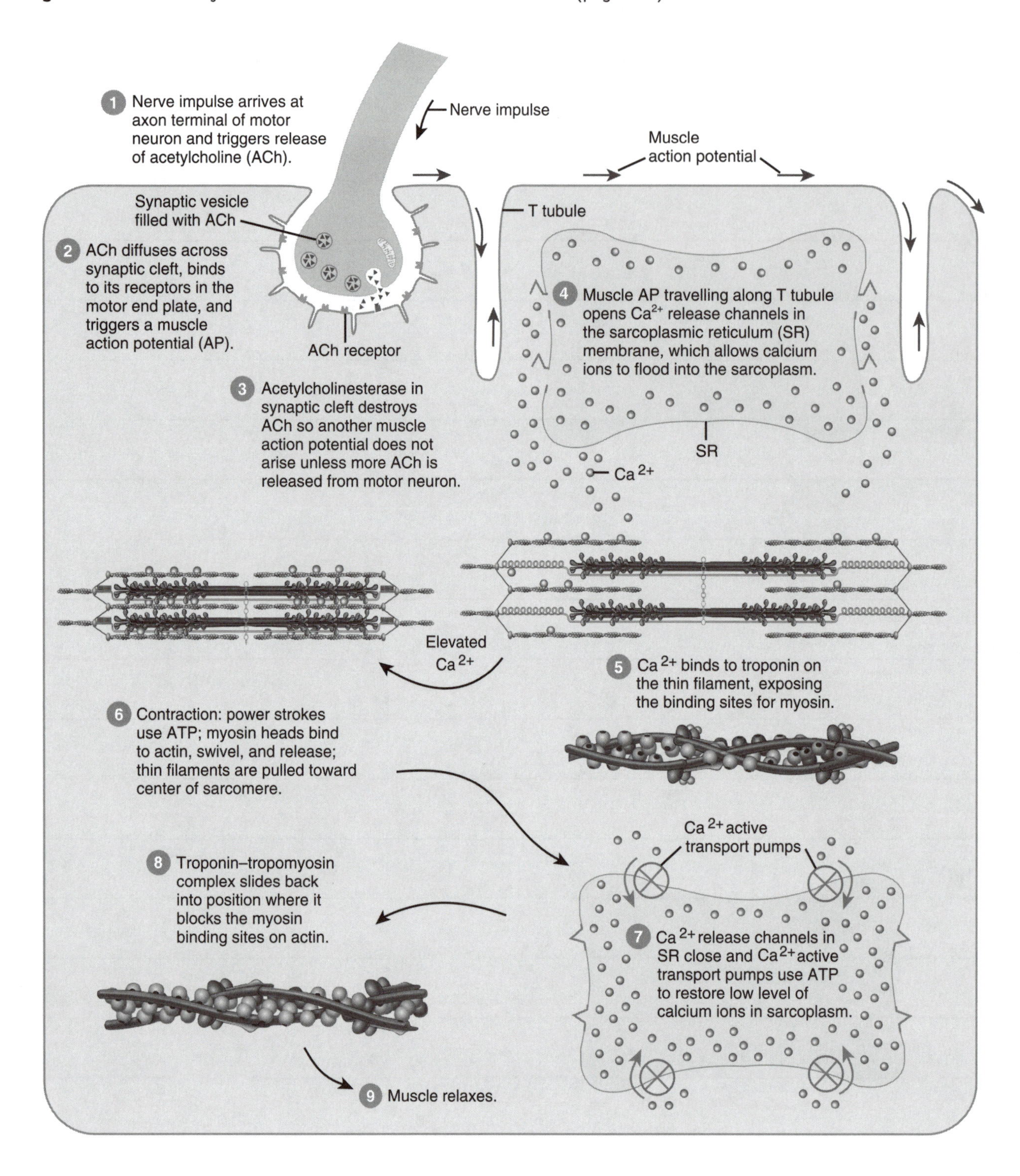

1 Nerve impulse arrives at axon terminal of motor neuron and triggers release of acetylcholine (ACh).
Nerve impulse
Muscle action potential
Synaptic vesicle filled with ACh
T tubule
2 ACh diffuses across synaptic cleft, binds to its receptors in the motor end plate, and triggers a muscle action potential (AP).
ACh receptor
4 Muscle AP travelling along T tubule opens Ca2+ release channels in the sarcoplasmic reticulum (SR) membrane, which allows calcium ions to flood into the sarcoplasm.
3 Acetylcholinesterase in synaptic cleft destroys ACh so another muscle action potential does not arise unless more ACh is released from motor neuron.
SR
Ca 2+
Elevated Ca 2+
5 Ca 2+ binds to troponin on the thin filament, exposing the binding sites for myosin.
6 Contraction: power strokes use ATP; myosin heads bind to actin, swivel, and release; thin filaments are pulled toward center of sarcomere.
Ca 2+ active transport pumps
8 Troponin–tropomyosin complex slides back into position where it blocks the myosin binding sites on actin.
7 Ca 2+ release channels in SR close and Ca 2+ active transport pumps use ATP to restore low level of calcium ions in sarcoplasm.
9 Muscle relaxes.

NOTES

10

Figure 10.13 Production of ATP for muscle contraction (page 284).

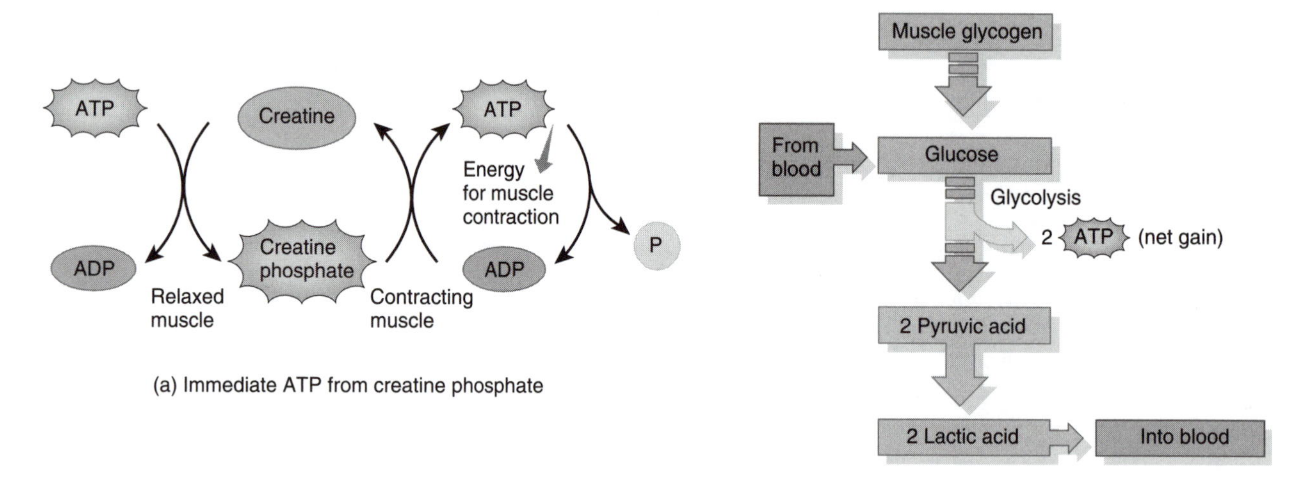
ATP
Creatine
ATP
Energy for muscle contraction
ADP
Creatine phosphate
ADP
P
Relaxed muscle
Contracting muscle
Muscle glycogen
From blood
Glucose
Glycolysis
2 ATP (net gain)
2 Pyruvic acid
2 Lactic acid
Into blood

(a) Immediate ATP from creatine phosphate

(b) Short-term ATP from anaerobic respiration

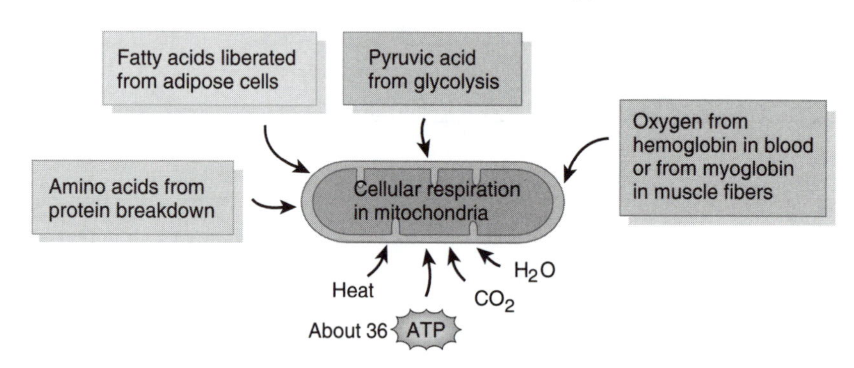
Fatty acids liberated from adipose cells
Pyruvic acid from glycolysis
Oxygen from hemoglobin in blood or from myoglobin in muscle fibers
Amino acids from protein breakdown
Cellular respiration in mitochondria
Heat
H_2O
CO_2
About 36 ATP

(c) Long-term ATP from aerobic cellular respiration

Figure 10.14 Motor units (page 286).

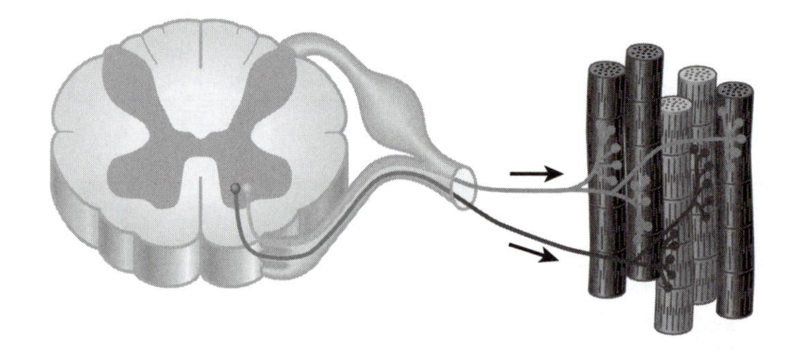

NOTES

10

Figure 10.15 Myogram of a twitch contraction (page 286).

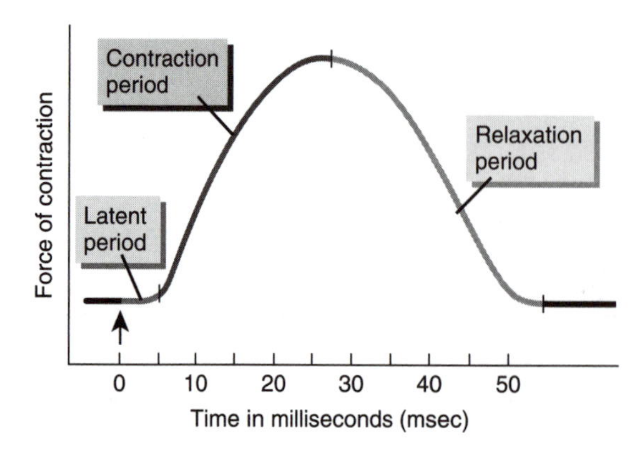
Contraction period
Relaxation period
Latent period
Force of contraction
0
10
20
30
40
50
Time in milliseconds (msec)

Figure 10.16 Myograms showing the effects of different frequencies of stimulation (page 287).

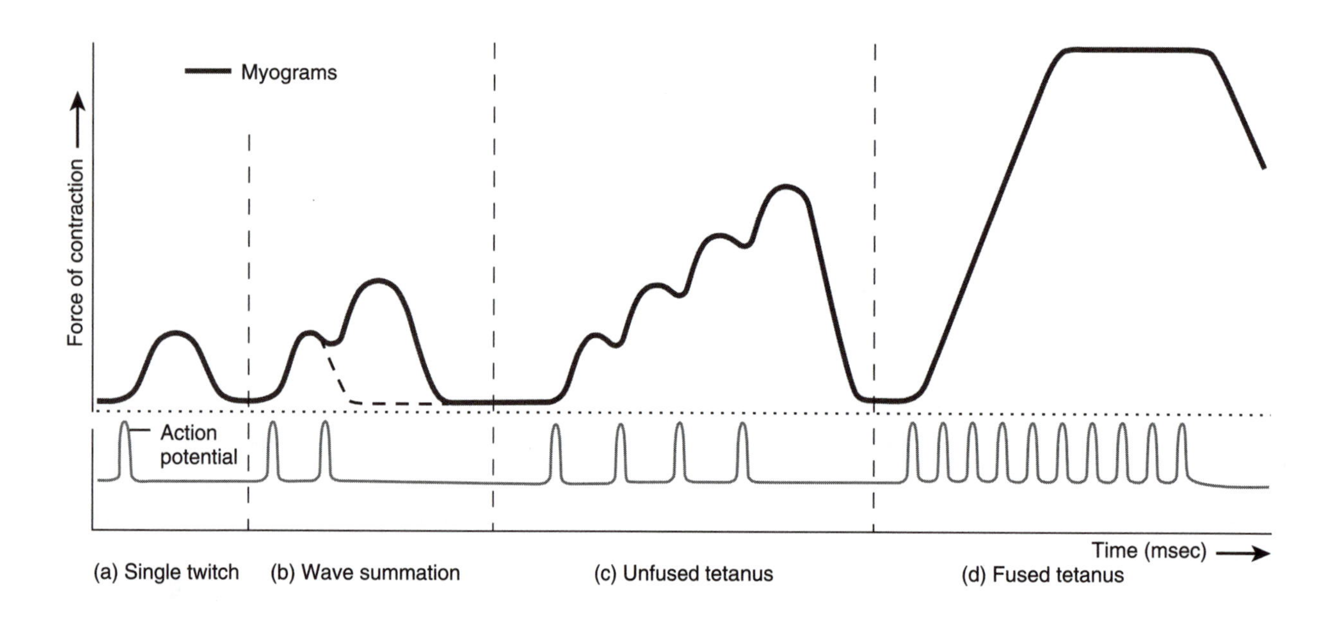
Myograms
Force of contraction
Action potential
Time (msec)
(a) Single twitch
(b) Wave summation
(c) Unfused tetanus
(d) Fused tetanus

NOTES

10

Figure 10.18 Two types of smooth muscle tissue (page 292).

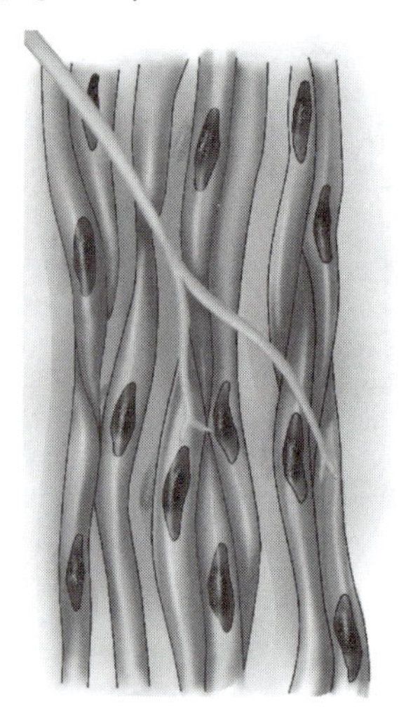

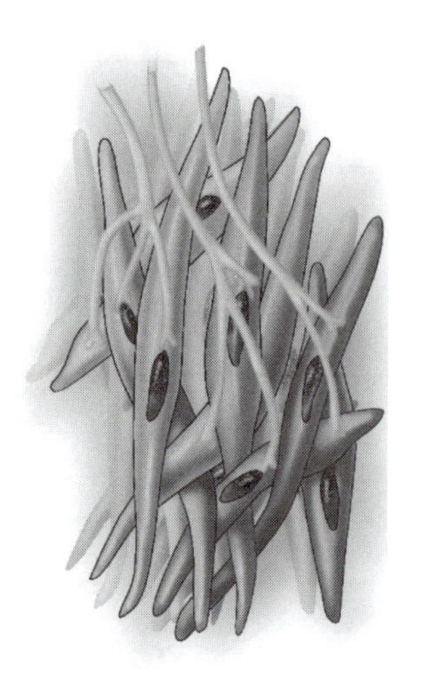

Figure 10.19 Microscopic anatomy of a smooth muscle fiber (page 293).

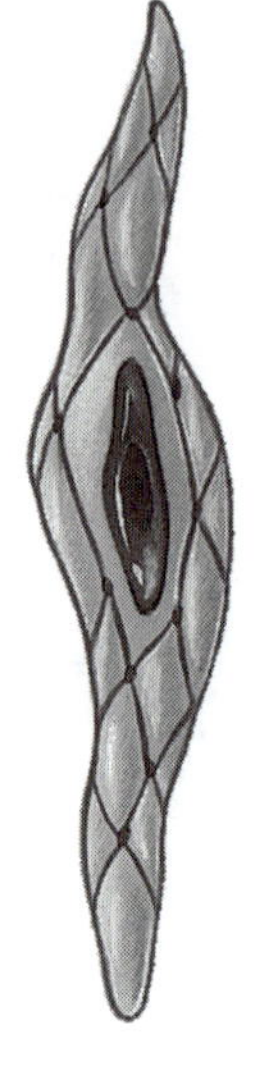

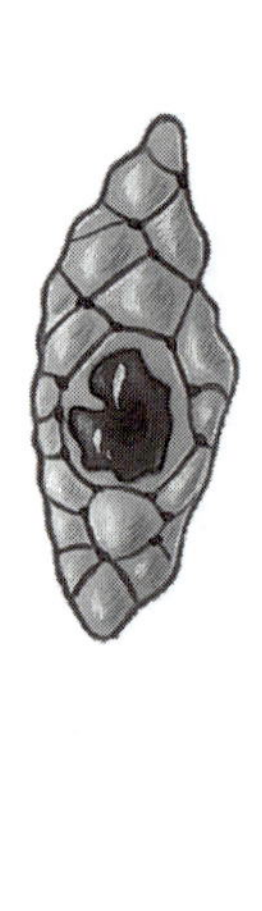

Figure 10.20 Location and structure of somites (page 295).

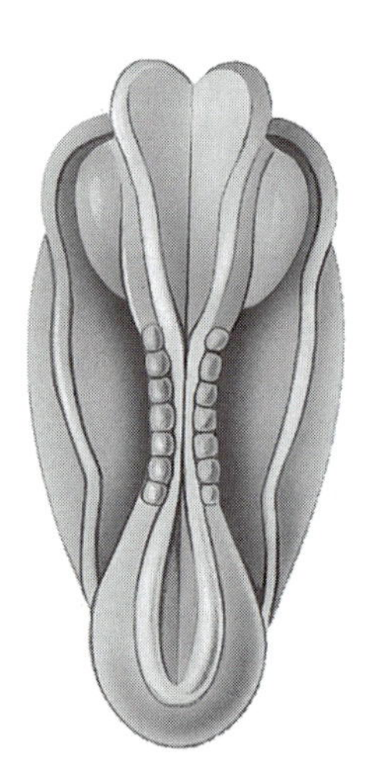

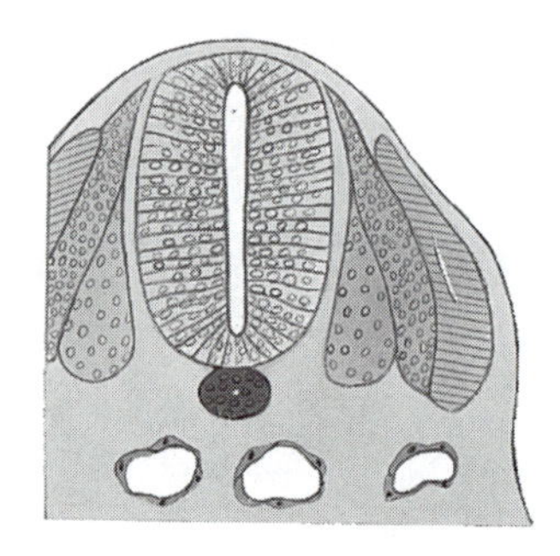

NOTES

10

Figure 11.1 Relationship of skeletal muscles to bones (page 303).

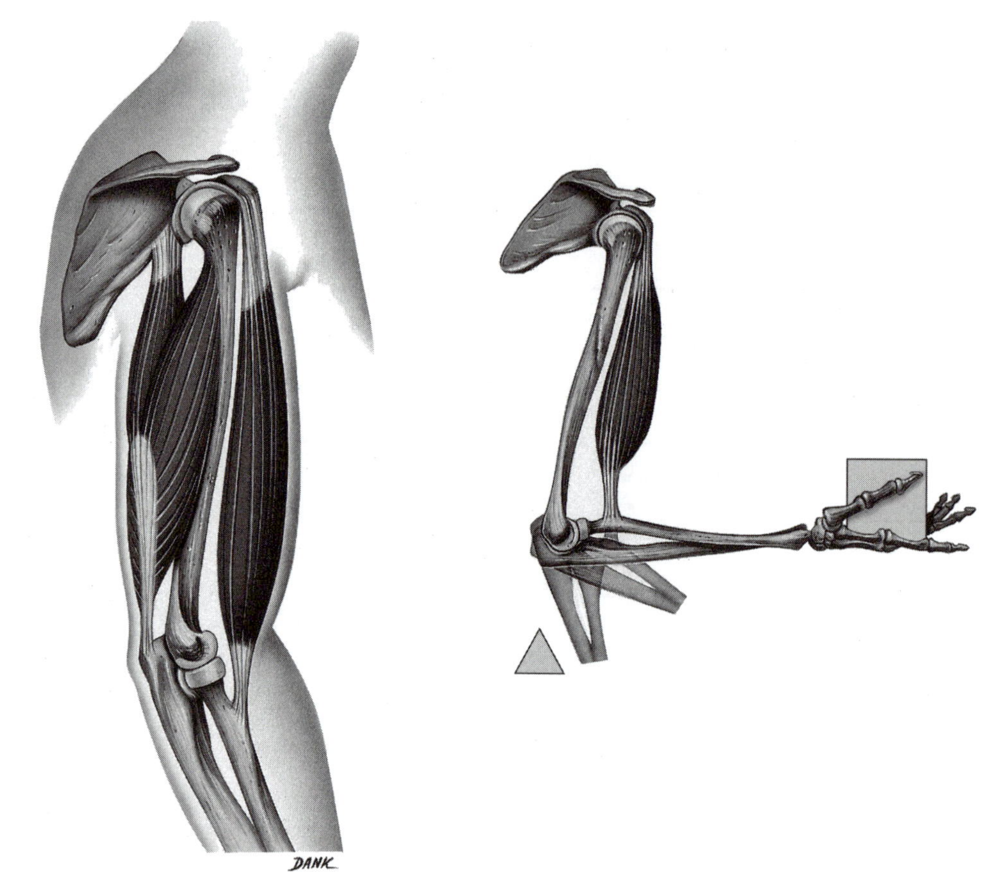
DANK

Figure 11.2 Types of levers (page 304).

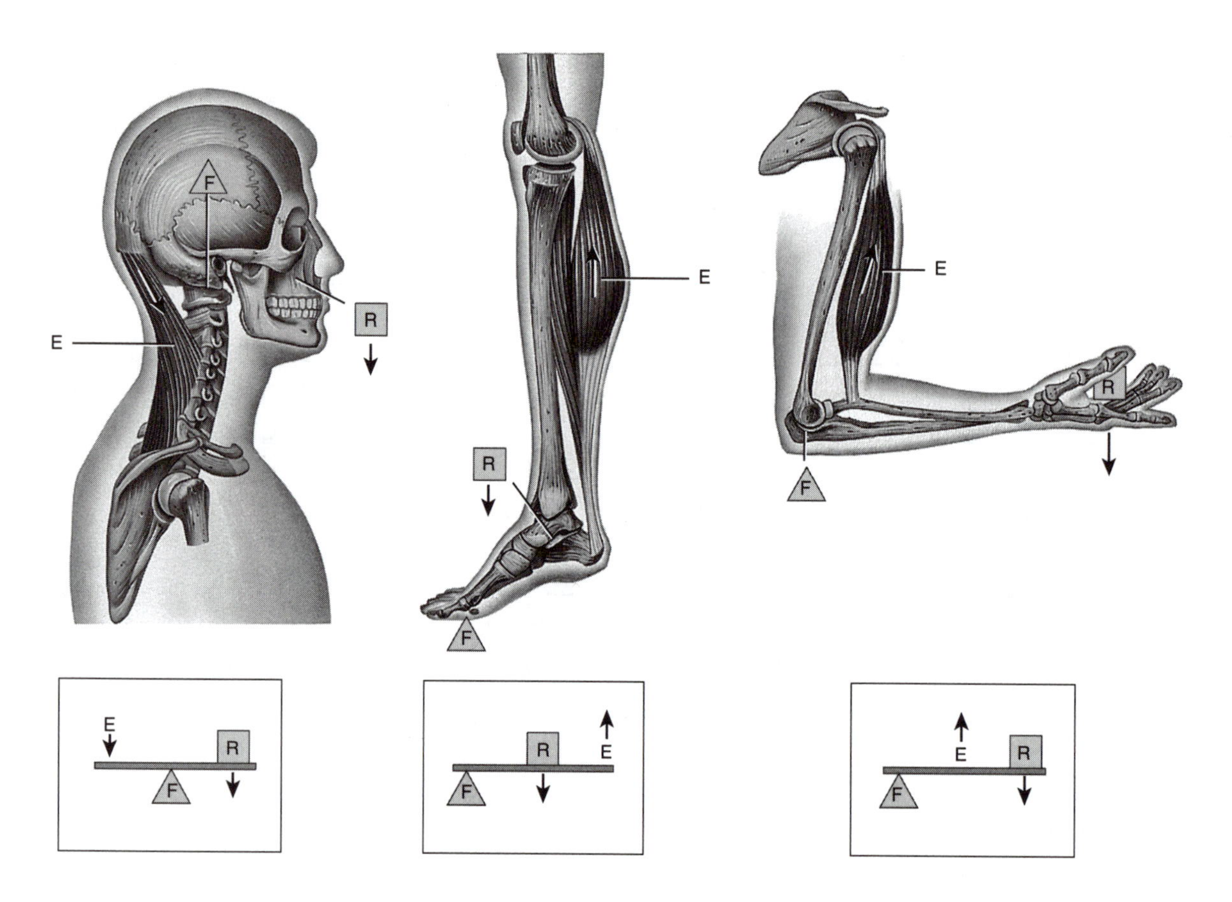
F
E
R
E
R
F
E
R
F
E
R
F
R
E
F
E
R
F

NOTES

11

Figure 11.3 Principal superficial skeletal muscles: anterior view (309).

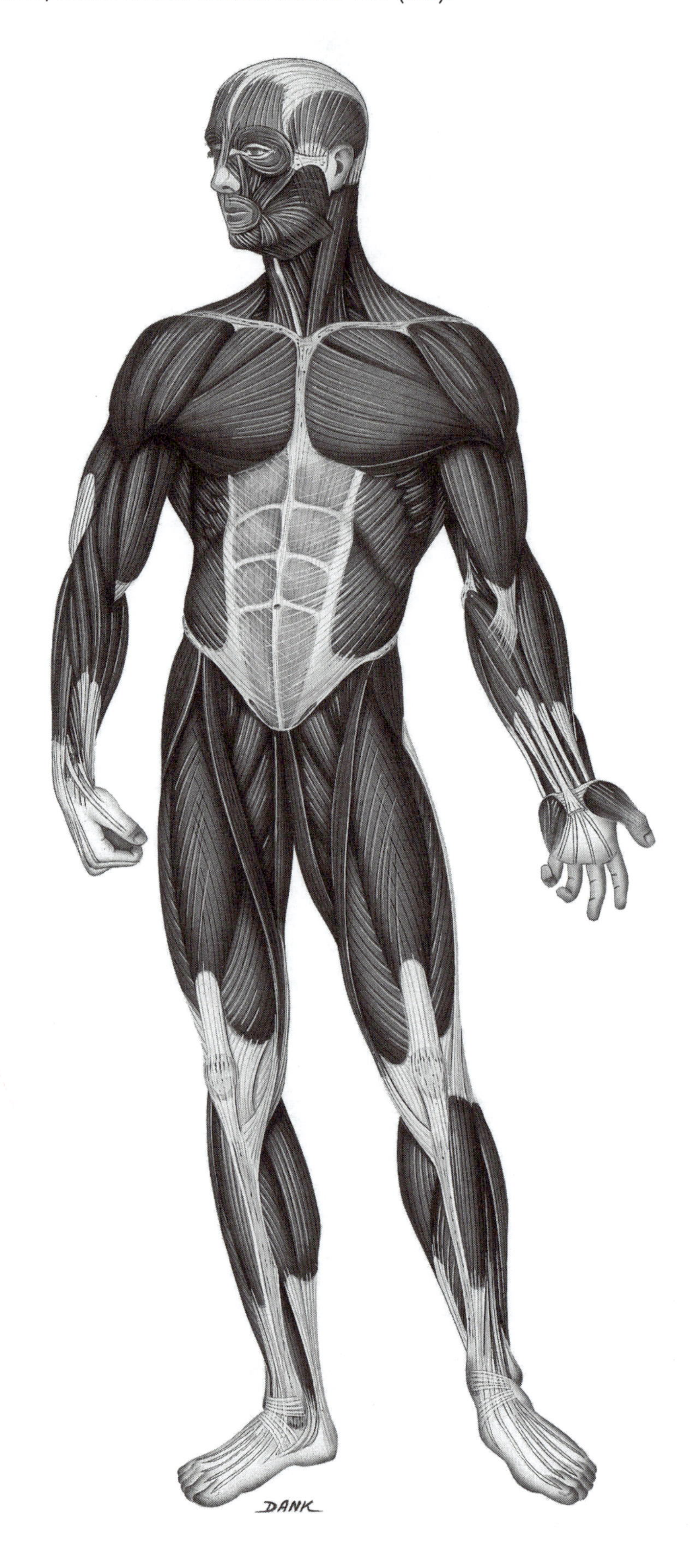
DANK

NOTES

11

Figure 11.3 Principal superficial skeletal muscles: posterior view (310).

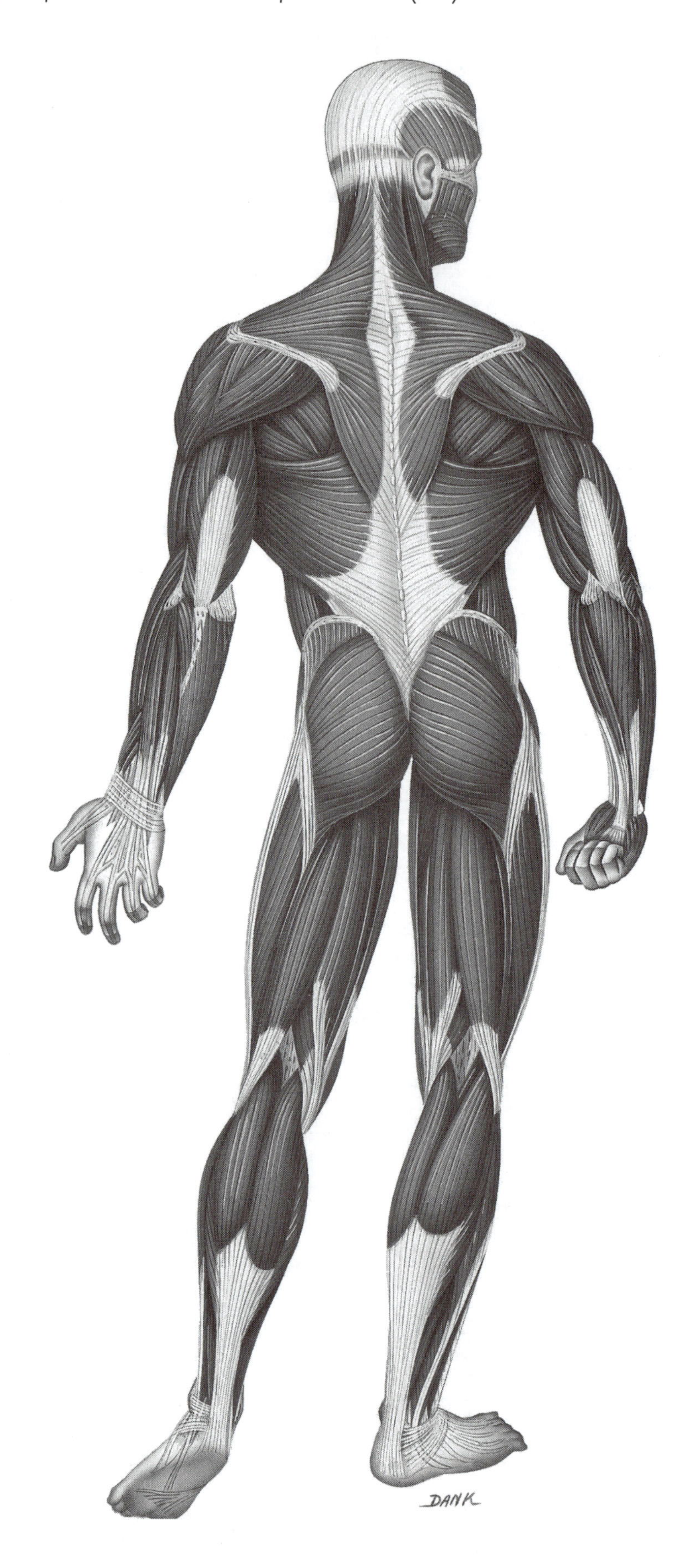
DANK

NOTES

Figure 11.4 Muscles of facial expression (pages 313, 314).

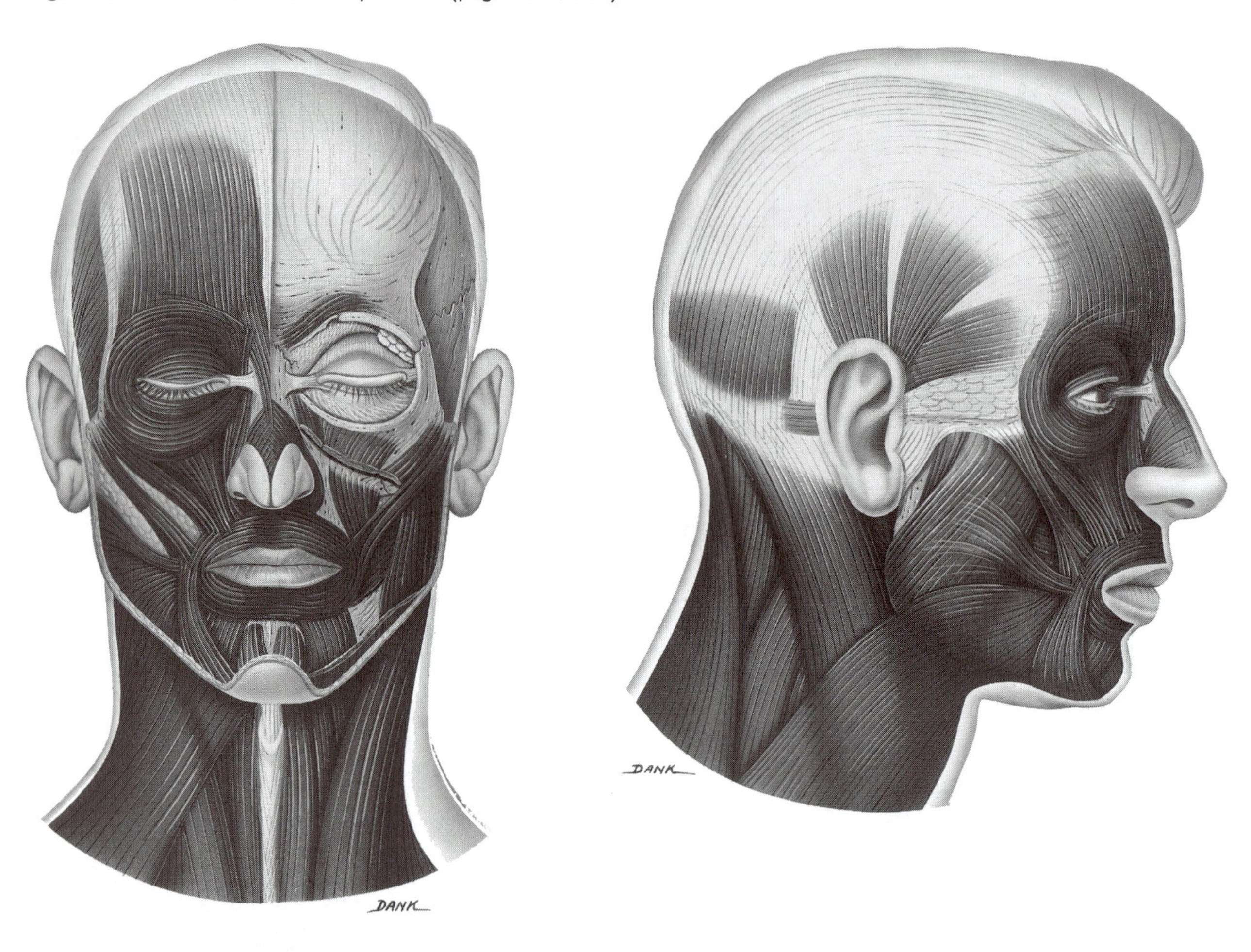
DANK
DANK

Figure 11.5 Extrinsic muscles of the eyeball (page 316).

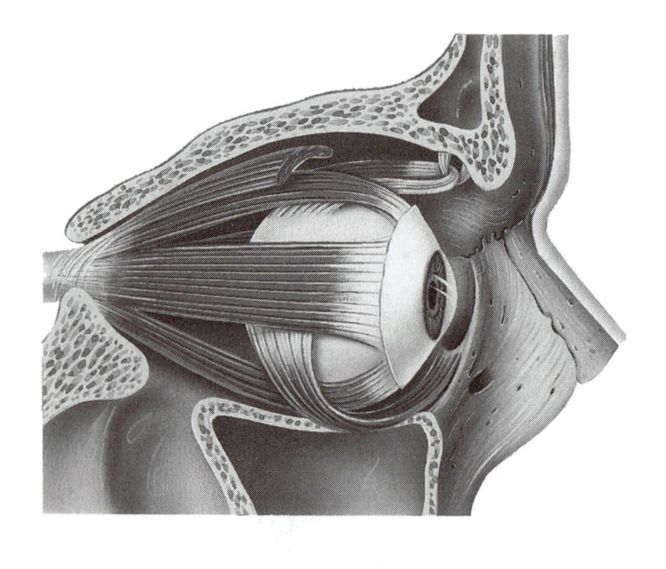

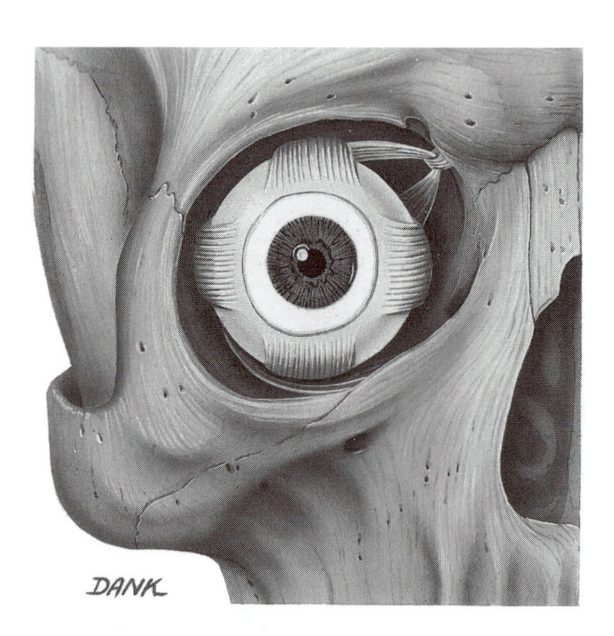
DANK

NOTES

11

Figure 11.6 Muscles that move the mandible (page 318).

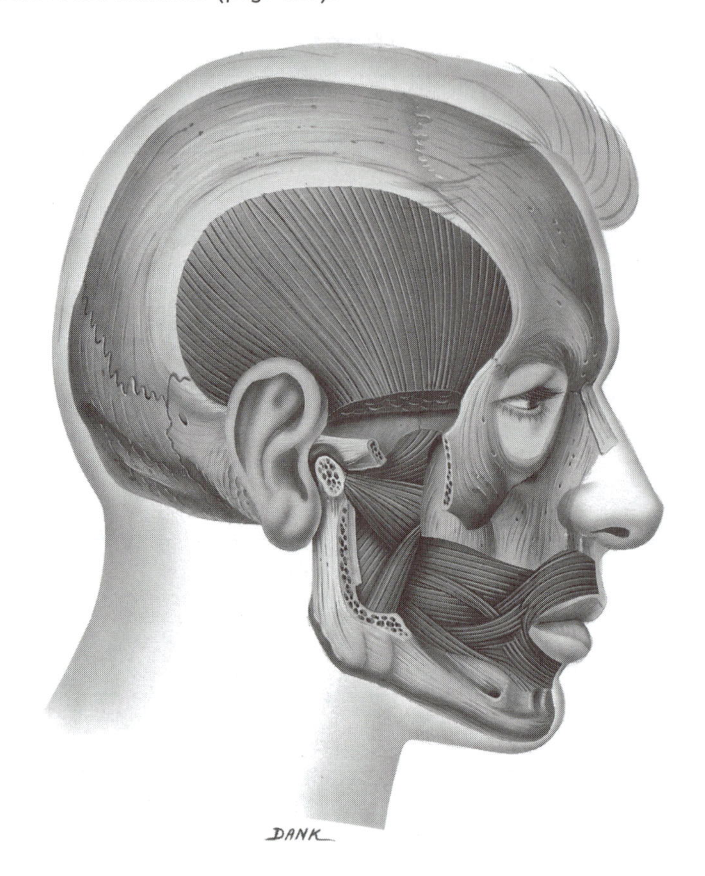
DANK

Figure 11.7 Muscles that move the tongue (page 320).

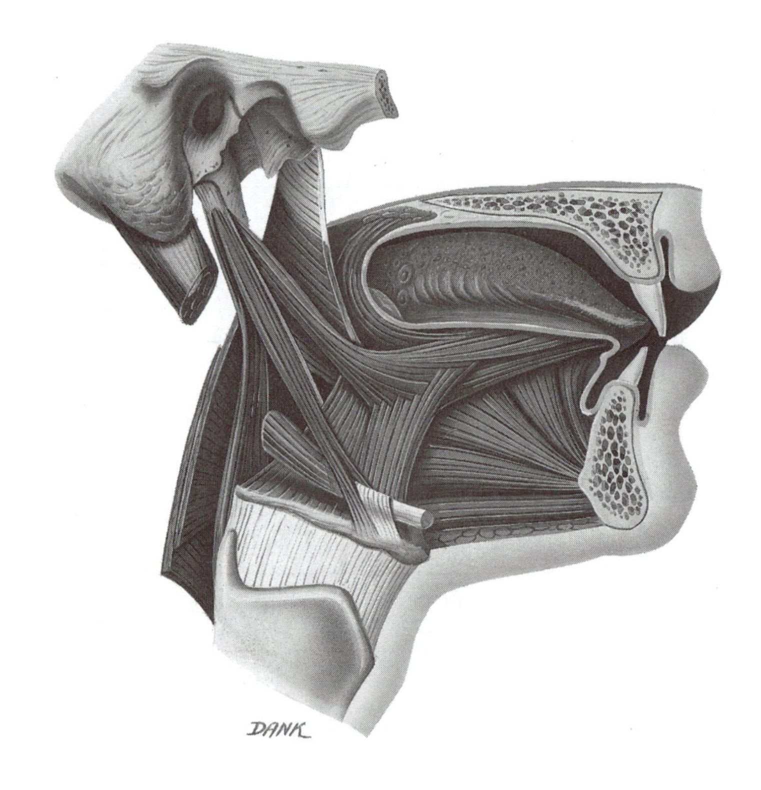
DANK

NOTES

Figure 11.8 Muscles of the floor of the oral cavity and front of the neck (page 322).

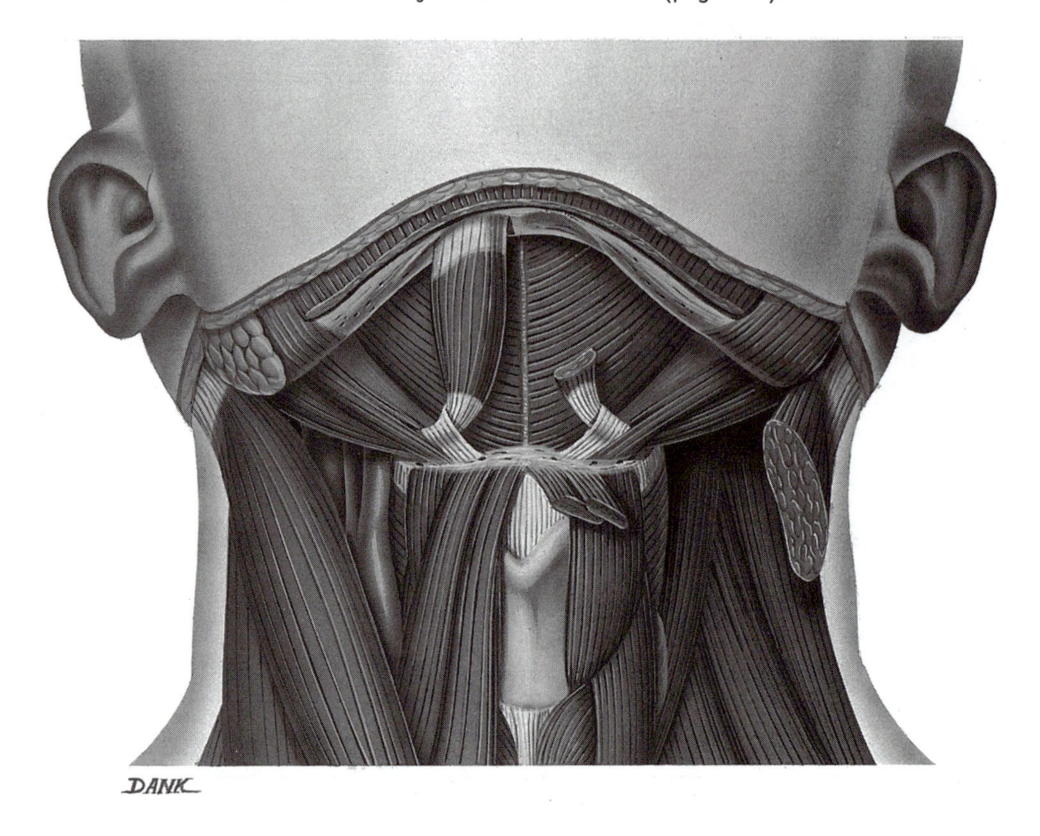
DANK

Figure 11.9 Triangles of the neck (page 324).

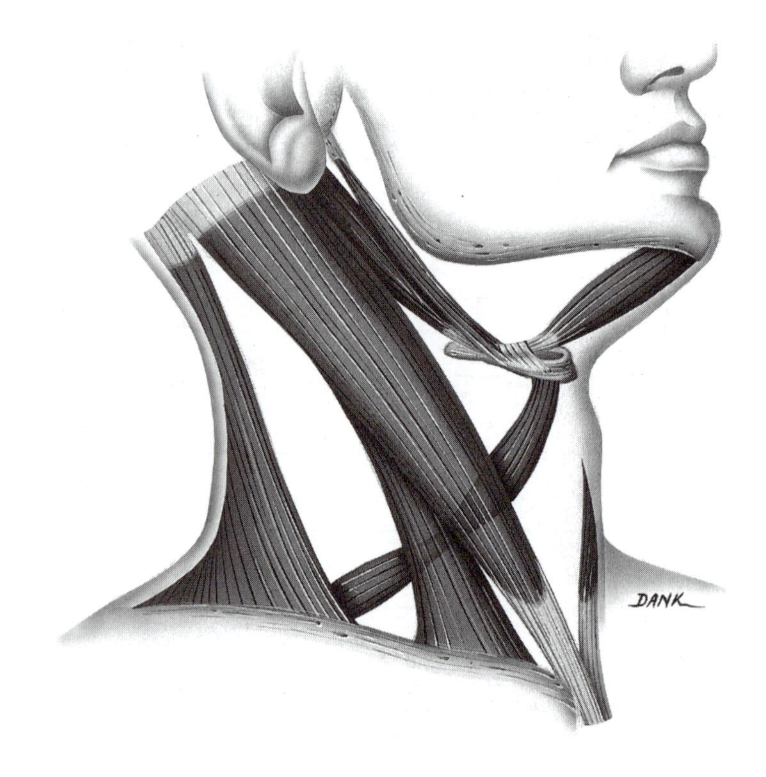
DANK

NOTES

11

Figure 11.10 Muscles of the male anterolateral abdominal wall (page 327).

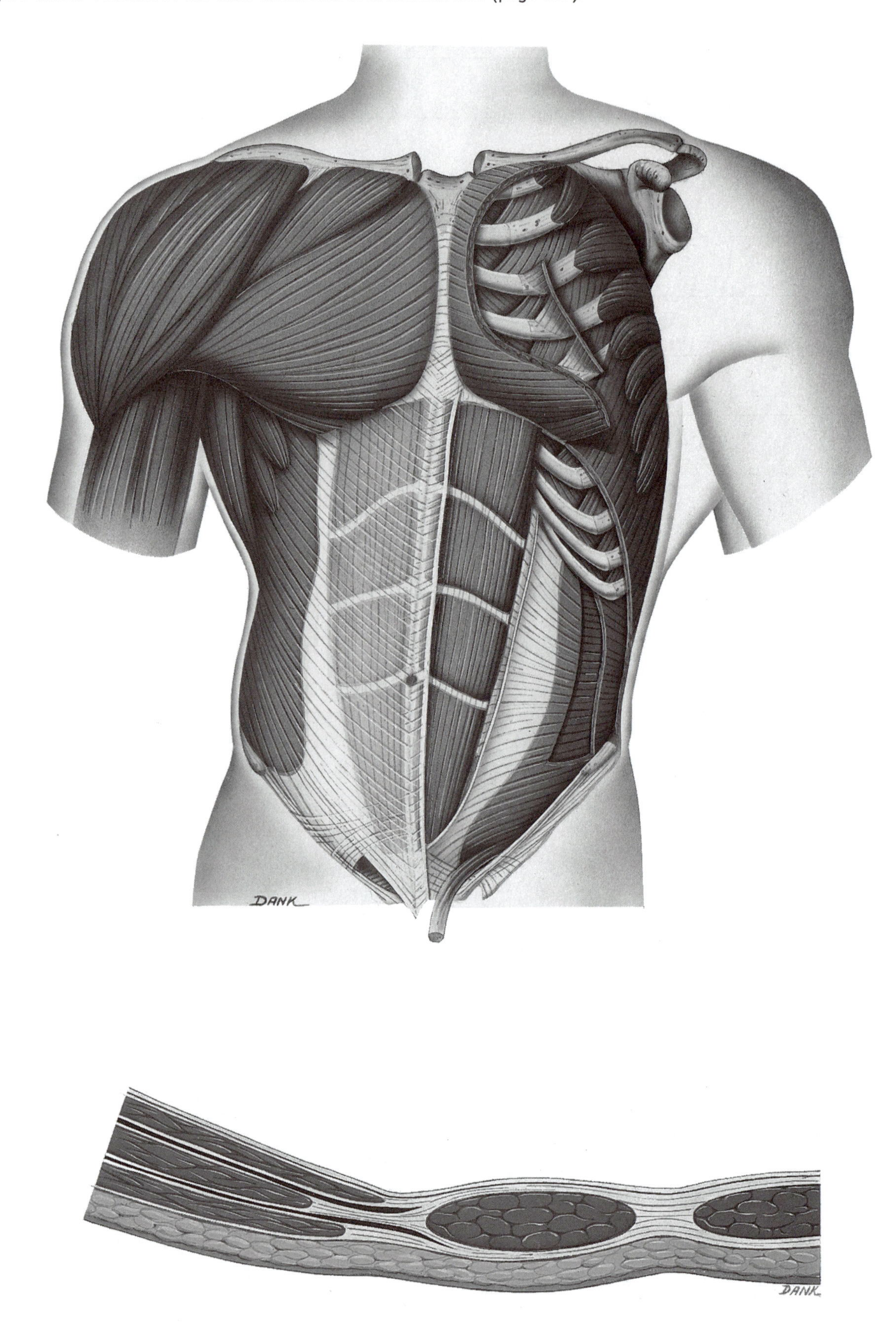
DANK
DANK

NOTES

11

Figure 11.11 Muscles used in breathing, as seen in a male (page 329).

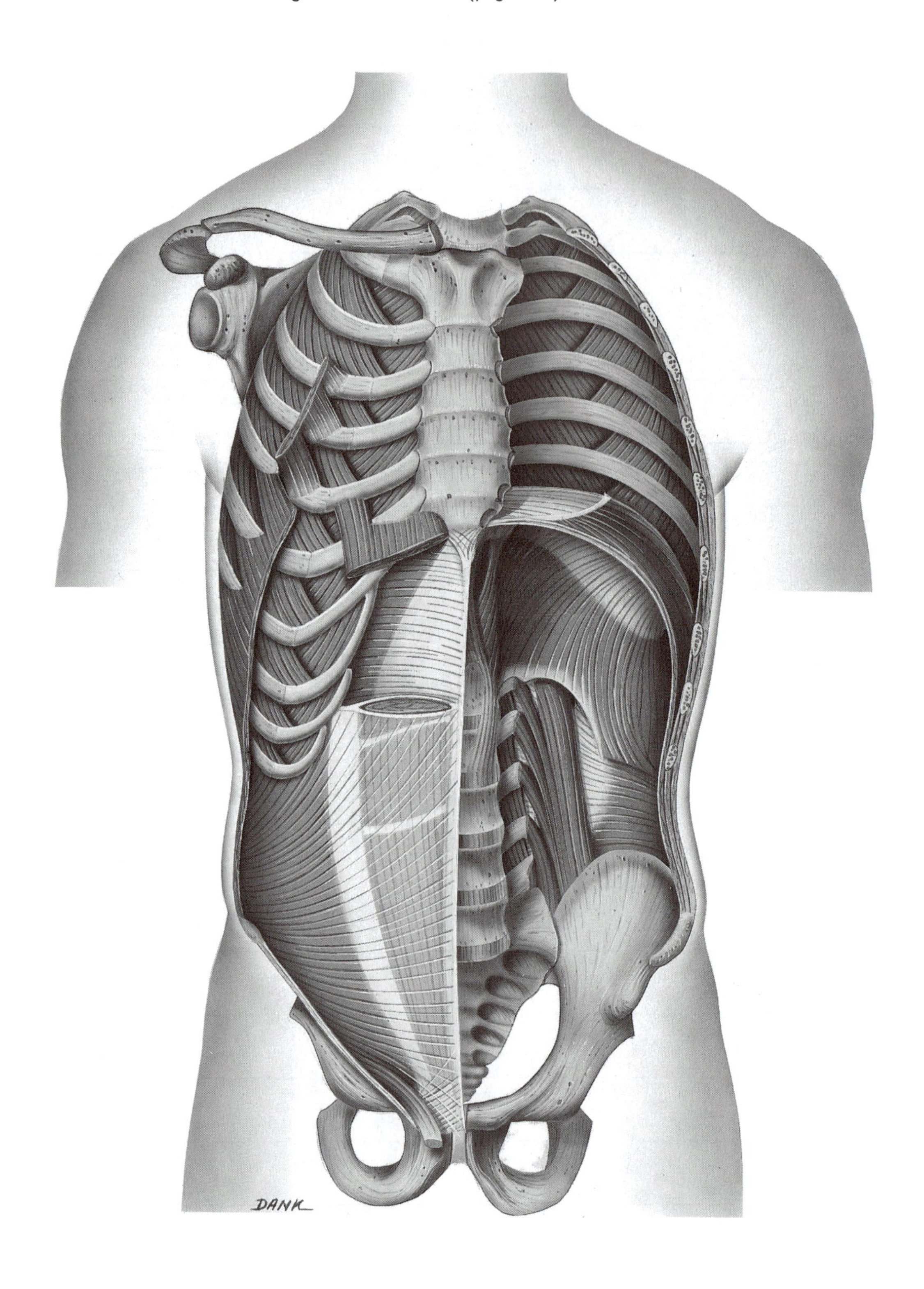
DANK

NOTES

11

Figure 11.12 Muscles of the pelvic floor, as seen in the female perineum (page 331).

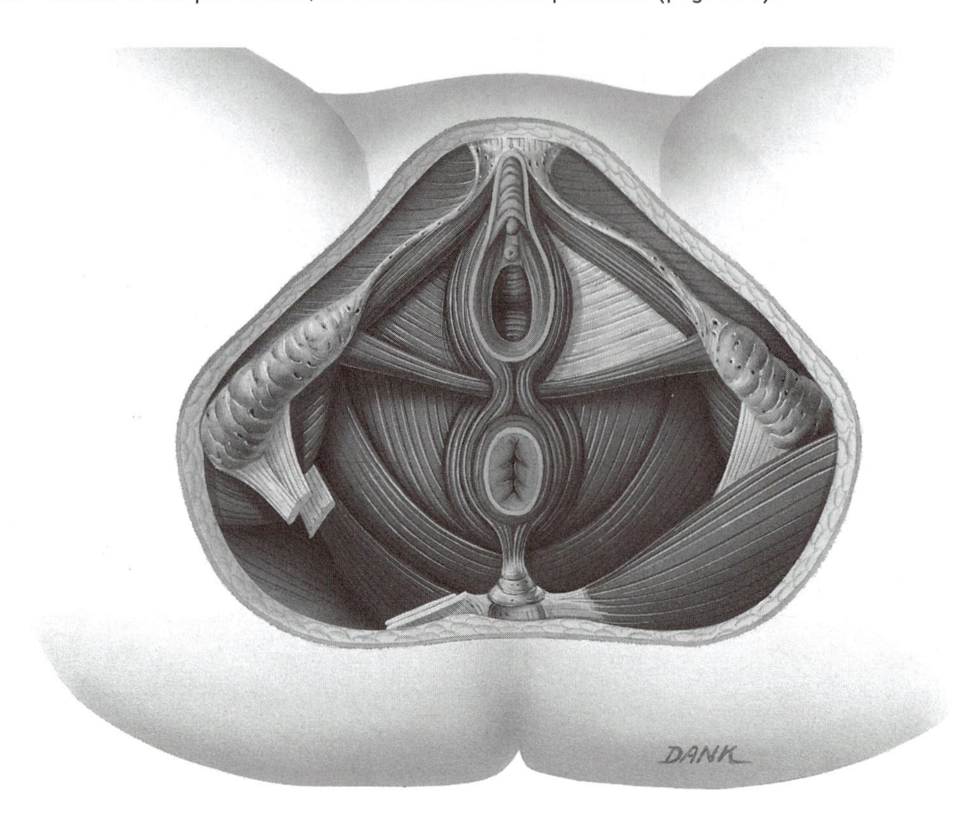
DANK

Figure 11.13 Muscles of the male perineum (page 333).

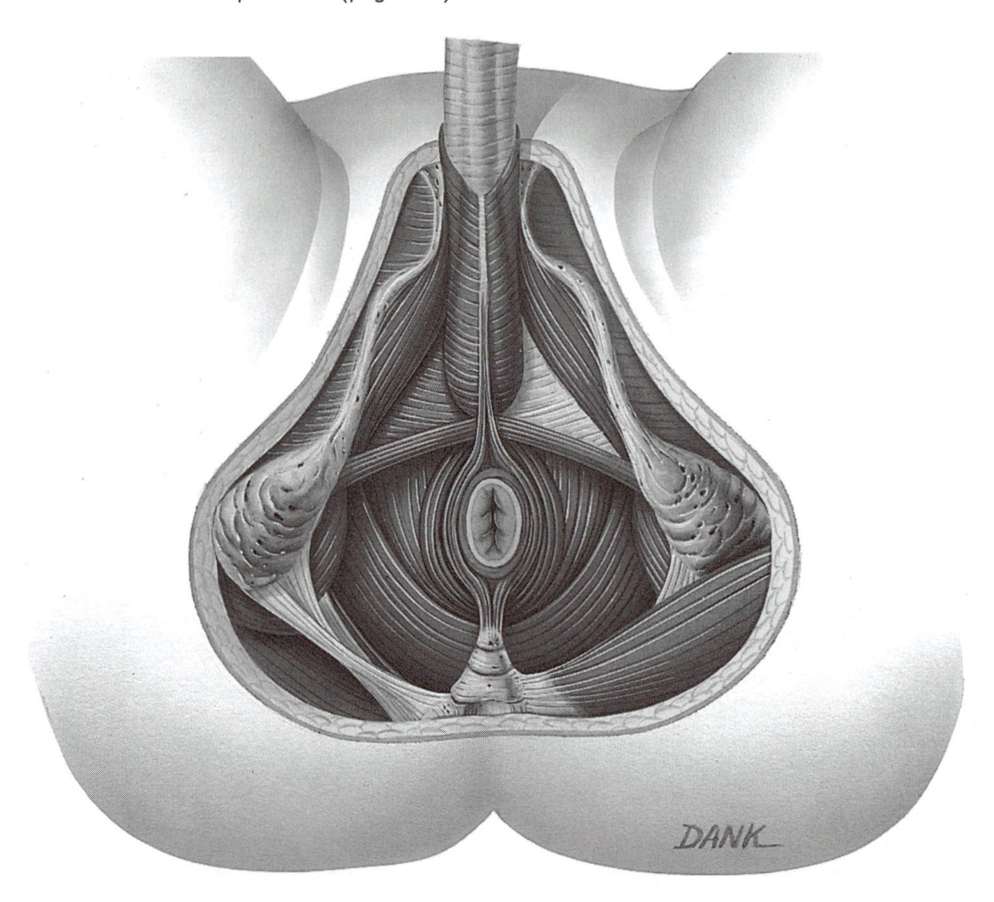
DANK

NOTES

11

Figure 11.14 Muscles that move the pectoral girdle (page 336).

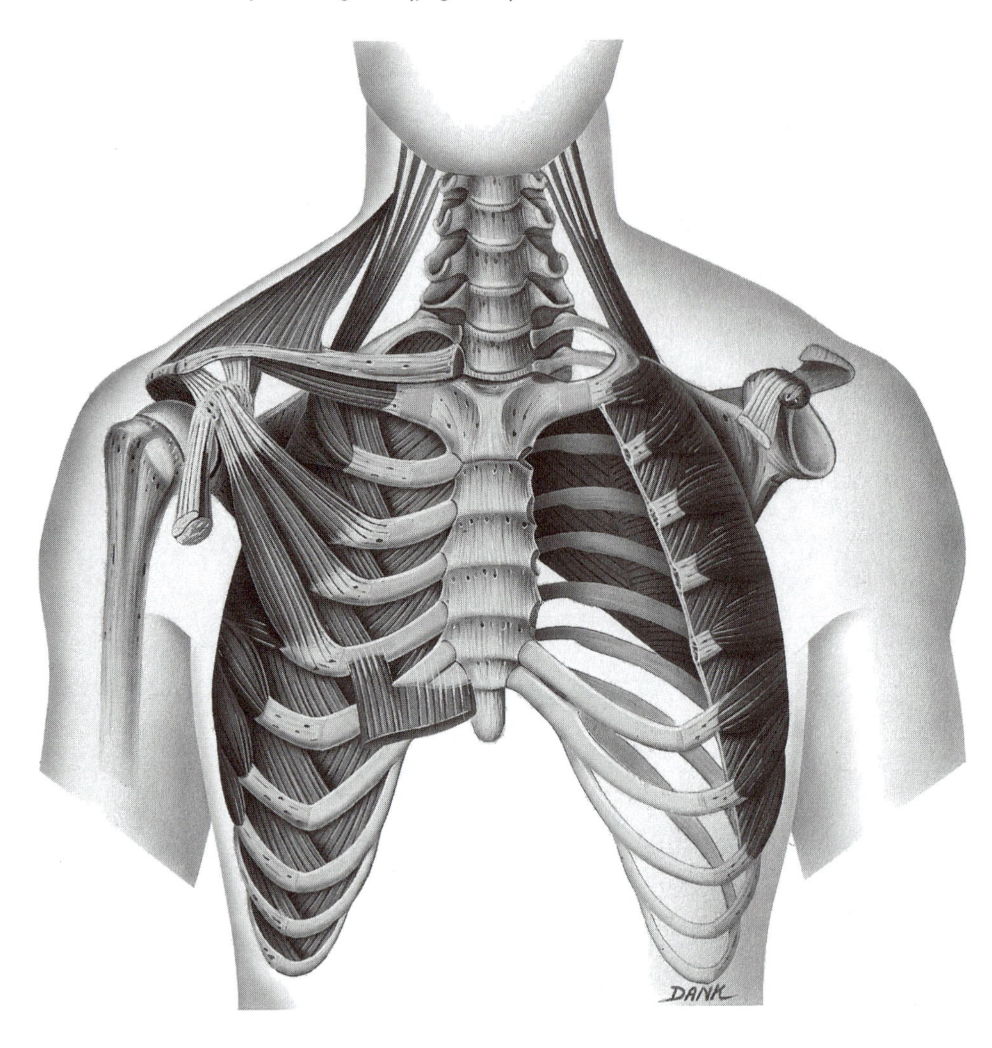
DANK

NOTES

11

Figure 11.15a Muscles that move the humerus (page 339).

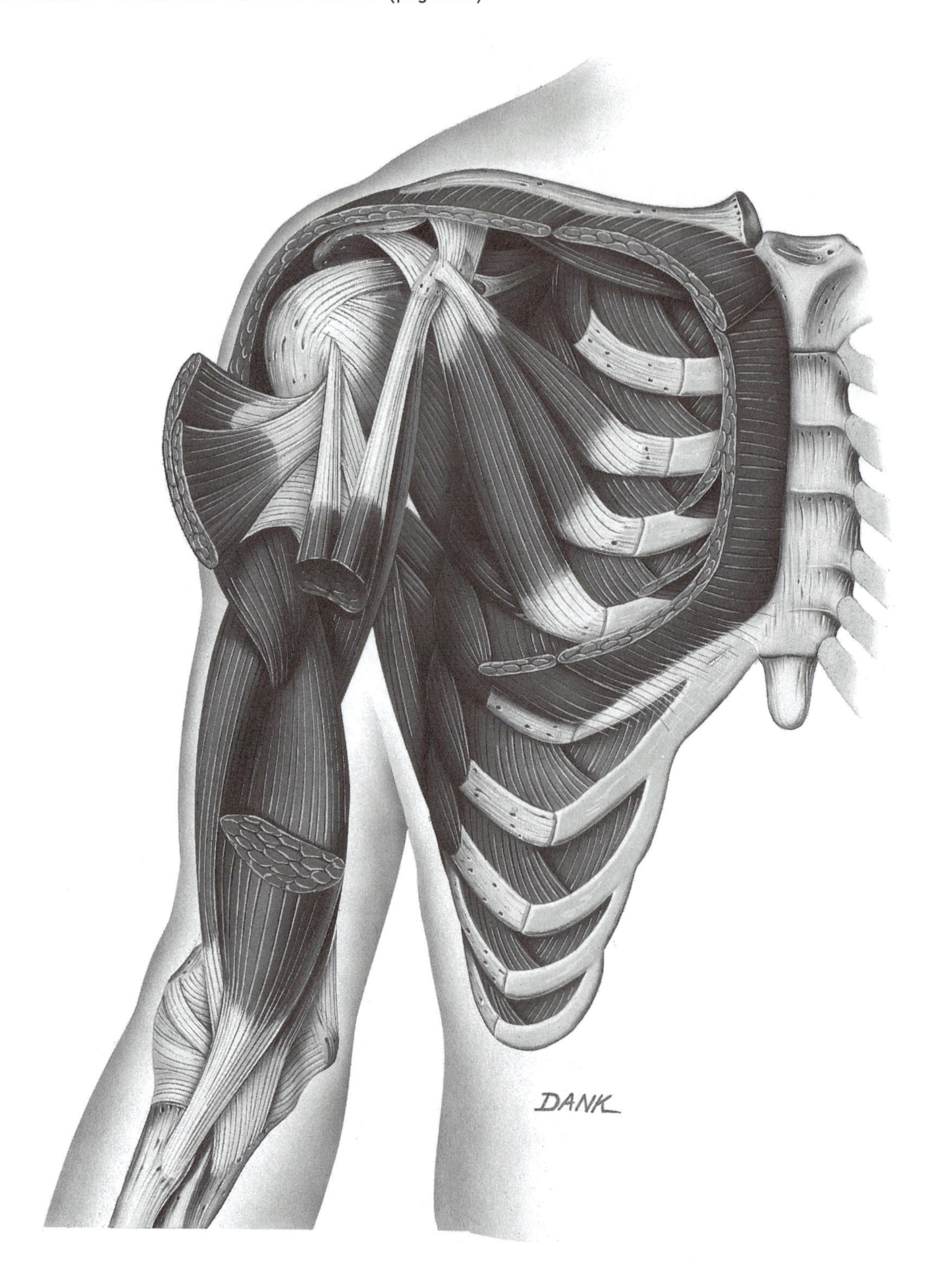
DANK

NOTES

11

Figure 11.15b-c Muscles that move the humerus (page 340).

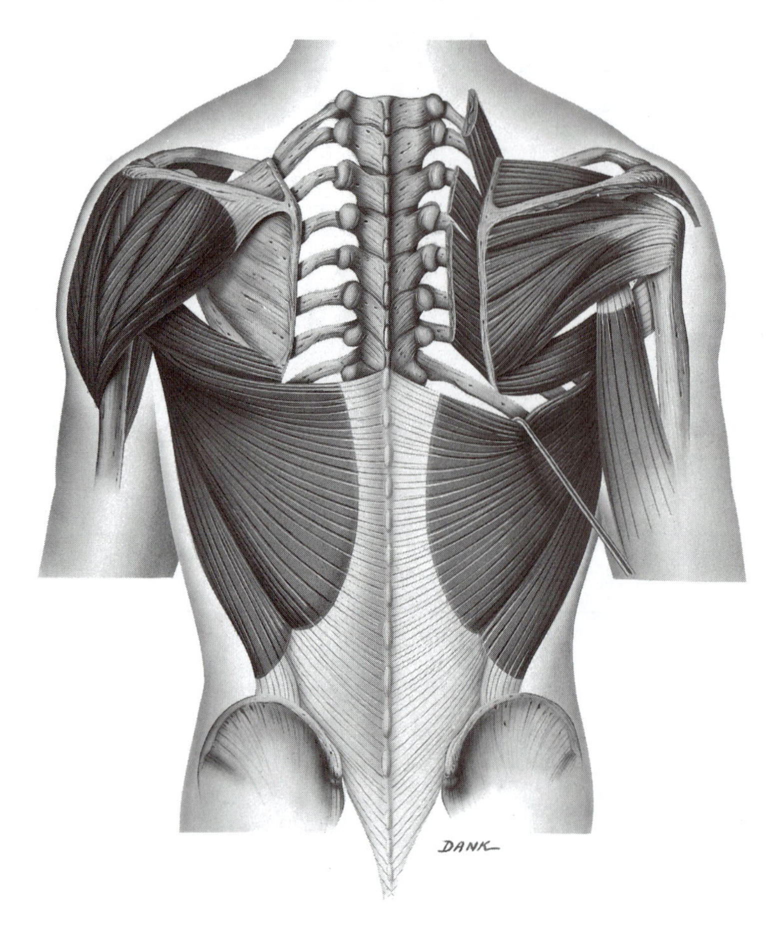
DANK

Figure 11.16a-b Muscles that move the radius and ulna (page 343).

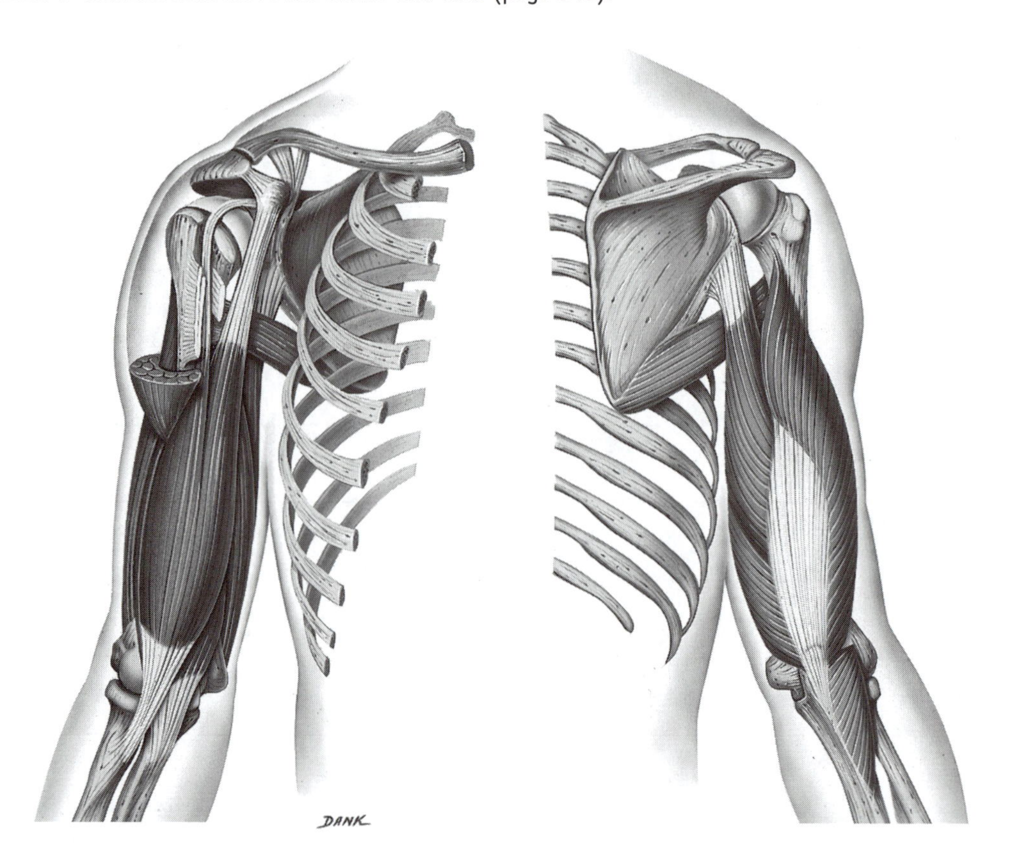
DANK

NOTES

11

Figure 11.16c Muscles that move the radius and ulna (page 344).

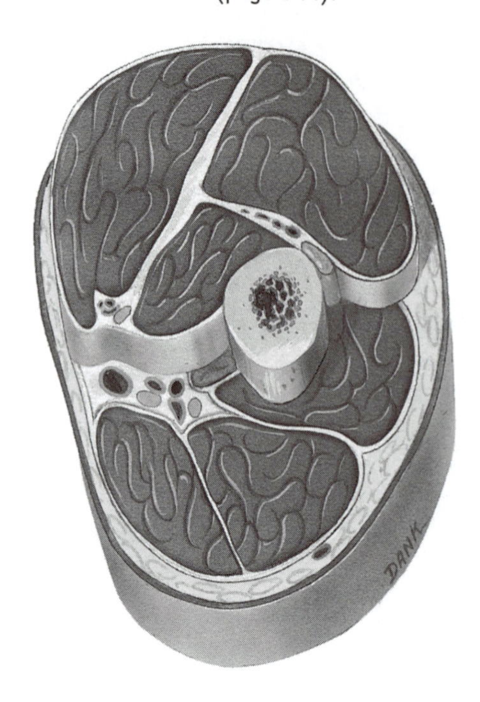

Figure 11.17a-b Muscles that move the wrist, hand, and digits (page 345).

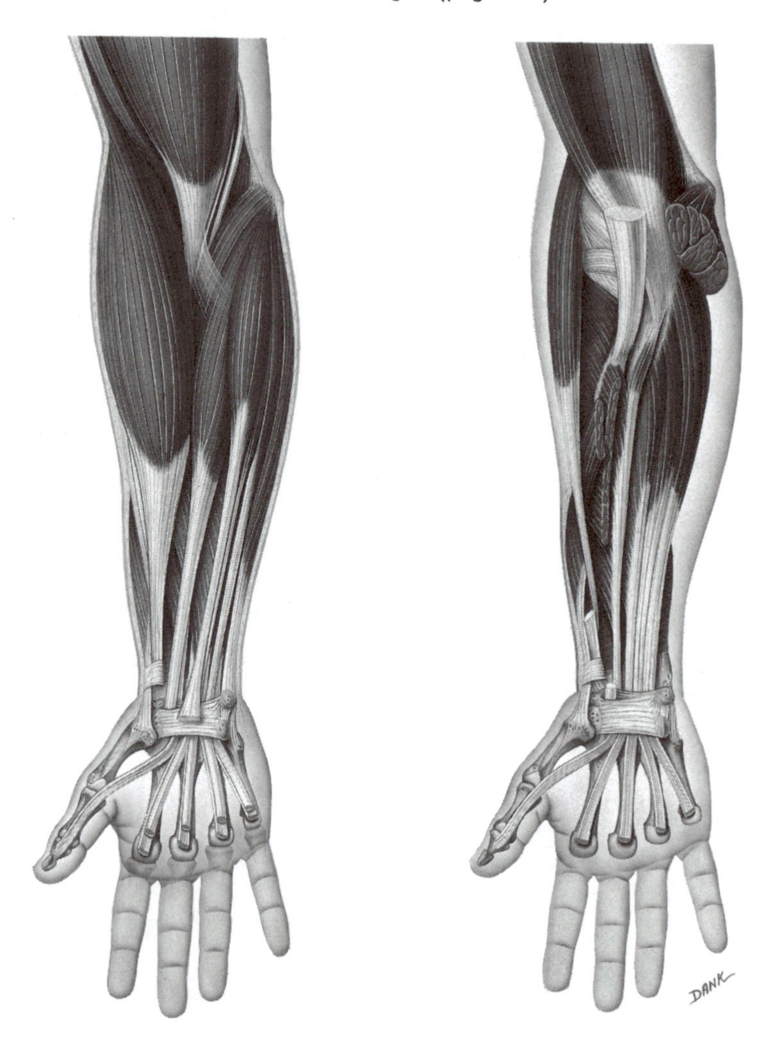

NOTES

11

Figure 11.17c-d Muscles that move the wrist, hand, and digits (page 346).

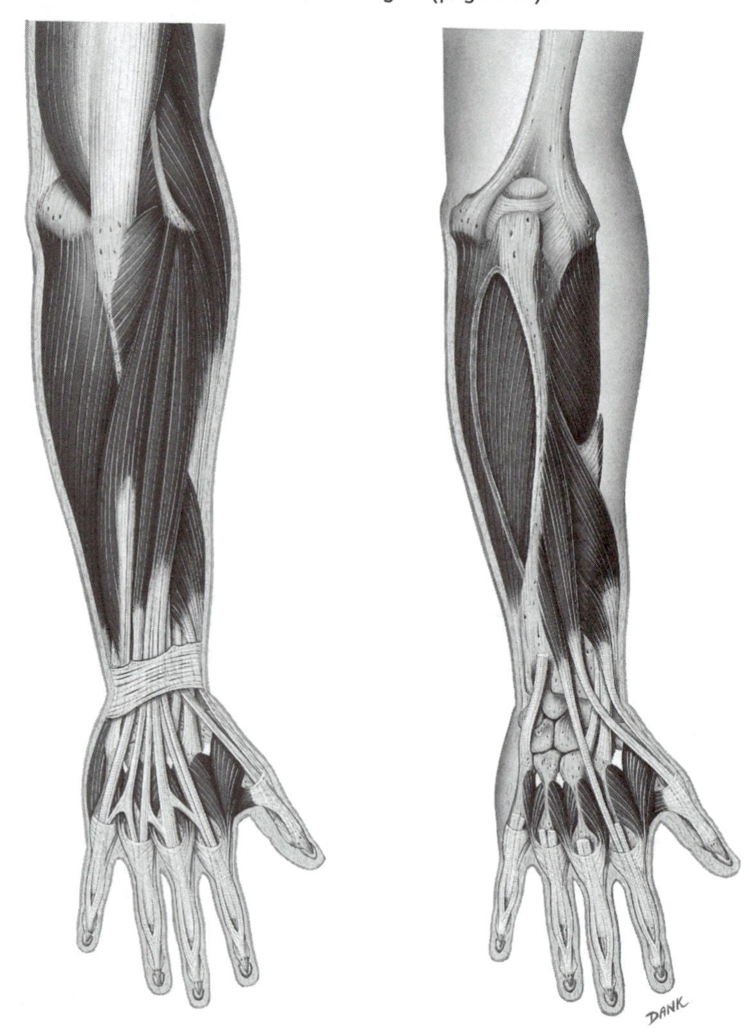
DANK

Figure 11.18 Intrinsic muscles of the hand (page 352).

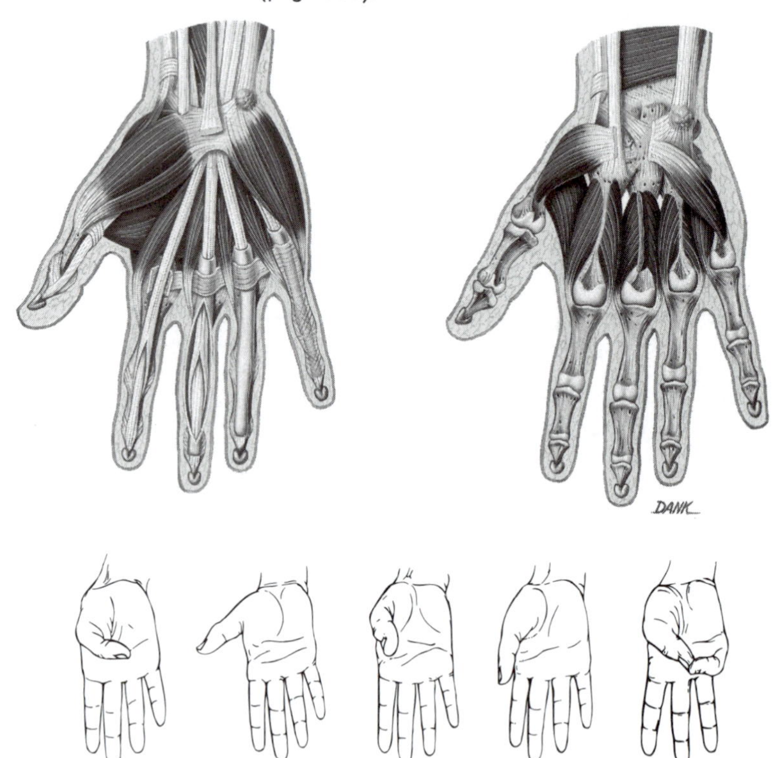
DANK

NOTES

11

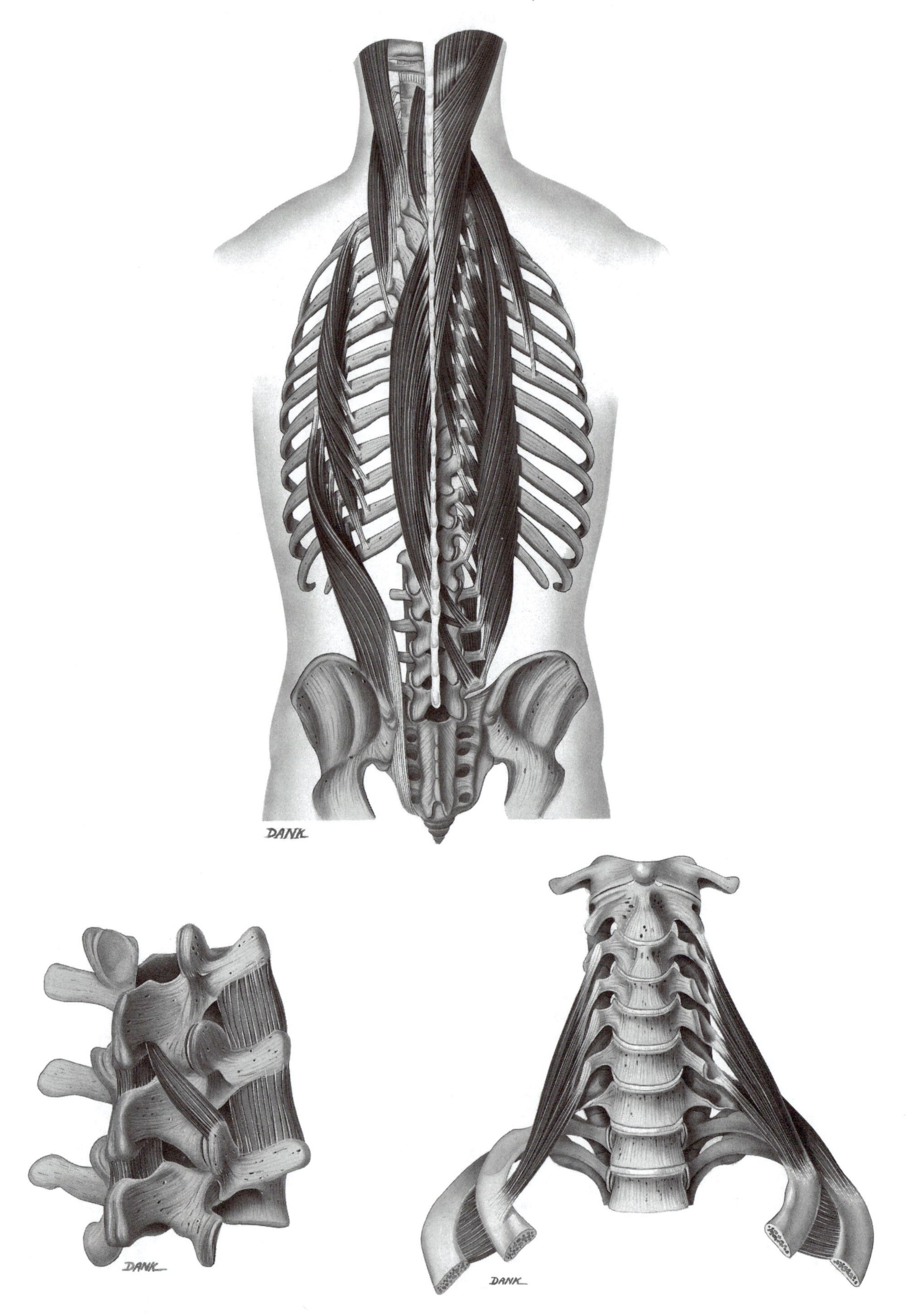
DANK
DANK
DANK

Figure 11.19 Muscles that move the vertebral column (pages 356, 357).

NOTES

11

Figure 11.20 Muscles that move the femur (pages 360, 361, 362).

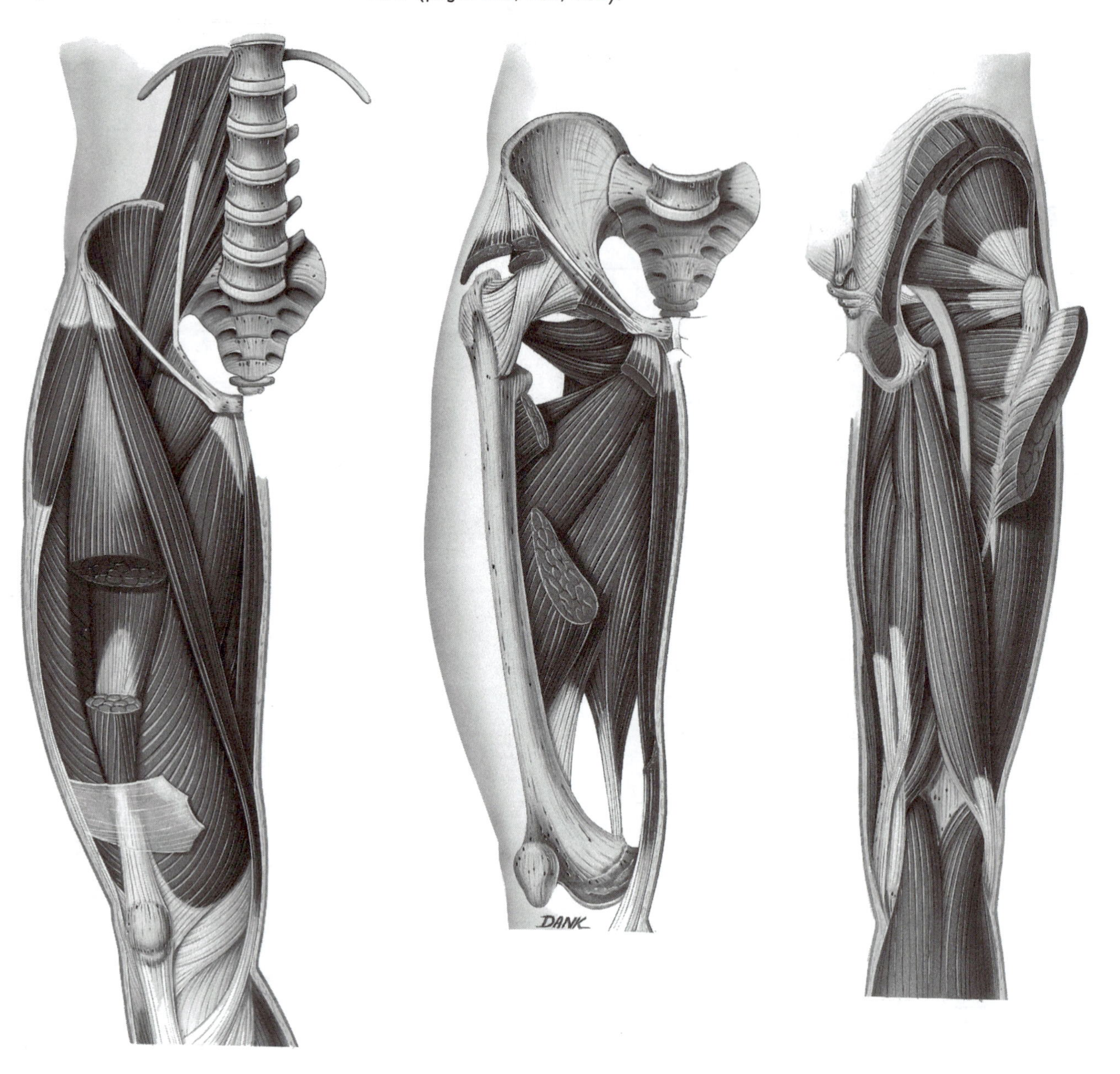
DANK

Figure 11.21 Muscles that act on the femur and tibia and fibula (page 365).

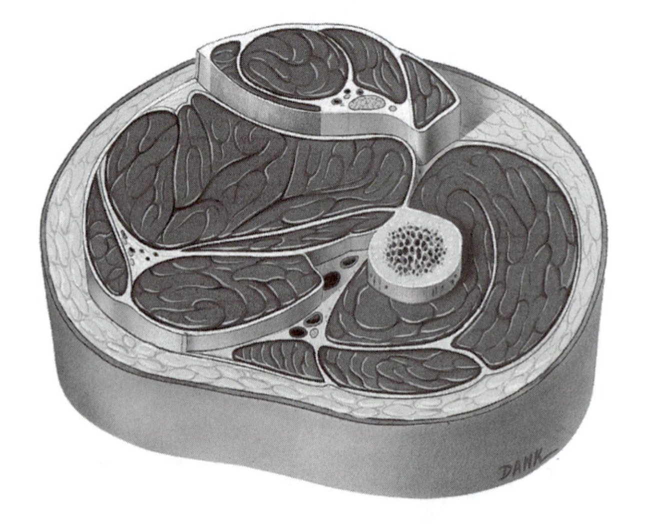
DANK

NOTES

11

Figure 11.22 Muscles that move the foot and toes (page 369, 370).

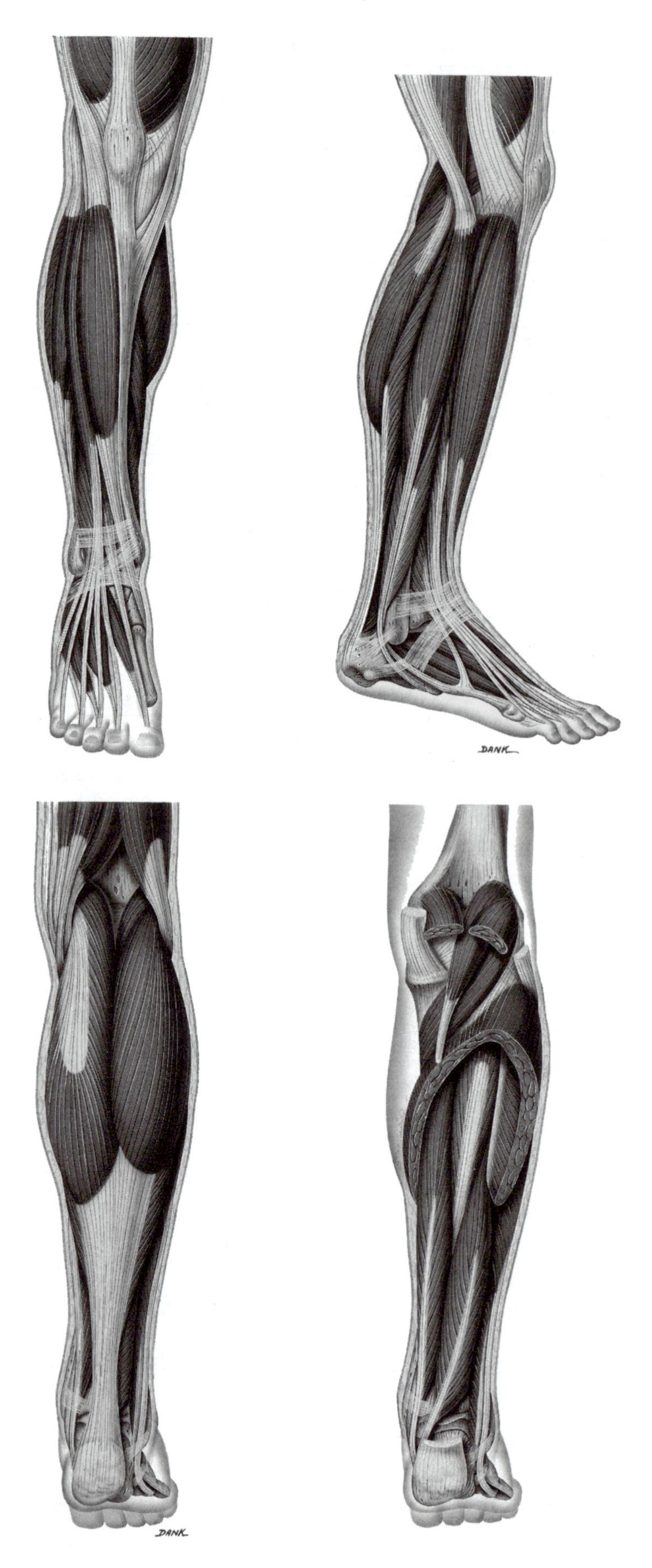
DANK
DANK

NOTES

11

Figure 11.23 Intrinsic muscles of the foot (page 373).

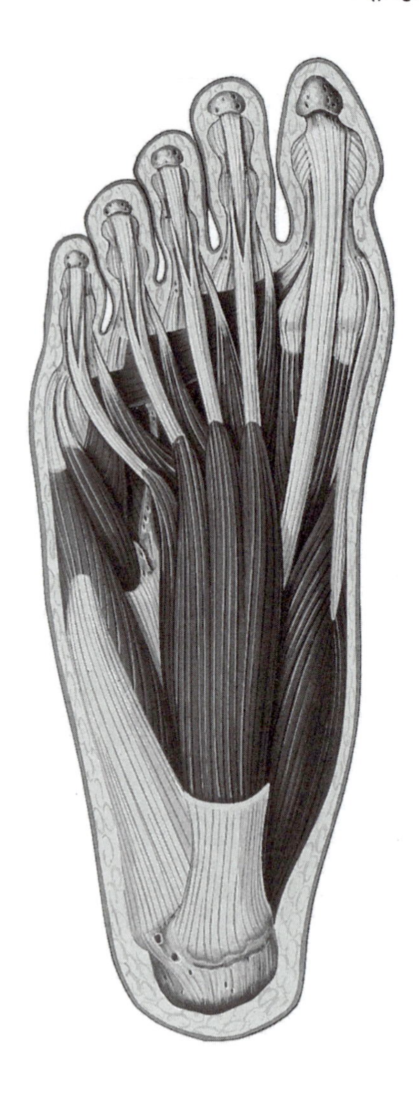

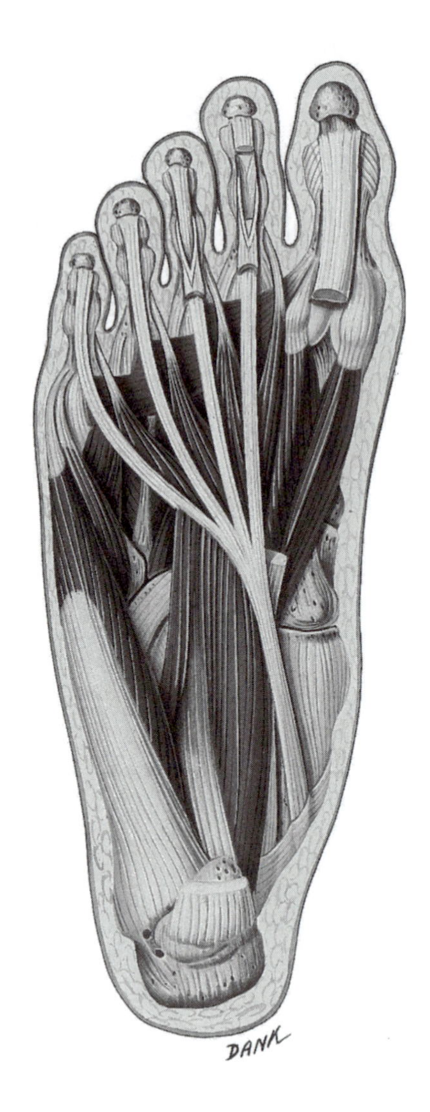
DANK

NOTES

Figure 12.1 Major structures of the nervous system (page 379).

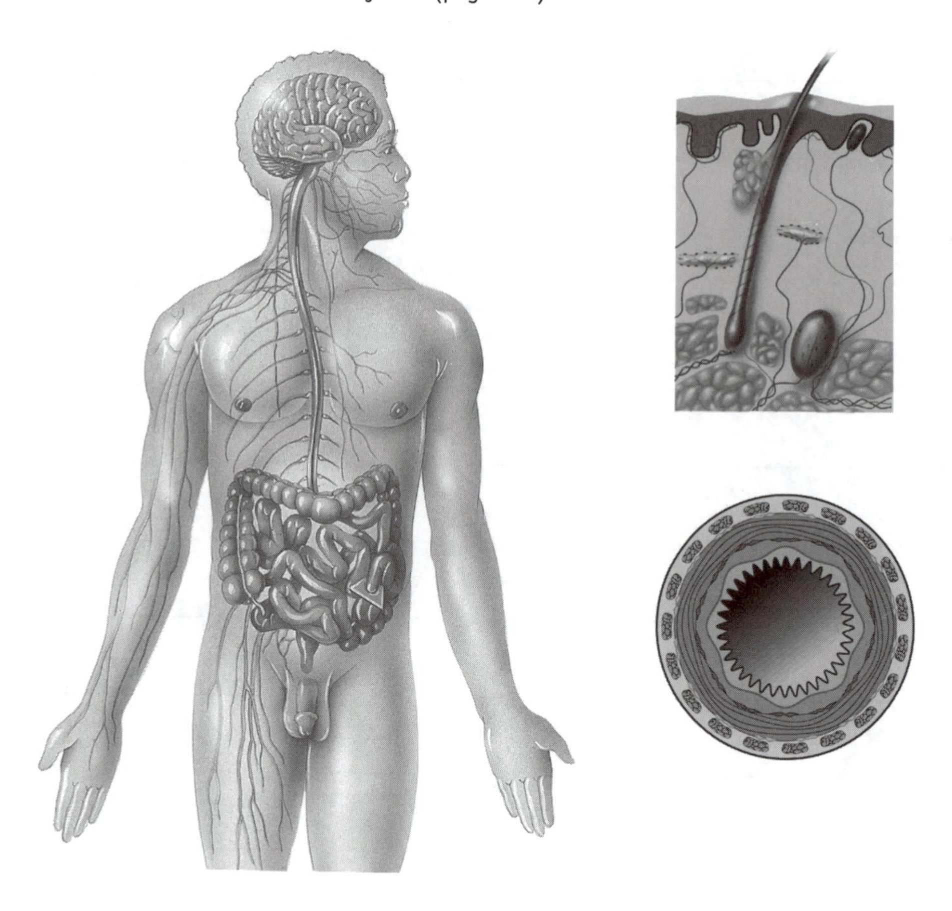

Figure 12.2 Organization of the nervous system (page 380).

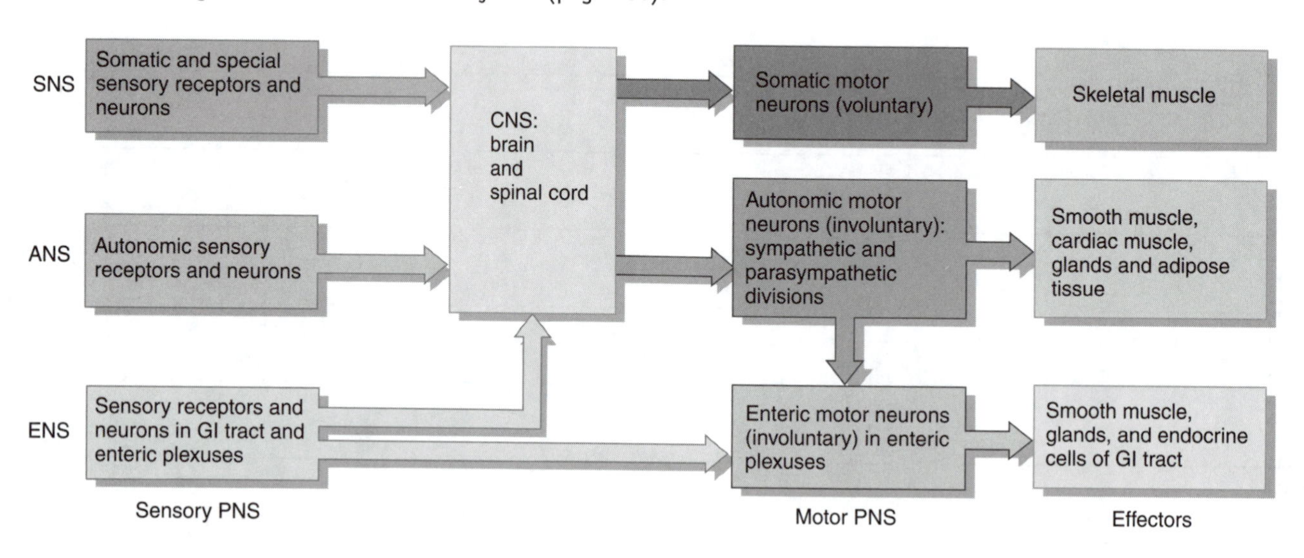

SNS
Somatic and special sensory receptors and neurons
ANS
Autonomic sensory receptors and neurons
ENS
Sensory receptors and neurons in GI tract and enteric plexuses
Sensory PNS
CNS: brain and spinal cord
Somatic motor neurons (voluntary)
Autonomic motor neurons (involuntary): sympathetic and parasympathetic divisions
Enteric motor neurons (involuntary) in enteric plexuses
Motor PNS
Skeletal muscle
Smooth muscle, cardiac muscle, glands and adipose tissue
Smooth muscle, glands, and endocrine cells of GI tract
Effectors

NOTES

12

Figure 12.3 Structure of a typical neuron (page 382).

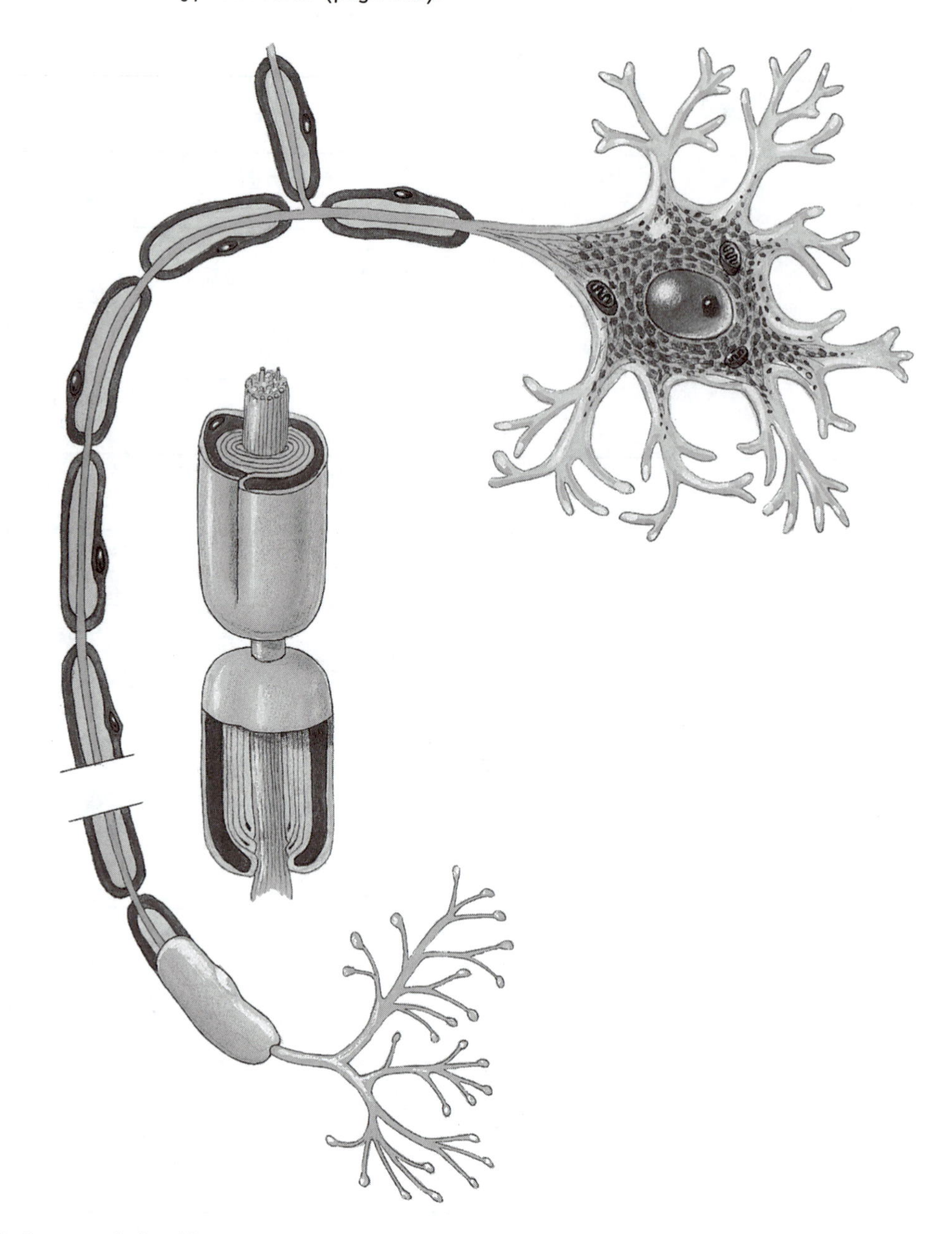

Figure 12.4 Structural classification of neurons (page 383).

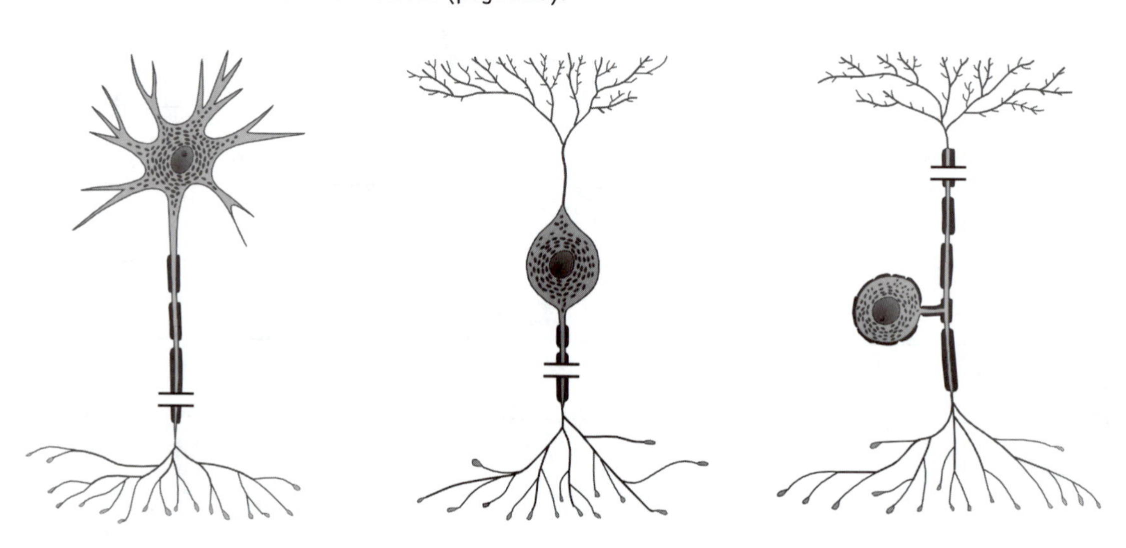

NOTES

12

Figure 12.5 Two examples of interneurons (page 384).

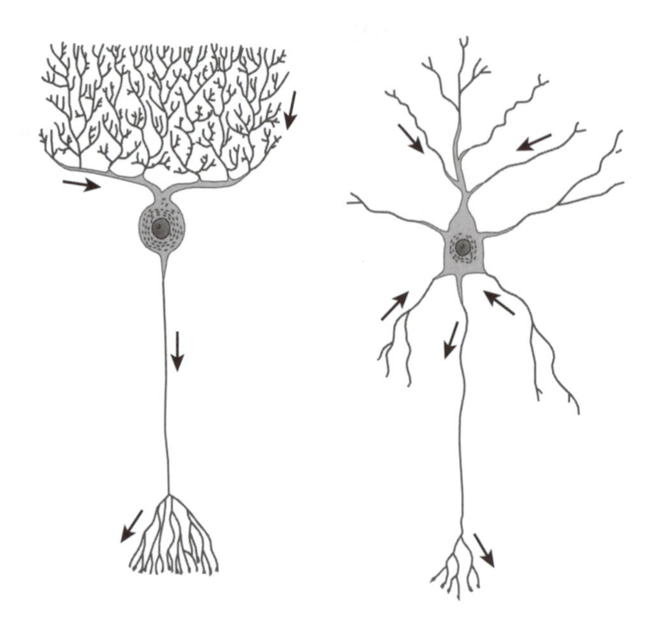

Figure 12.6 Myelinated and unmyelinated axons (page 385).

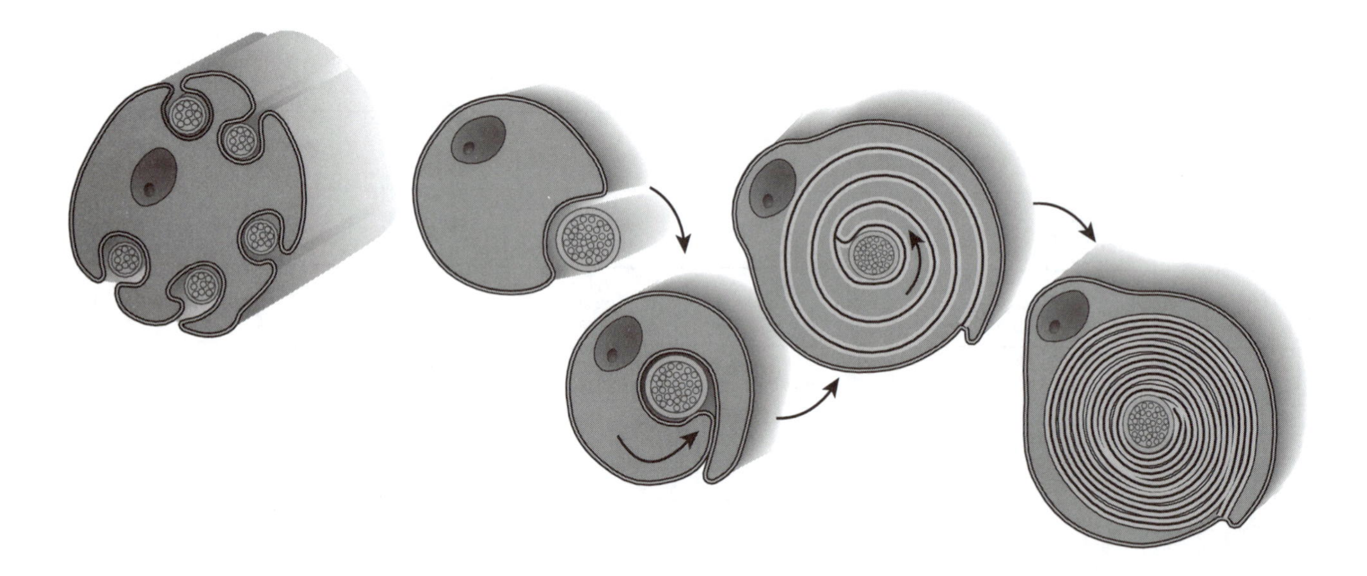

NOTES

12

Figure 12.7 Distribution of gray and white matter in the spinal cord and brain (page 388).

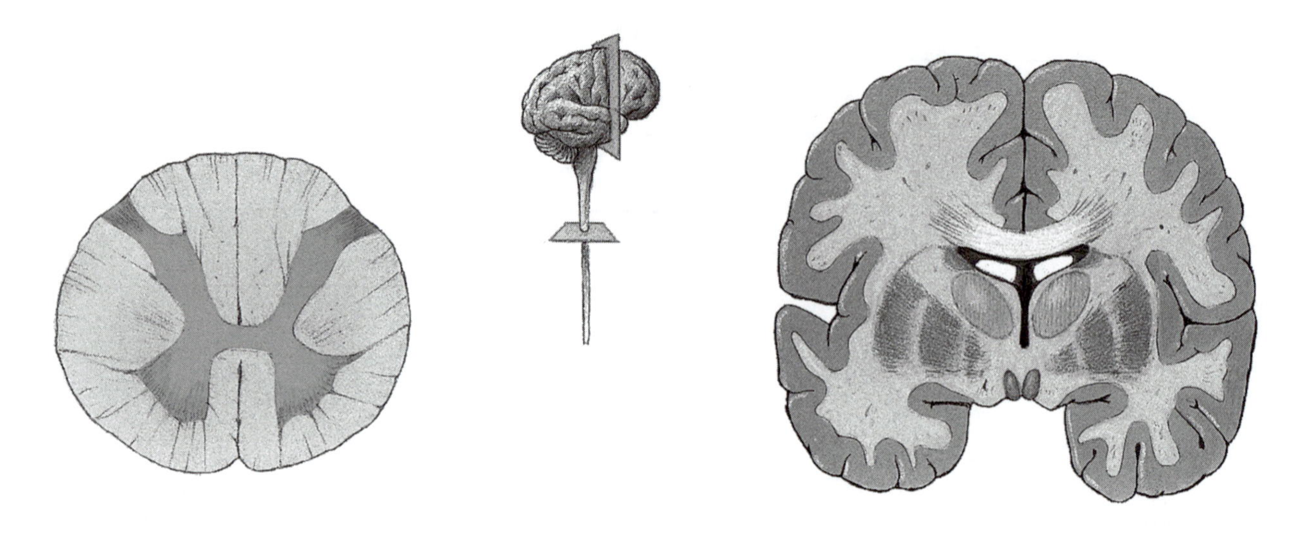

Figure 12.8 Voltage-gated and ligand-gated ion channels in the plasma membrane (page 388).

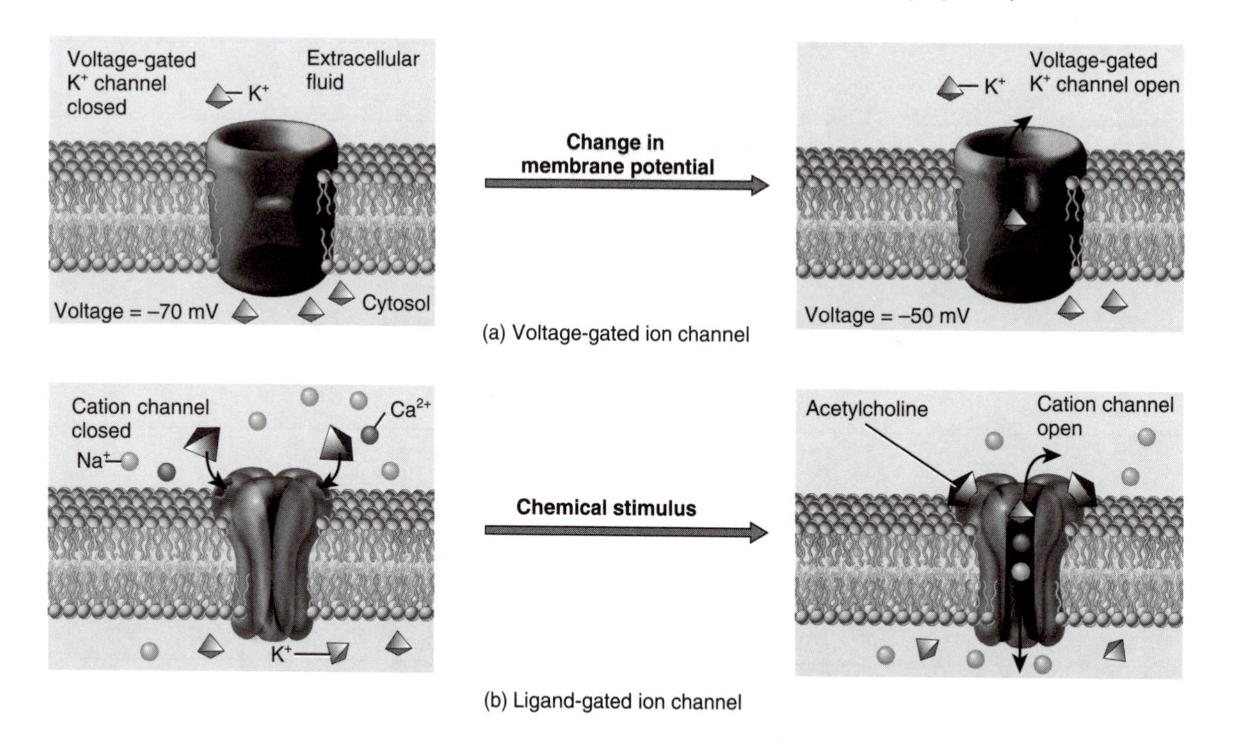
Voltage-gated
K⁺ channel
closed
Extracellular
fluid
K⁺
Voltage = −70 mV
Cytosol
Change in
membrane potential
Voltage-gated
K⁺ channel open
K⁺
Voltage = −50 mV
(a) Voltage-gated ion channel
Cation channel
closed
Na⁺
Ca²⁺
K⁺
Chemical stimulus
Acetylcholine
Cation channel
open
(b) Ligand-gated ion channel

NOTES

12

Figure 12.9 The distributions of charges and ions that produce the resting membrane potential (page 389).

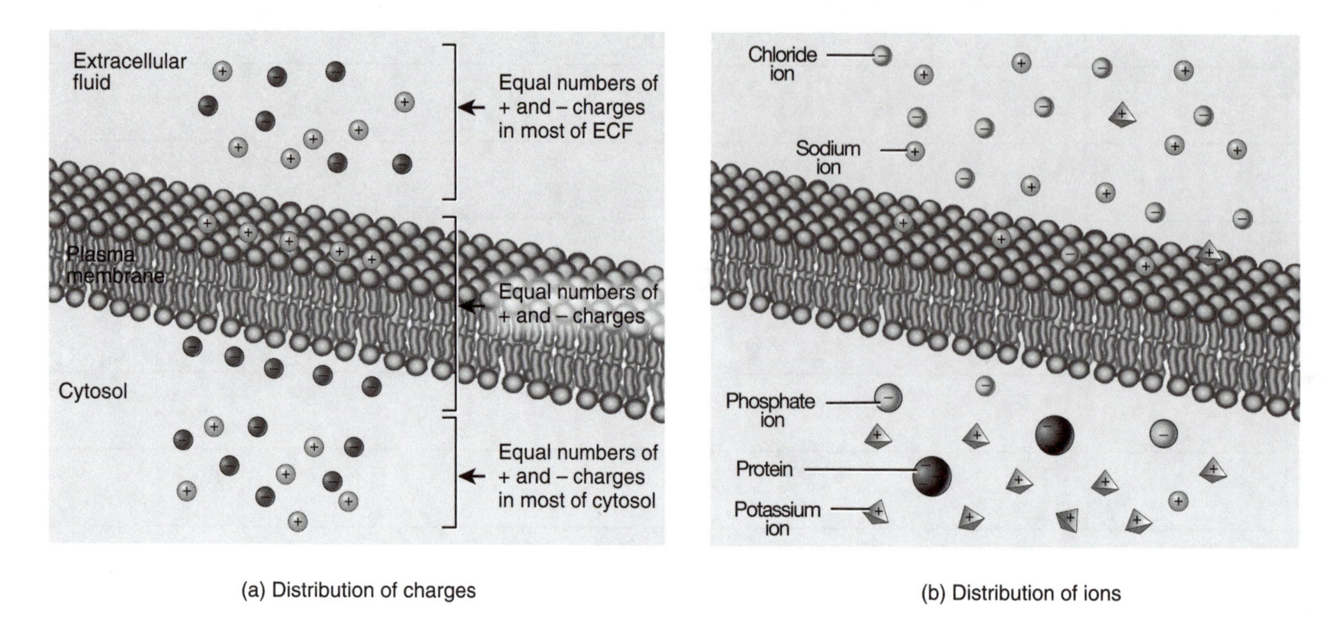
Extracellular fluid
Equal numbers of + and – charges in most of ECF
Plasma membrane
Equal numbers of + and – charges
Cytosol
Equal numbers of + and – charges in most of cytosol
(a) Distribution of charges
Chloride ion
Sodium ion
Phosphate ion
Protein
Potassium ion
(b) Distribution of ions

Figure 12.10 Graded potentials (page 391).

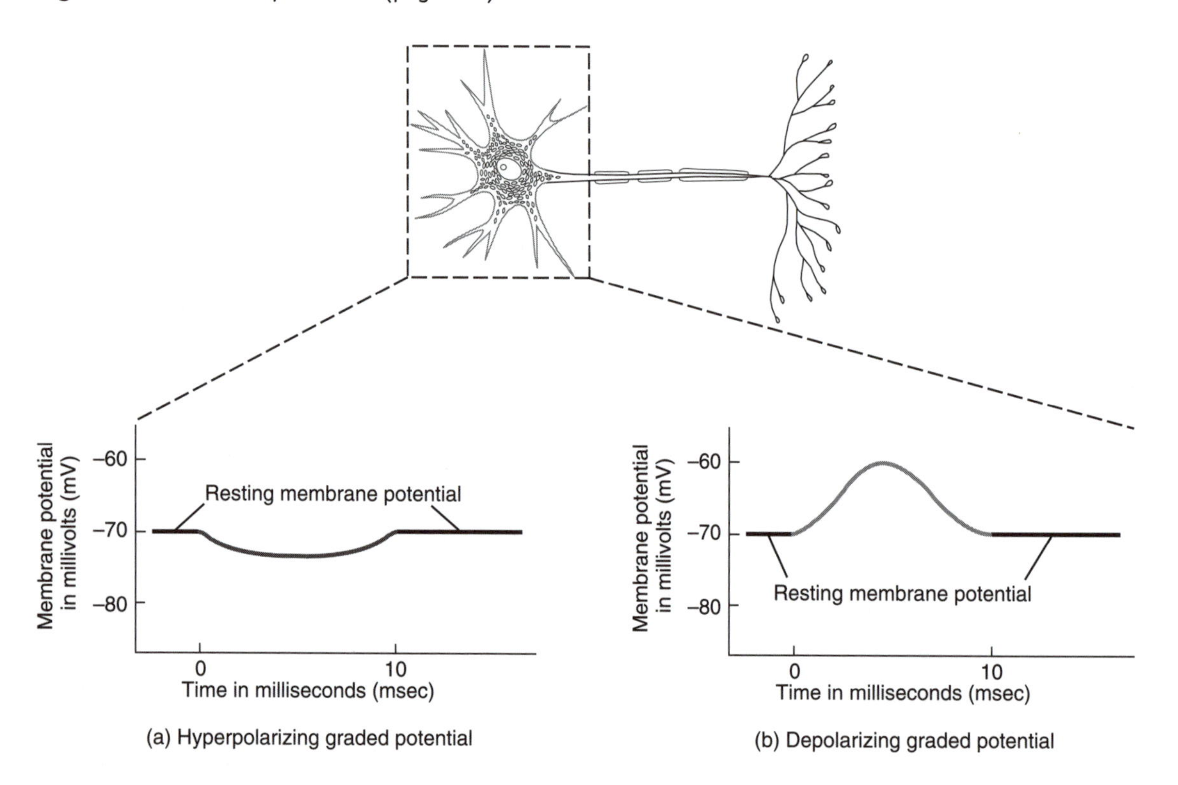
Membrane potential in millivolts (mV)
-60
-70
-80
Resting membrane potential
0
10
Time in milliseconds (msec)
(a) Hyperpolarizing graded potential
Membrane potential in millivolts (mV)
-60
-70
-80
Resting membrane potential
0
10
Time in milliseconds (msec)
(b) Depolarizing graded potential

NOTES

12

Figure 12.11 Action potential or impulse (page 392).

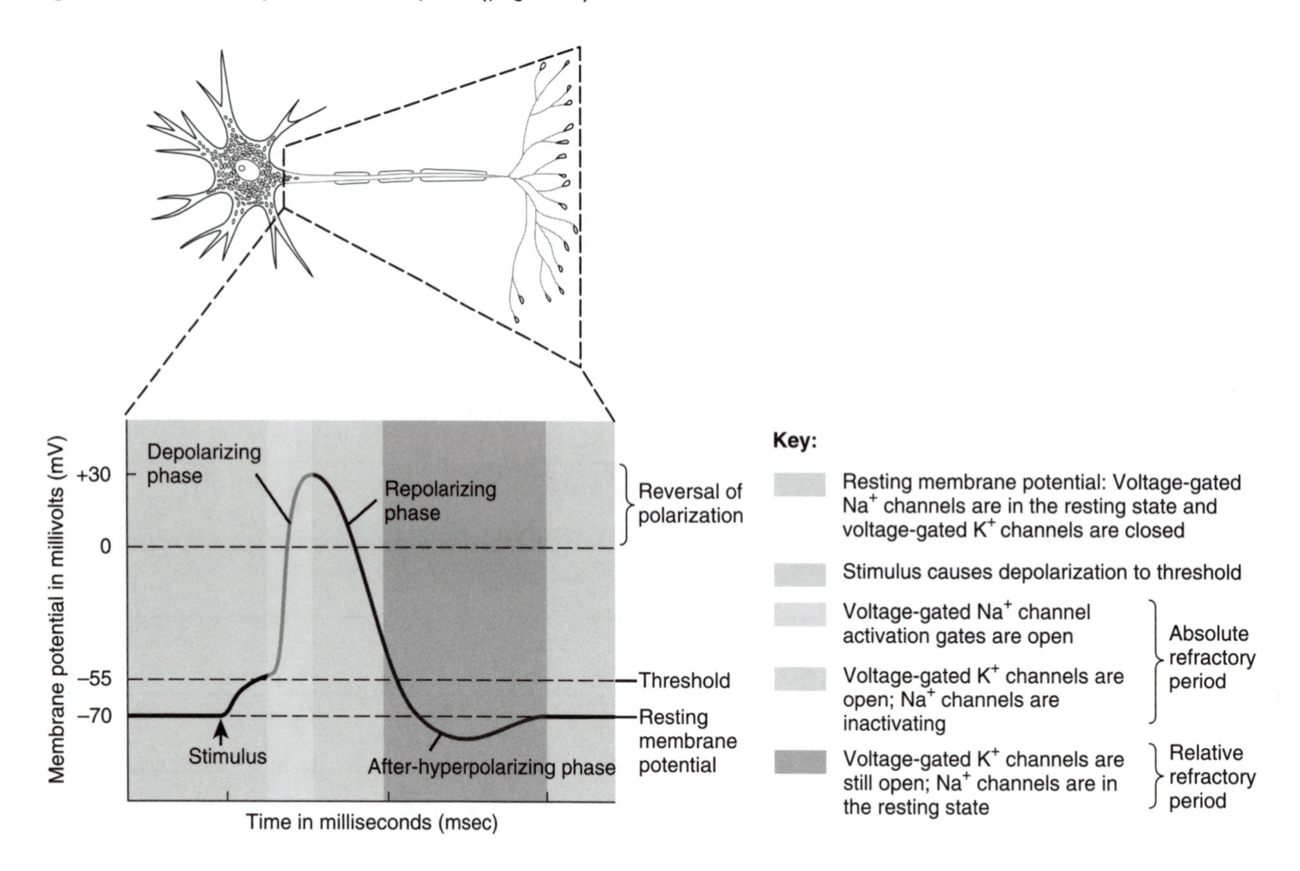

Membrane potential in millivolts (mV)
+30
0
–55
–70
Depolarizing phase
Repolarizing phase
Reversal of polarization
Threshold
Resting membrane potential
Stimulus
After-hyperpolarizing phase
Time in milliseconds (msec)
Key:
Resting membrane potential: Voltage-gated Na+ channels are in the resting state and voltage-gated K+ channels are closed
Stimulus causes depolarization to threshold
Voltage-gated Na+ channel activation gates are open
Voltage-gated K+ channels are open; Na+ channels are inactivating
Absolute refractory period
Voltage-gated K+ channels are still open; Na+ channels are in the resting state
Relative refractory period

NOTES

12

Figure 12.12 Changes in ion flow (page 393).

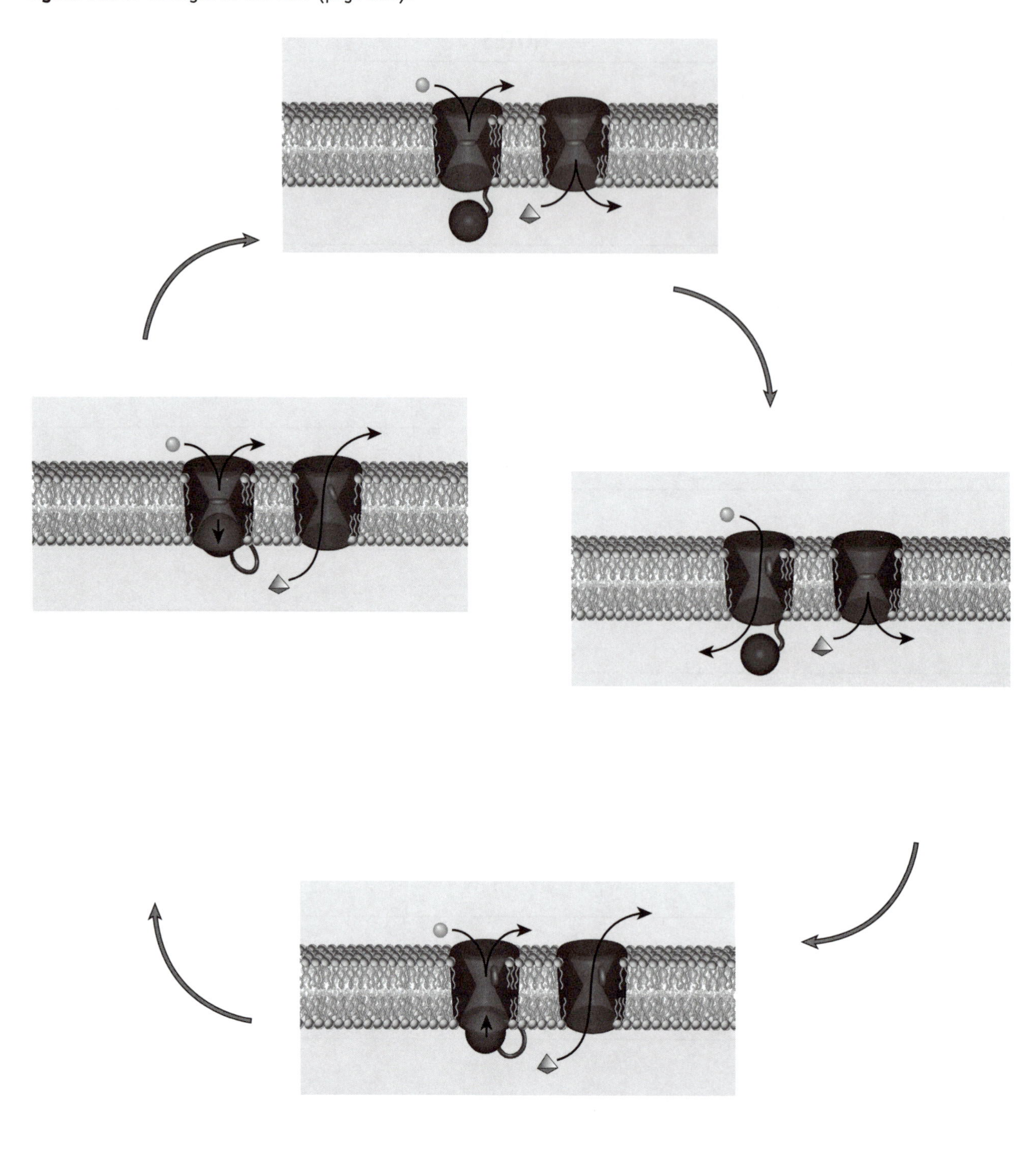

Adapted from Becker et al., *The World of the Cell,* 3e, F22.18, p732 (Menlo Park, CA: Benjamin/Cummings, 1996).

NOTES

12

Figure 12.13 Propagation of a nerve impulse (page 395).

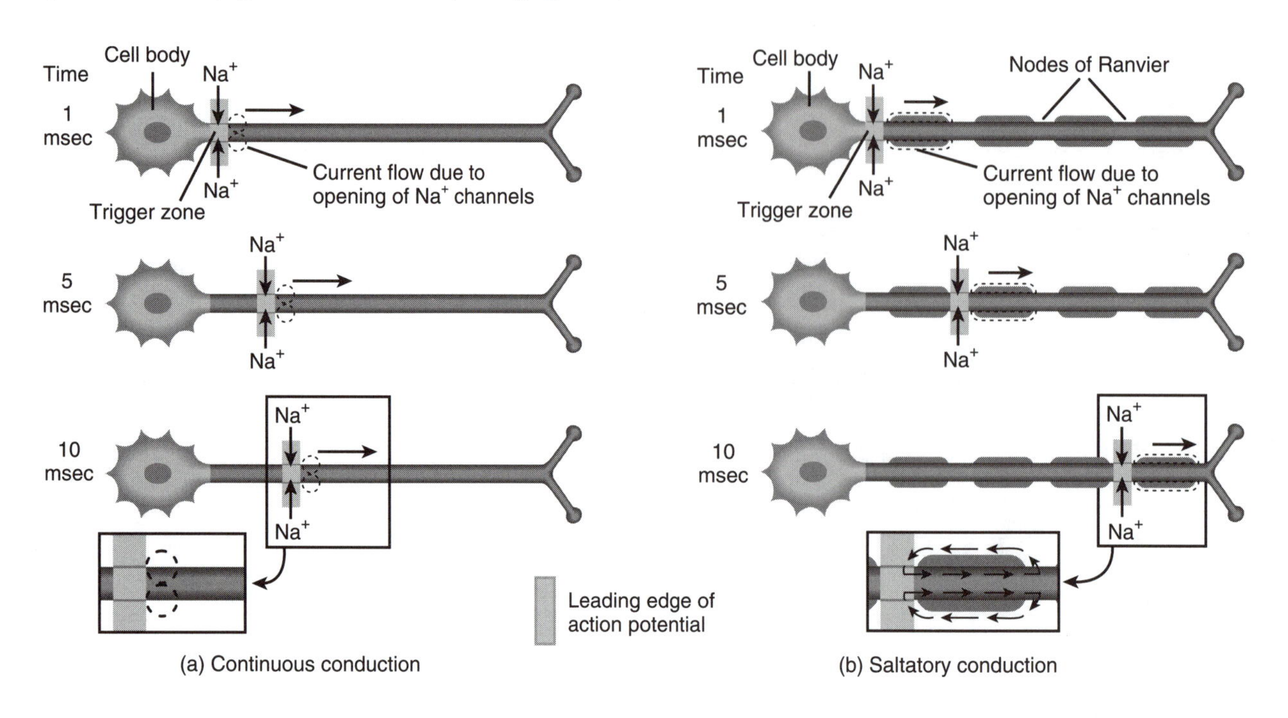

Time
Cell body
Na^+
1 msec
Current flow due to opening of Na^+ channels
Trigger zone
5 msec
10 msec
Nodes of Ranvier
Leading edge of action potential
(a) Continuous conduction
(b) Saltatory conduction

Figure 12.14 Signal transmission at a chemical synapse (page 398).

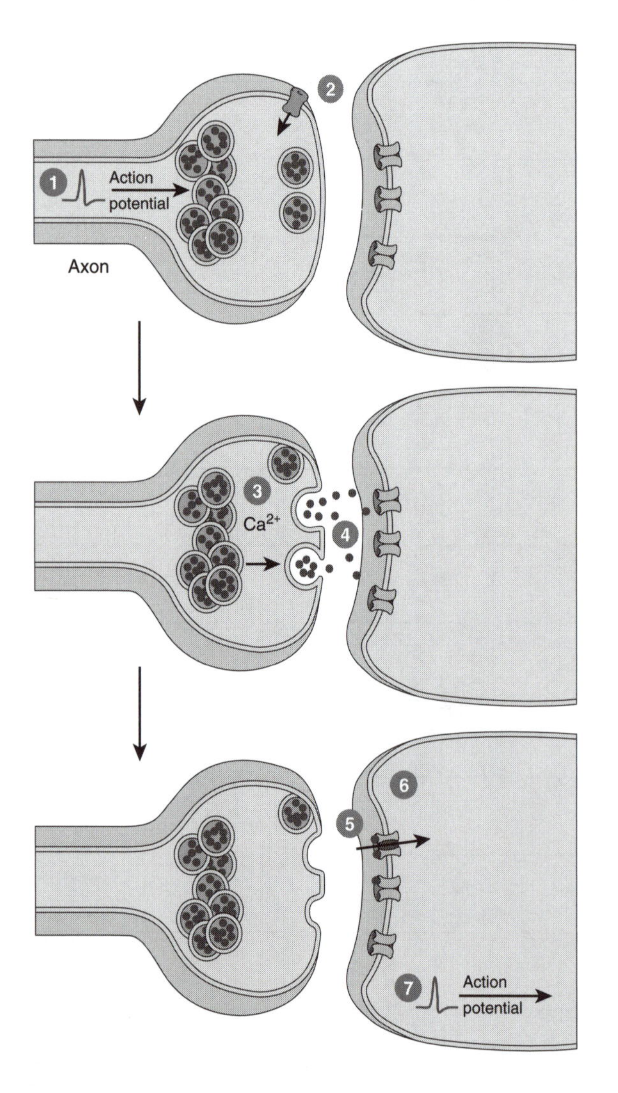

1
Action potential
2
Axon
3
Ca^{2+}
4
5
6
7
Action potential

Adapted from Becker et al., *The World of the Cell,* 3e, F22.28, p741 (Menlo Park, CA: Benjamin/Cummings, 1996). © 1996 The Benjamin/Cummings Publishing Company.

NOTES

Figure 12.15 Spatial and temporal summation (page 400).

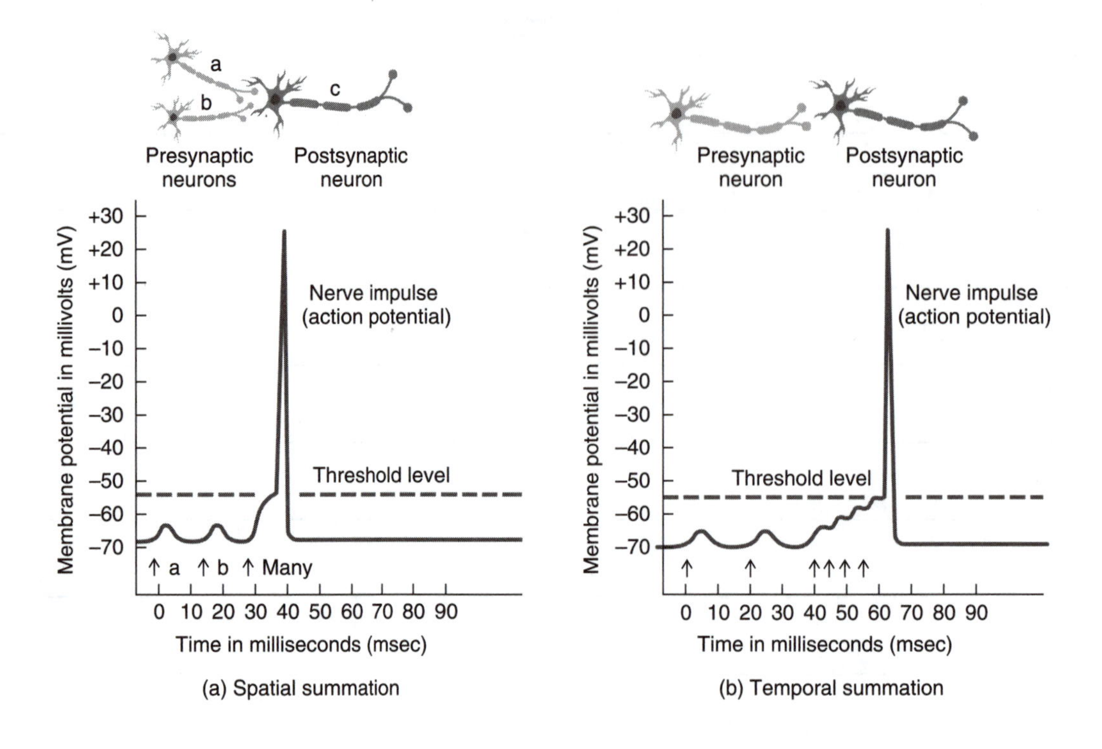
a
b
c
Presynaptic neurons
Postsynaptic neuron
Membrane potential in millivolts (mV)
+30
+20
+10
0
−10
−20
−30
−40
−50
−60
−70
Nerve impulse (action potential)
Threshold level
↑ a ↑ b ↑ Many
0 10 20 30 40 50 60 70 80 90
Time in milliseconds (msec)
(a) Spatial summation
Presynaptic neuron
Postsynaptic neuron
Membrane potential in millivolts (mV)
Nerve impulse (action potential)
Threshold level
0 10 20 30 40 50 60 70 80 90
Time in milliseconds (msec)
(b) Temporal summation

Figure 12.16 Examples of neuronal circuits (page 404).

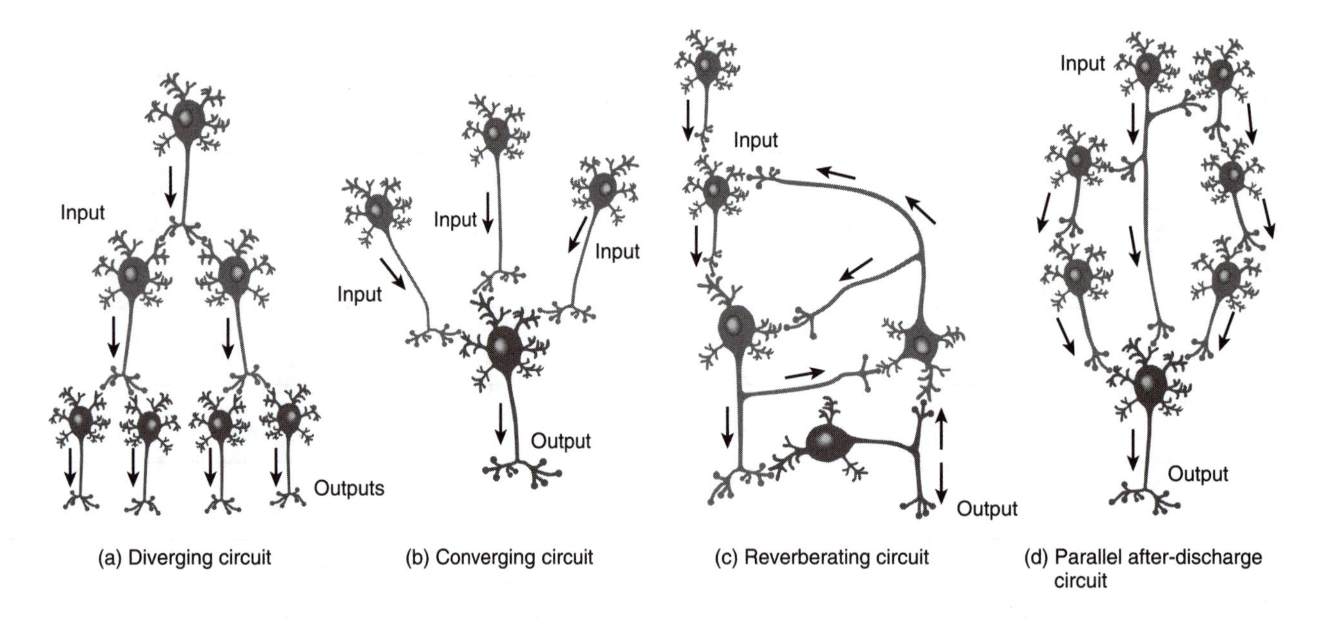
Input
Outputs
(a) Diverging circuit
Input
Input
Input
Output
(b) Converging circuit
Input
Output
(c) Reverberating circuit
Input
Output
(d) Parallel after-discharge circuit

NOTES

12

Figure 12.17 Damage and repair of a neuron in the PNS (page 405).

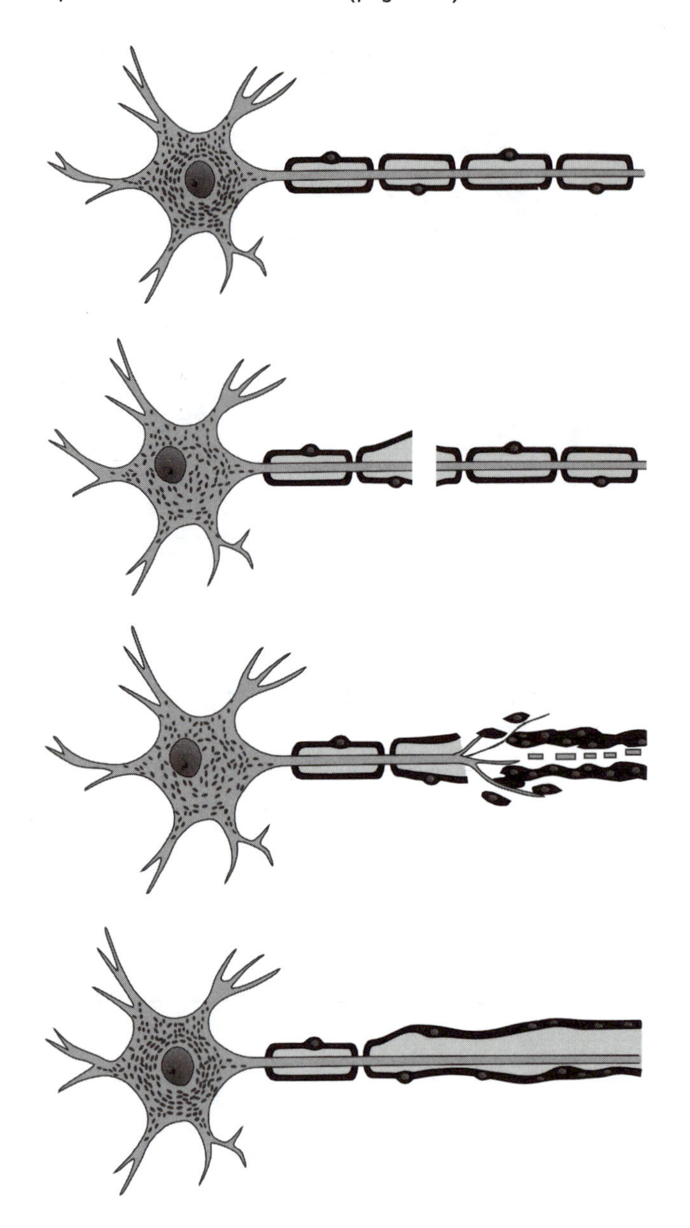

NOTES

12

Figure 13.1 Gross anatomy of the spinal cord (page 413).

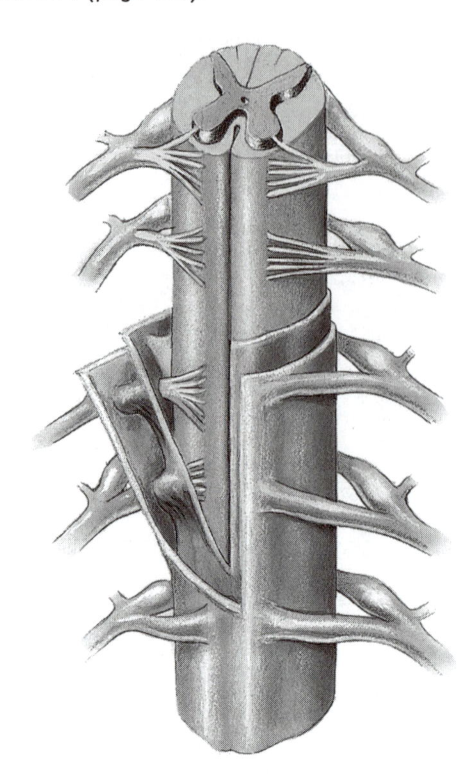

Figure 13.2 External anatomy of the spinal cord and the spinal nerves (page 415).

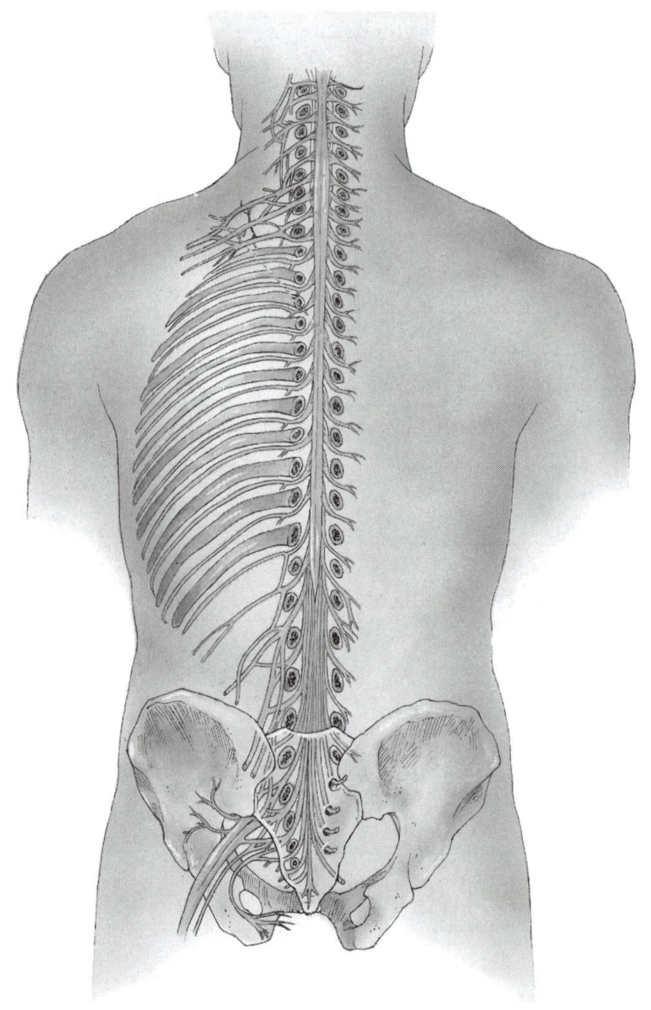

NOTES

Figure 13.3 Internal anatomy of the spinal cord (page 416).

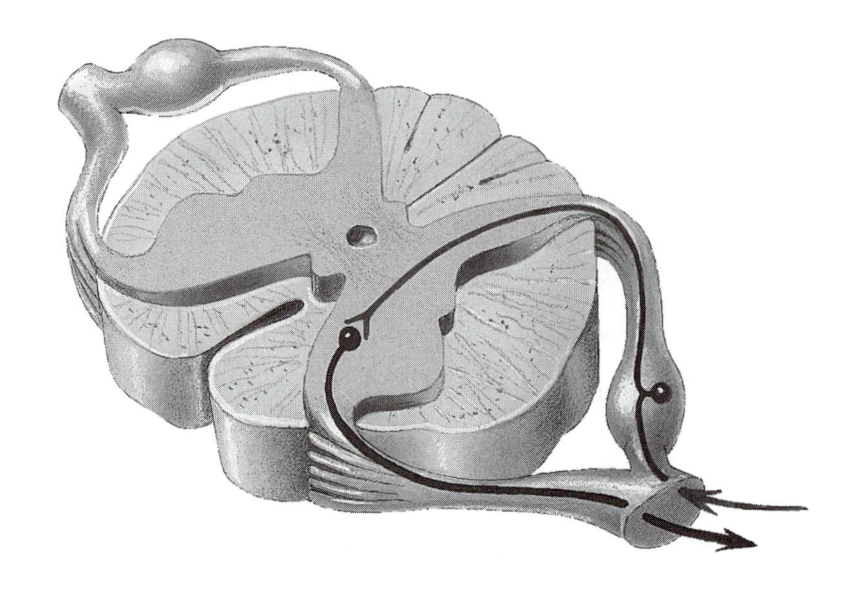

Figure 13.4 The locations of selected sensory and motor tracts (page 418).

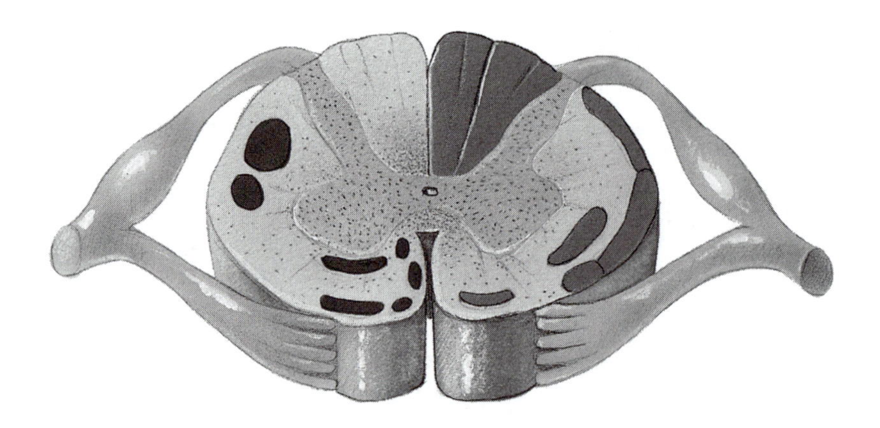

NOTES

Figure 13.5 General components of a reflex arc (page 419).

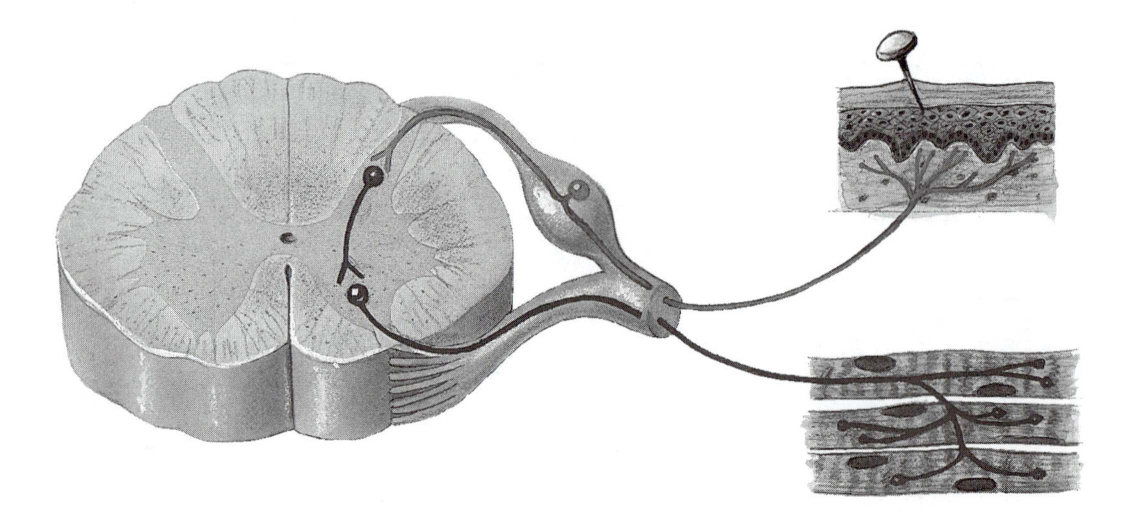

Figure 13.6 Stretch reflex (page 421).

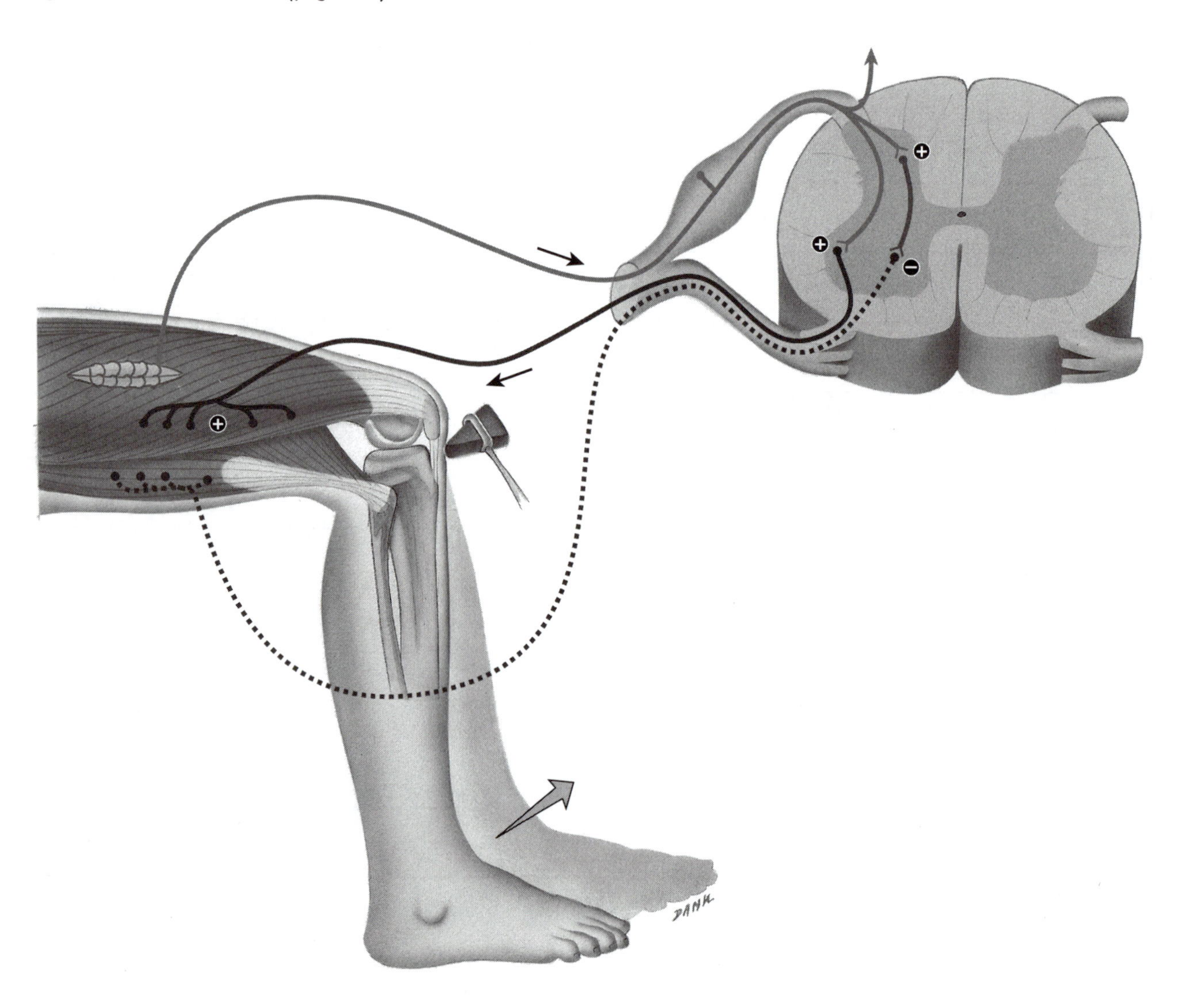

NOTES

13

Figure 13.7 Tendon reflex (page 422).

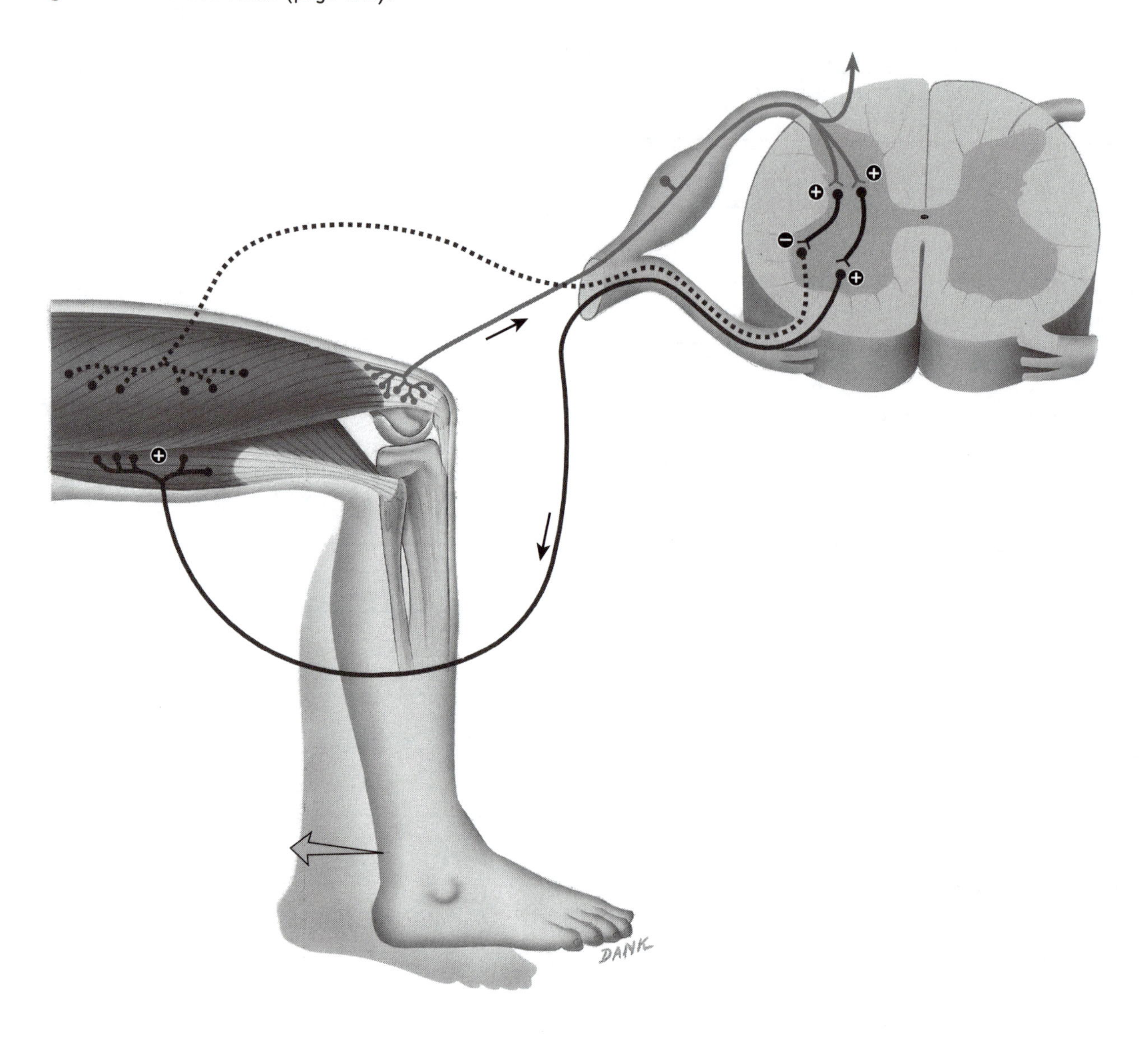
DANK

NOTES

13

Figure 13.8 Flexor (withdrawal) reflex (page 424).

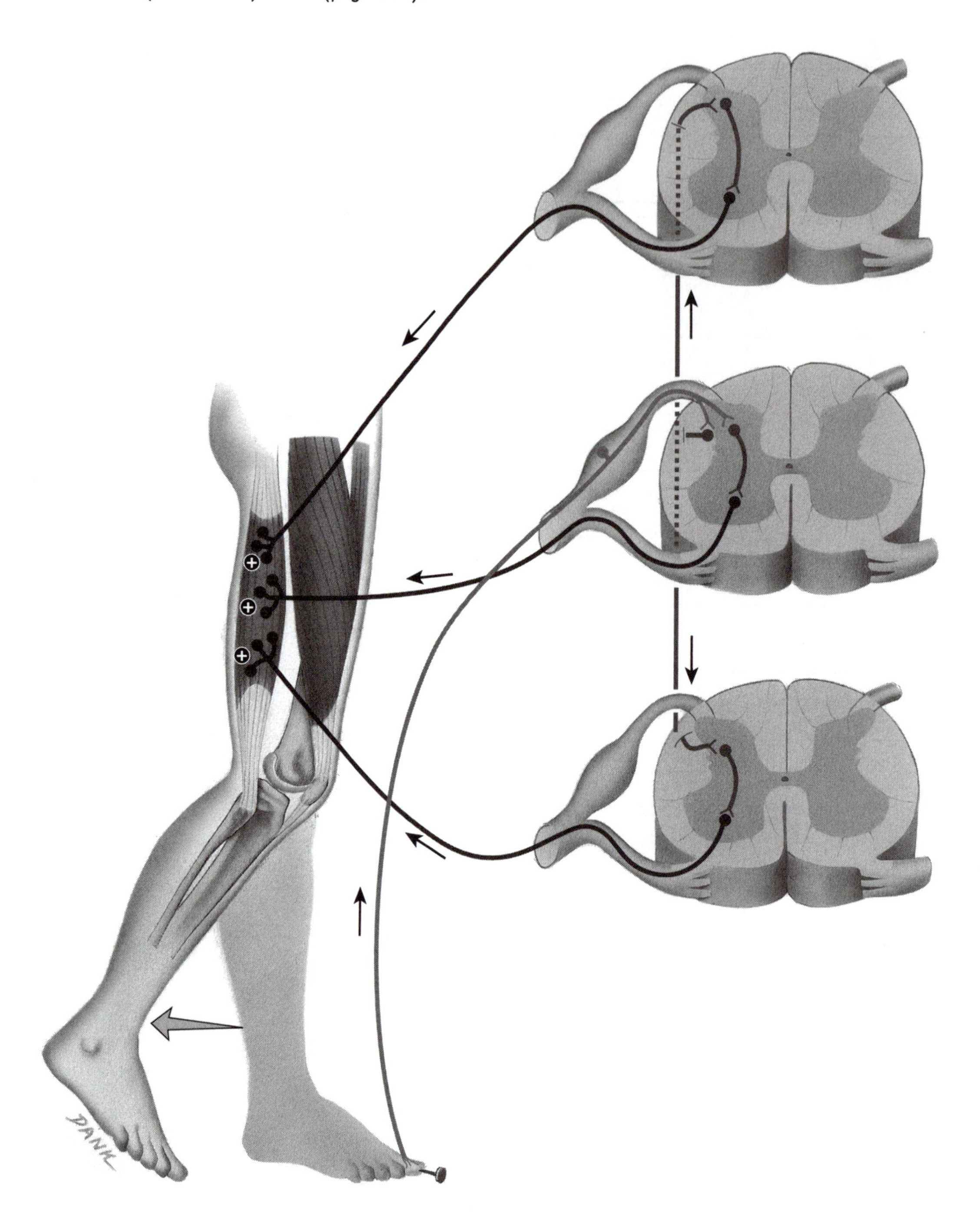

NOTES

Figure 13.9 Crossed extensor reflex (page 425).

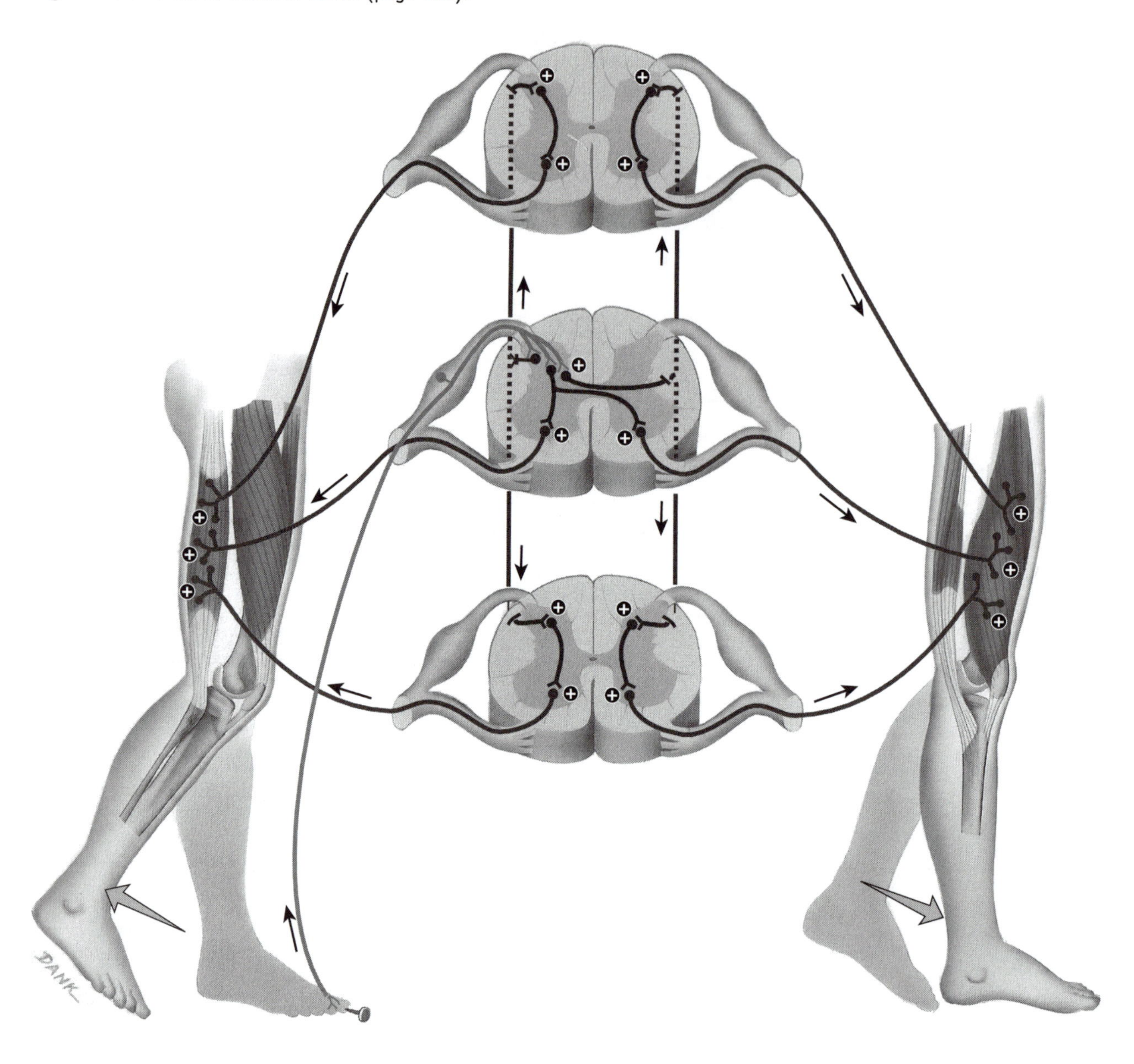

Figure 13.10 Organization and connective tissue coverings of a spinal nerve (page 426).

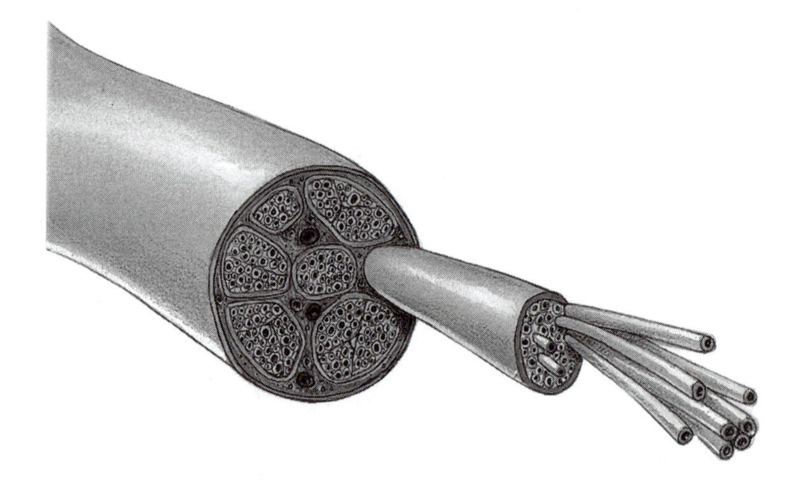

NOTES

13

Figure 13.11 Branches of a typical spinal nerve (page 428).

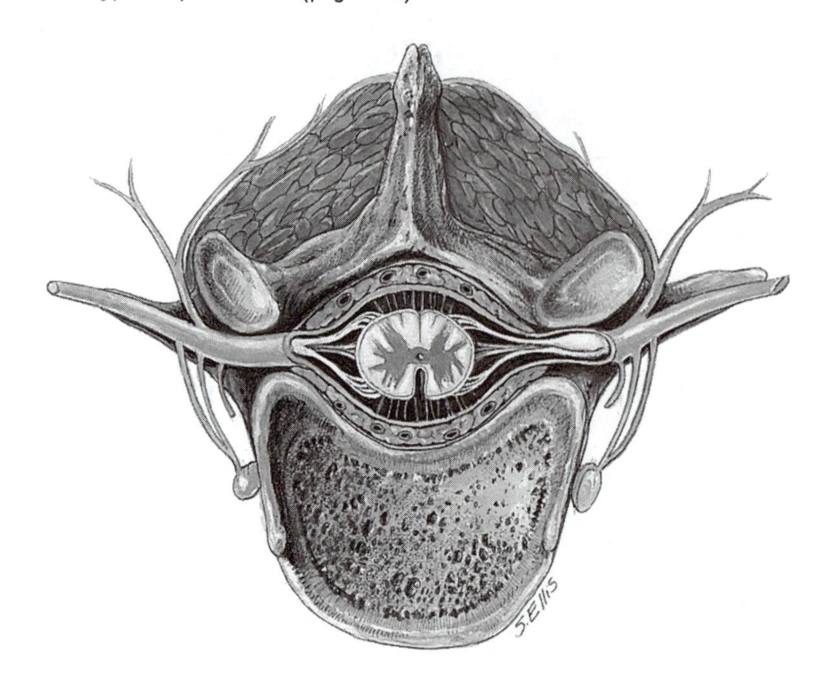
S.Ellis

Figure 13.12 Cervical plexus (page 430).

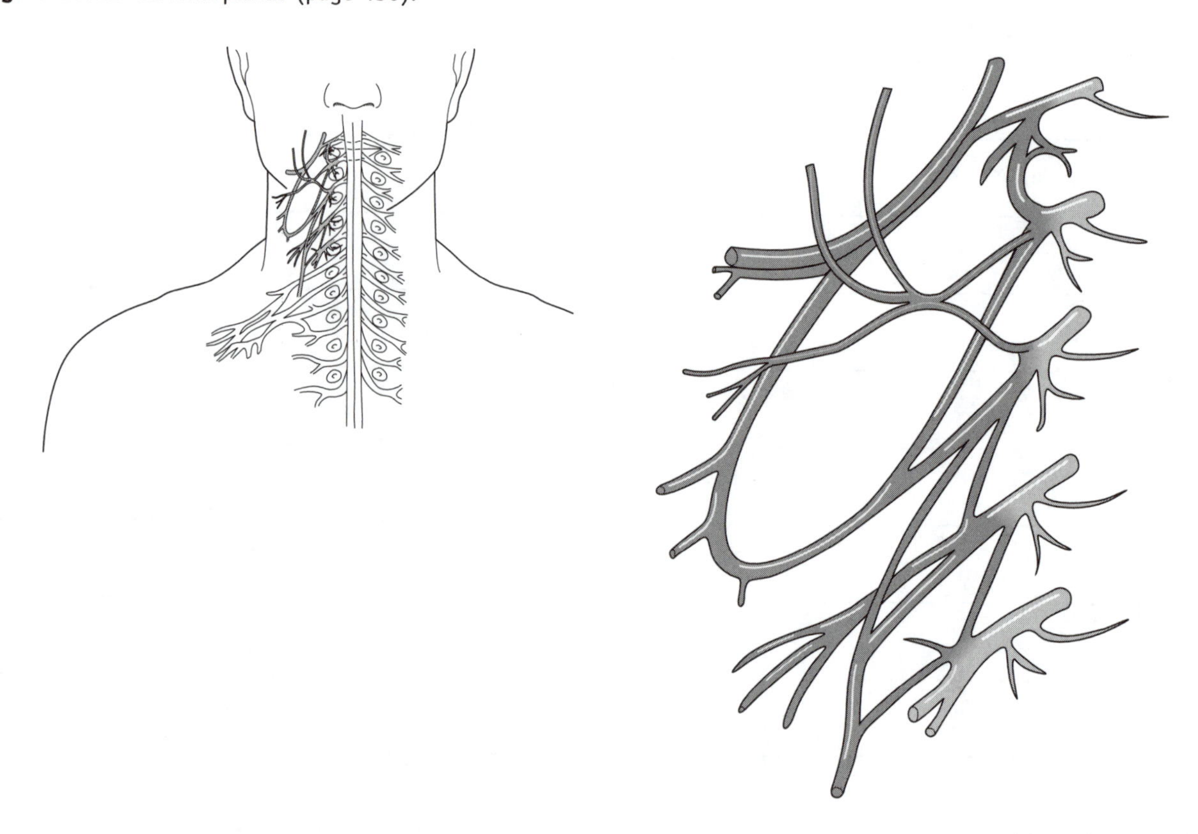

NOTES

13

Figure 13.13a Brachial plexus (page 433).

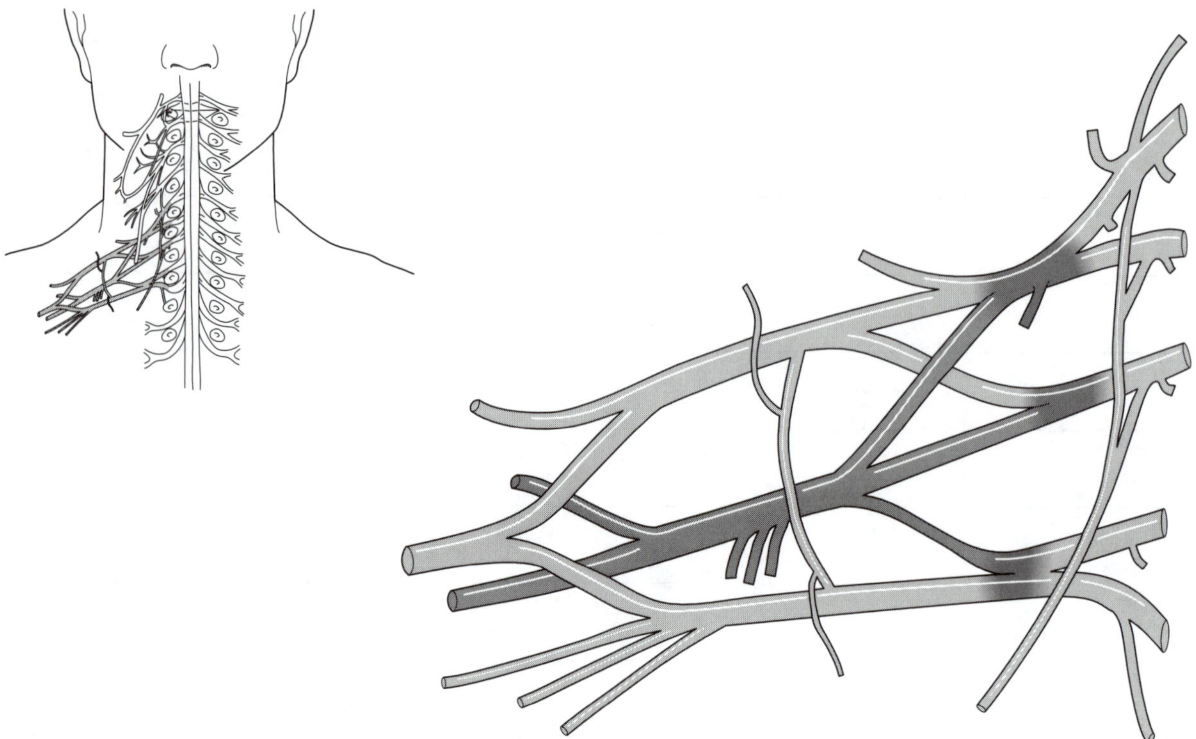

Figure 13.13b Brachial plexus (page 434).

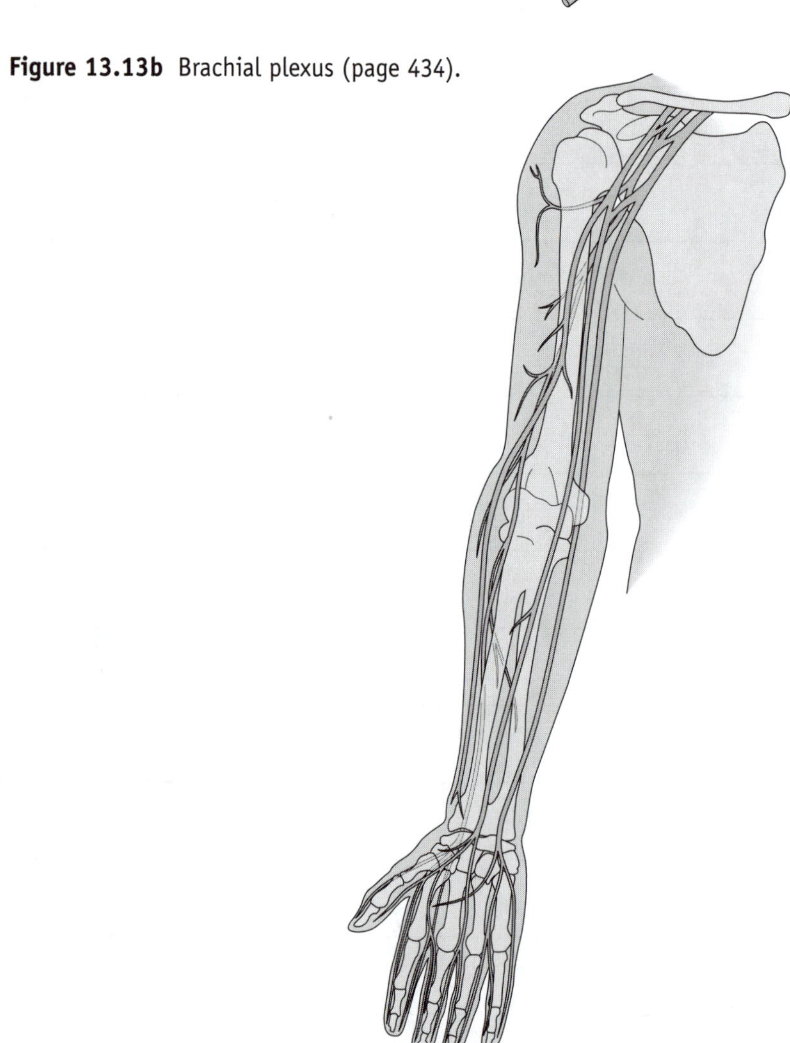

NOTES

13

Figure 13.14 Injuries to the brachial plexus (page 435).

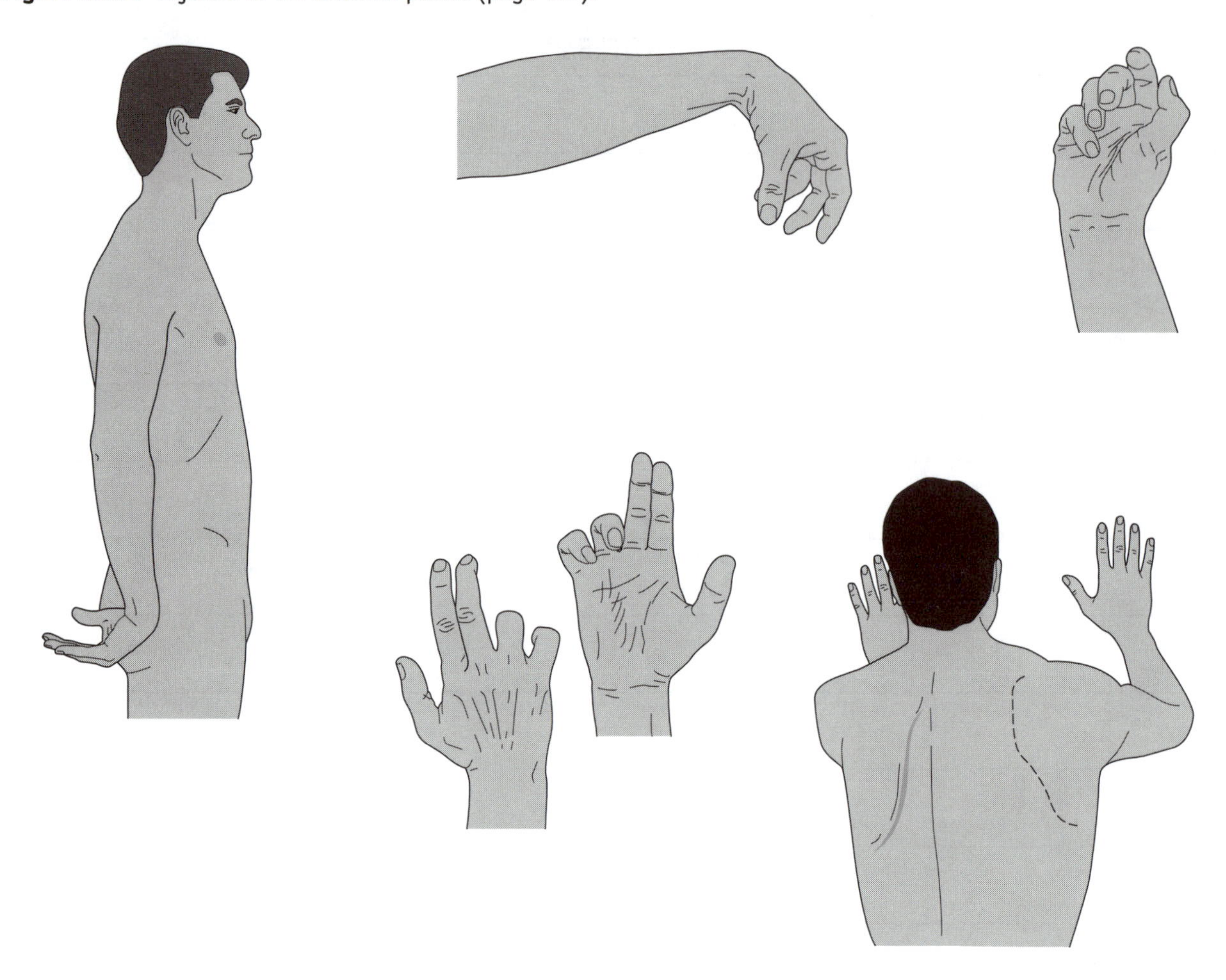

Figure 13.15a Lumbar plexus (page 437).

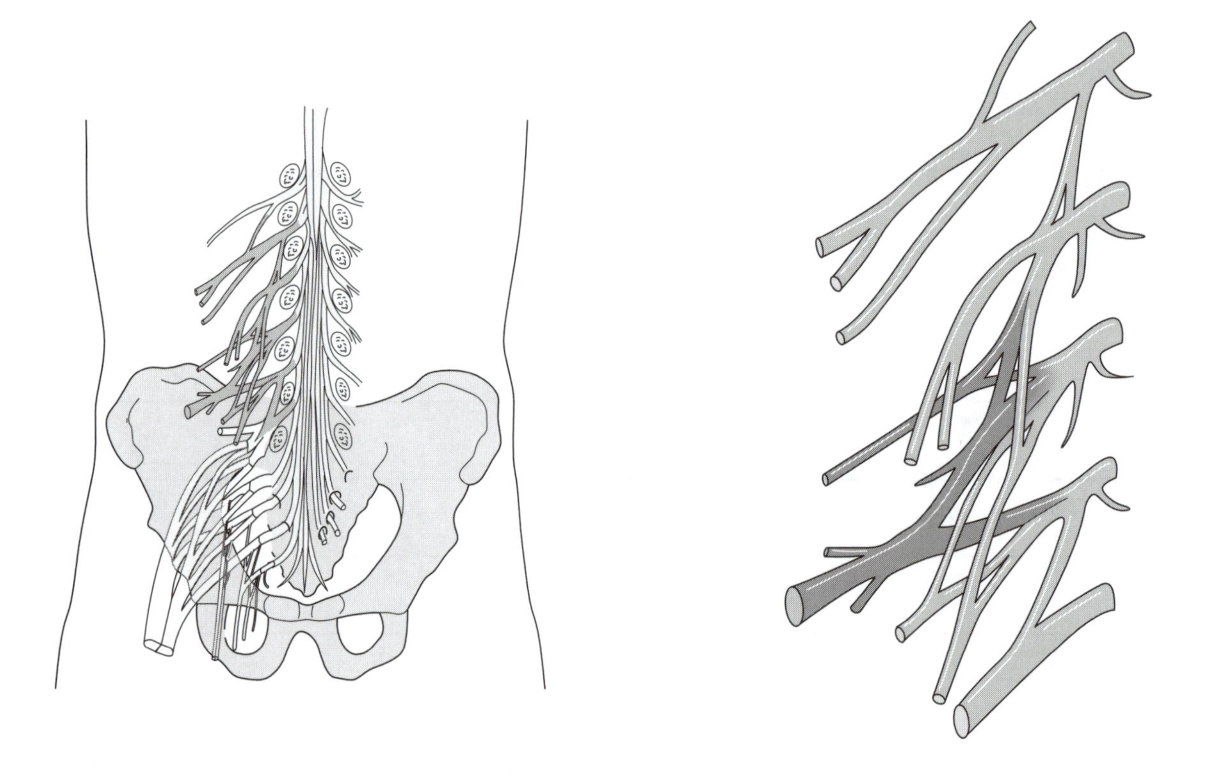

NOTES

Figure 13.15b Lumbar plexus (page 438).

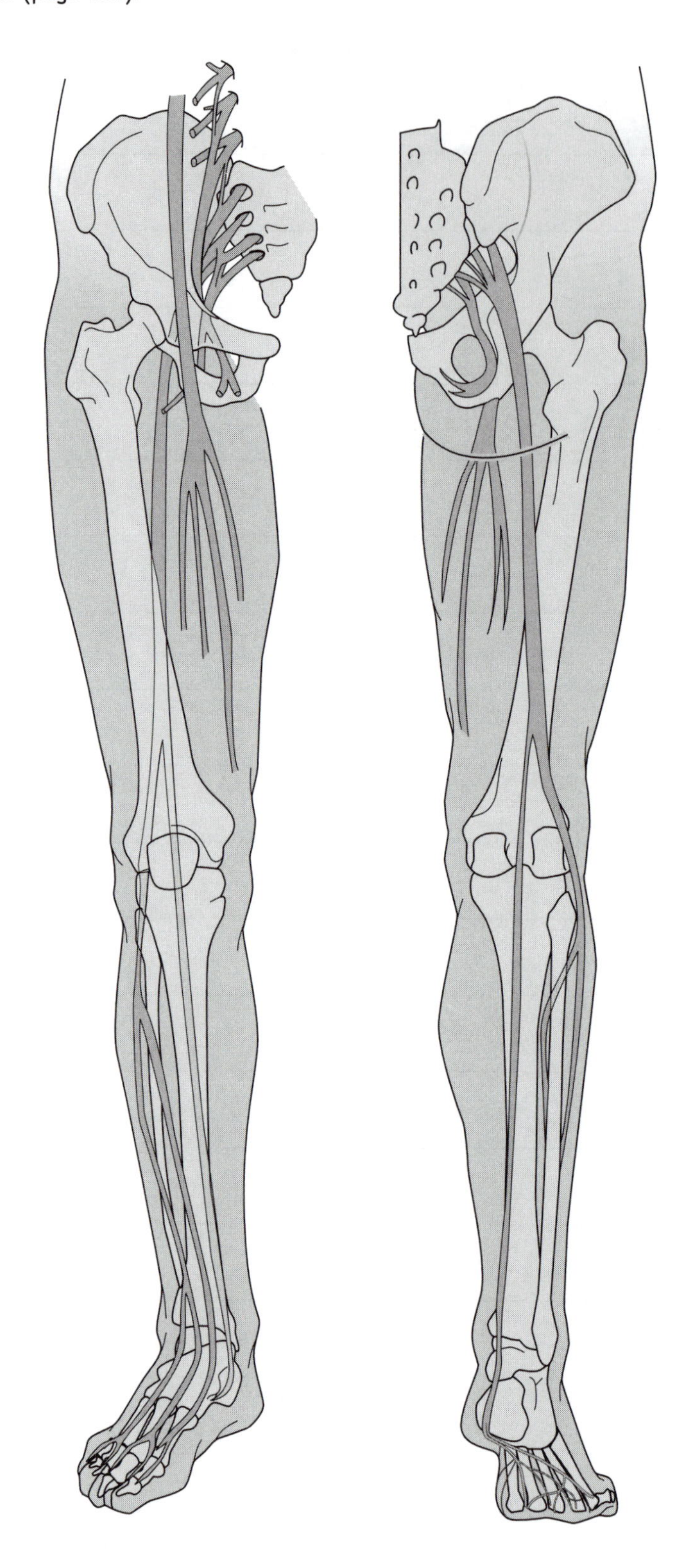

NOTES

13

Figure 13.16 Sacral plexus (page 440).

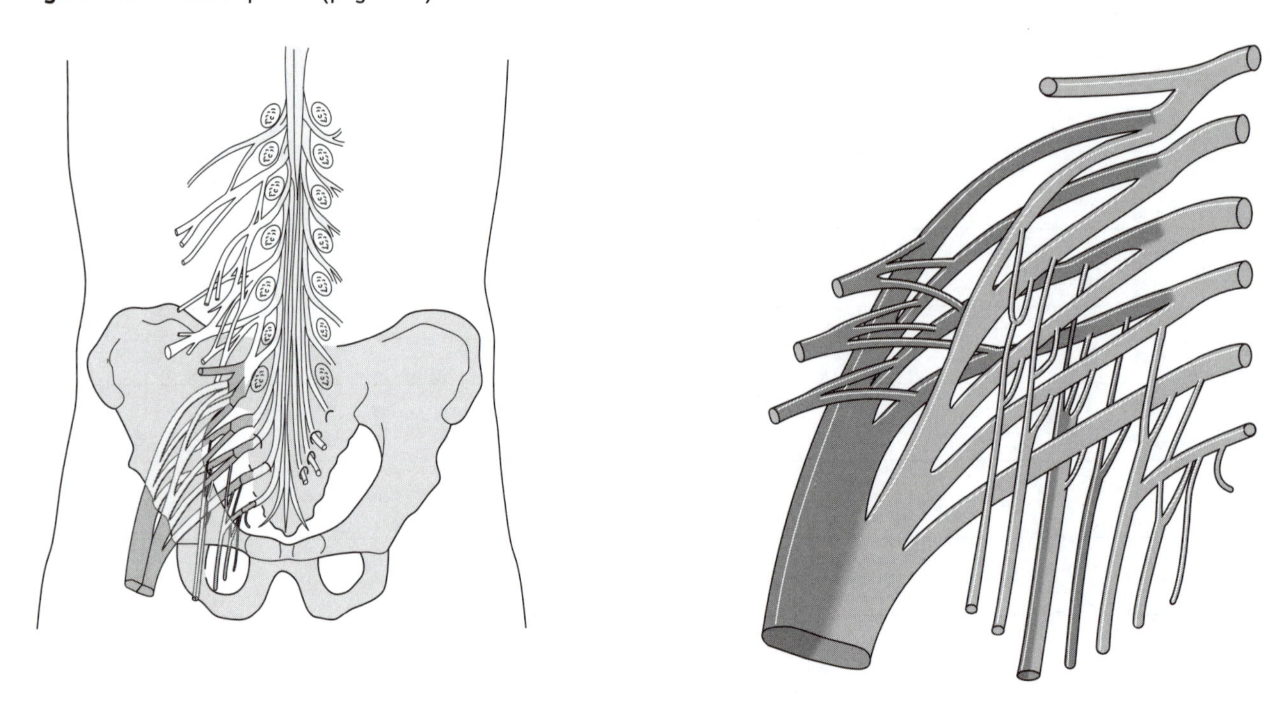

Figure 13.17 Distribution of dermatomes (page 441).

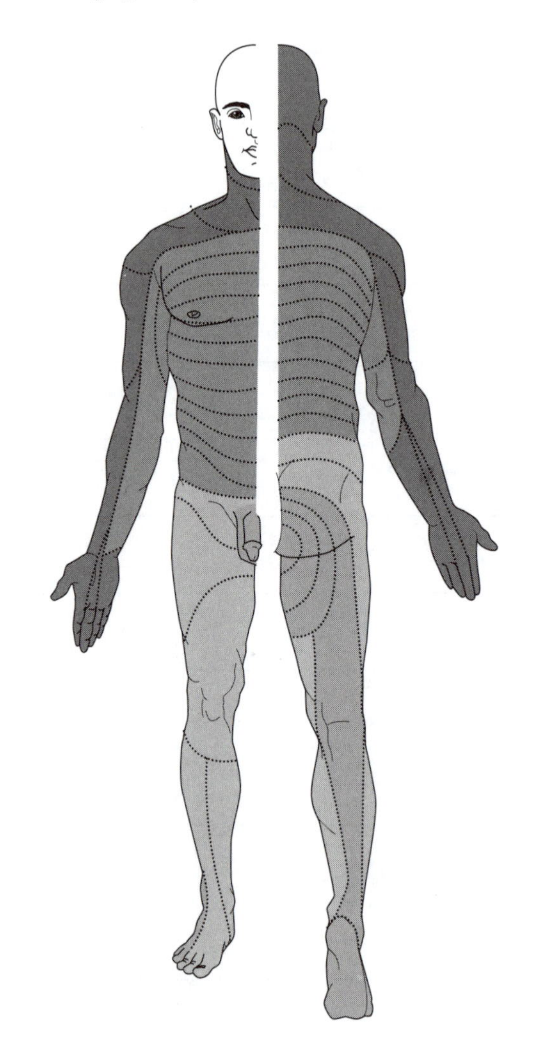

NOTES

13

Figure 14.1 The brain (page 447).

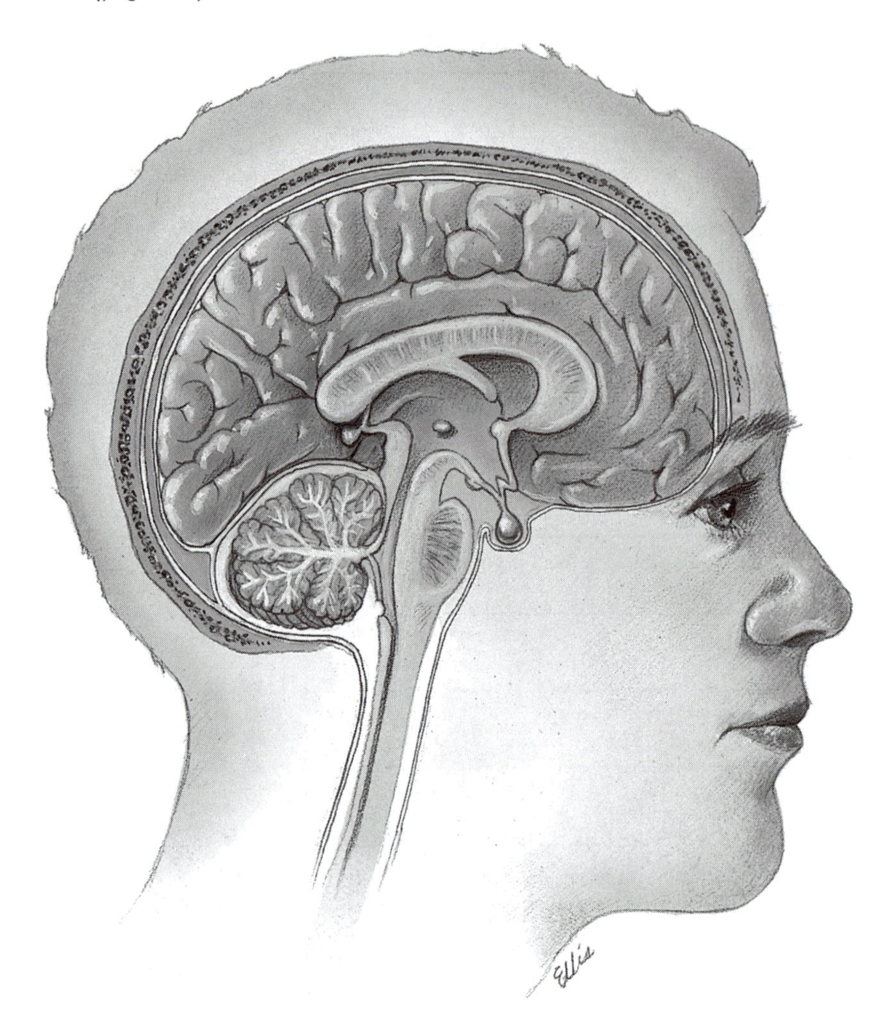

Figure 14.2 The protective coverings of the brain (page 449).

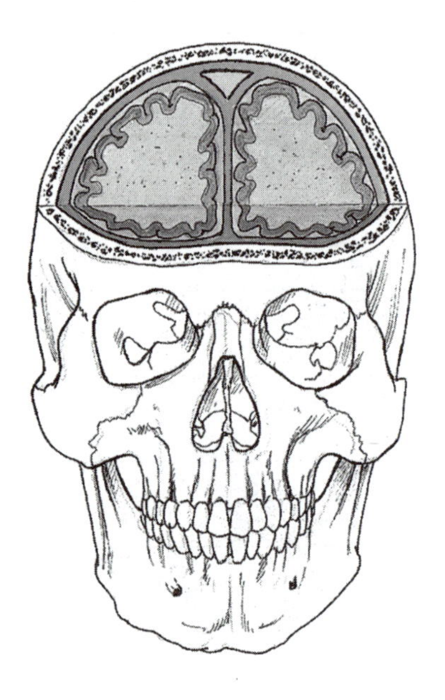

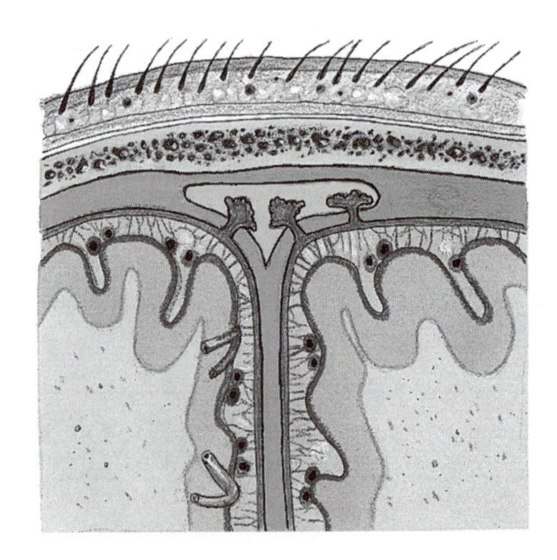

NOTES

14

Figure 14.3 Locations of ventricles (page 450).

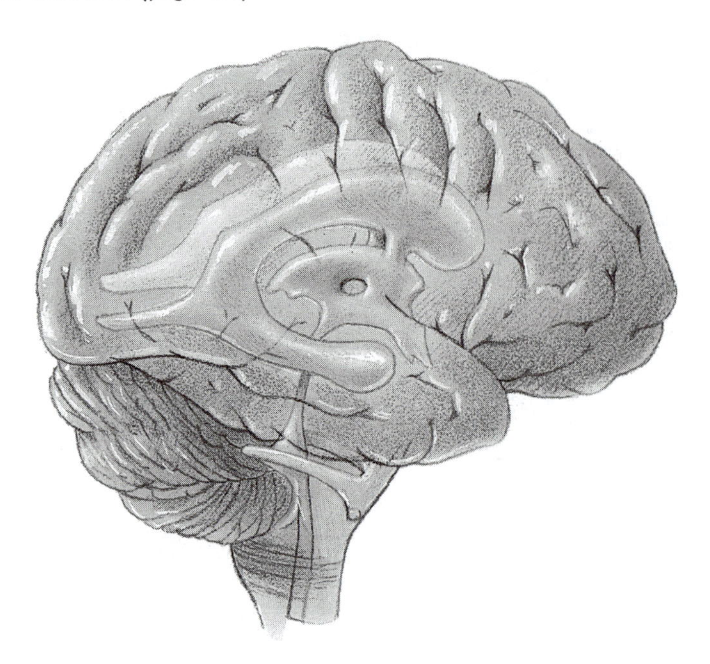

Figure 14.4a Pathways of circulating cerebrospinal fluid (page 451).

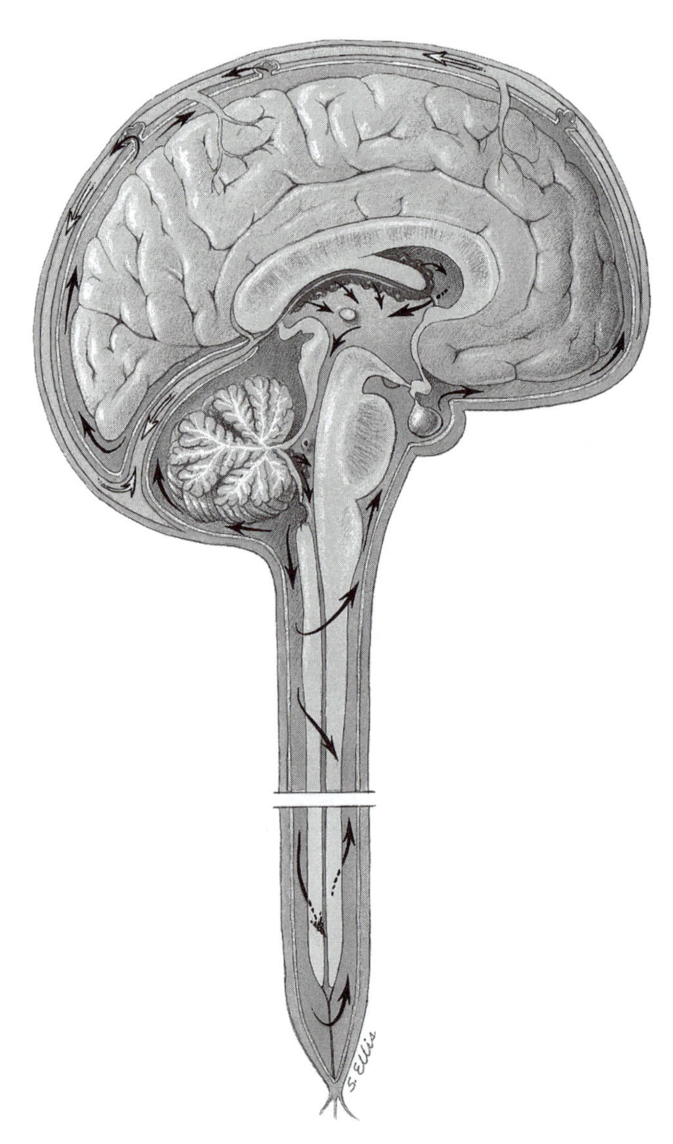
S. Ellis

NOTES

Figure 14.4b Frontal section of brain and spinal cord (page 452).

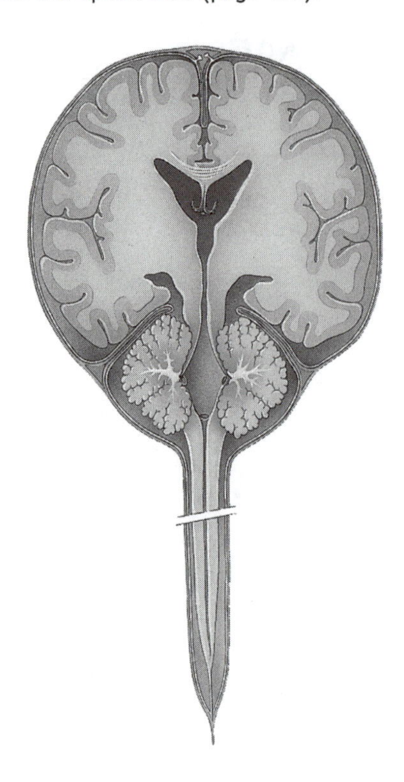

Figure 14.4c Summary of the formation, circulation, and absorption of cerebrospinal fluid (CSF) (page 453).

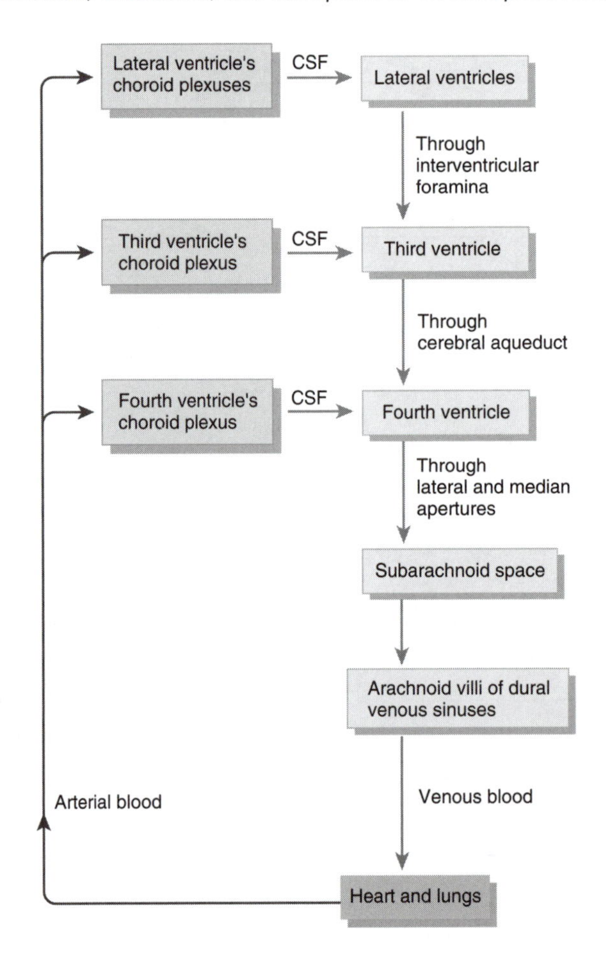

Lateral ventricle's choroid plexuses
CSF
Lateral ventricles
Through interventricular foramina
Third ventricle's choroid plexus
CSF
Third ventricle
Through cerebral aqueduct
Fourth ventricle's choroid plexus
CSF
Fourth ventricle
Through lateral and median apertures
Subarachnoid space
Arachnoid villi of dural venous sinuses
Arterial blood
Venous blood
Heart and lungs

NOTES

14

Figure 14.5 Medulla oblongata in relation to the rest of the brain stem (page 454).

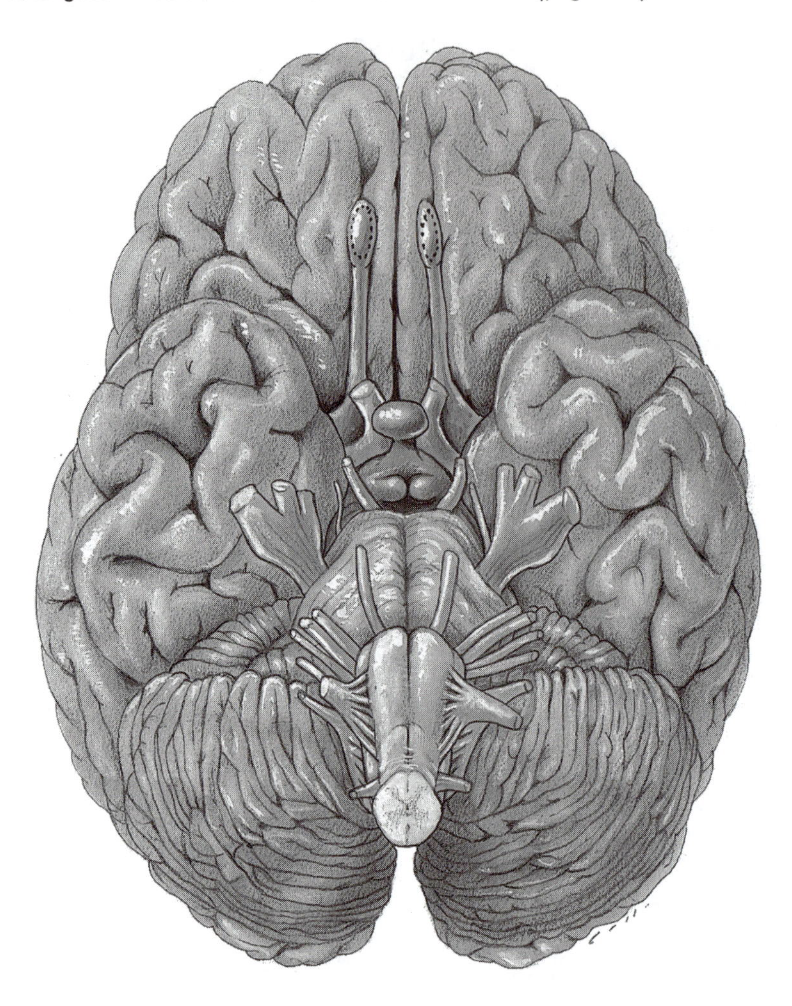

Figure 14.6 Internal anatomy of the medulla oblongata (page 455).

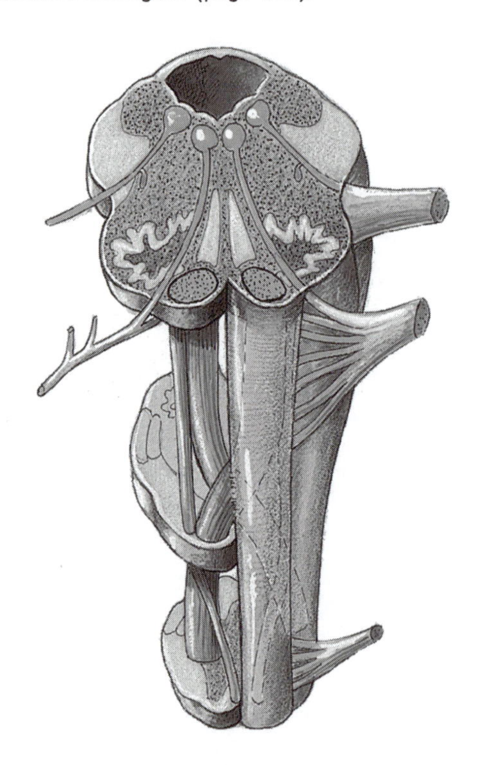

NOTES

14

Figure 14.7 Midbrain (page 456).

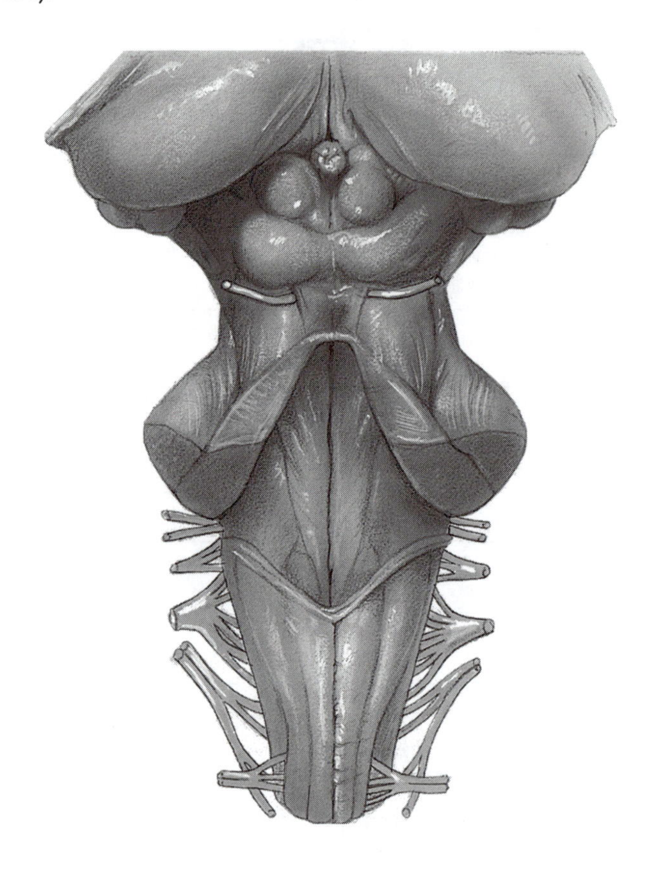

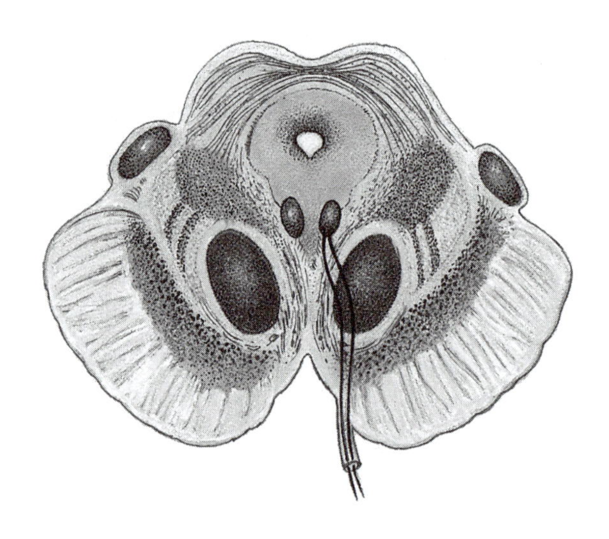

NOTES

14

Figure 14.8 Cerebellum (page 459).

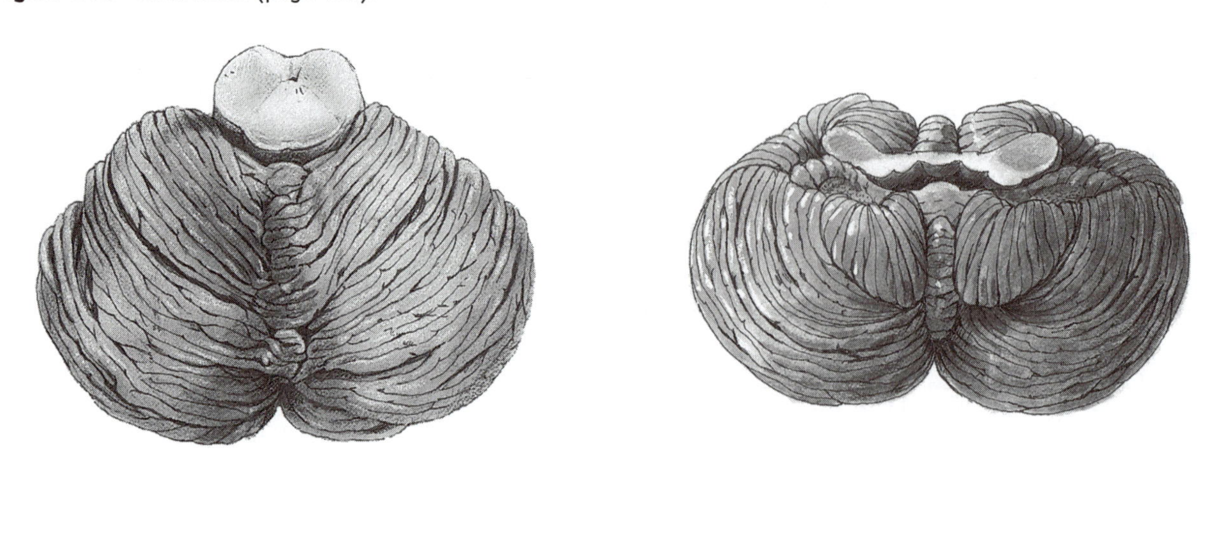

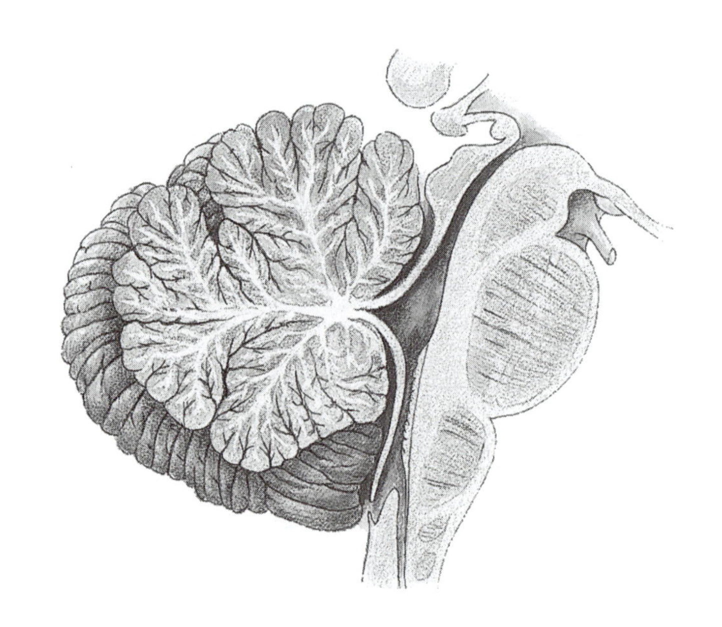

Figure 14.9 Thalamus (page 461).

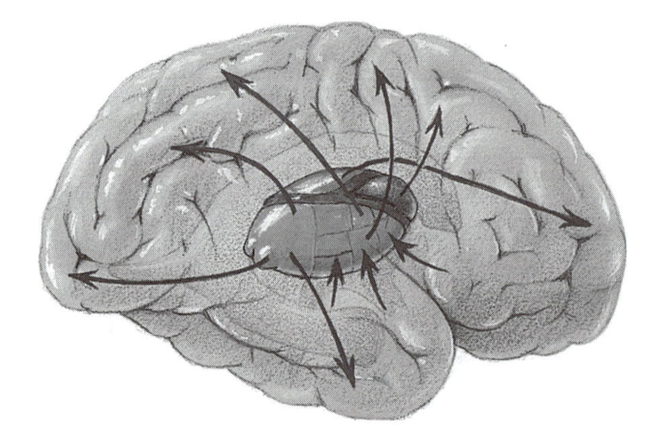

NOTES

14

Figure 14.10 Hypothalamus (page 462).

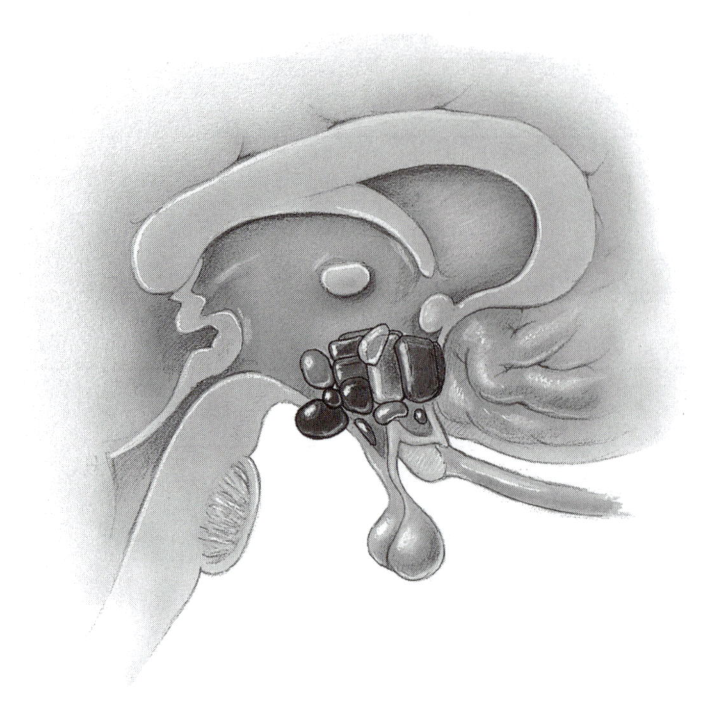

Figure 14.11 Cerebrum (page 464).

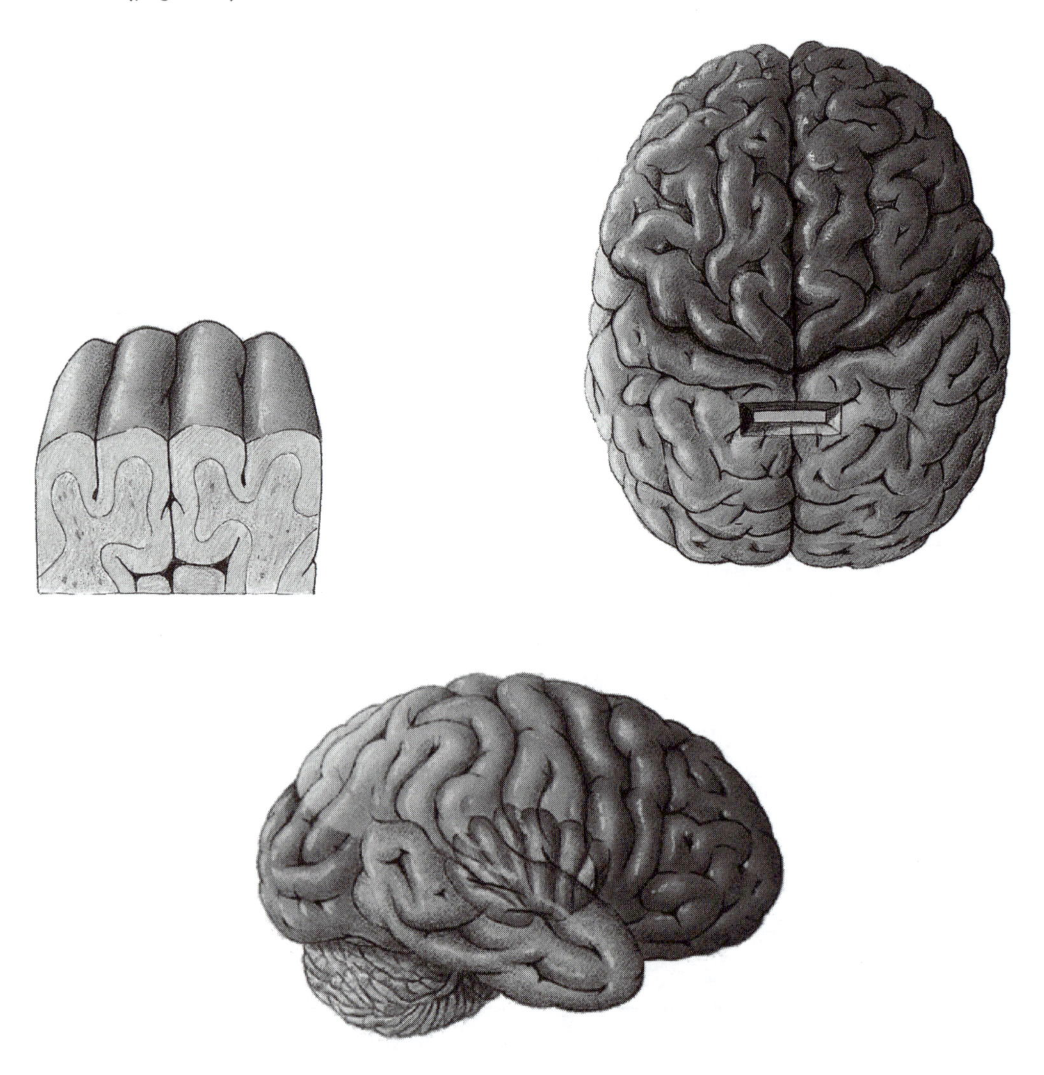

NOTES

14

Figure 14.13 Basal ganglia (page 466).

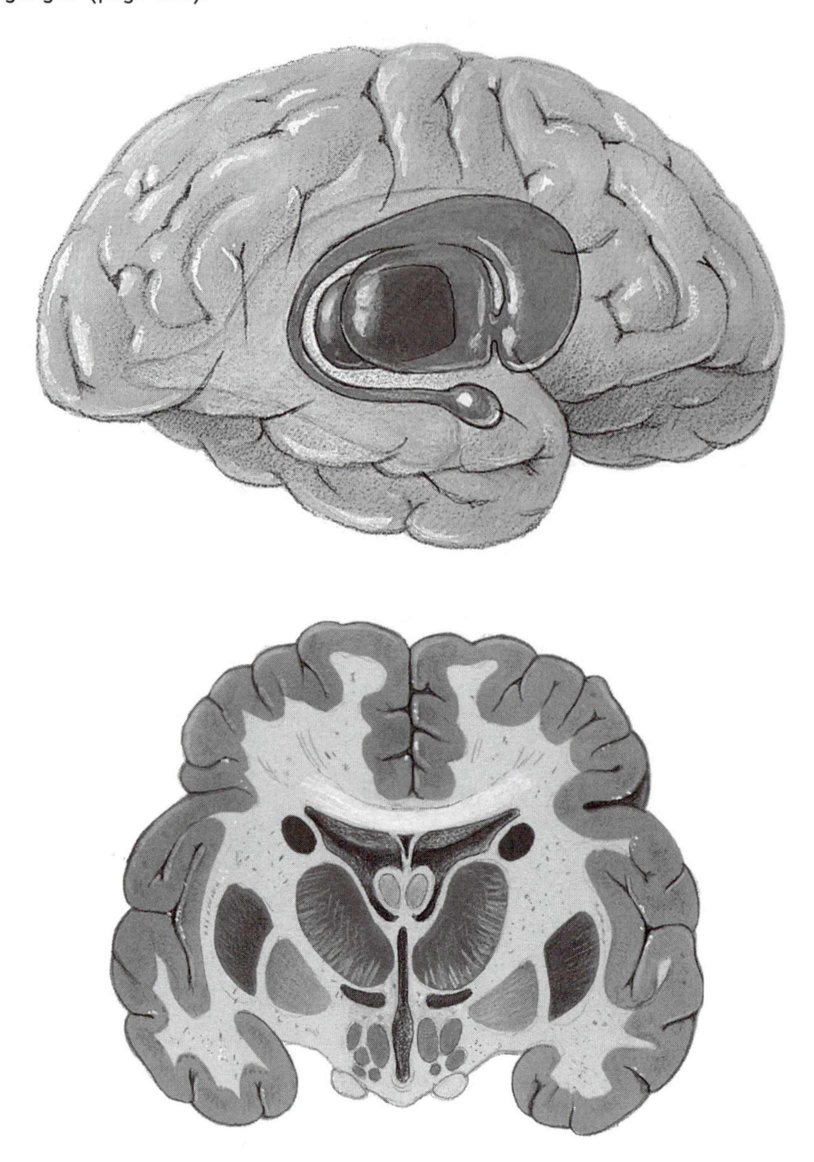

Figure 14.14 Components of the limbic system and surrounding structures (page 467).

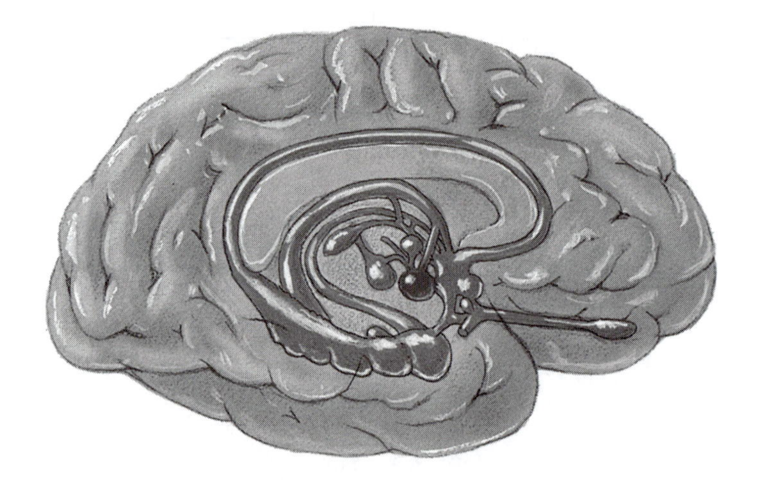

NOTES

14

Figure 14.15 Functional areas of the cerebrum (page 469).

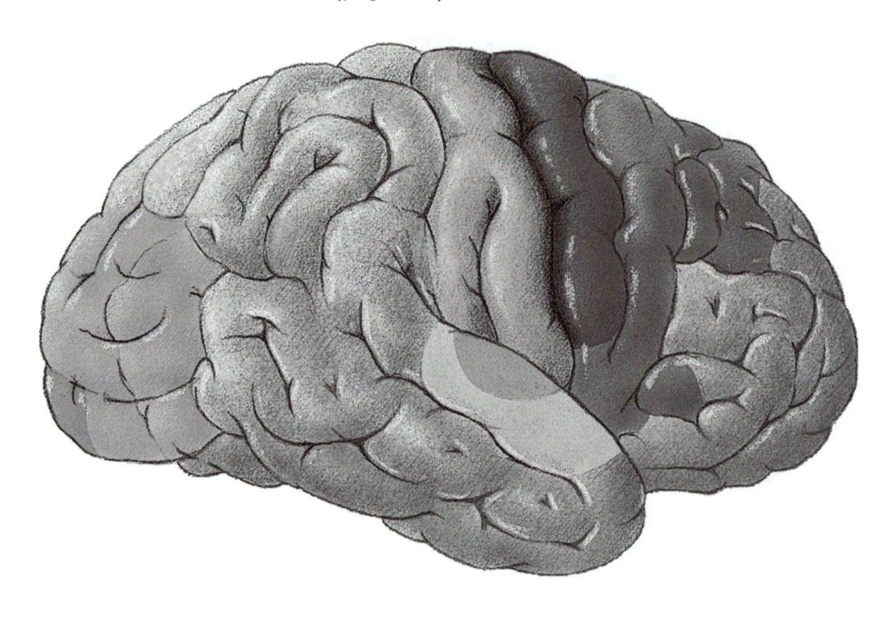

Figure 14.16 Summary of the principal functional differences between the left and right cerebral hemispheres (page 471).

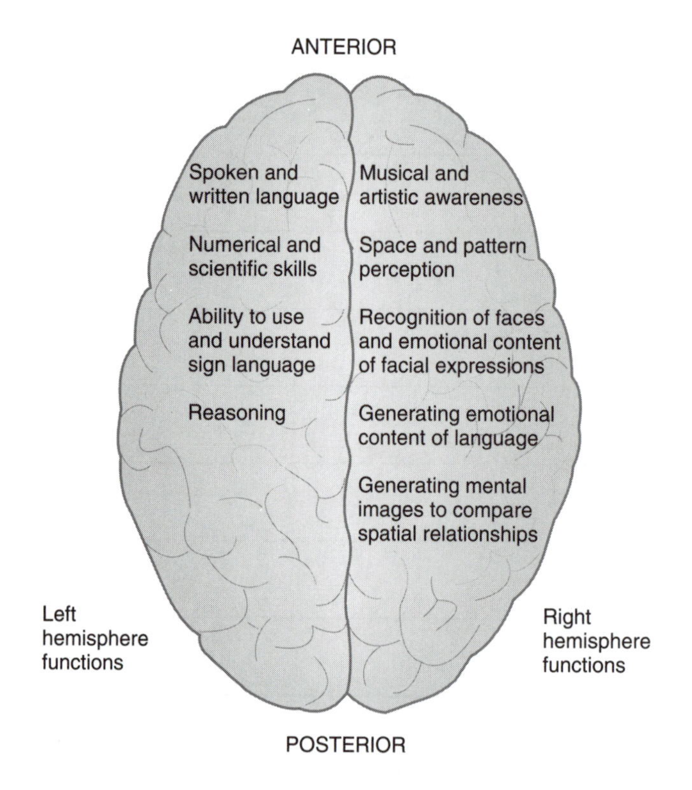
ANTERIOR
Spoken and written language
Numerical and scientific skills
Ability to use and understand sign language
Reasoning
Musical and artistic awareness
Space and pattern perception
Recognition of faces and emotional content of facial expressions
Generating emotional content of language
Generating mental images to compare spatial relationships
Left hemisphere functions
Right hemisphere functions
POSTERIOR

NOTES

14

Figure 14.17 Types of brain waves recorded in an electroencephalogram (page 471).

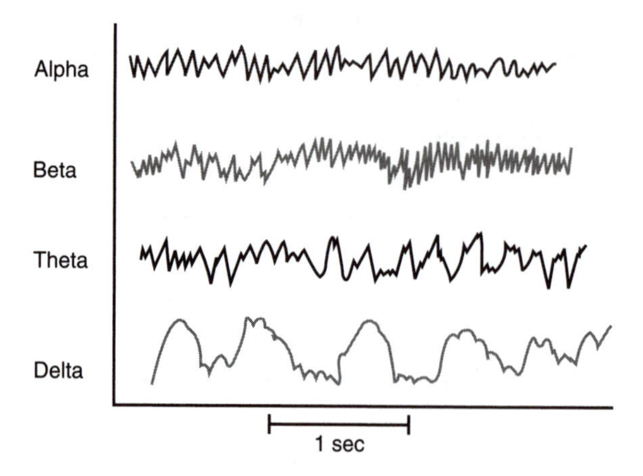
Alpha
Beta
Theta
Delta
1 sec

Figure 14.18 Origin of the nervous system (page 473).

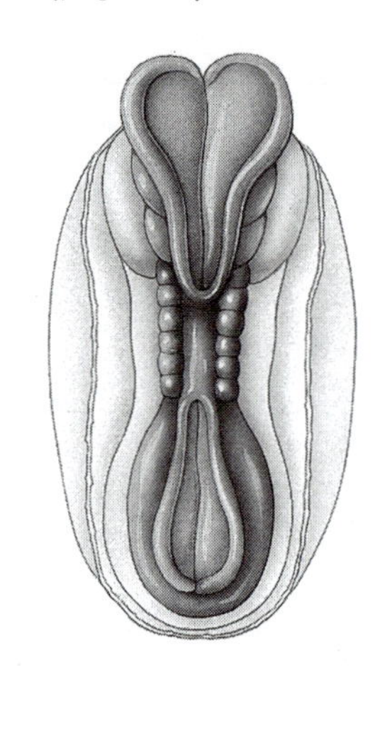

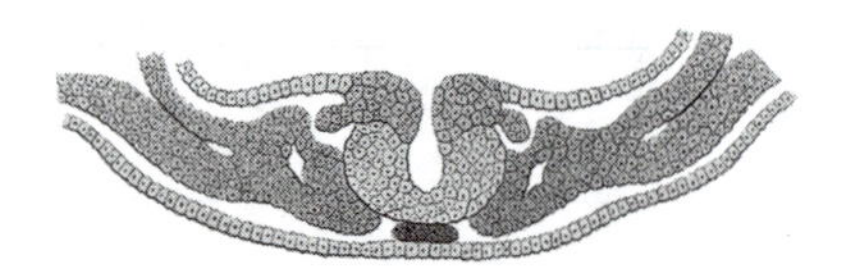

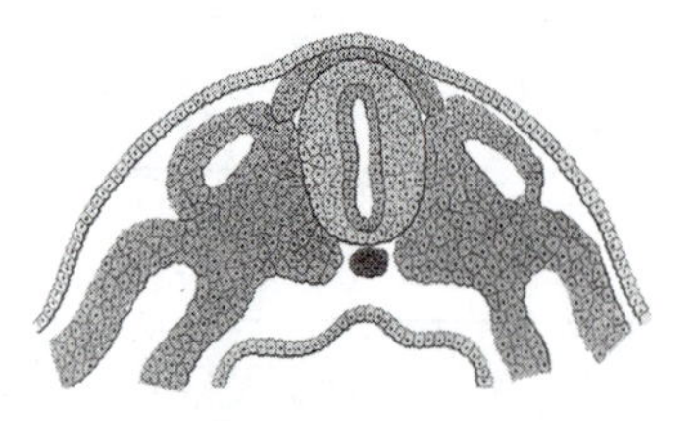

NOTES

14

Figure 14.19 Development of the brain and spinal cord (page 474).

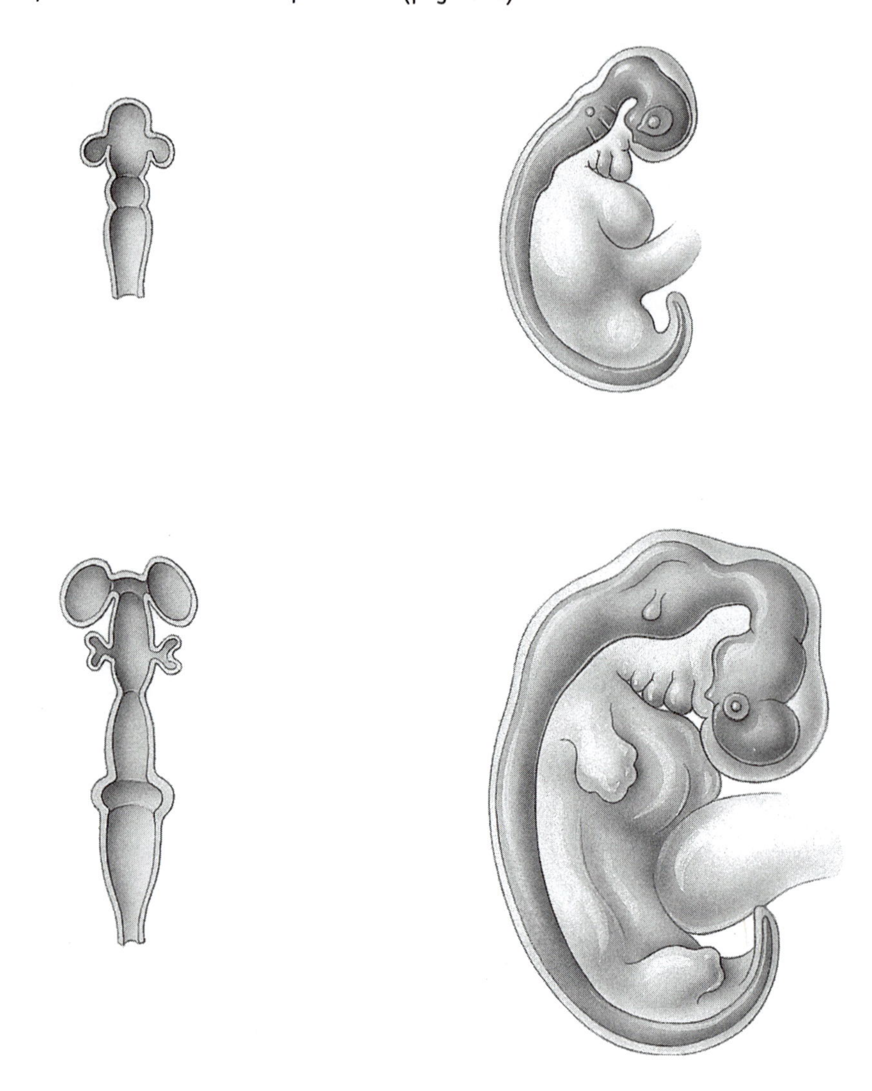

NOTES

14

Figure 15.1 Types of sensory receptors and their relationship to first-order sensory neurons (page 487).

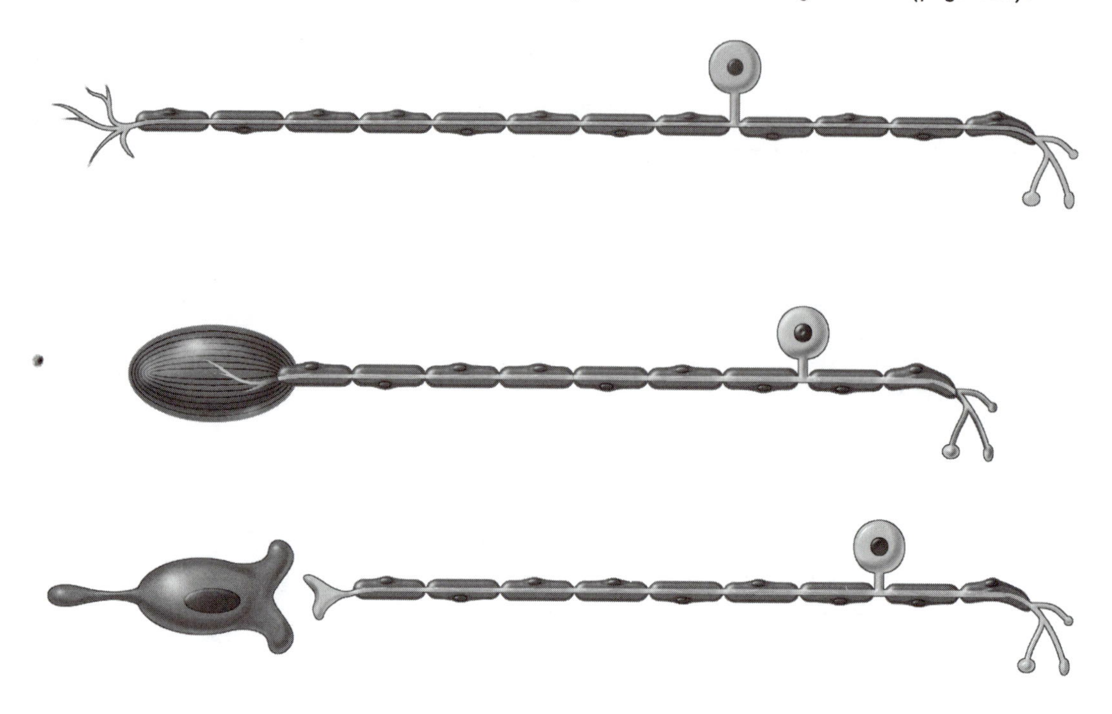

Figure 15.2 Structure and location of sensory receptors in the skin (page 490).

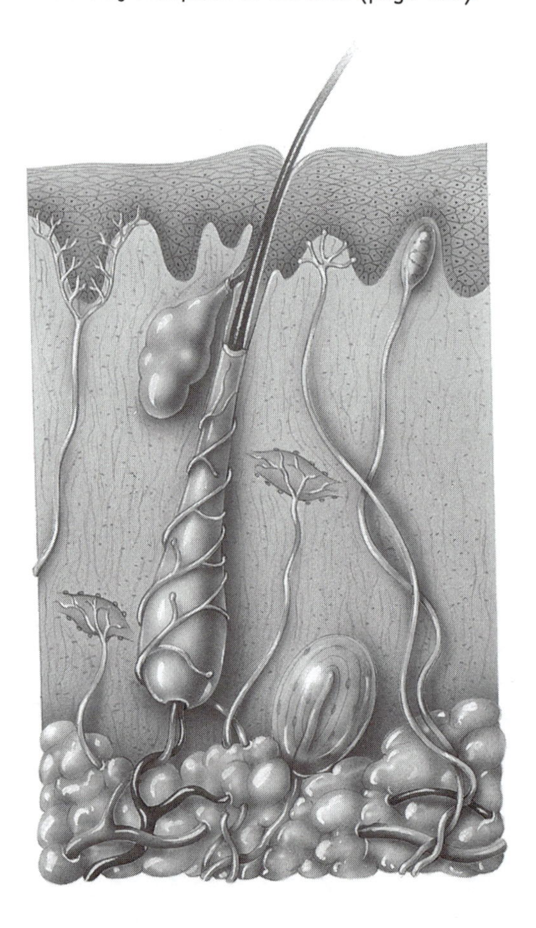

NOTES

Figure 15.3 Distribution of referred pain (page 492).

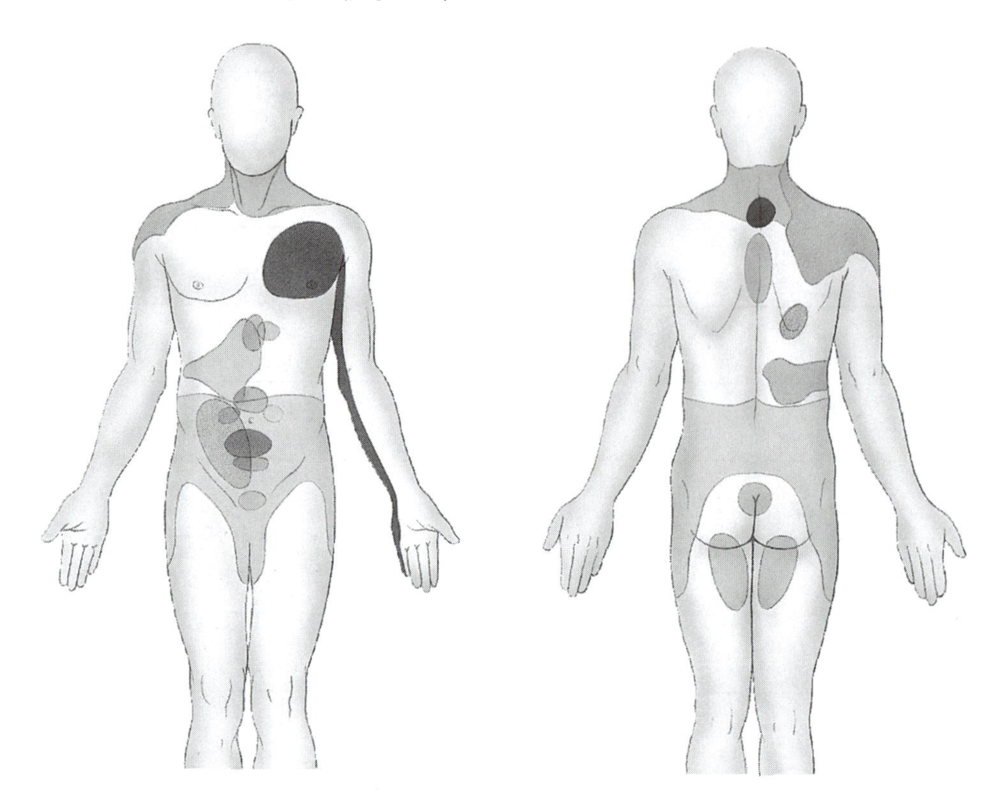

Figure 15.4 Two types of proprioceptors (page 493).

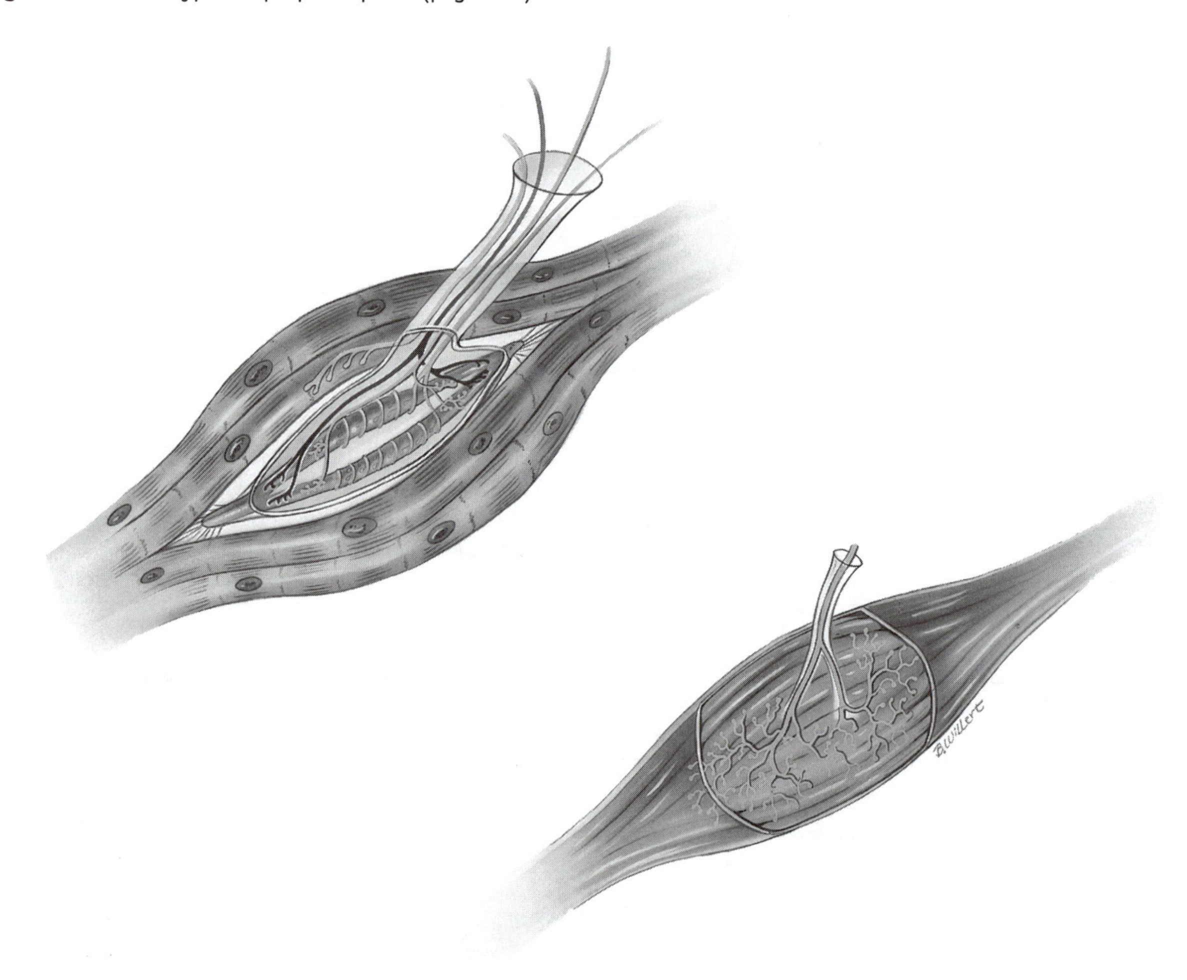

15

NOTES

Figure 15.5 Somatic sensory pathways (page 496).

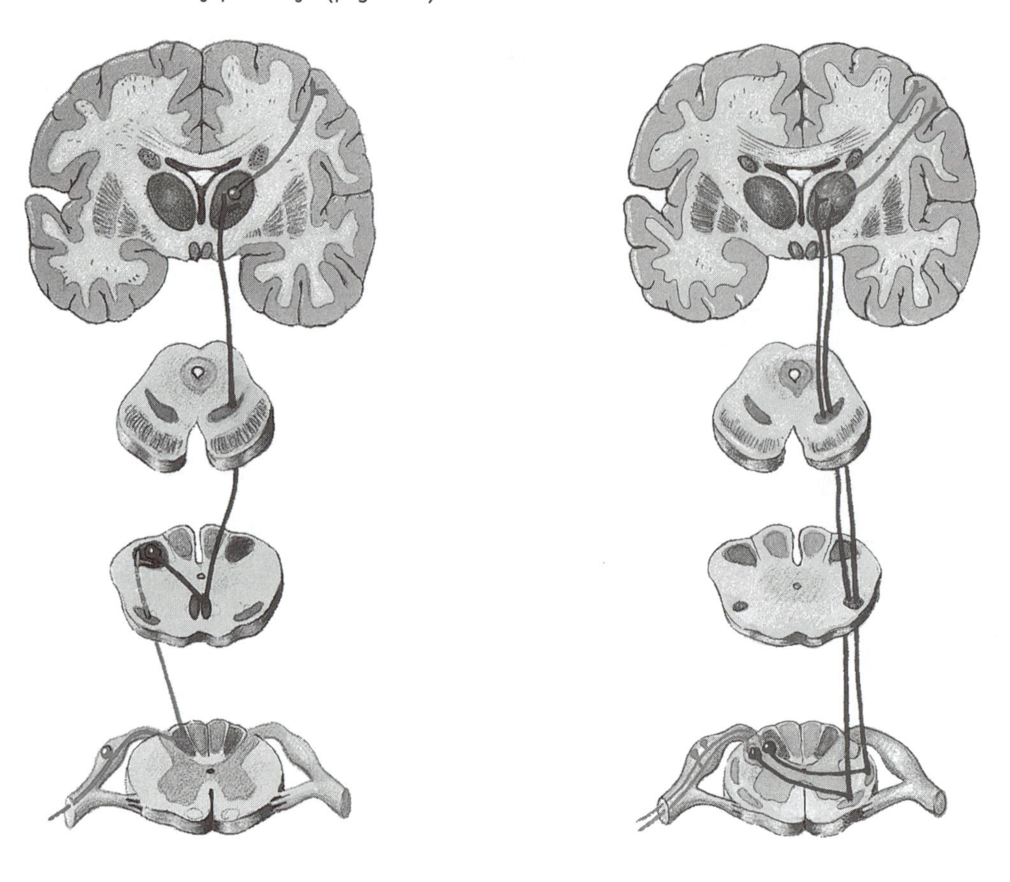

Figure 15.6 Primary somatosensory area and primary motor area (page 497).

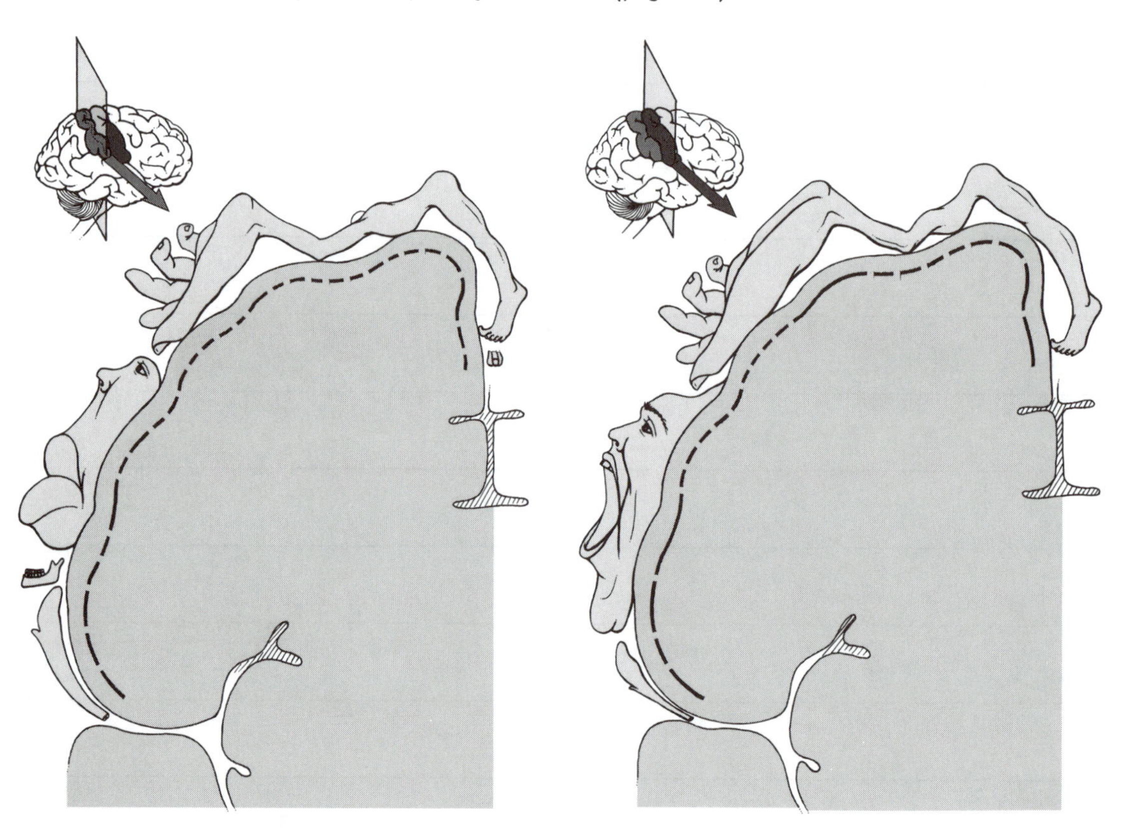

NOTES

Figure 15.7 Lateral and anterior corticospinal tract (page 499).

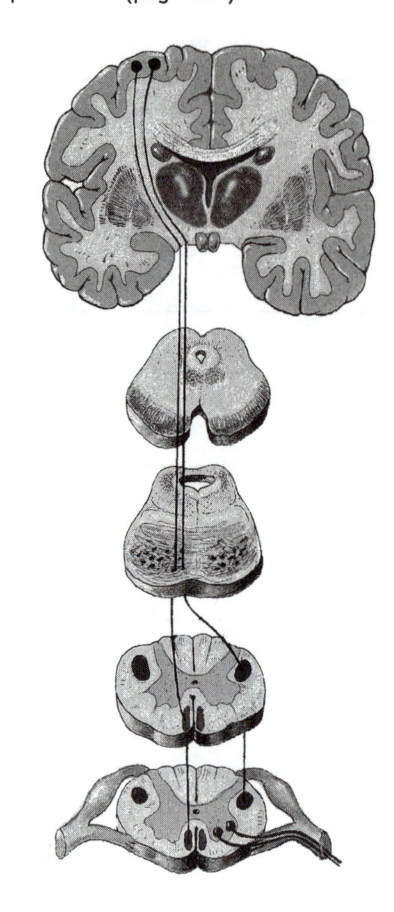

Figure 15.8 Indirect motor pathways for coordination and control of movement (page 500).

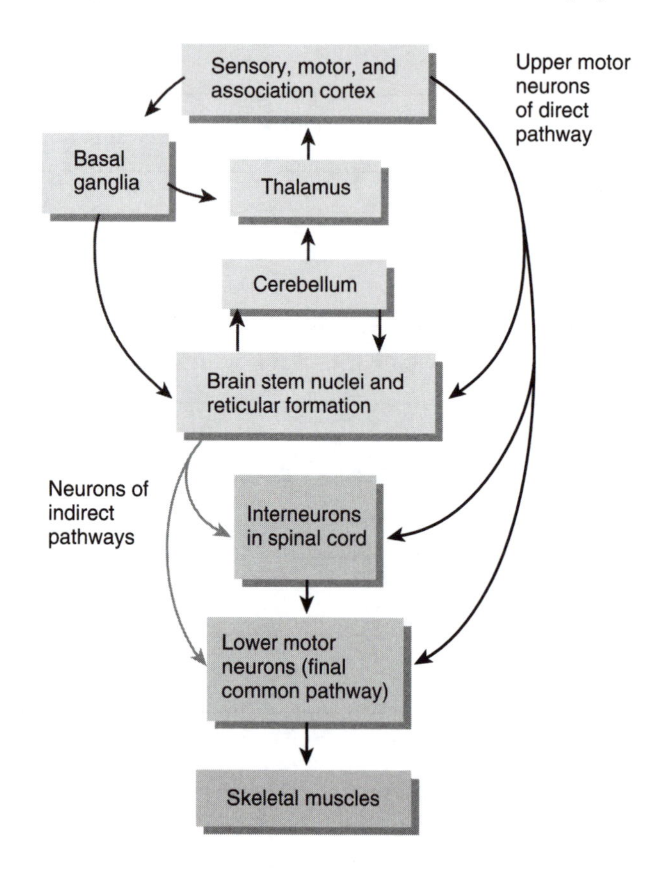
Sensory, motor, and association cortex
Upper motor neurons of direct pathway
Basal ganglia
Thalamus
Cerebellum
Brain stem nuclei and reticular formation
Neurons of indirect pathways
Interneurons in spinal cord
Lower motor neurons (final common pathway)
Skeletal muscles

NOTES

Figure 15.9 Input to and output from the cerebellum (page 503).

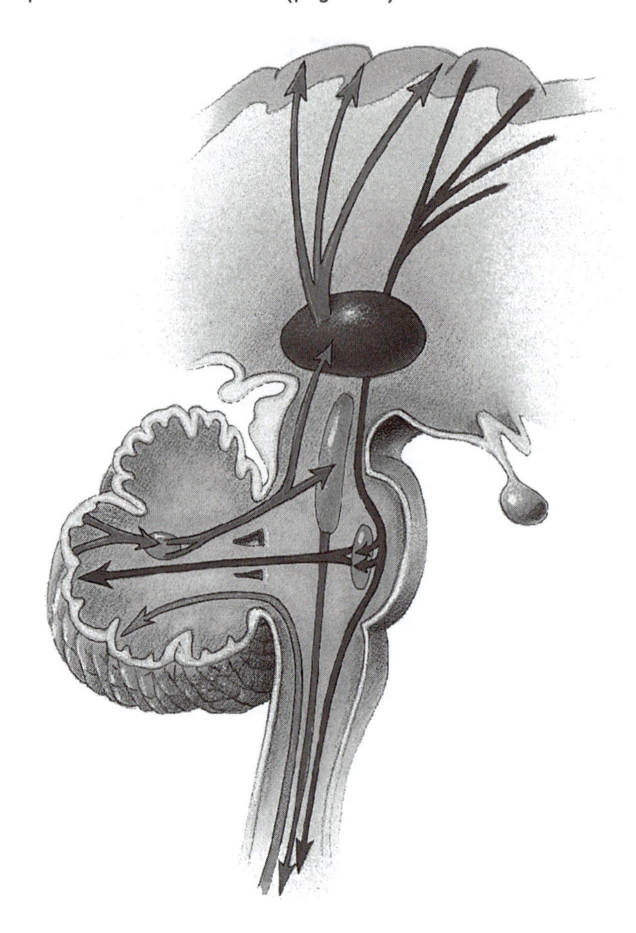

Figure 15.10 The reticular activating system (page 504).

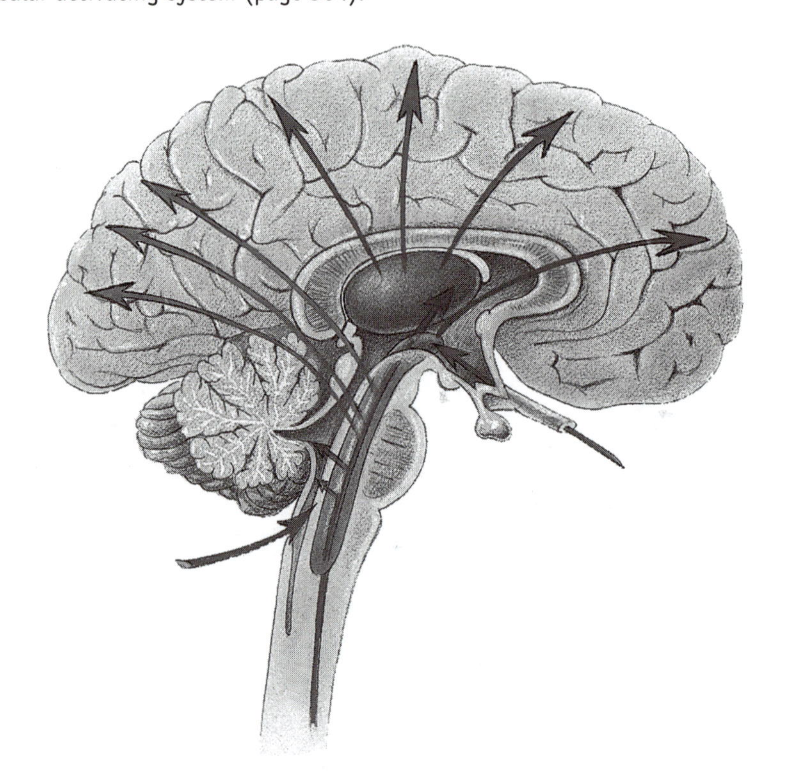

NOTES

Figure 15.11 The stages of sleep (page 505).

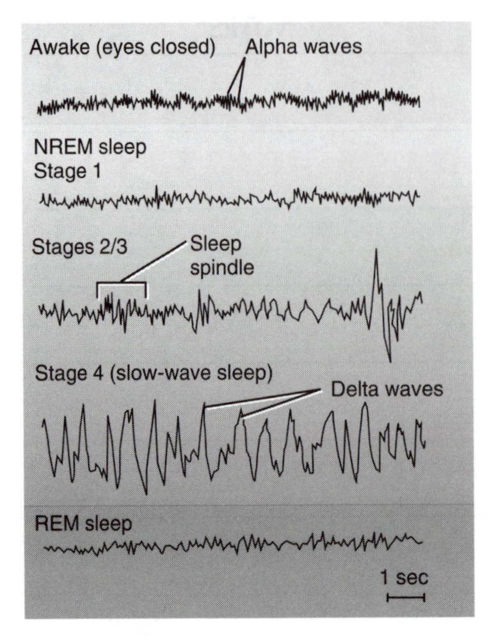
Awake (eyes closed)
Alpha waves
NREM sleep
Stage 1
Stages 2/3
Sleep spindle
Stage 4 (slow-wave sleep)
Delta waves
REM sleep
1 sec

(a) EEG waves during sleep stages

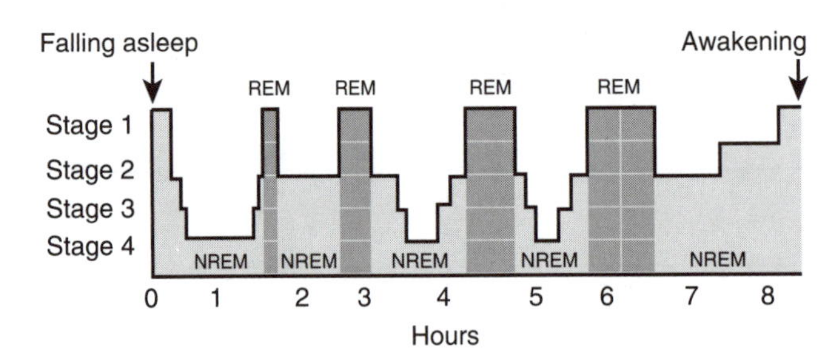
Falling asleep
Awakening
REM
REM
REM
REM
Stage 1
Stage 2
Stage 3
Stage 4
NREM
NREM
NREM
NREM
NREM
0
1
2
3
4
5
6
7
8
Hours

(b) Pattern of NREM and REM sleep over one sleep period

NOTES

Figure 16.1 Olfactory epithelium and olfactory receptors (page 513).

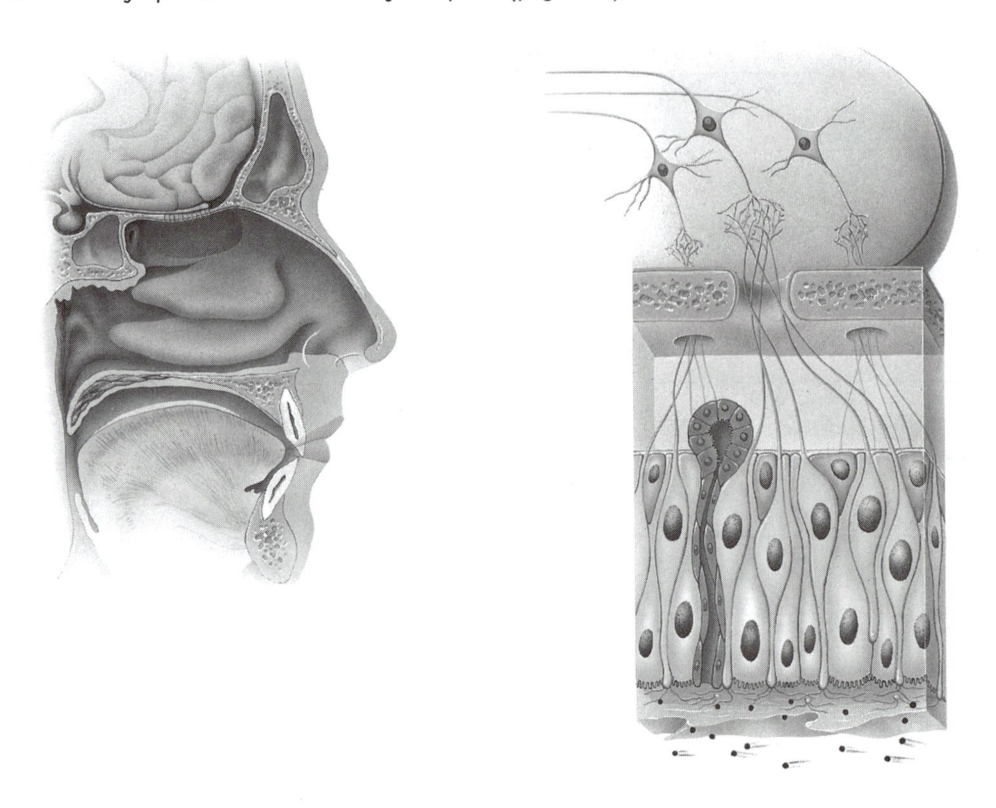

Figure 16.2 The relationship of gustatory receptors in taste buds to tongue papillae (page 515).

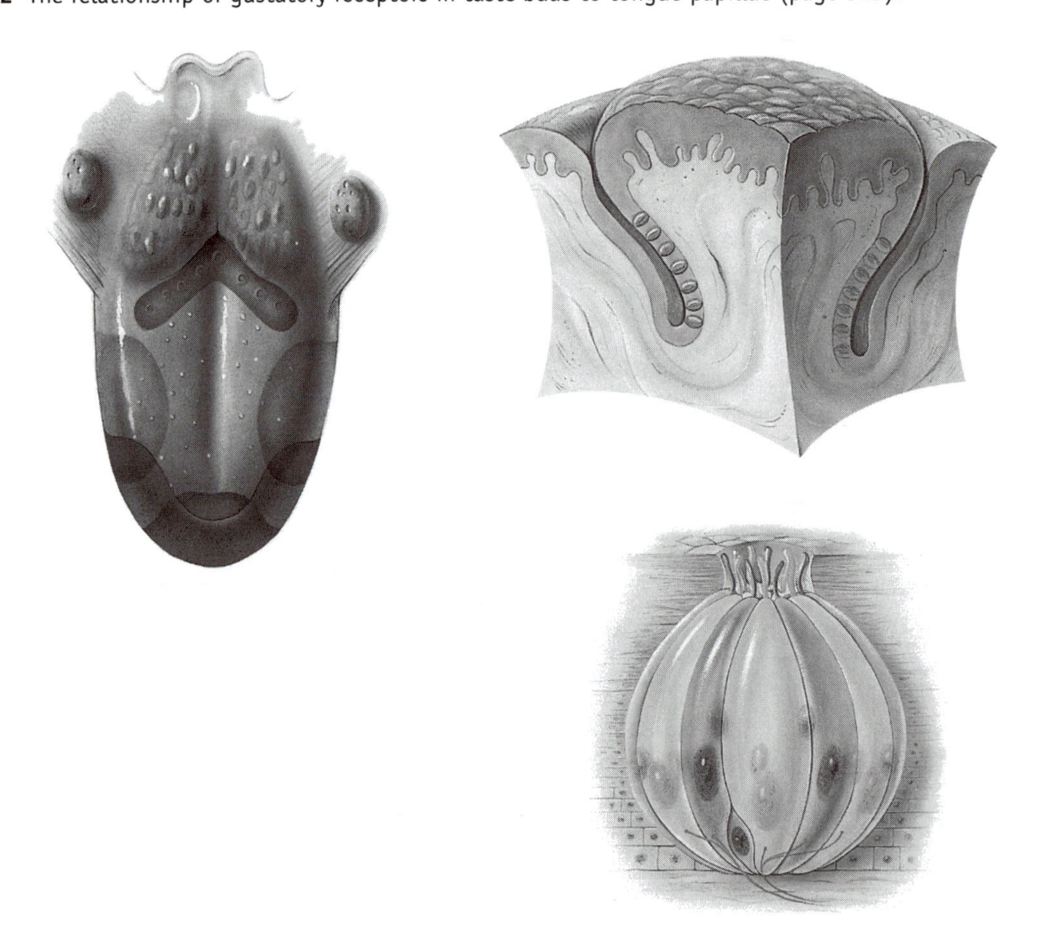

NOTES

16

Figure 16.4 Accessory structures of the eye (page 518).

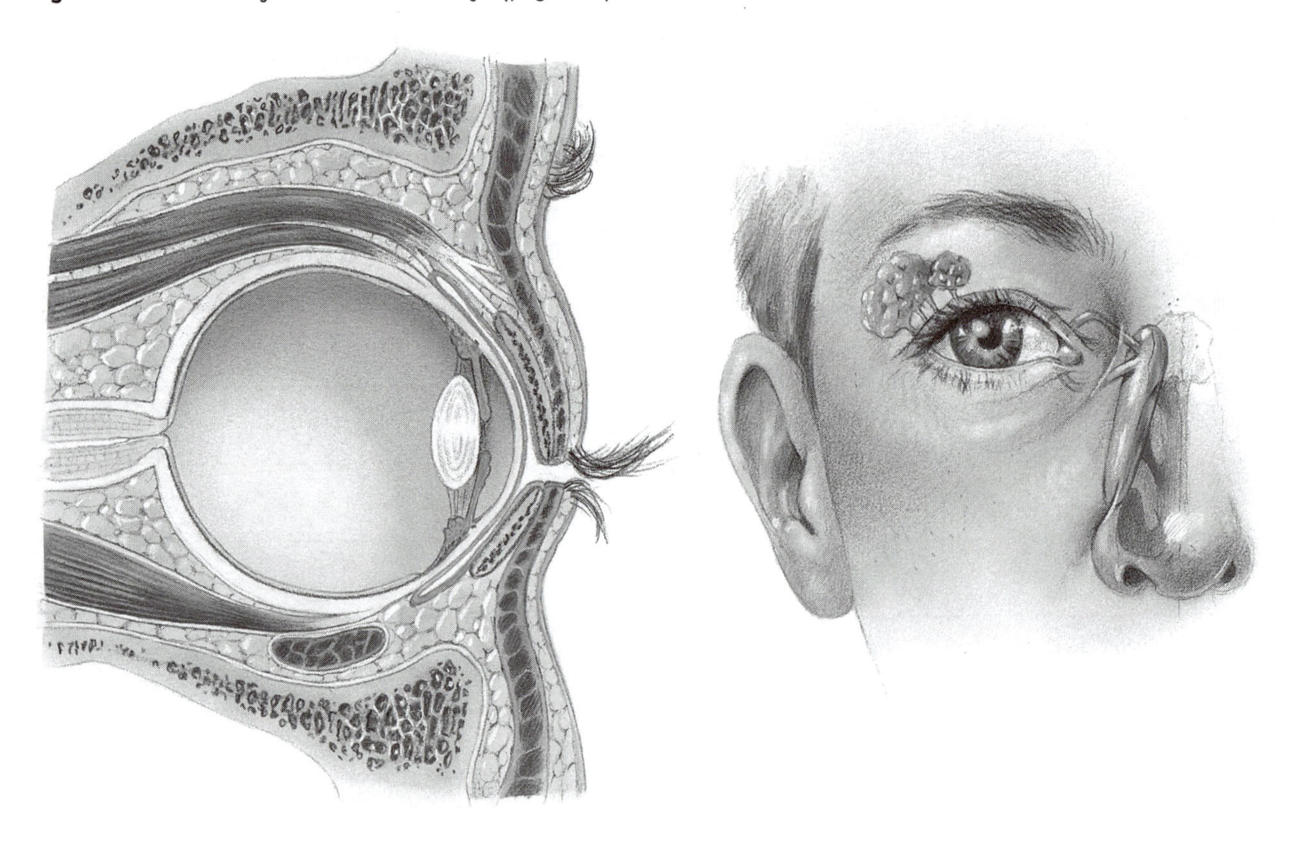

Figure 16.5 Gross structure of the eyeball (page 519).

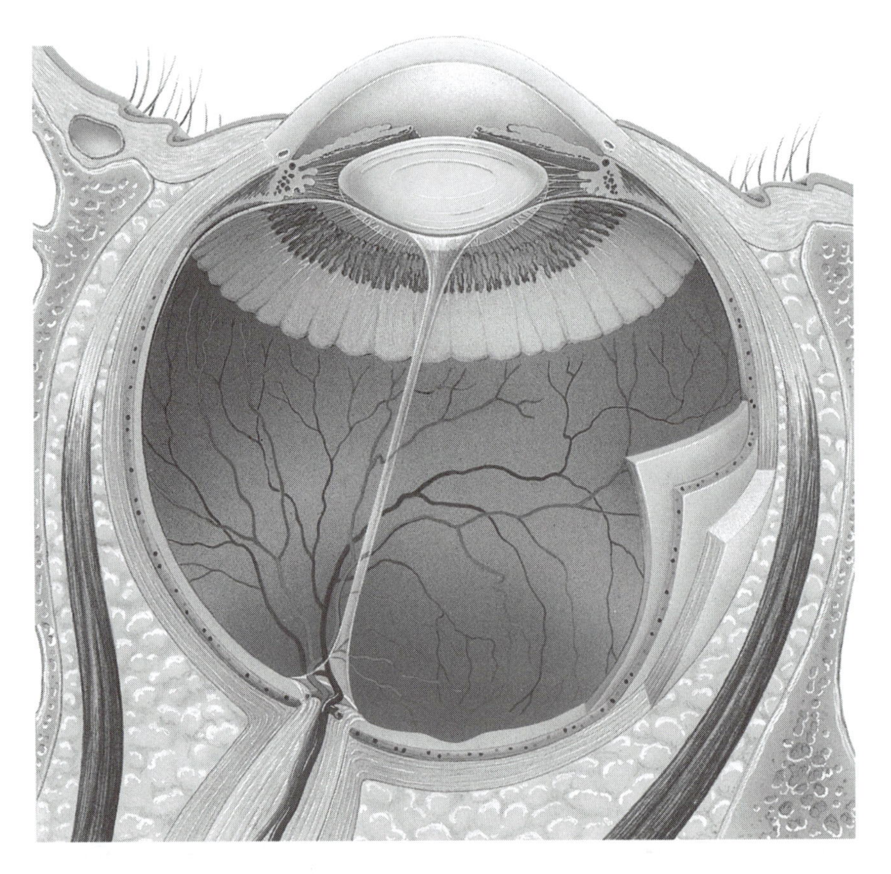

NOTES

Figure 16.6 Responses of the pupil to light of varying brightness (page 520).

Figure 16.8 Microscopic structure of the retina (page 521).

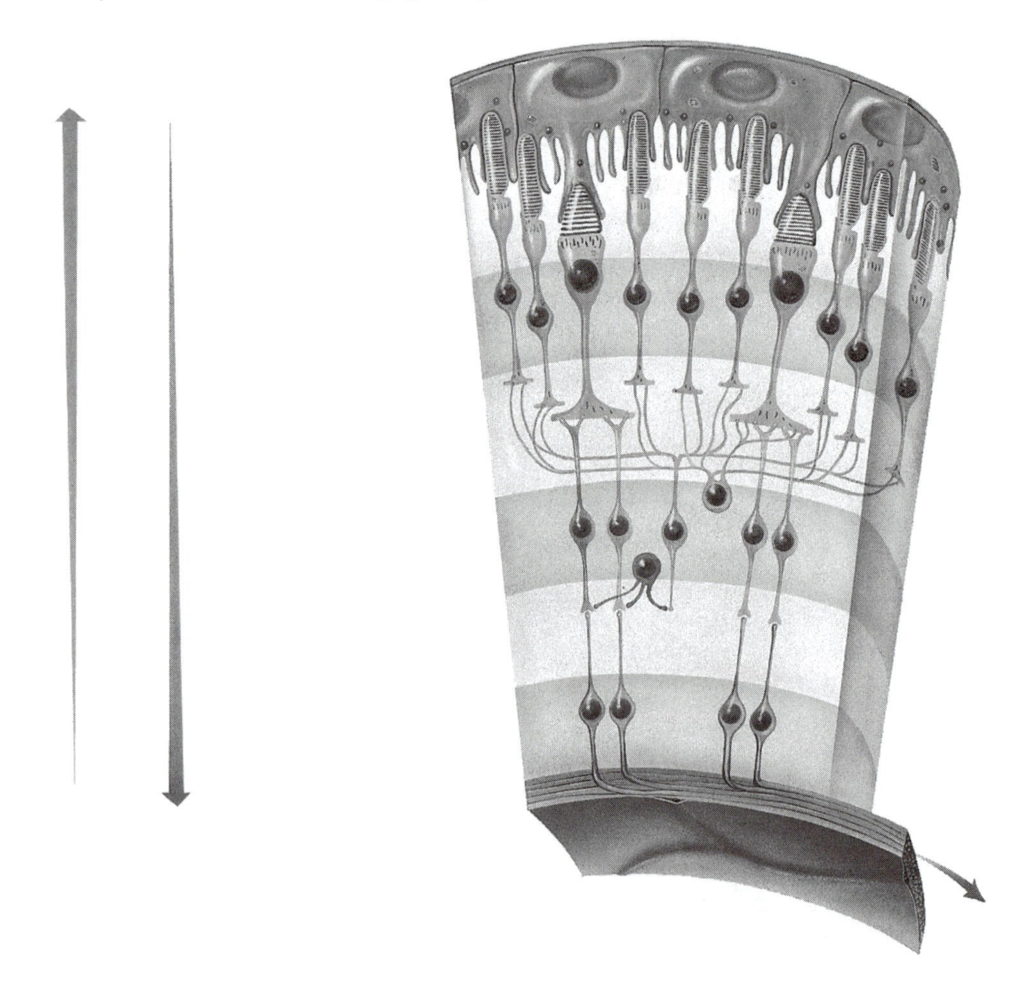

NOTES

16

Figure 16.9 The anterior and posterior chambers of the eye (page 522).

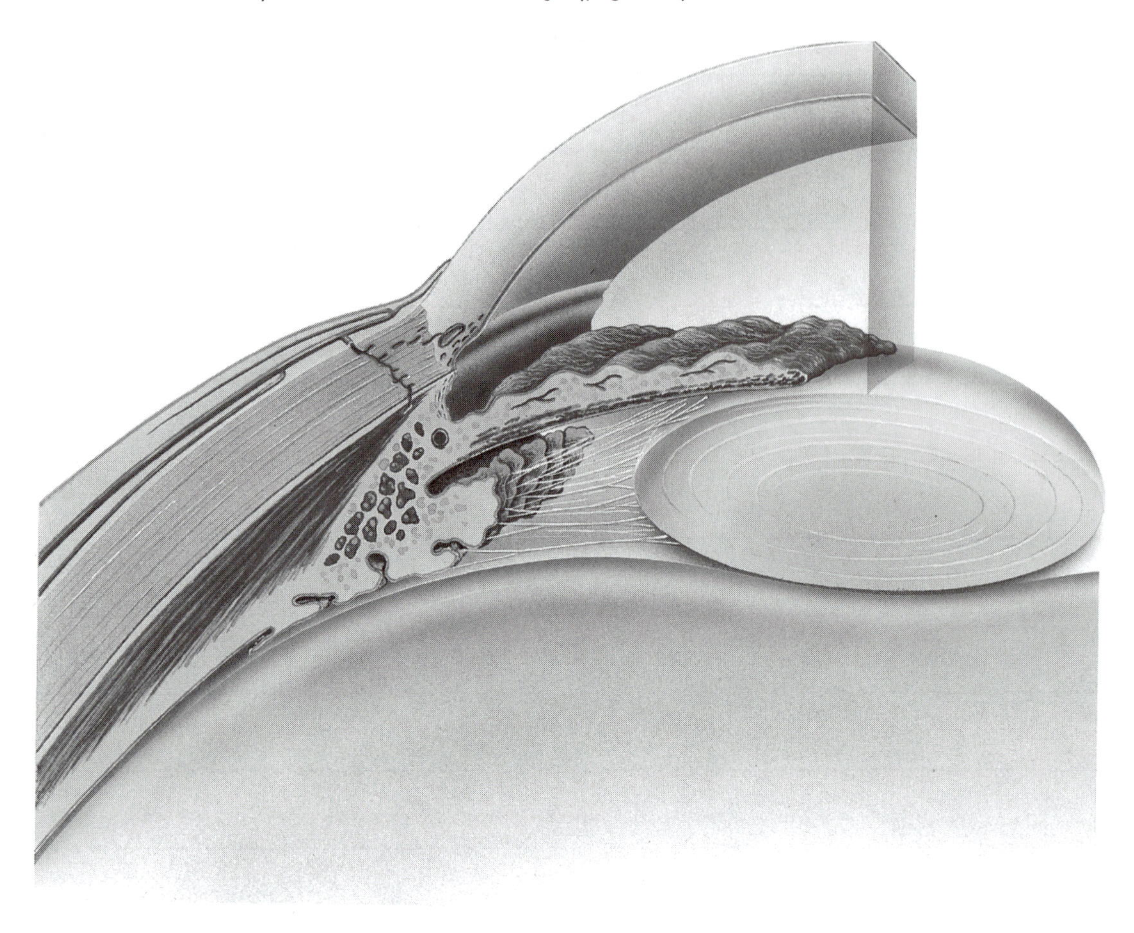

Figure 16.10 Refraction of light rays (page 524).

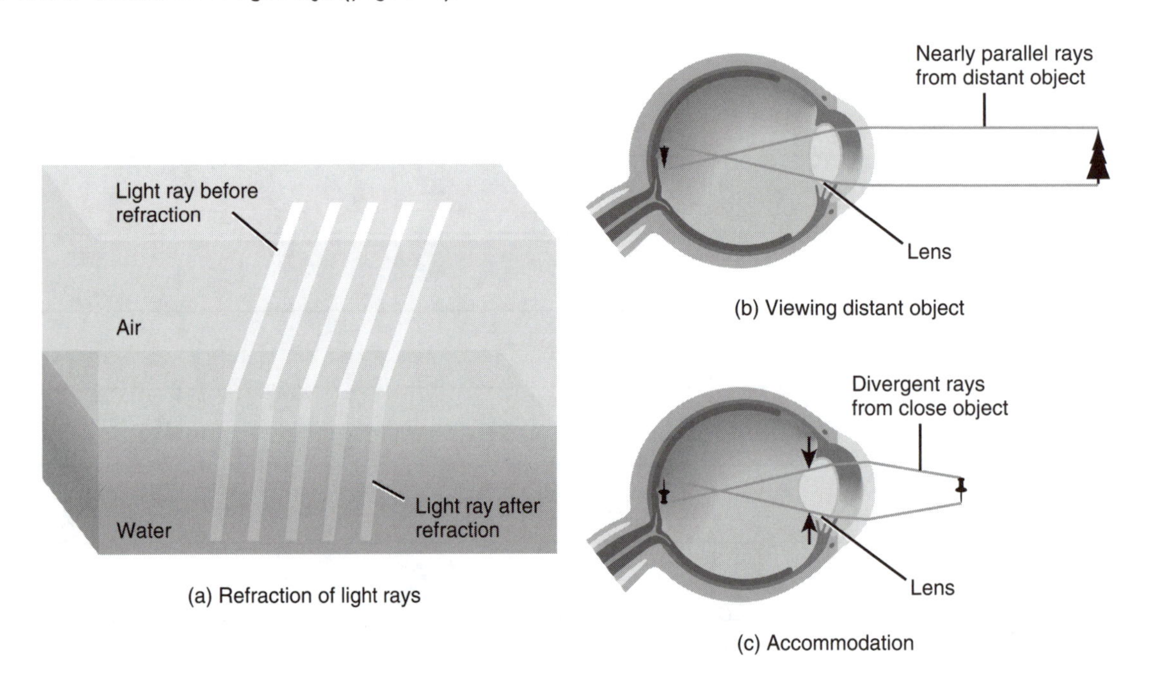

Light ray before refraction
Air
Water
Light ray after refraction
(a) Refraction of light rays
Nearly parallel rays from distant object
Lens
(b) Viewing distant object
Divergent rays from close object
Lens
(c) Accommodation

NOTES

16

Figure 16.11 Refraction abnormalities in the eyeball (page 525).

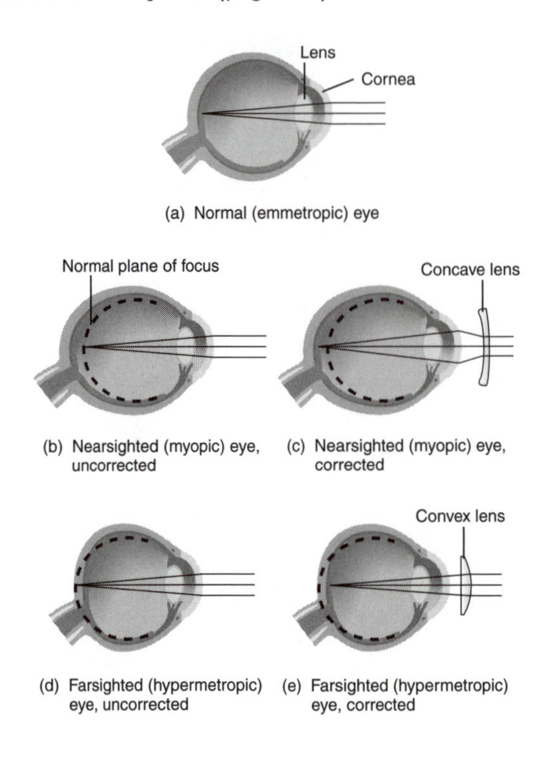
Lens
Cornea
(a) Normal (emmetropic) eye
Normal plane of focus
Concave lens
(b) Nearsighted (myopic) eye, uncorrected
(c) Nearsighted (myopic) eye, corrected
Convex lens
(d) Farsighted (hypermetropic) eye, uncorrected
(e) Farsighted (hypermetropic) eye, corrected

Figure 16.12 Structure of rod and cone photoreceptors (page 526).

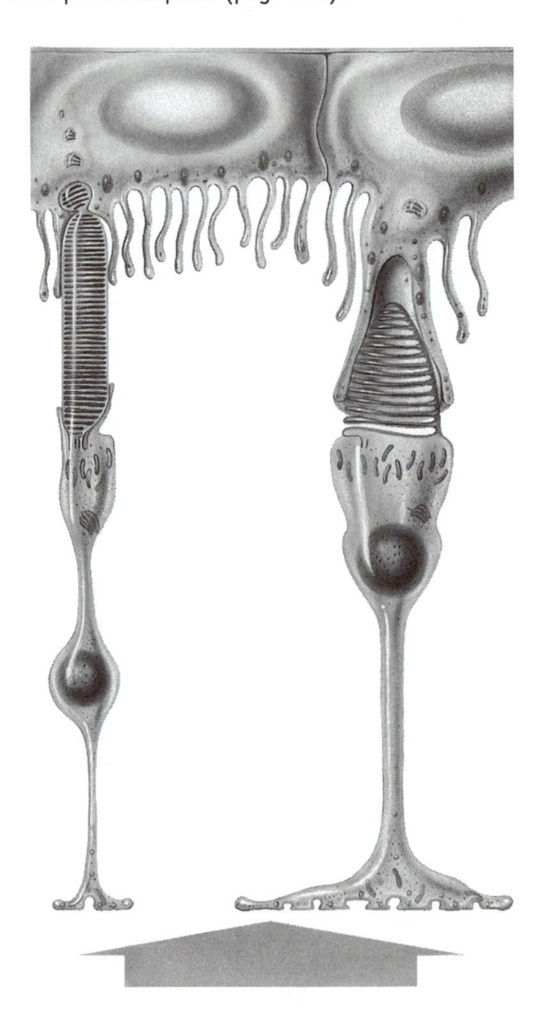

NOTES

Figure 16.13 The cyclical bleaching and regeneration of photopigment (page 527).

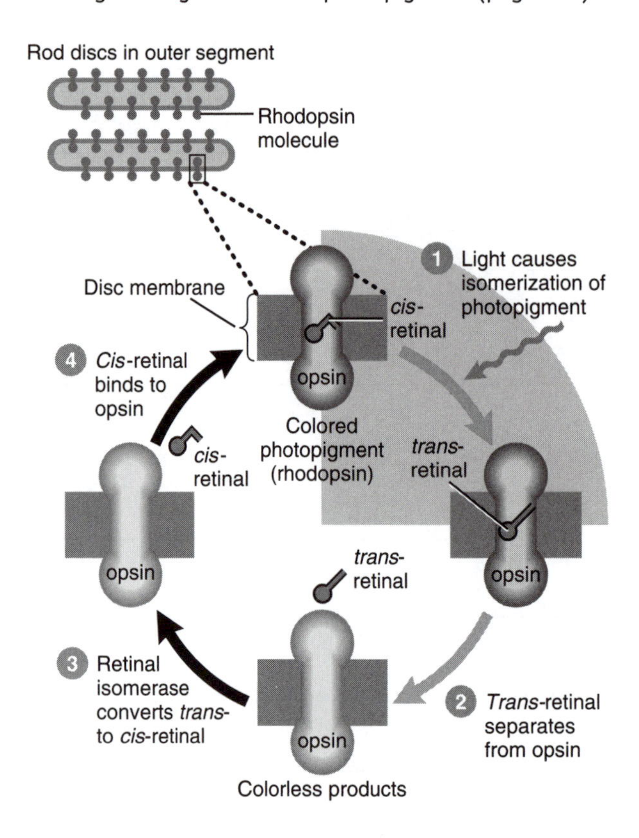
Rod discs in outer segment
Rhodopsin molecule
Disc membrane
1 Light causes isomerization of photopigment
cis-retinal
opsin
Colored photopigment (rhodopsin)
trans-retinal
opsin
4 Cis-retinal binds to opsin
cis-retinal
opsin
trans-retinal
3 Retinal isomerase converts trans- to cis-retinal
opsin
Colorless products
2 Trans-retinal separates from opsin

Figure 16.14 Operation of rod photoreceptors (page 528).

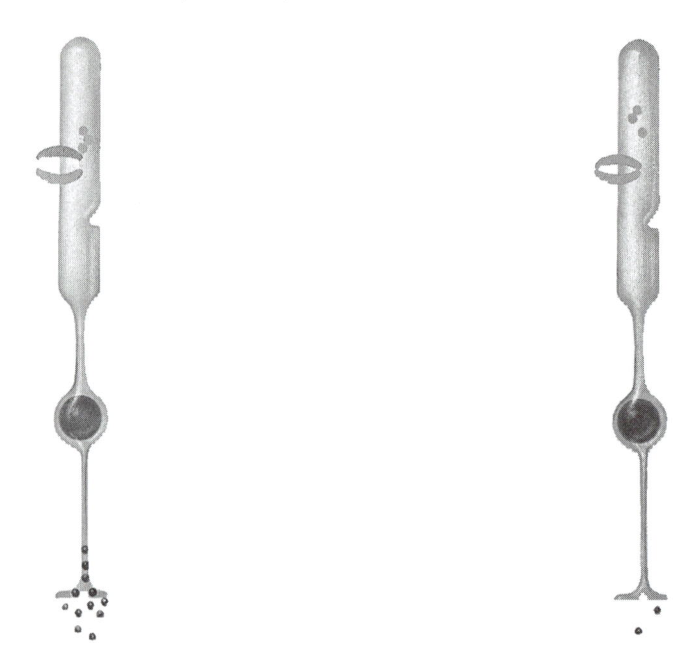

NOTES

16

Figure 16.15b-d The visual pathway (page 530).

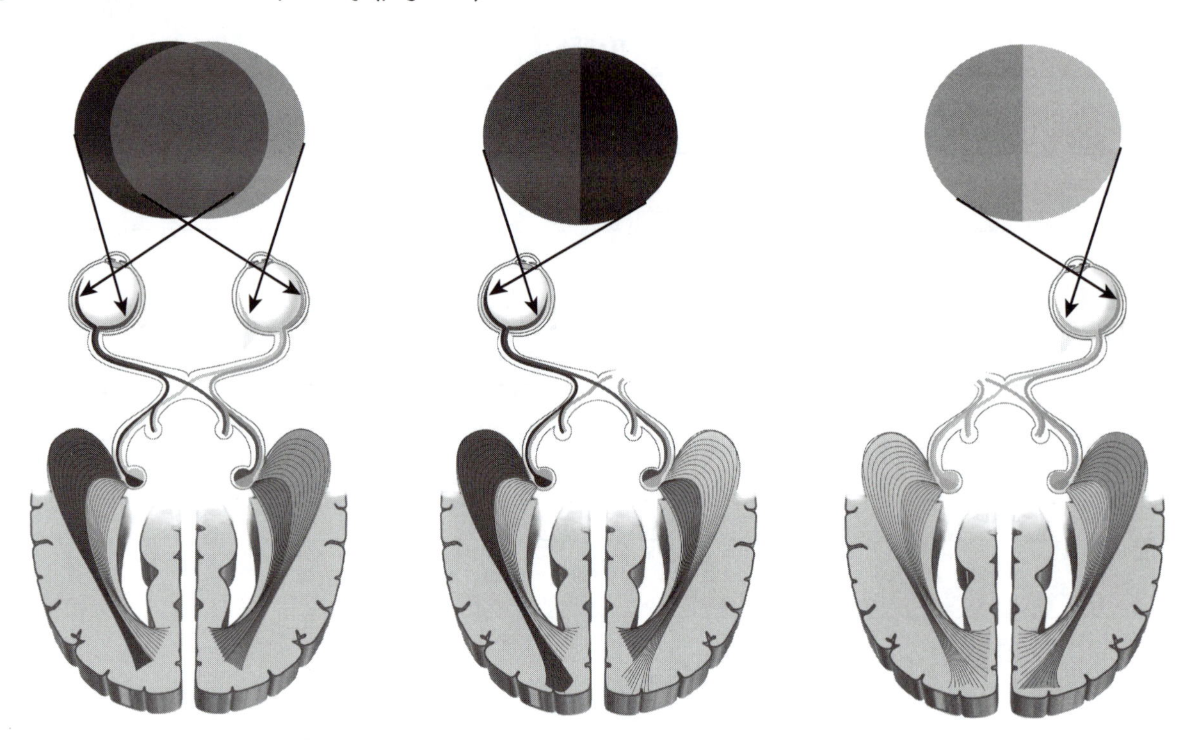

Adapted from Seeley et al., *Anatomy and Physiology* 4e, F15.22, p480; (New York: WCB McGraw-Hill, 1998) ©1998 The McGraw-Hill Companies.

Figure 16.16 Structure of the ear (page 531).

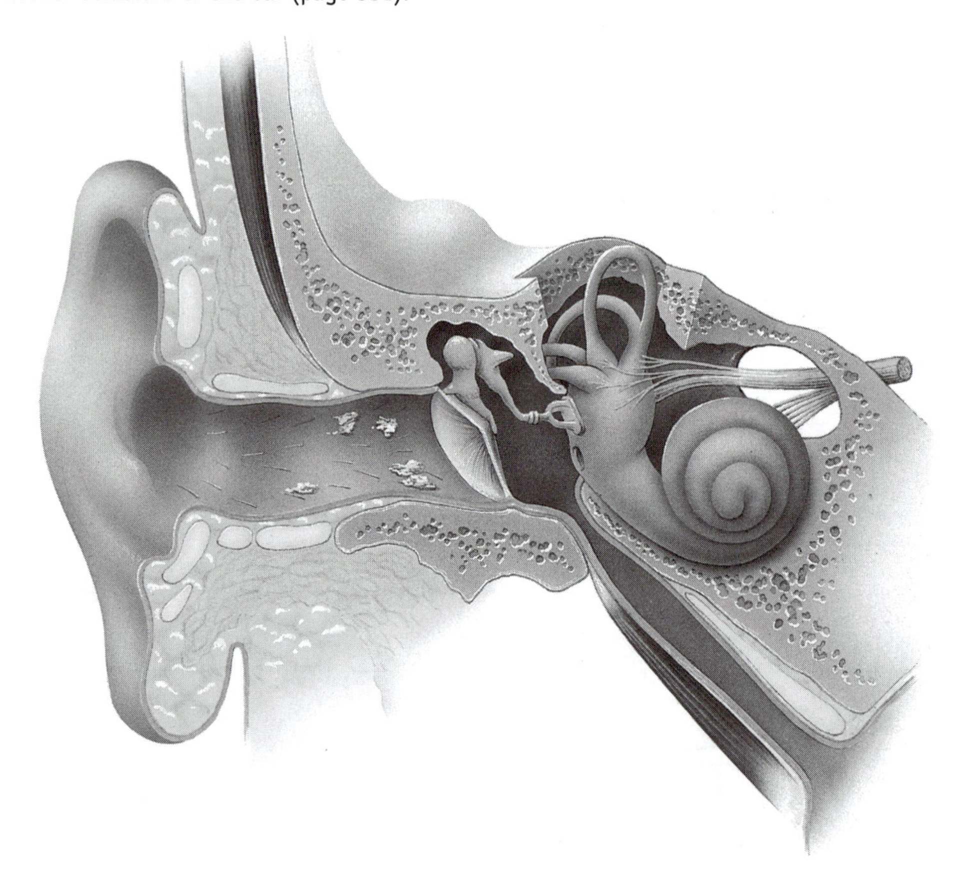

NOTES

16

Figure 16.17 The right middle ear containing the auditory ossicles (page 532).

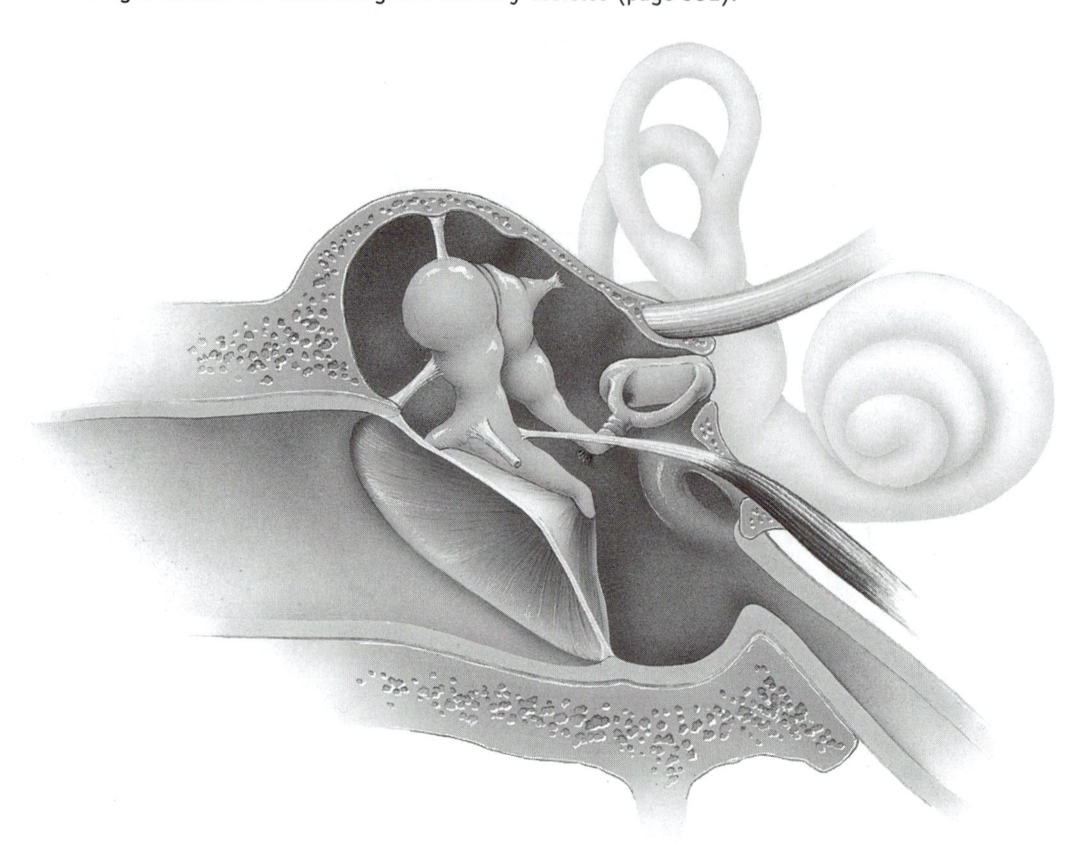

Figure 16.18 The right internal ear (page 533).

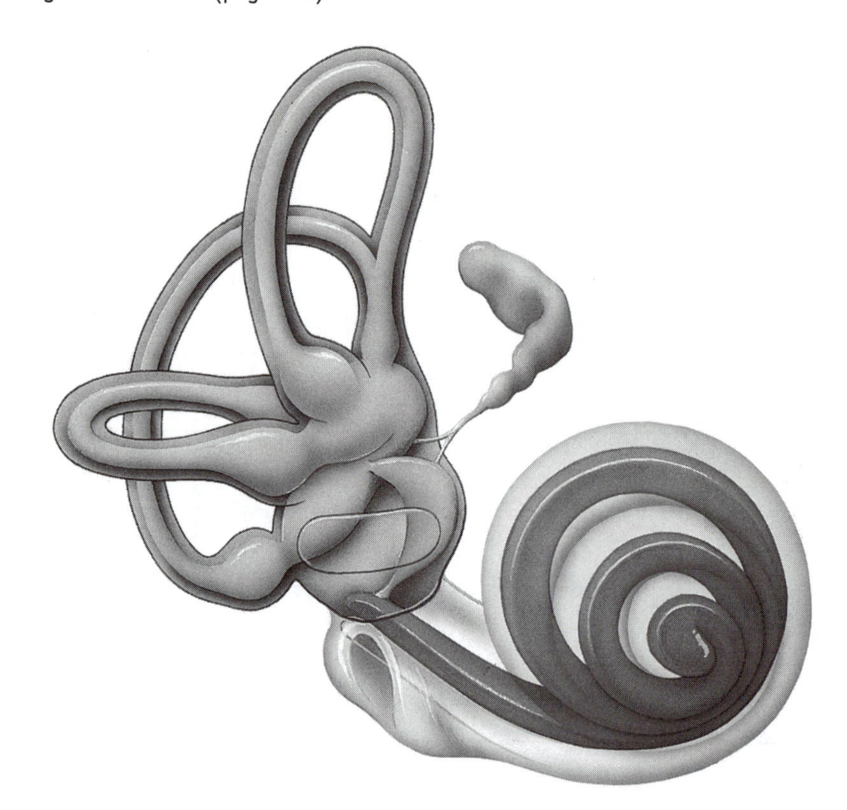

NOTES

16

Figure 16.19a-b Semicircular canals, vestibule, and cochlea of the right ear (page 536).

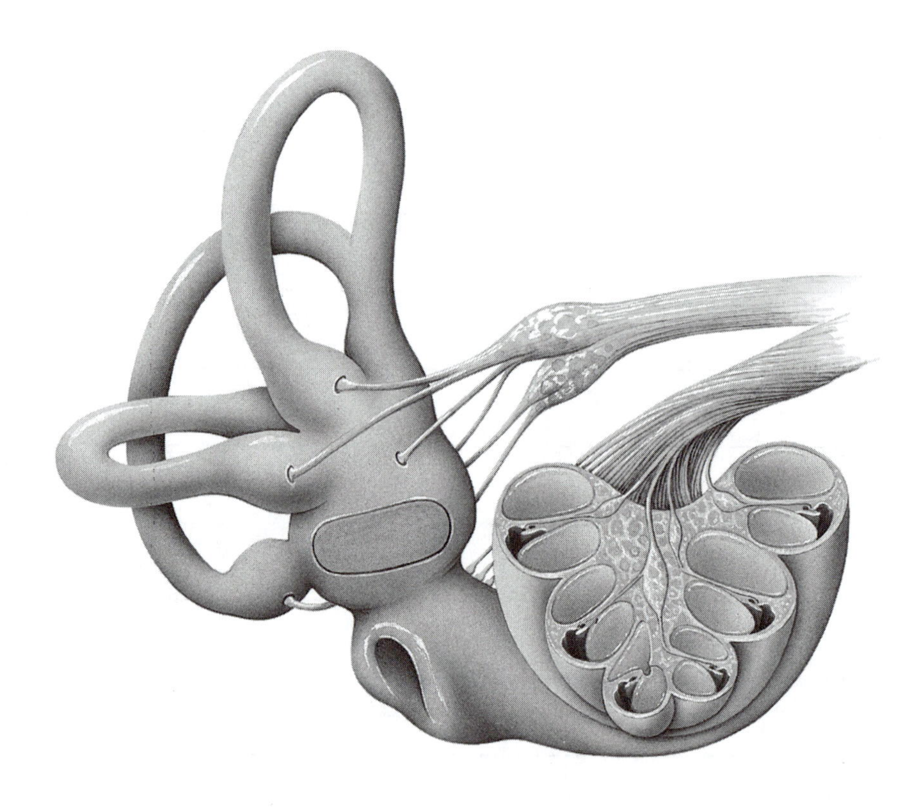

NOTES

16

Figure 16.19c-d Semicircular canals, vestibule, and cochlea of the right ear (page 537).

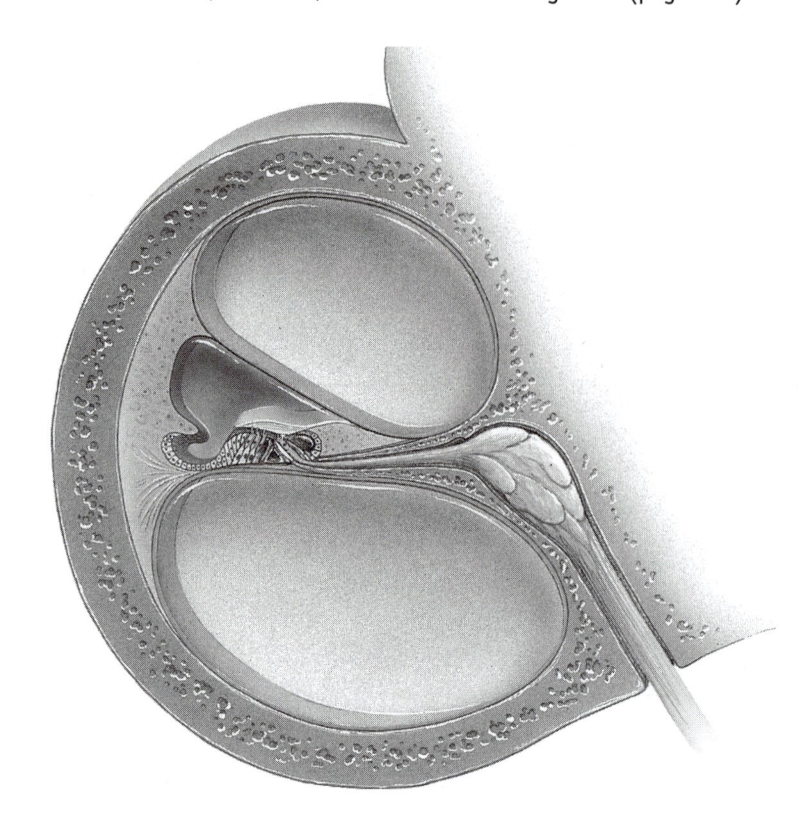

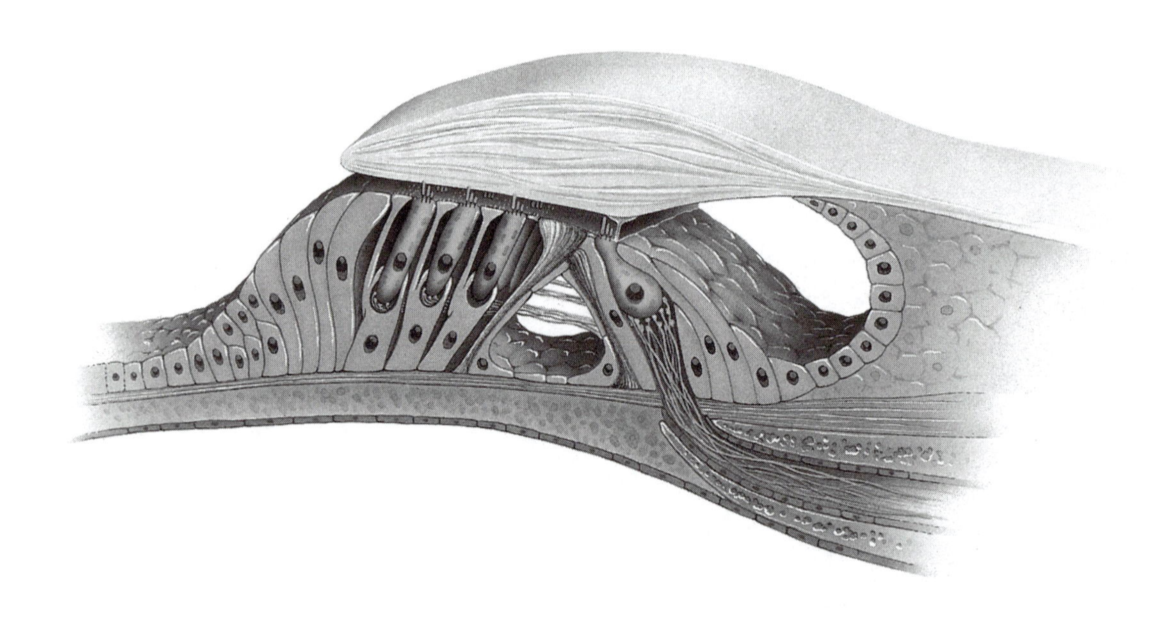

NOTES

16

Figure 16.20 Events in the stimulation of auditory receptors in the right ear (page 538).

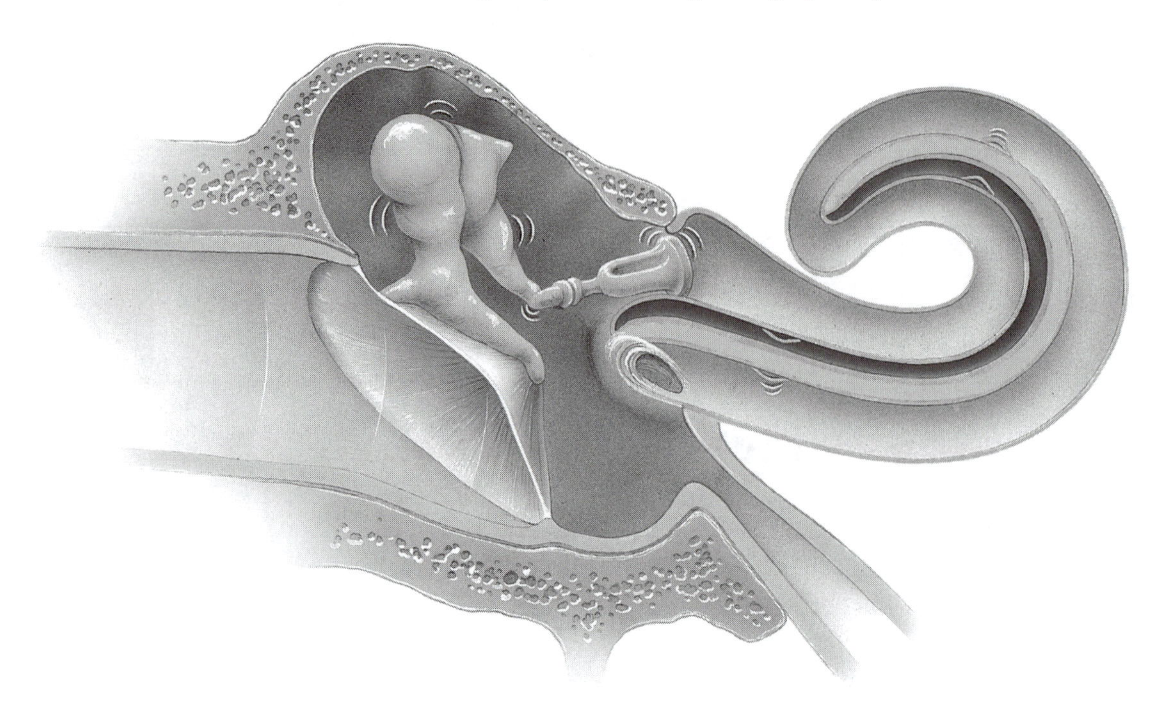

Figure 16.21a-b Location and structure of receptors in the masculae (page 539).

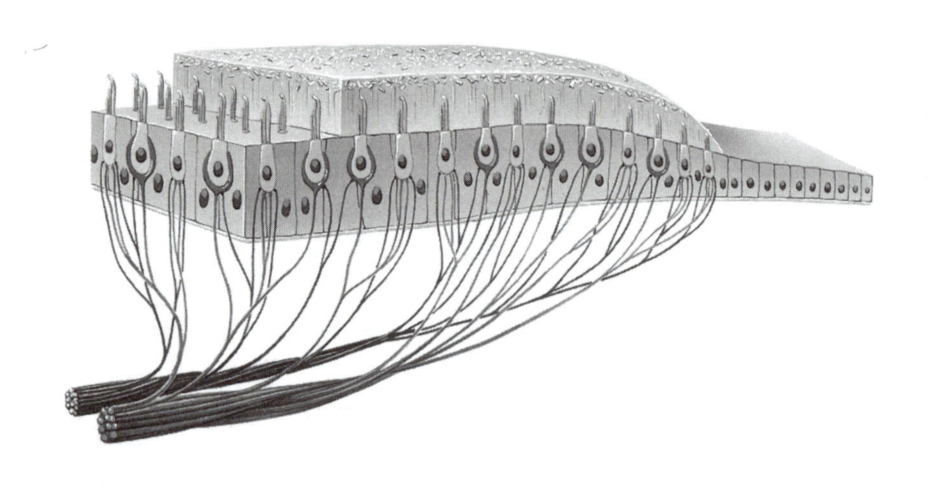

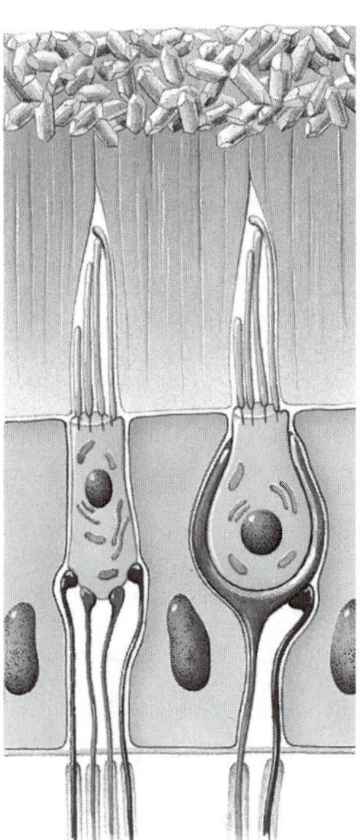

NOTES

Figure 16.22 Location and structure of the membranous semicircular ducts of the right ear (page 540).

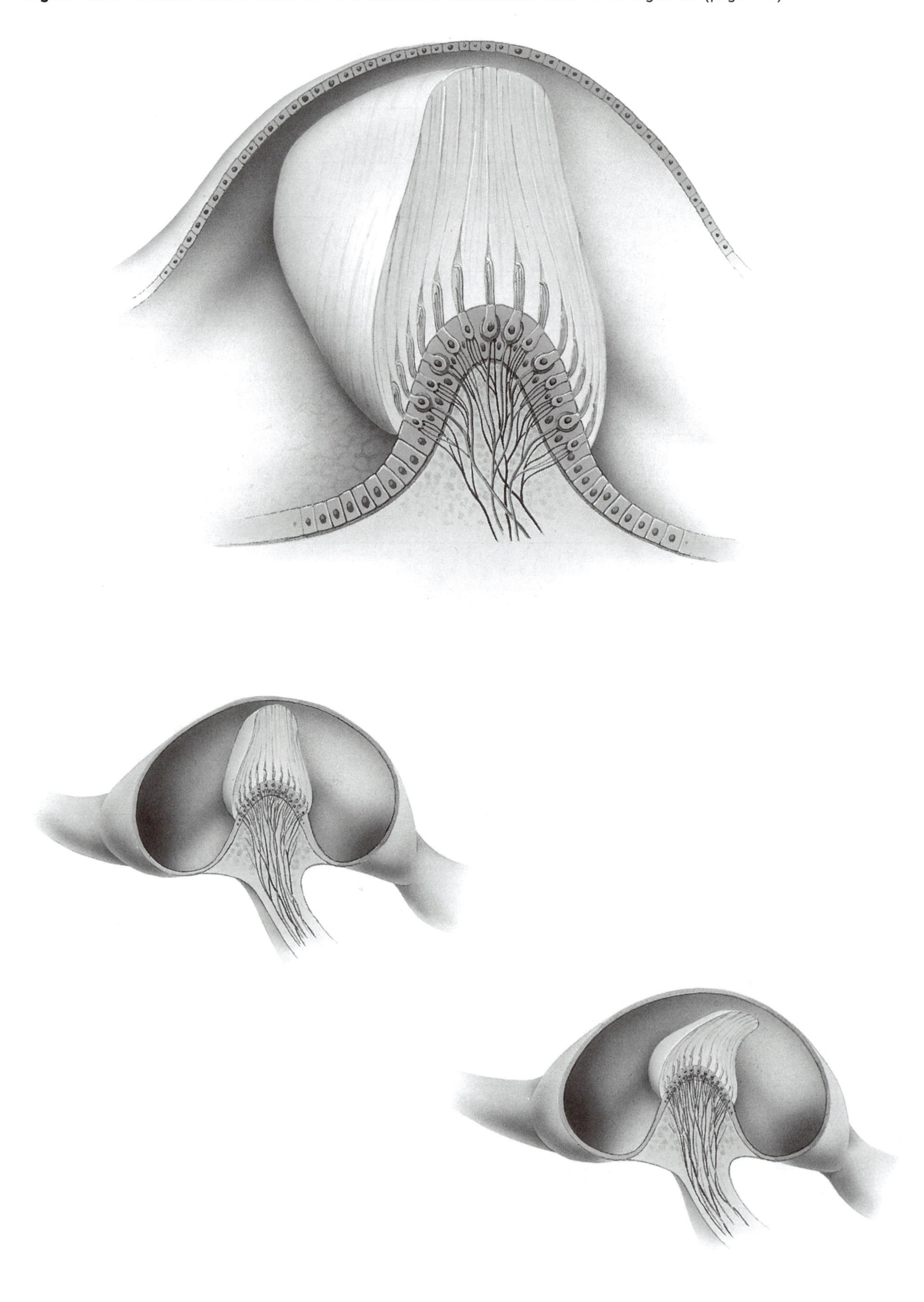

NOTES

Figure 17.1 Motor neuron pathways in the ANS (page 549).

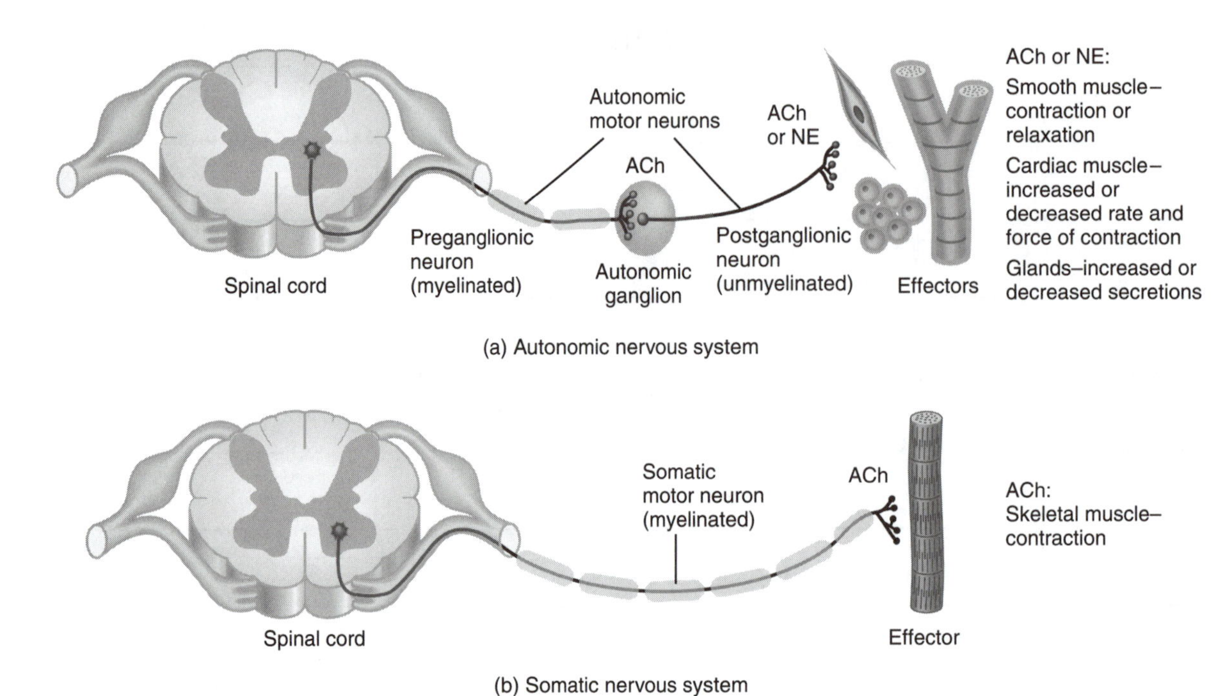
Autonomic motor neurons
ACh
ACh or NE
ACh or NE:
Smooth muscle–contraction or relaxation
Cardiac muscle–increased or decreased rate and force of contraction
Glands–increased or decreased secretions
Spinal cord
Preganglionic neuron (myelinated)
Autonomic ganglion
Postganglionic neuron (unmyelinated)
Effectors
(a) Autonomic nervous system
Somatic motor neuron (myelinated)
ACh
ACh:
Skeletal muscle–contraction
Spinal cord
Effector
(b) Somatic nervous system

NOTES

17

Figure 17.2 Structure of the sympathetic and parasympathetic divisions of the autonomic nervous system (page 551).

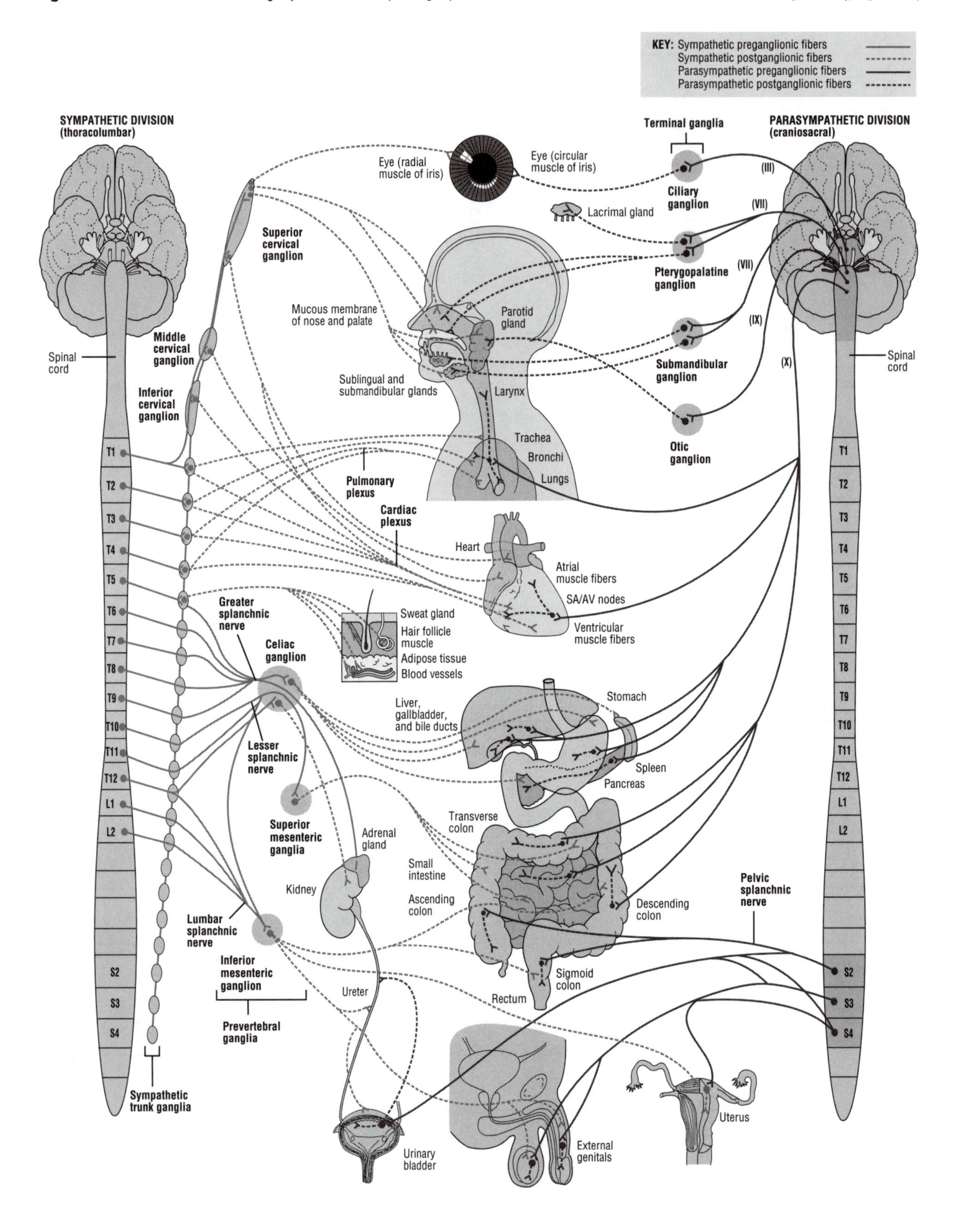

KEY: Sympathetic preganglionic fibers
Sympathetic postganglionic fibers
Parasympathetic preganglionic fibers
Parasympathetic postganglionic fibers
SYMPATHETIC DIVISION (thoracolumbar)
PARASYMPATHETIC DIVISION (craniosacral)
Terminal ganglia
Eye (radial muscle of iris)
Eye (circular muscle of iris)
Ciliary ganglion
Lacrimal gland
(III)
(VII)
(VII)
(IX)
(X)
Superior cervical ganglion
Pterygopalatine ganglion
Mucous membrane of nose and palate
Parotid gland
Submandibular ganglion
Spinal cord
Middle cervical ganglion
Inferior cervical ganglion
Sublingual and submandibular glands
Larynx
Otic ganglion
Trachea
Bronchi
Lungs
Pulmonary plexus
Cardiac plexus
Heart
Atrial muscle fibers
SA/AV nodes
Ventricular muscle fibers
Greater splanchnic nerve
Celiac ganglion
Sweat gland
Hair follicle muscle
Adipose tissue
Blood vessels
Liver, gallbladder, and bile ducts
Stomach
Spleen
Pancreas
Lesser splanchnic nerve
Superior mesenteric ganglia
Adrenal gland
Transverse colon
Small intestine
Kidney
Ascending colon
Descending colon
Pelvic splanchnic nerve
Lumbar splanchnic nerve
Inferior mesenteric ganglion
Prevertebral ganglia
Ureter
Sigmoid colon
Rectum
Sympathetic trunk ganglia
Uterus
Urinary bladder
External genitals
T1 T2 T3 T4 T5 T6 T7 T8 T9 T10 T11 T12 L1 L2 S2 S3 S4

NOTES

17

Figure 17.3 Autonomic plexuses in the thorax, abdomen, and pelvis (page 552).

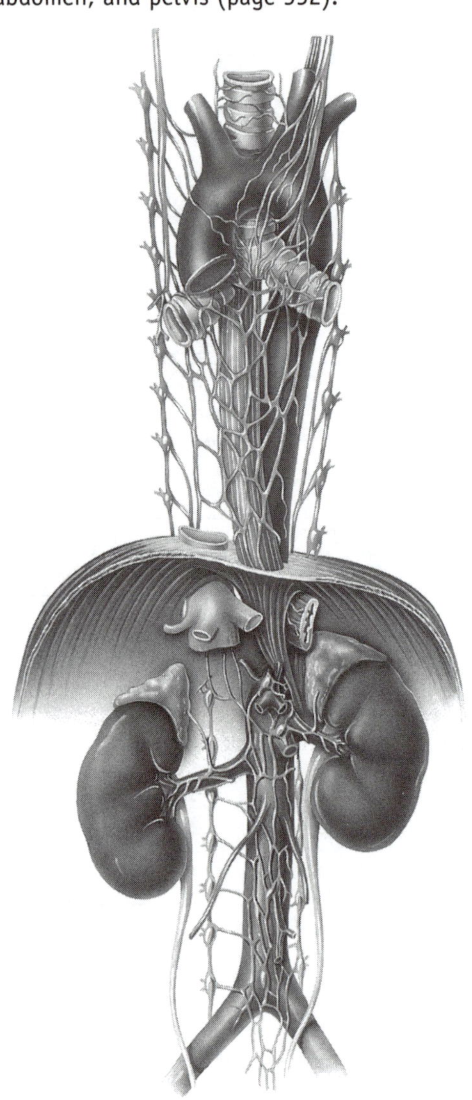

Figure 17.4 Types of connections between ganglia and postganglionic neurons (page 554).

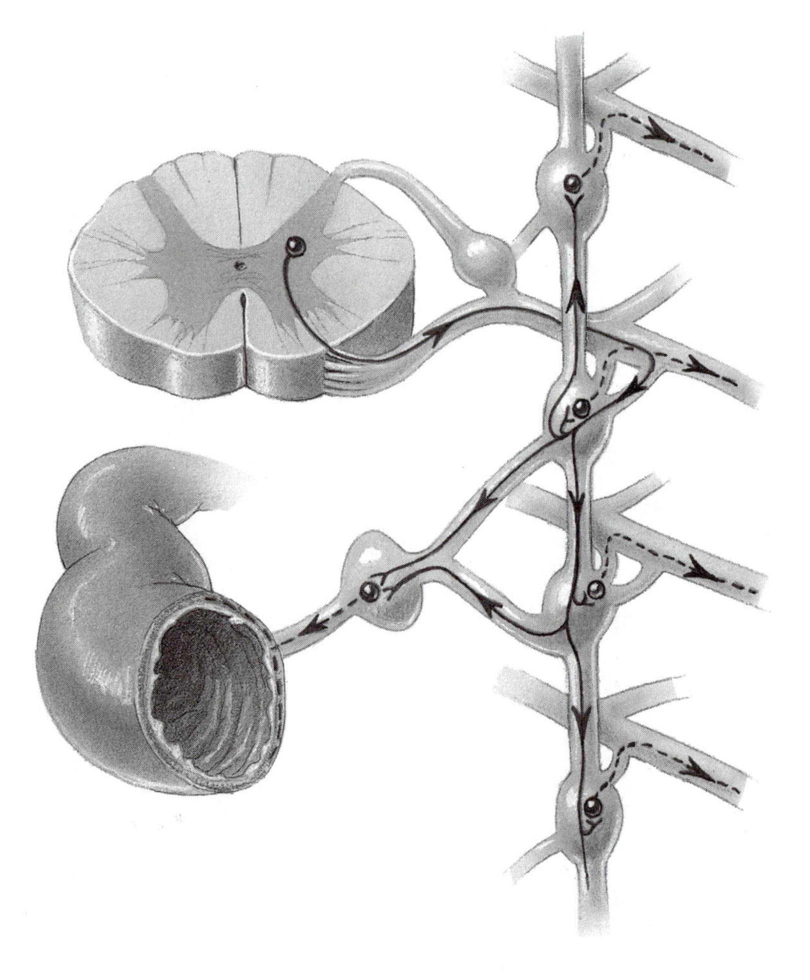

NOTES

17

Figure 17.5 Cholinergic neurons and adrenergic neurons (page 557).

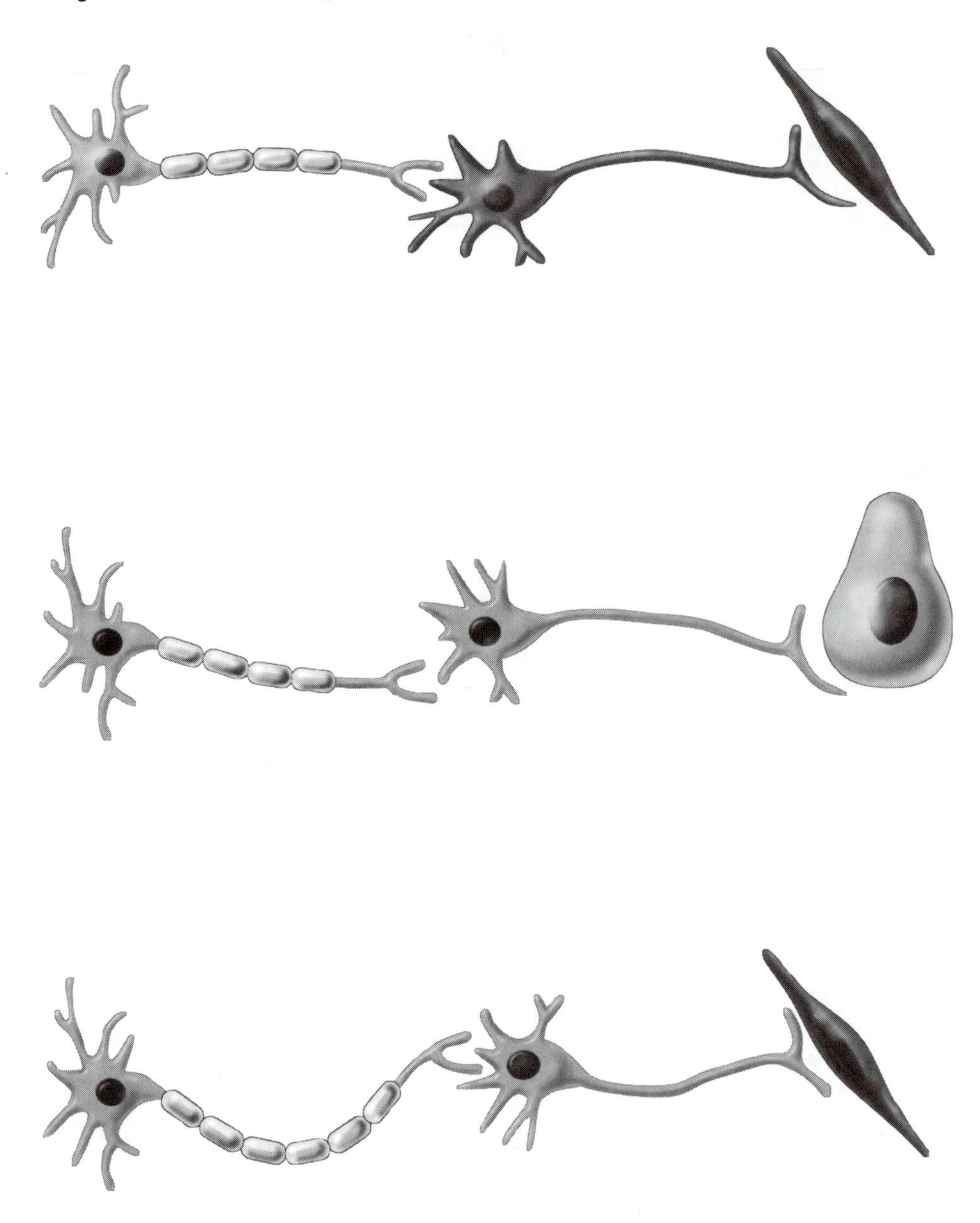

NOTES

17

Figure 18.1 Location of many endocrine glands (page 567).

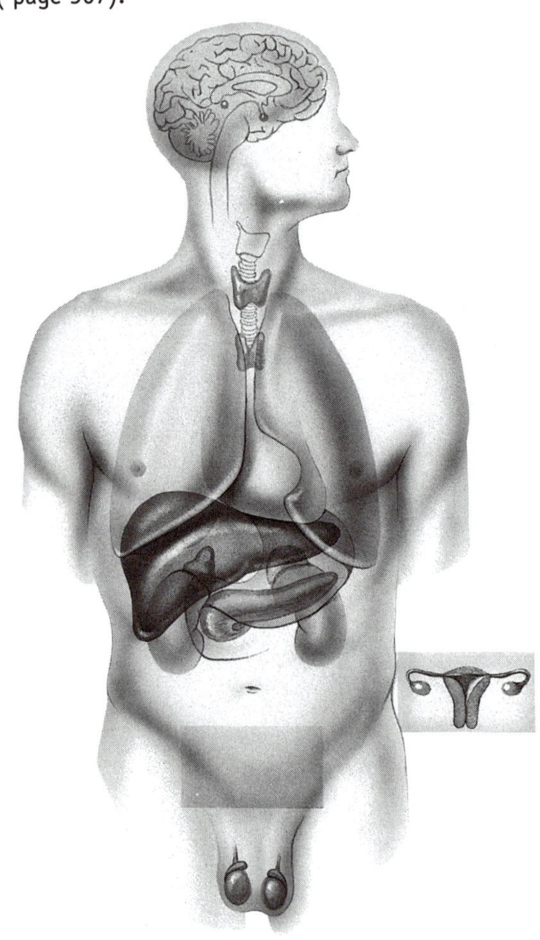

Figure 18.2 Comparison of circulating hormones and local hormones (page 568).

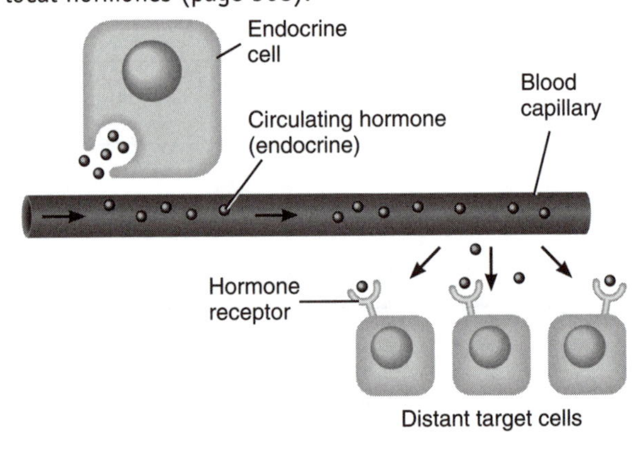
Endocrine cell
Blood capillary
Circulating hormone (endocrine)
Hormone receptor
Distant target cells

(a) Circulating hormones (endocrines)

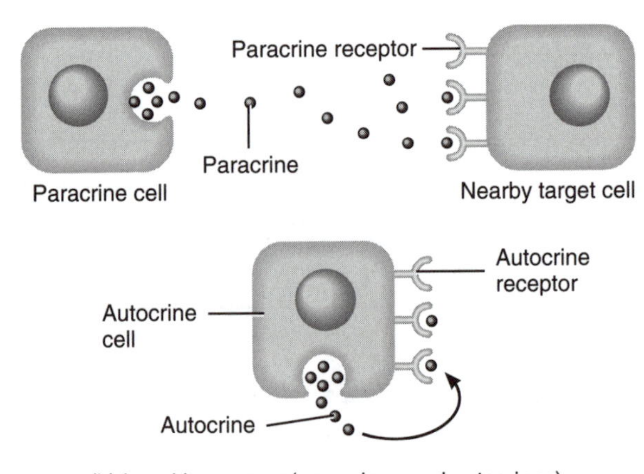
Paracrine receptor
Paracrine
Paracrine cell
Nearby target cell
Autocrine receptor
Autocrine cell
Autocrine

(b) Local hormones (paracrines and autocrines)

NOTES

18

Figure 18.4 Mechanism of action of the water-soluble hormones (page 571).

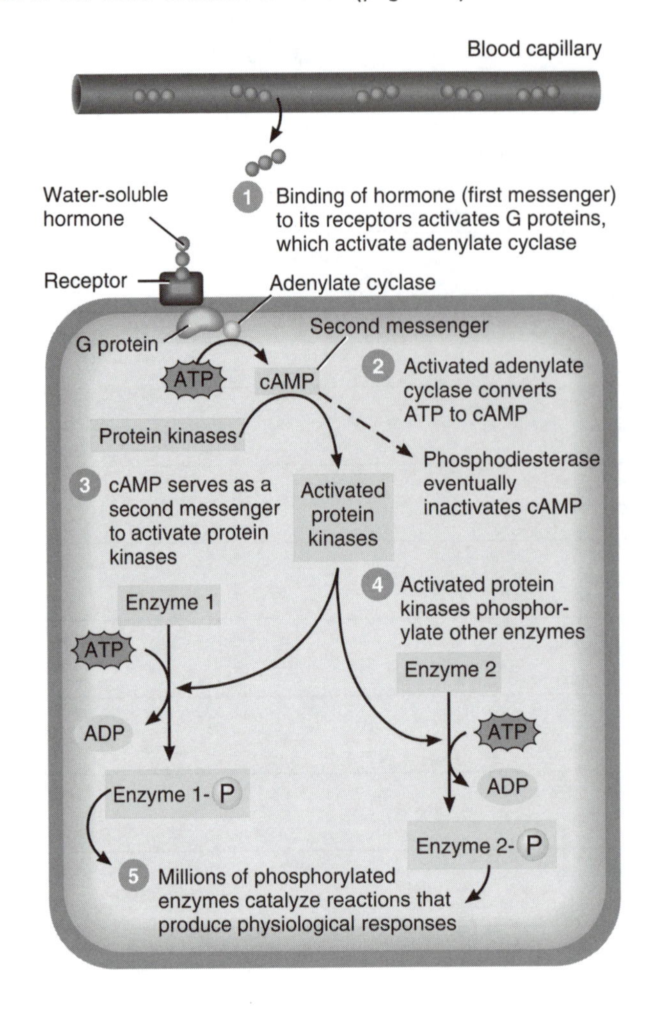
Blood capillary
Water-soluble hormone
1 Binding of hormone (first messenger) to its receptors activates G proteins, which activate adenylate cyclase
Receptor
Adenylate cyclase
G protein
Second messenger
ATP
cAMP
2 Activated adenylate cyclase converts ATP to cAMP
Protein kinases
Phosphodiesterase eventually inactivates cAMP
3 cAMP serves as a second messenger to activate protein kinases
Activated protein kinases
4 Activated protein kinases phosphorylate other enzymes
Enzyme 1
ATP
ADP
Enzyme 2
ATP
ADP
Enzyme 1- P
Enzyme 2- P
5 Millions of phosphorylated enzymes catalyze reactions that produce physiological responses

Figure 18.5 Hypothalamus and pituitary gland (page 574).

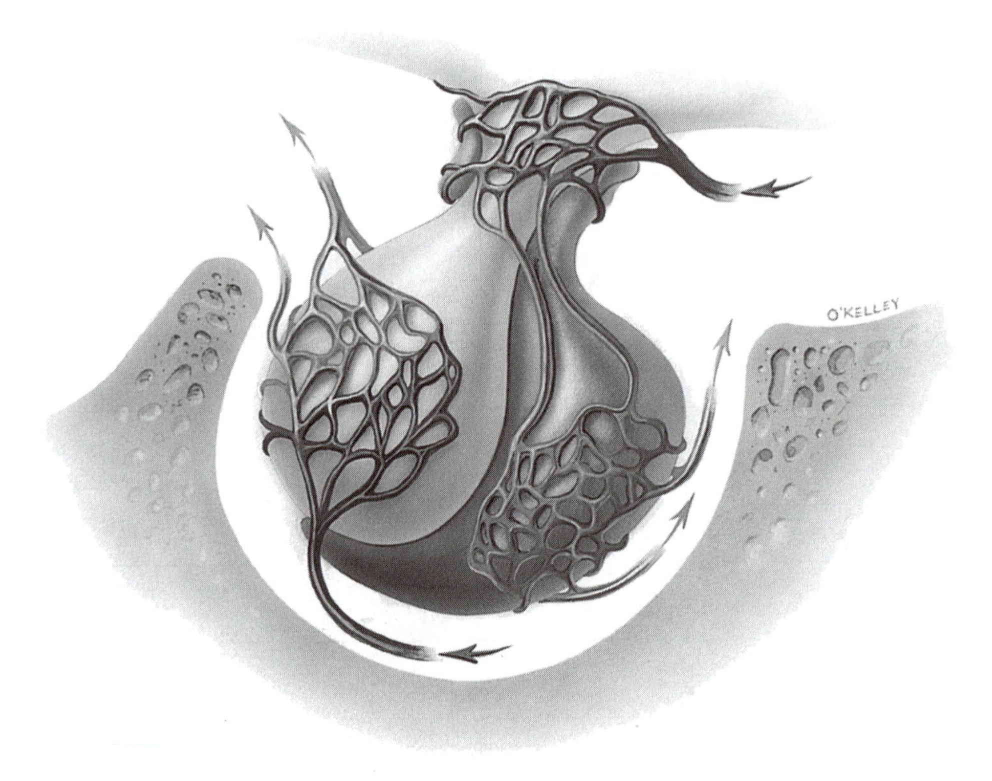
O'KELLEY

NOTES

18

Figure 18.6 Negative feedback regulation of hypothalamic neurosecretory cells and anterior pituitary corticotrophs (page 576).

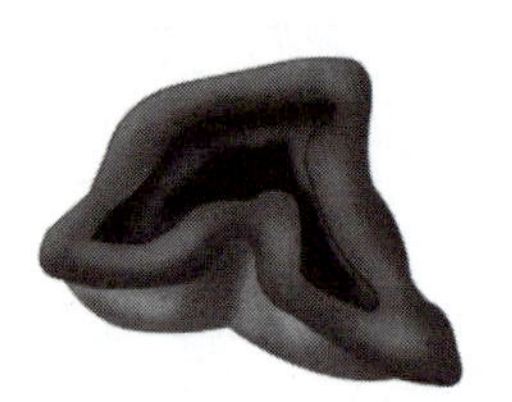

Figure 18.7 Effects of human growth hormone and insulinlike growth factors (page 576).

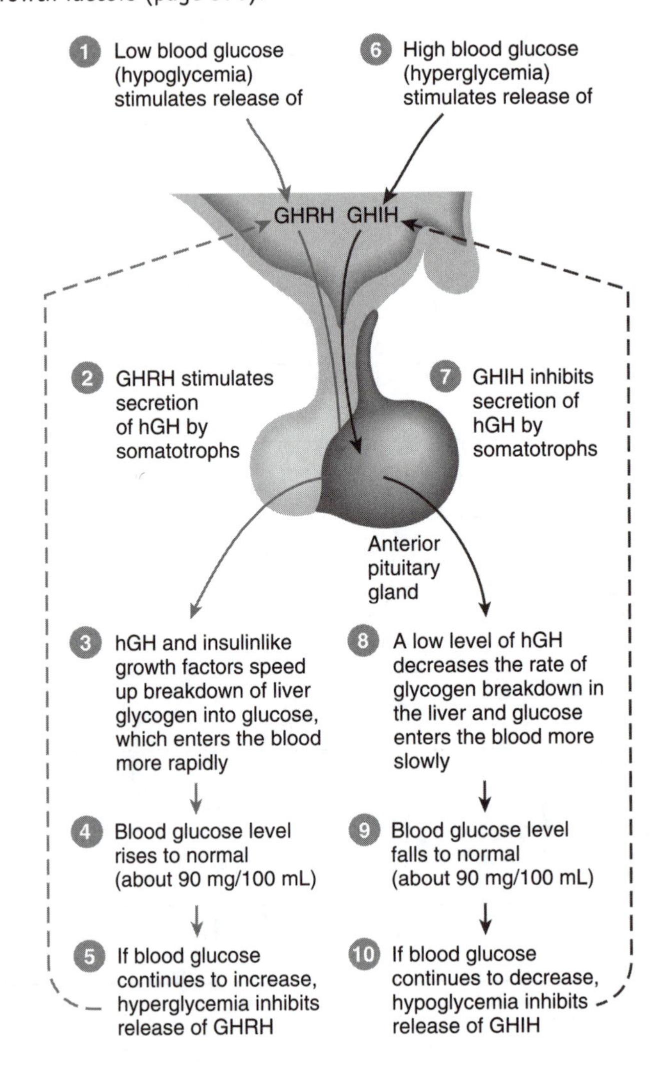

1 Low blood glucose (hypoglycemia) stimulates release of
6 High blood glucose (hyperglycemia) stimulates release of
GHRH GHIH
2 GHRH stimulates secretion of hGH by somatotrophs
7 GHIH inhibits secretion of hGH by somatotrophs
Anterior pituitary gland
3 hGH and insulinlike growth factors speed up breakdown of liver glycogen into glucose, which enters the blood more rapidly
8 A low level of hGH decreases the rate of glycogen breakdown in the liver and glucose enters the blood more slowly
4 Blood glucose level rises to normal (about 90 mg/100 mL)
9 Blood glucose level falls to normal (about 90 mg/100 mL)
5 If blood glucose continues to increase, hyperglycemia inhibits release of GHRH
10 If blood glucose continues to decrease, hypoglycemia inhibits release of GHIH

NOTES

18

Figure 18.8 Axons of hypothalamic neurosecretory cells form the hypothalamohypophyseal tract (page 579).

Figure 18.9 Regulation of secretion and actions of antidiuretic hormone (page 580).

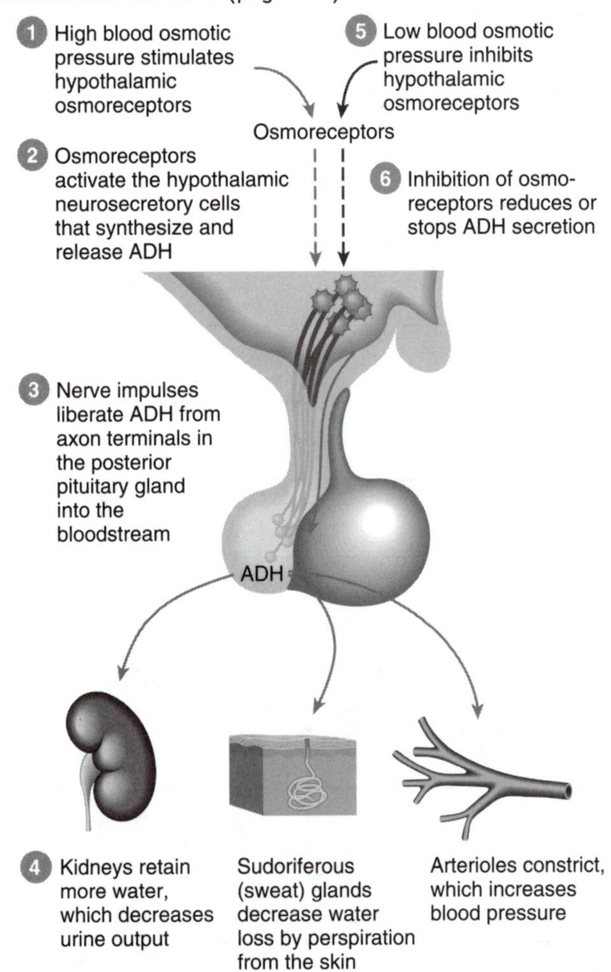

1 High blood osmotic pressure stimulates hypothalamic osmoreceptors
5 Low blood osmotic pressure inhibits hypothalamic osmoreceptors
Osmoreceptors
2 Osmoreceptors activate the hypothalamic neurosecretory cells that synthesize and release ADH
6 Inhibition of osmoreceptors reduces or stops ADH secretion
3 Nerve impulses liberate ADH from axon terminals in the posterior pituitary gland into the bloodstream
ADH
4 Kidneys retain more water, which decreases urine output
Sudoriferous (sweat) glands decrease water loss by perspiration from the skin
Arterioles constrict, which increases blood pressure

NOTES

18

Figure 18.10 Location, blood supply, and histology of the thyroid gland (page 582).

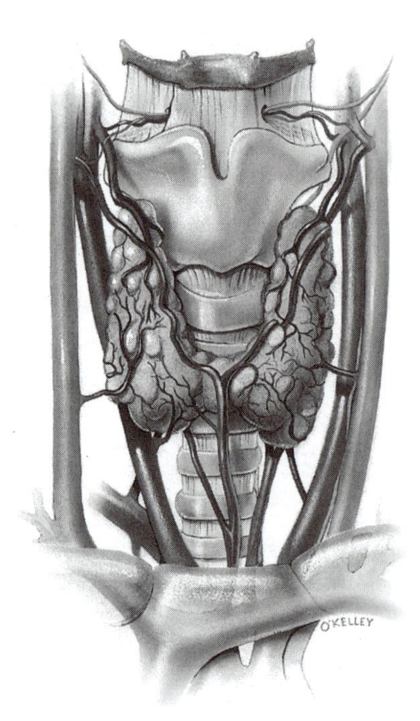
O'KELLEY

Figure 18.11 Steps in the synthesis and secretion of thyroid hormones (page 583).

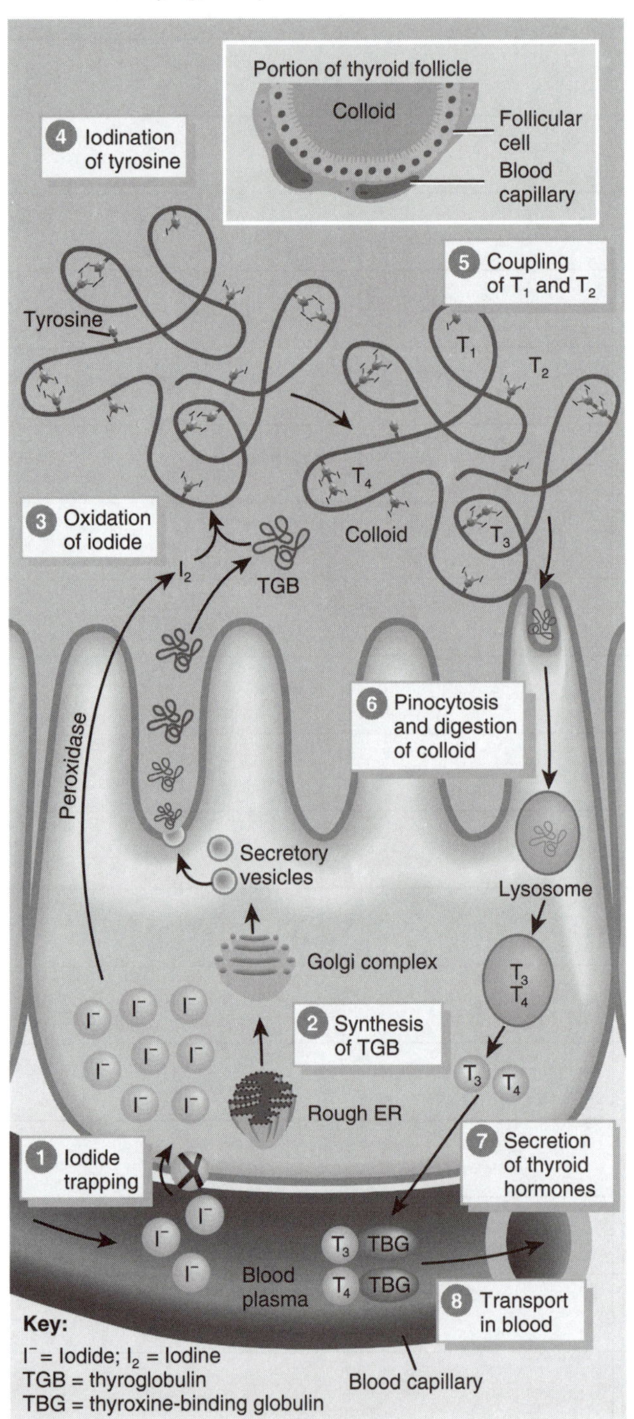
Portion of thyroid follicle
Colloid
Follicular cell
Blood capillary
4 Iodination of tyrosine
5 Coupling of T_1 and T_2
Tyrosine
T_1
T_2
T_4
T_3
3 Oxidation of iodide
I_2
TGB
Colloid
Peroxidase
6 Pinocytosis and digestion of colloid
Secretory vesicles
Lysosome
Golgi complex
T_3 T_4
2 Synthesis of TGB
I^-
Rough ER
1 Iodide trapping
7 Secretion of thyroid hormones
T_3 TBG
T_4 TBG
Blood plasma
8 Transport in blood
Key:
I^- = Iodide; I_2 = Iodine
TGB = thyroglobulin
TBG = thyroxine-binding globulin
Blood capillary

NOTES

18

Figure 18.12 Negative feedback regulation of thyroid hormone secretion (page 584).

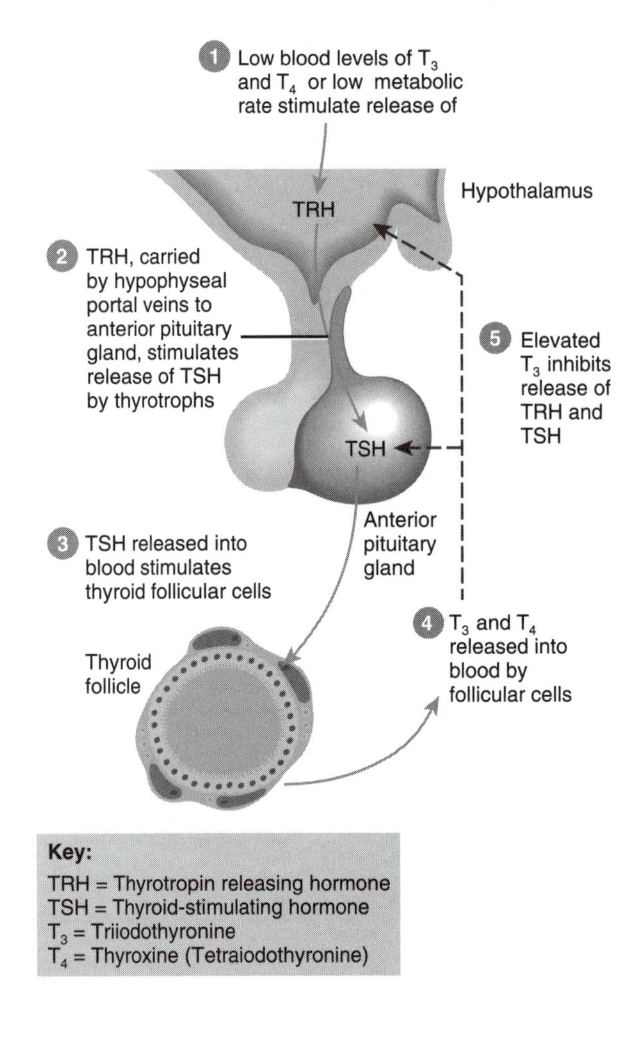

1 Low blood levels of T_3 and T_4 or low metabolic rate stimulate release of
TRH
Hypothalamus
2 TRH, carried by hypophyseal portal veins to anterior pituitary gland, stimulates release of TSH by thyrotrophs
5 Elevated T_3 inhibits release of TRH and TSH
TSH
Anterior pituitary gland
3 TSH released into blood stimulates thyroid follicular cells
4 T_3 and T_4 released into blood by follicular cells
Thyroid follicle
Key:
TRH = Thyrotropin releasing hormone
TSH = Thyroid-stimulating hormone
T_3 = Triiodothyronine
T_4 = Thyroxine (Tetraiodothyronine)

Figure 18.13 Location, blood supply, and histology of the parathyroid glands (page 586).

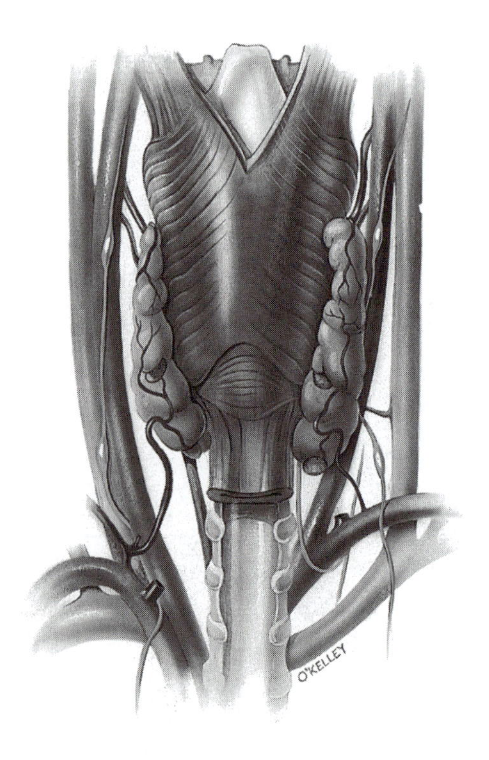

O'KELLEY

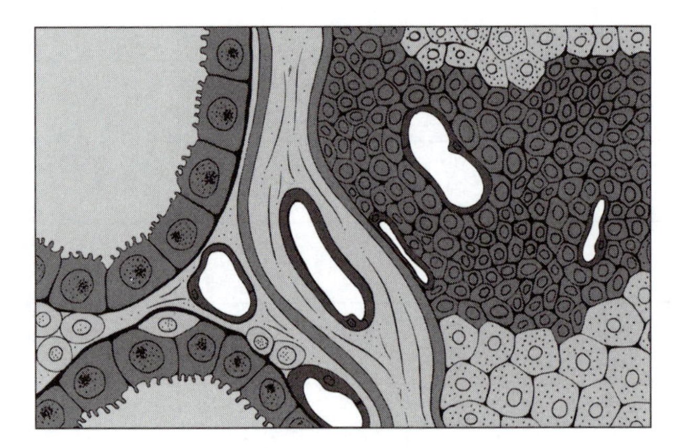

NOTES

18

Figure 18.14 The roles of calcitonin, parathyroid hormone, and calcitriol in calcium homeostasis (page 587).

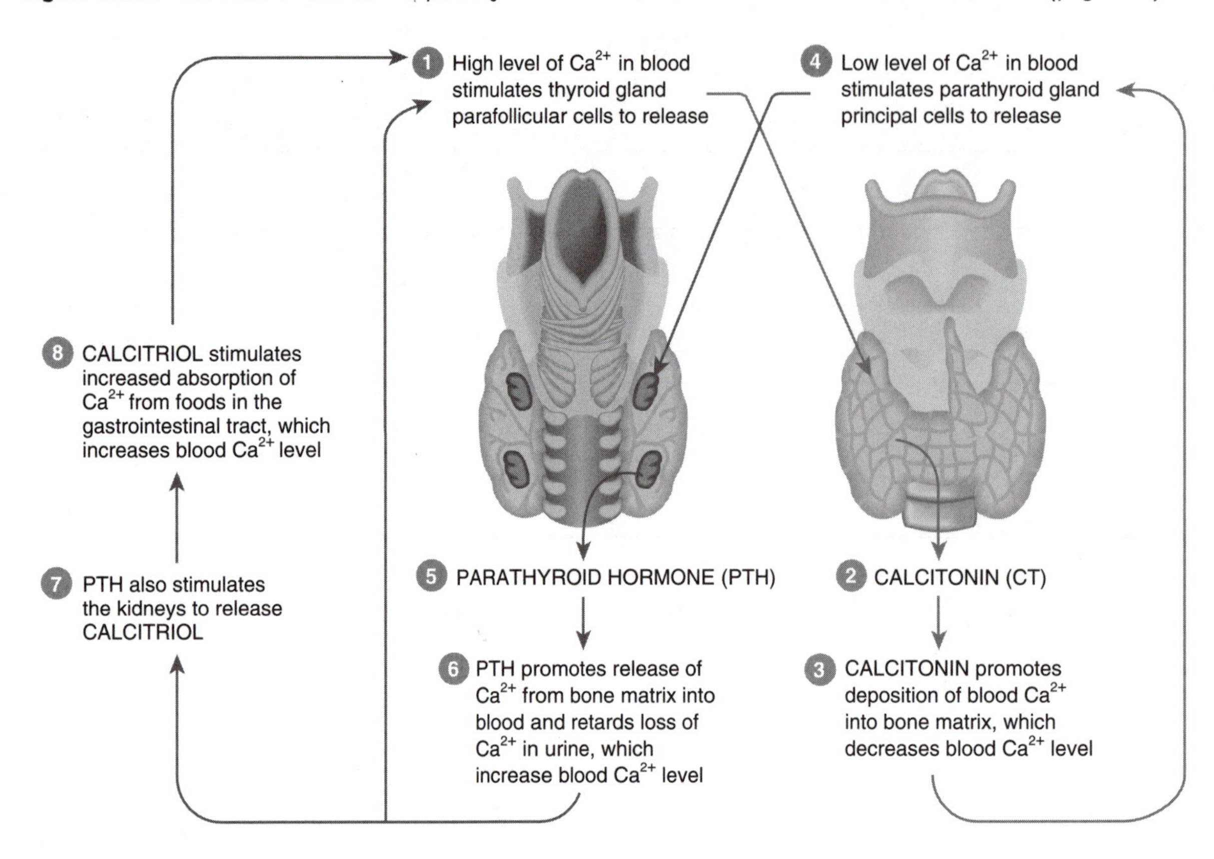

1 High level of Ca^{2+} in blood stimulates thyroid gland parafollicular cells to release
2 CALCITONIN (CT)
3 CALCITONIN promotes deposition of blood Ca^{2+} into bone matrix, which decreases blood Ca^{2+} level
4 Low level of Ca^{2+} in blood stimulates parathyroid gland principal cells to release
5 PARATHYROID HORMONE (PTH)
6 PTH promotes release of Ca^{2+} from bone matrix into blood and retards loss of Ca^{2+} in urine, which increase blood Ca^{2+} level
7 PTH also stimulates the kidneys to release CALCITRIOL
8 CALCITRIOL stimulates increased absorption of Ca^{2+} from foods in the gastrointestinal tract, which increases blood Ca^{2+} level

Figure 18.15 Location, blood supply, and histology of the adrenal (suprarenal) glands (page 588).

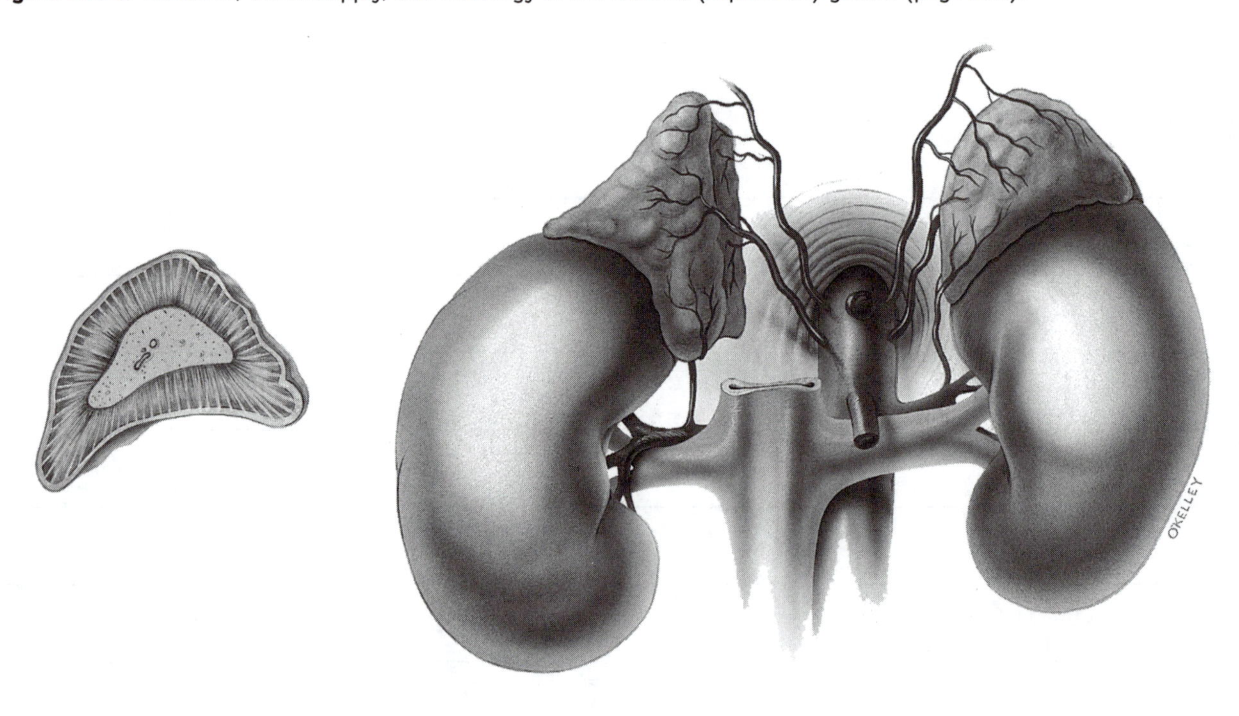

O'KELLEY

NOTES

18

Figure 18.16 Regulation of aldosterone secretion by the renin–angiotensin pathway (page 590).

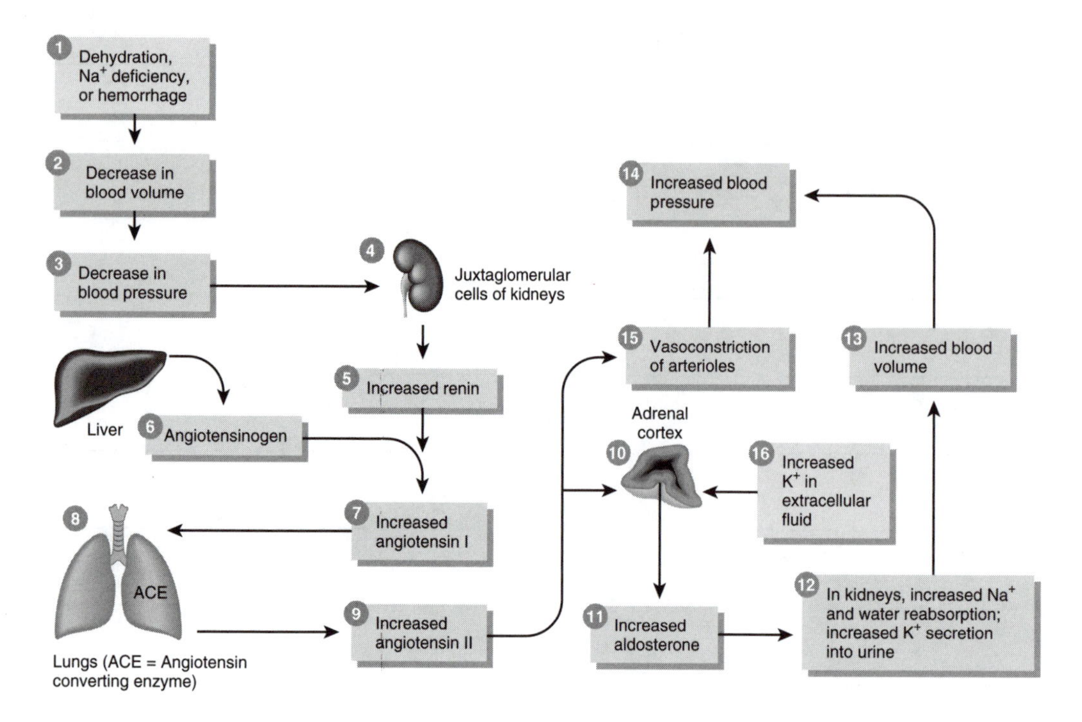
1 Dehydration, Na^+ deficiency, or hemorrhage
2 Decrease in blood volume
3 Decrease in blood pressure
4 Juxtaglomerular cells of kidneys
5 Increased renin
Liver
6 Angiotensinogen
7 Increased angiotensin I
8 ACE
Lungs (ACE = Angiotensin converting enzyme)
9 Increased angiotensin II
Adrenal cortex
10
11 Increased aldosterone
12 In kidneys, increased Na^+ and water reabsorption; increased K^+ secretion into urine
13 Increased blood volume
14 Increased blood pressure
15 Vasoconstriction of arterioles
16 Increased K^+ in extracellular fluid

Figure 18.17 Negative feedback regulation of glucocorticoid secretion (page 591).

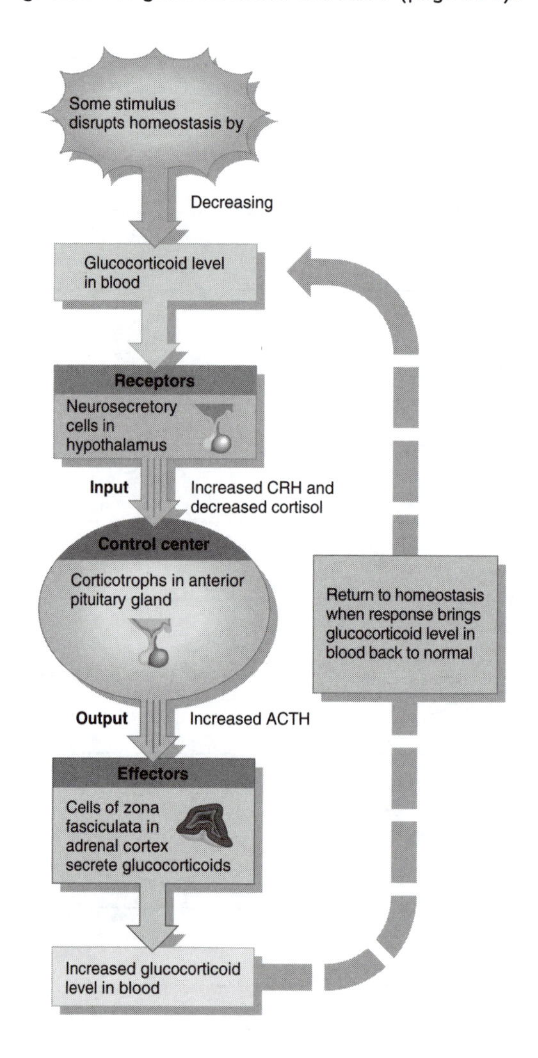
Some stimulus disrupts homeostasis by
Decreasing
Glucocorticoid level in blood
Receptors
Neurosecretory cells in hypothalamus
Input
Increased CRH and decreased cortisol
Control center
Corticotrophs in anterior pituitary gland
Output
Increased ACTH
Effectors
Cells of zona fasciculata in adrenal cortex secrete glucocorticoids
Increased glucocorticoid level in blood
Return to homeostasis when response brings glucocorticoid level in blood back to normal

NOTES

18

Figure 18.18 Location, blood supply, and histology of the pancreas (page 593).

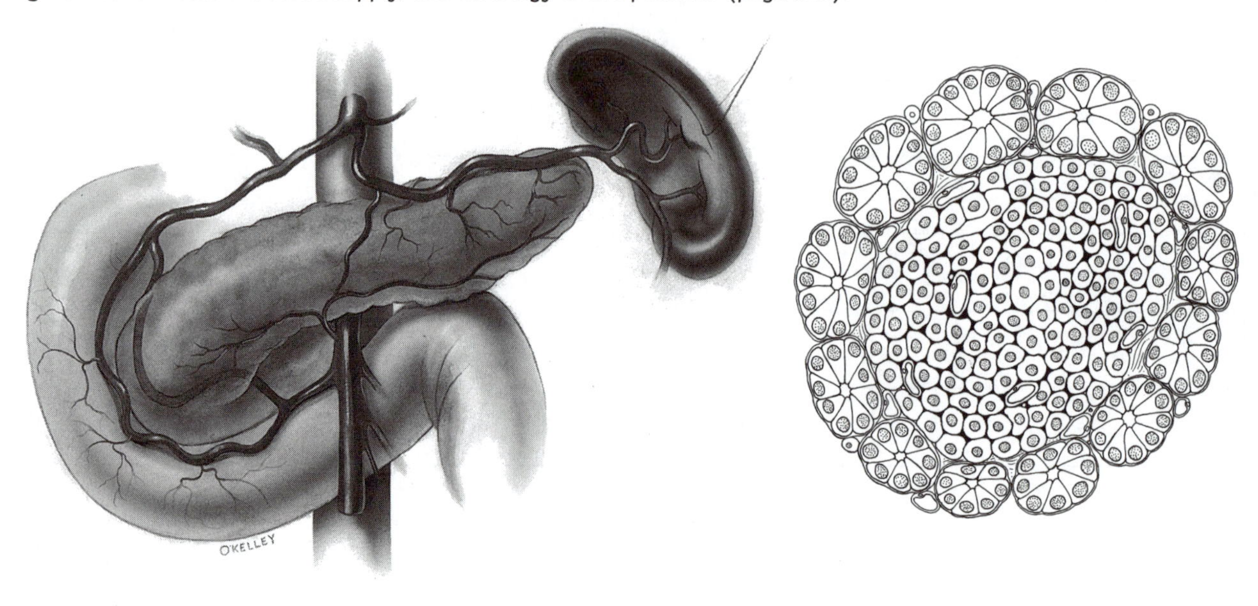
O'KELLEY

Figure 18.19 Negative feedback regulation of the secretion of glucagon and insulin (page 594).

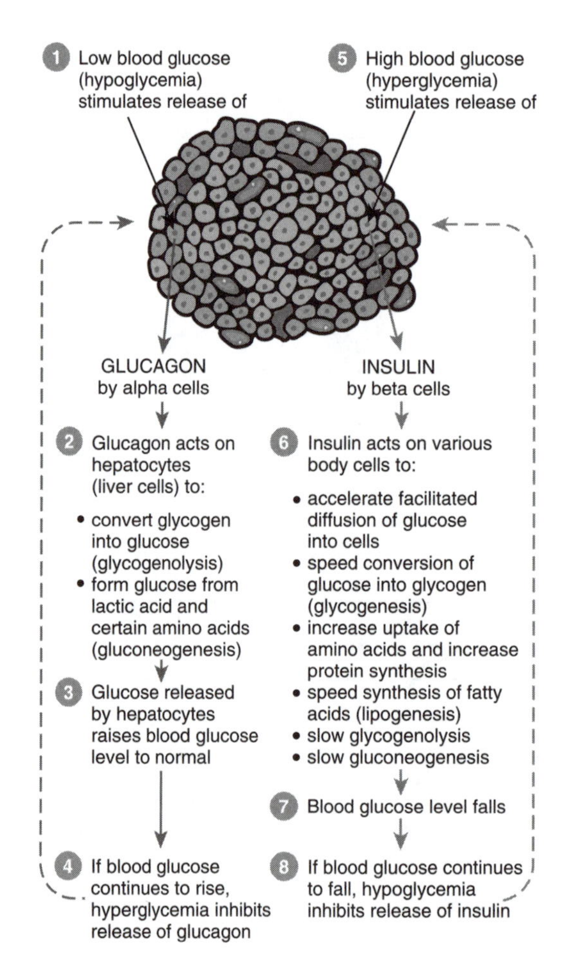
1 Low blood glucose (hypoglycemia) stimulates release of
5 High blood glucose (hyperglycemia) stimulates release of
GLUCAGON by alpha cells
INSULIN by beta cells
2 Glucagon acts on hepatocytes (liver cells) to:
• convert glycogen into glucose (glycogenolysis)
• form glucose from lactic acid and certain amino acids (gluconeogenesis)
3 Glucose released by hepatocytes raises blood glucose level to normal
4 If blood glucose continues to rise, hyperglycemia inhibits release of glucagon
6 Insulin acts on various body cells to:
• accelerate facilitated diffusion of glucose into cells
• speed conversion of glucose into glycogen (glycogenesis)
• increase uptake of amino acids and increase protein synthesis
• speed synthesis of fatty acids (lipogenesis)
• slow glycogenolysis
• slow gluconeogenesis
7 Blood glucose level falls
8 If blood glucose continues to fall, hypoglycemia inhibits release of insulin

NOTES

18

Figure 18.20 Pathway whereby light slows release of melatonin from the pineal gland (page 597).

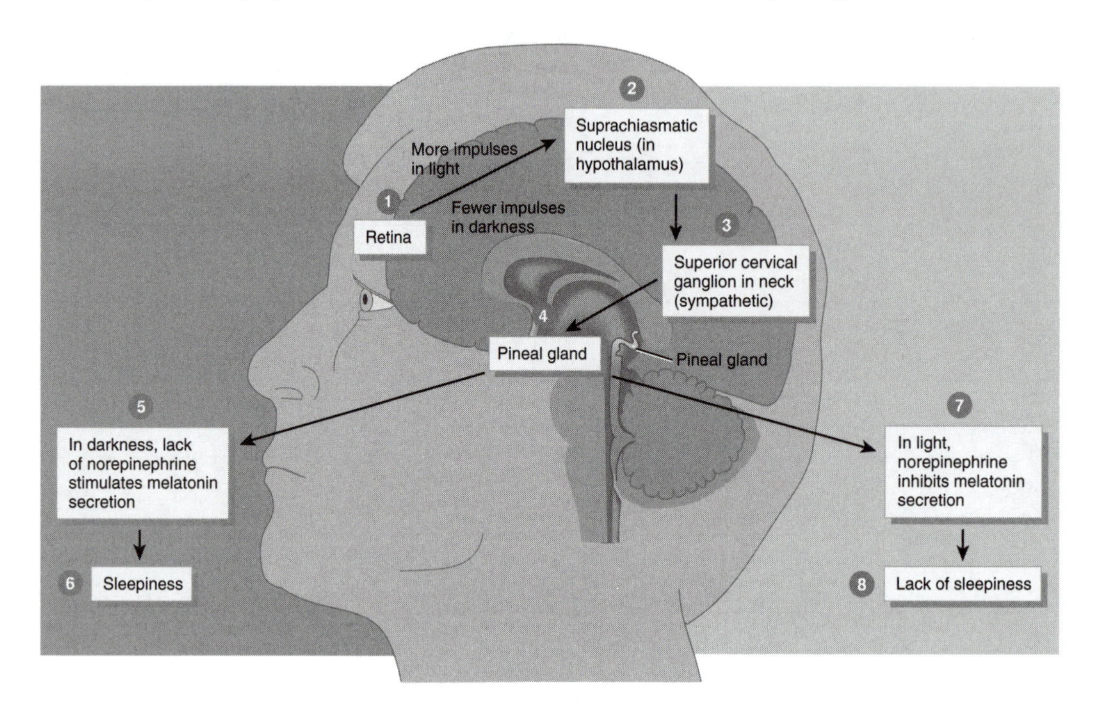

1 Retina
More impulses in light
Fewer impulses in darkness
2 Suprachiasmatic nucleus (in hypothalamus)
3 Superior cervical ganglion in neck (sympathetic)
4 Pineal gland
Pineal gland
5 In darkness, lack of norepinephrine stimulates melatonin secretion
6 Sleepiness
7 In light, norepinephrine inhibits melatonin secretion
8 Lack of sleepiness

Figure 18.21 Responses to stressors during the general adaptation syndrome (page 600).

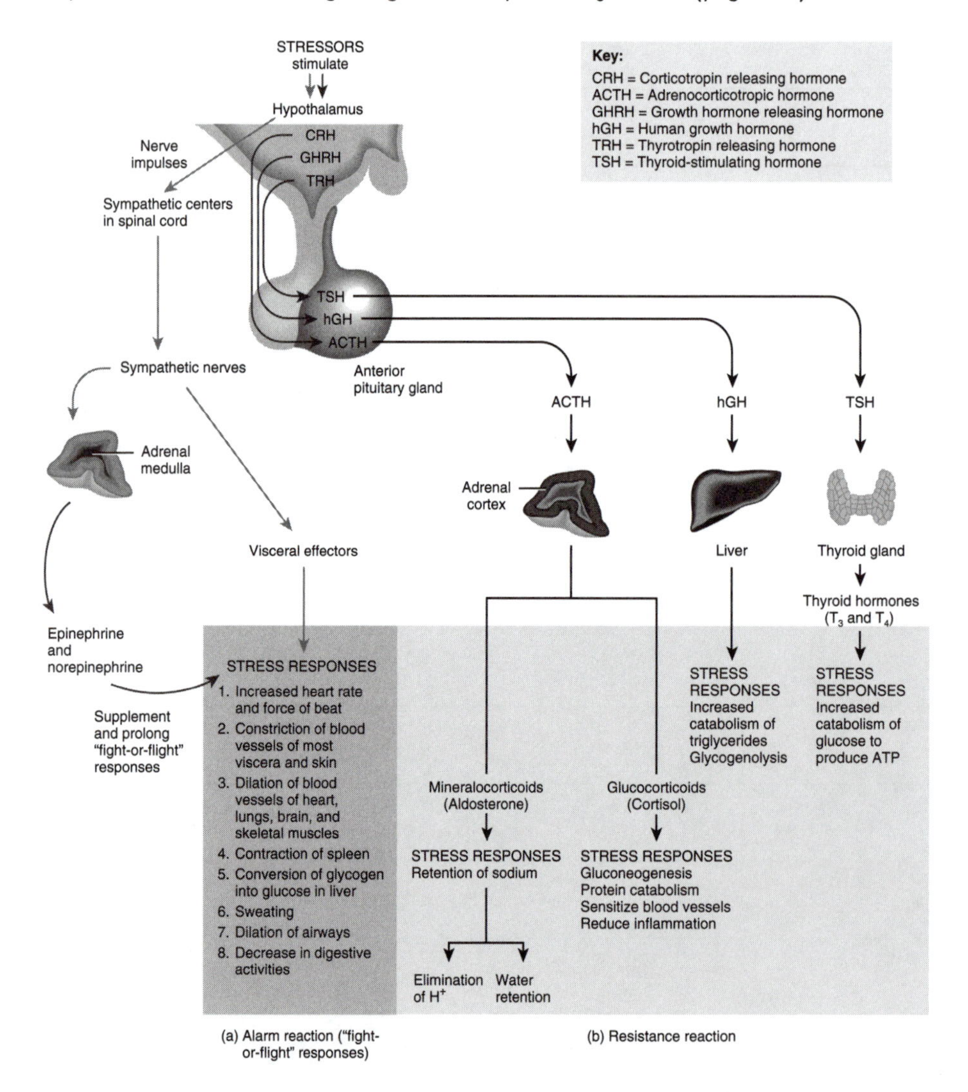

STRESSORS stimulate
Hypothalamus
Key:
CRH = Corticotropin releasing hormone
ACTH = Adrenocorticotropic hormone
GHRH = Growth hormone releasing hormone
hGH = Human growth hormone
TRH = Thyrotropin releasing hormone
TSH = Thyroid-stimulating hormone
Nerve impulses
CRH
GHRH
TRH
Sympathetic centers in spinal cord
TSH
hGH
ACTH
Sympathetic nerves
Anterior pituitary gland
ACTH
hGH
TSH
Adrenal medulla
Adrenal cortex
Visceral effectors
Liver
Thyroid gland
Thyroid hormones (T_3 and T_4)
Epinephrine and norepinephrine
Supplement and prolong "fight-or-flight" responses
STRESS RESPONSES
1. Increased heart rate and force of beat
2. Constriction of blood vessels of most viscera and skin
3. Dilation of blood vessels of heart, lungs, brain, and skeletal muscles
4. Contraction of spleen
5. Conversion of glycogen into glucose in liver
6. Sweating
7. Dilation of airways
8. Decrease in digestive activities
STRESS RESPONSES Increased catabolism of triglycerides Glycogenolysis
STRESS RESPONSES Increased catabolism of glucose to produce ATP
Mineralocorticoids (Aldosterone)
Glucocorticoids (Cortisol)
STRESS RESPONSES Retention of sodium
STRESS RESPONSES Gluconeogenesis Protein catabolism Sensitize blood vessels Reduce inflammation
Elimination of H^+
Water retention
(a) Alarm reaction ("fight-or-flight" responses)
(b) Resistance reaction

NOTES

18

Figure 18.22 Development of the endocrine system (page 602).

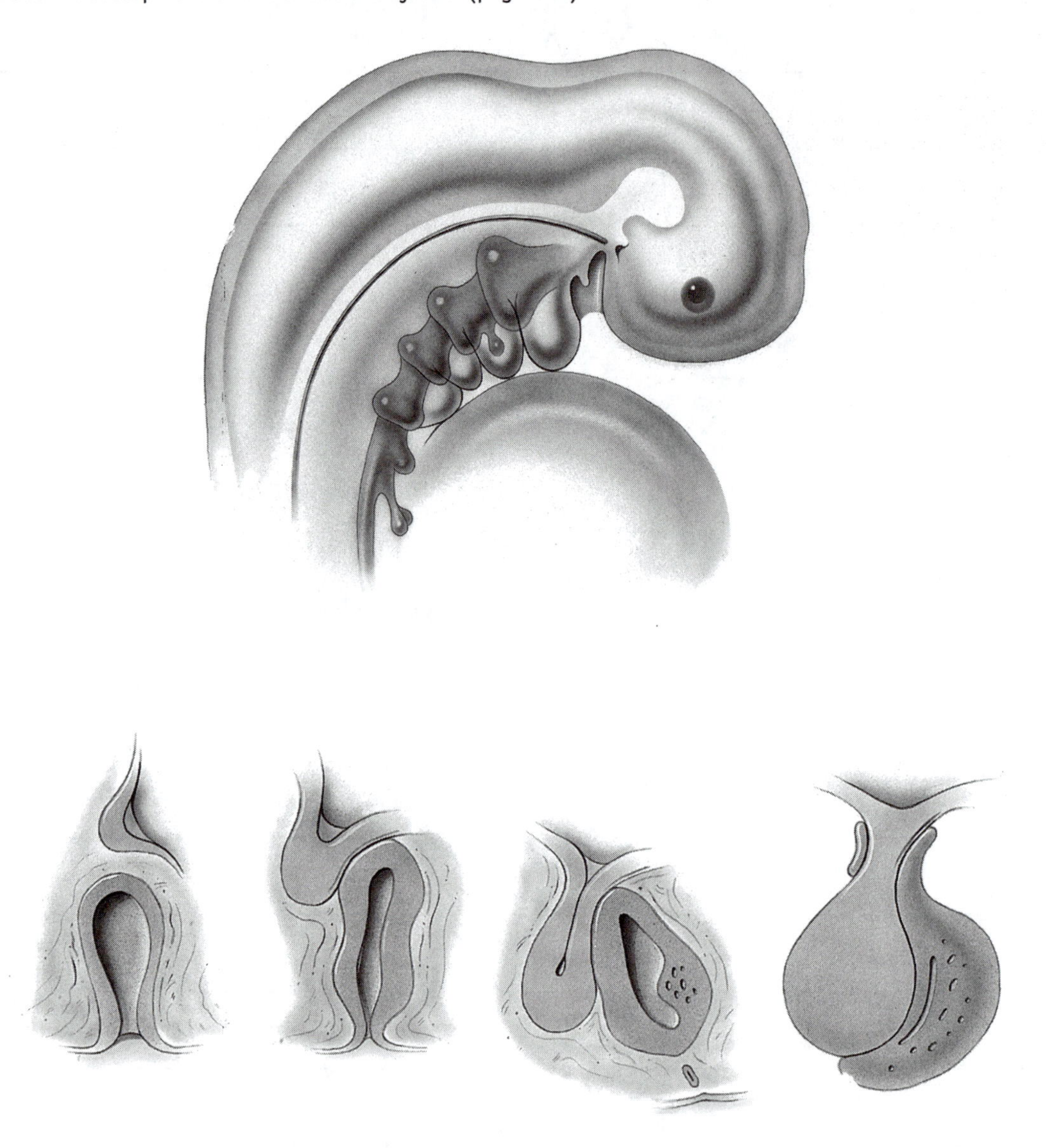

NOTES

18

Figure 19.1 Components of blood in a normal adult (page 612).

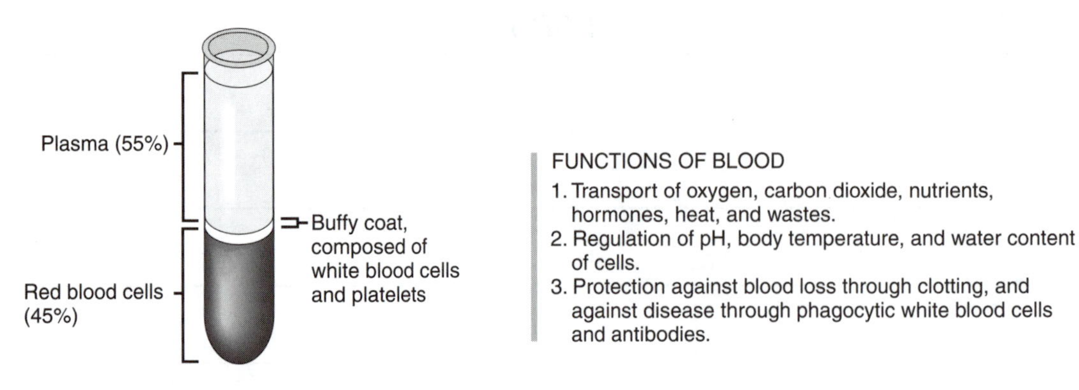

Plasma (55%)
Buffy coat, composed of white blood cells and platelets
Red blood cells (45%)
FUNCTIONS OF BLOOD
1. Transport of oxygen, carbon dioxide, nutrients, hormones, heat, and wastes.
2. Regulation of pH, body temperature, and water content of cells.
3. Protection against blood loss through clotting, and against disease through phagocytic white blood cells and antibodies.

(a) Appearance of centrifuged blood

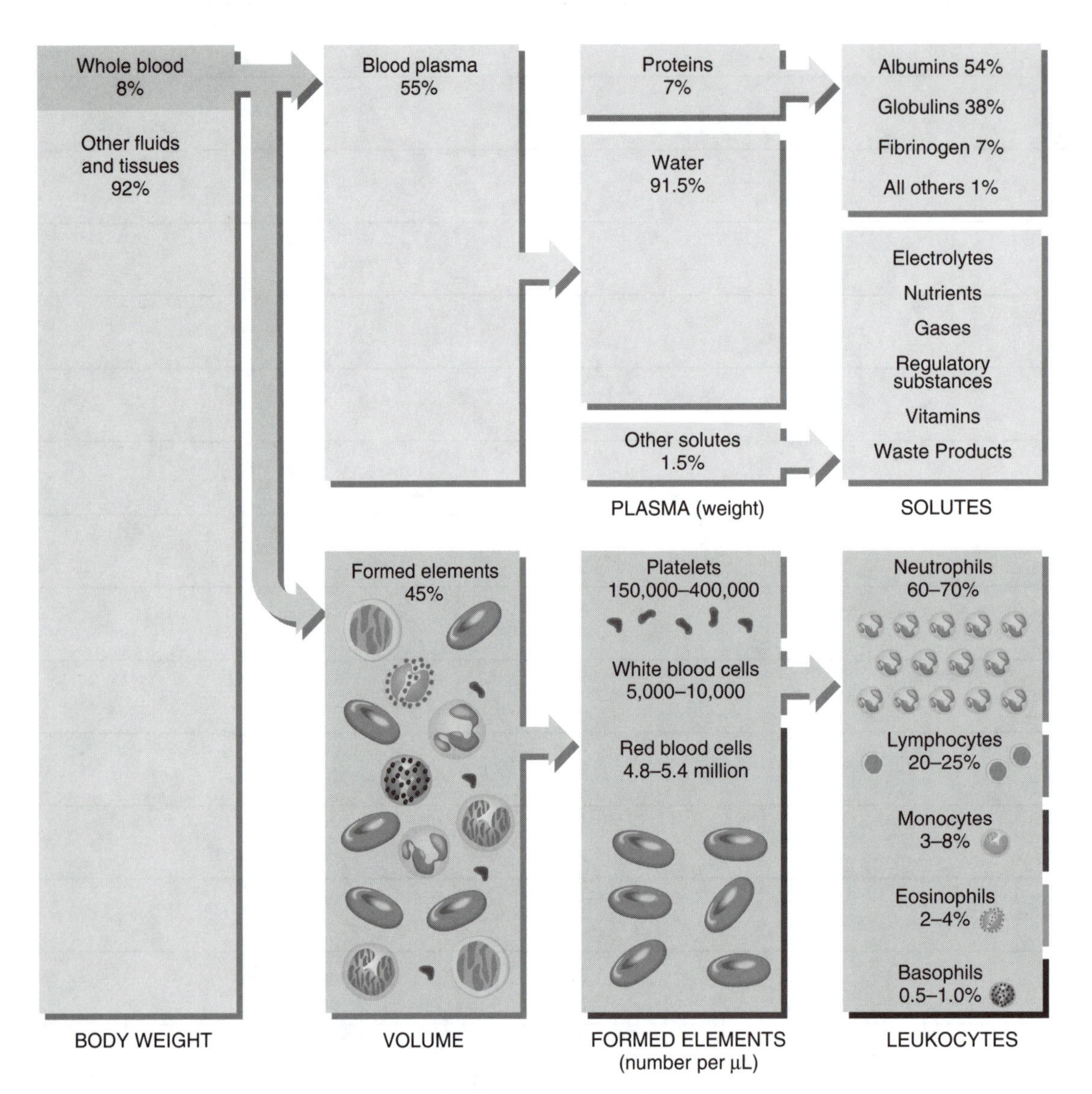

Whole blood 8%
Other fluids and tissues 92%
Blood plasma 55%
Proteins 7%
Water 91.5%
Other solutes 1.5%
Albumins 54%
Globulins 38%
Fibrinogen 7%
All others 1%
Electrolytes
Nutrients
Gases
Regulatory substances
Vitamins
Waste Products
PLASMA (weight)
SOLUTES
Formed elements 45%
Platelets 150,000–400,000
White blood cells 5,000–10,000
Red blood cells 4.8–5.4 million
Neutrophils 60–70%
Lymphocytes 20–25%
Monocytes 3–8%
Eosinophils 2–4%
Basophils 0.5–1.0%
BODY WEIGHT
VOLUME
FORMED ELEMENTS (number per μL)
LEUKOCYTES

(b) Components of blood

NOTES

19

Figure 19.3 Origin, development, and structure of blood cells (page 614).

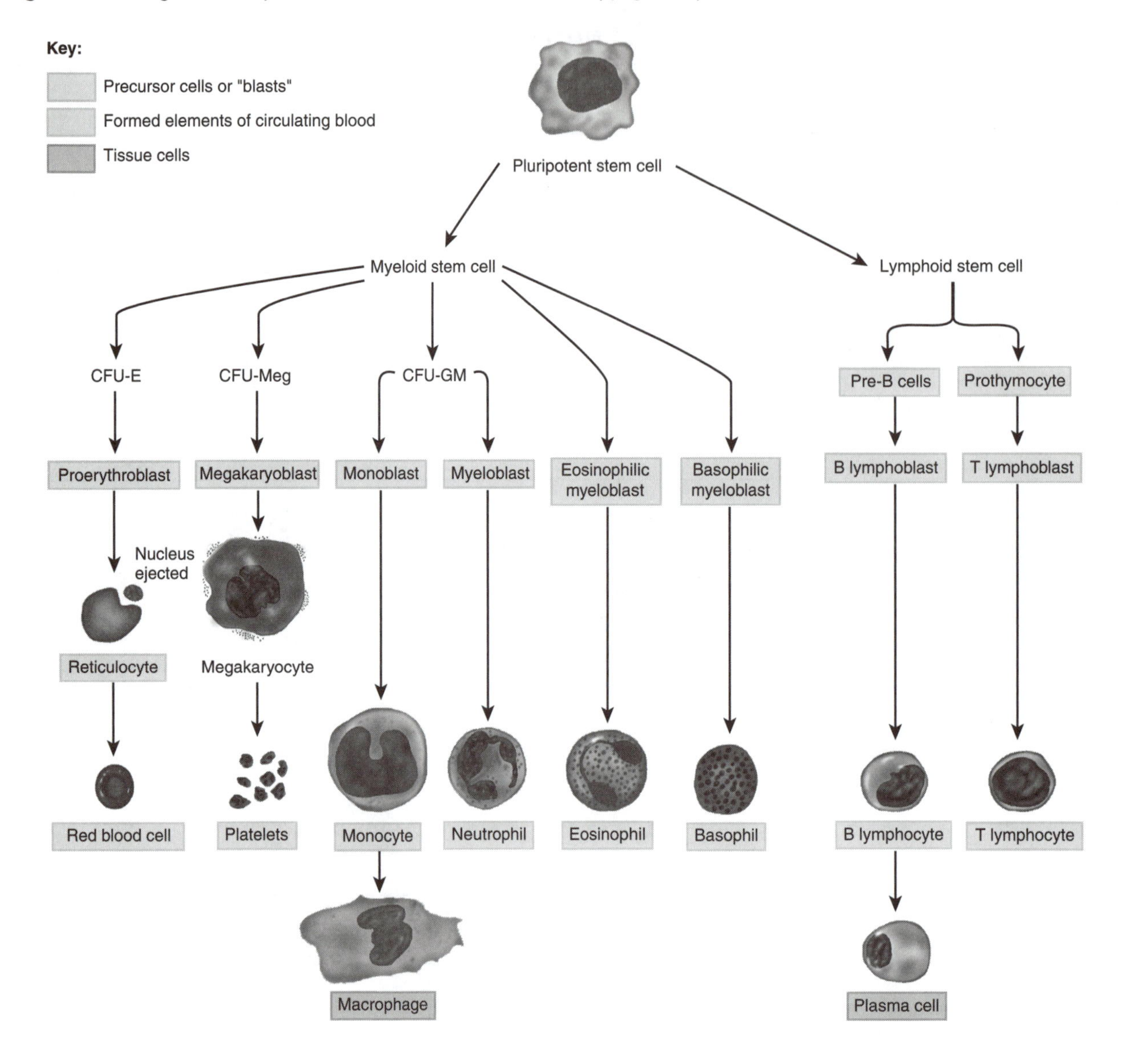

Key:
Precursor cells or "blasts"
Formed elements of circulating blood
Tissue cells
Pluripotent stem cell
Myeloid stem cell
Lymphoid stem cell
CFU-E
CFU-Meg
CFU-GM
Pre-B cells
Prothymocyte
Proerythroblast
Megakaryoblast
Monoblast
Myeloblast
Eosinophilic myeloblast
Basophilic myeloblast
B lymphoblast
T lymphoblast
Nucleus ejected
Reticulocyte
Megakaryocyte
Red blood cell
Platelets
Monocyte
Neutrophil
Eosinophil
Basophil
B lymphocyte
T lymphocyte
Macrophage
Plasma cell

Figure 19.4 The shapes of a red blood cell (page 616).

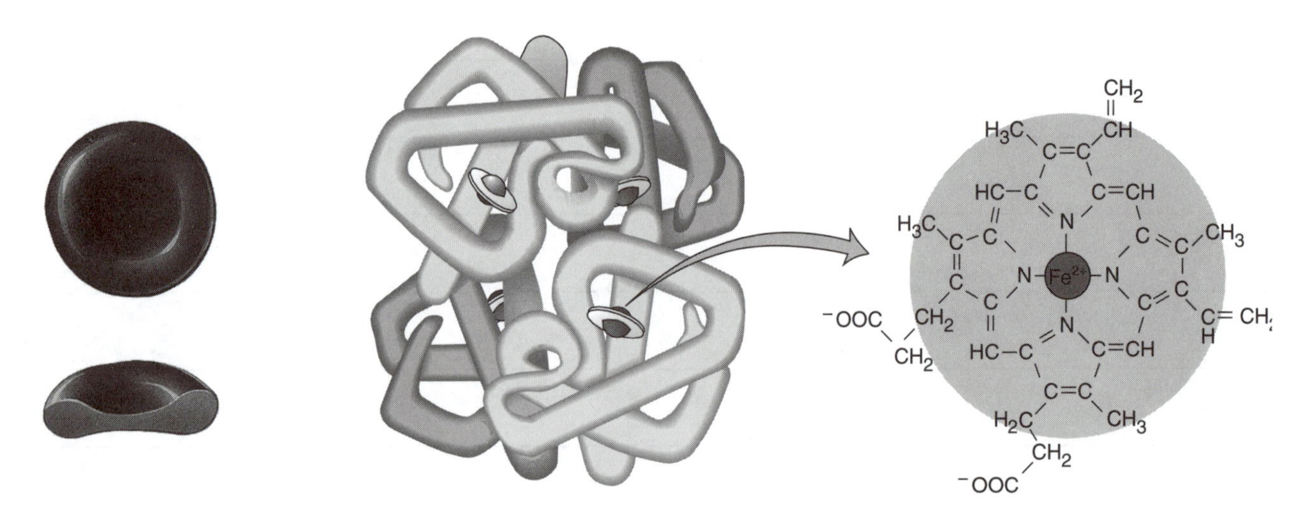

CH2
H3C
CH
C=C
HC–C
C=CH
N
H3C
C
C
CH3
N
Fe2+
N
-OOC
CH2
CH2
N
HC–C
C=CH
C=C
H2C
CH3
CH2
-OOC
H

NOTES

19

Figure 19.5 Formation and destruction of red blood cells, and the recycling of hemoglobin components (page 617).

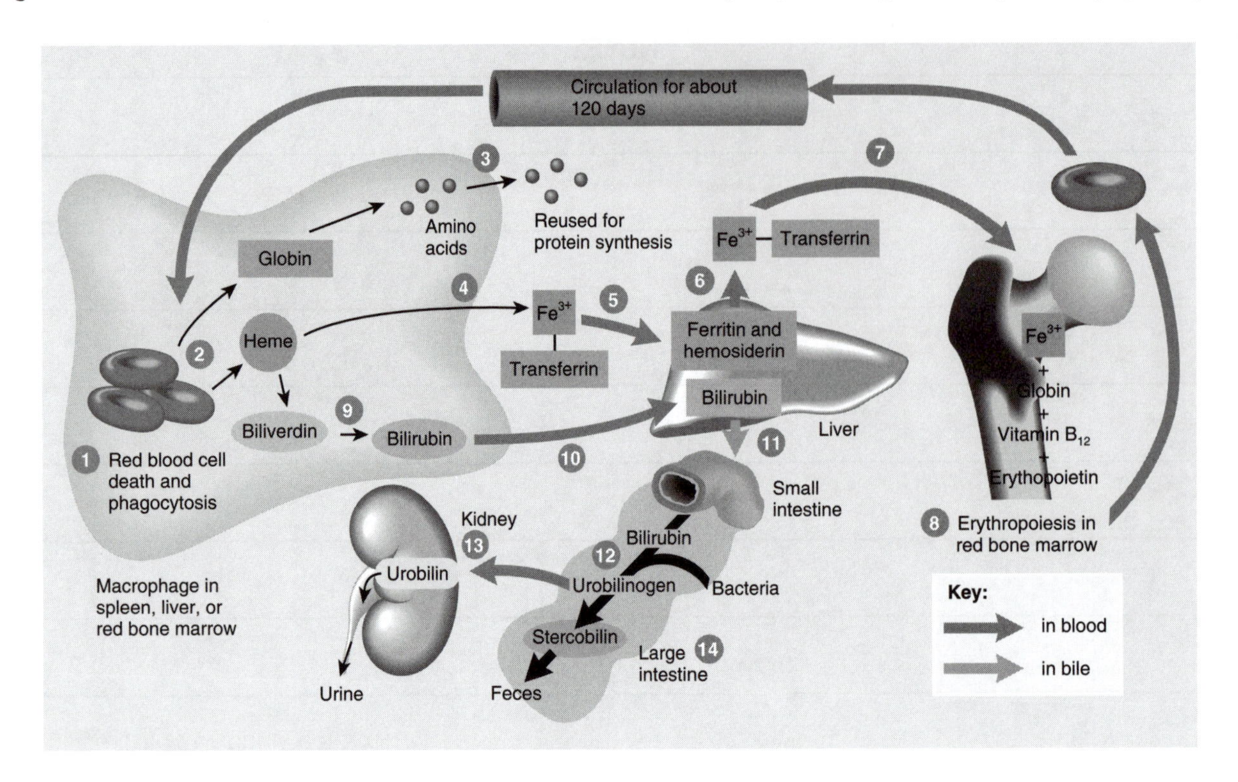
Circulation for about 120 days
3
Amino acids
Reused for protein synthesis
7
Globin
4
Fe^{3+}
5
Fe^{3+} — Transferrin
6
Heme
2
Transferrin
Ferritin and hemosiderin
Fe^{3+}
+
Globin
+
Vitamin B_{12}
+
Erythopoietin
Biliverdin
9
Bilirubin
Bilirubin
Liver
1 Red blood cell death and phagocytosis
10
11
Small intestine
Kidney
13
Bilirubin
12
8 Erythropoiesis in red bone marrow
Macrophage in spleen, liver, or red bone marrow
Urobilin
Urobilinogen
Bacteria
Key:
in blood
Stercobilin
Large intestine
14
in bile
Urine
Feces

Figure 19.6 Negative feedback regulation of erythropoiesis (page 618).

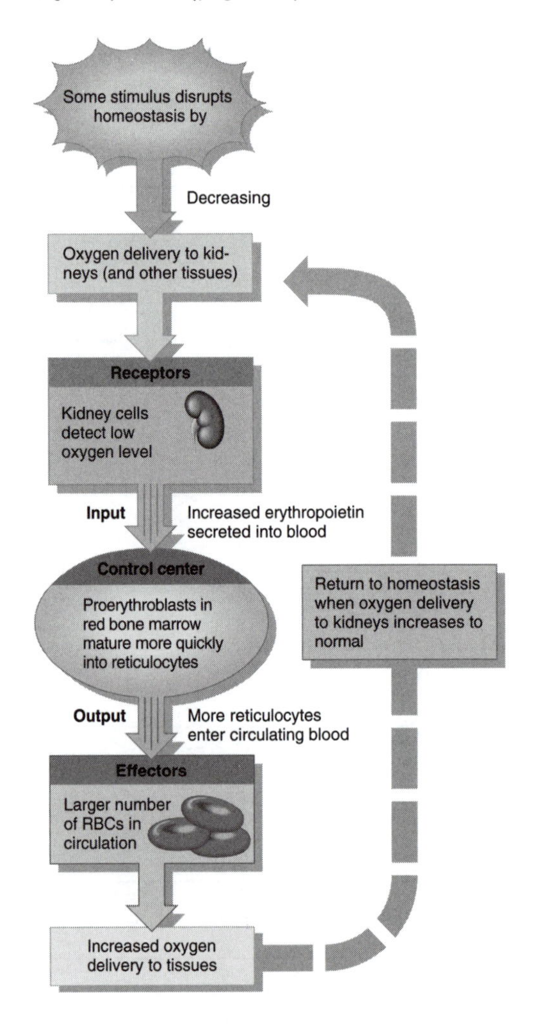
Some stimulus disrupts homeostasis by
Decreasing
Oxygen delivery to kidneys (and other tissues)
Receptors
Kidney cells detect low oxygen level
Input
Increased erythropoietin secreted into blood
Control center
Proerythroblasts in red bone marrow mature more quickly into reticulocytes
Return to homeostasis when oxygen delivery to kidneys increases to normal
Output
More reticulocytes enter circulating blood
Effectors
Larger number of RBCs in circulation
Increased oxygen delivery to tissues

NOTES

19

Figure 19.8 Emigration of white blood cells (page 620).

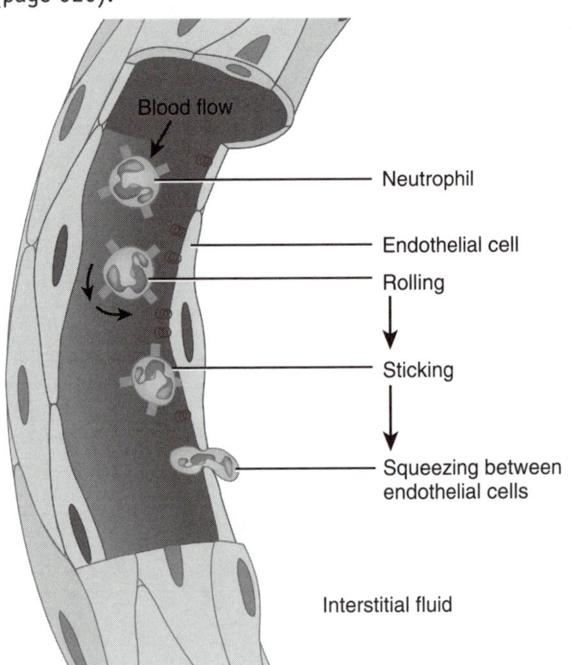
Blood flow
Neutrophil
Endothelial cell
Rolling
Sticking
Squeezing between endothelial cells
Interstitial fluid

Key:

Selectins on endothelial cells

Integrins on neutrophil

Figure 19.9 Platelet plug formation (page 624).

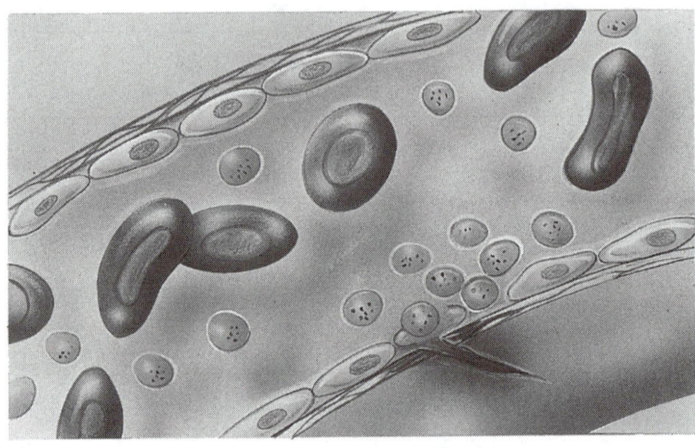

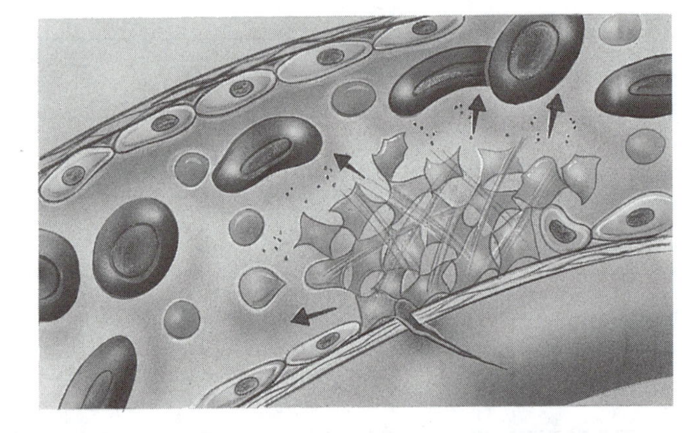

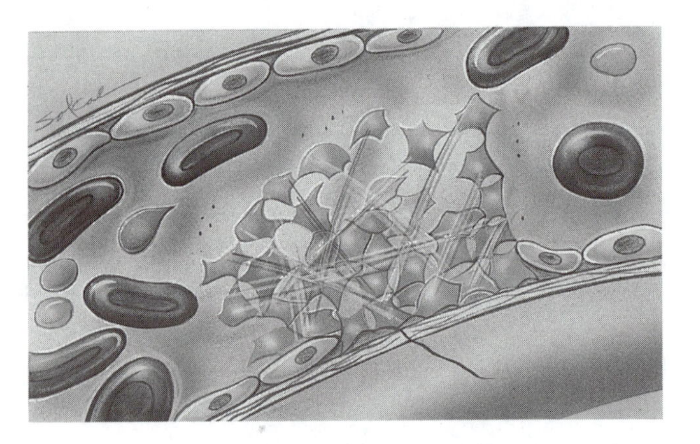

NOTES

19

Figure 19.11 The blood clotting cascade (page 625).

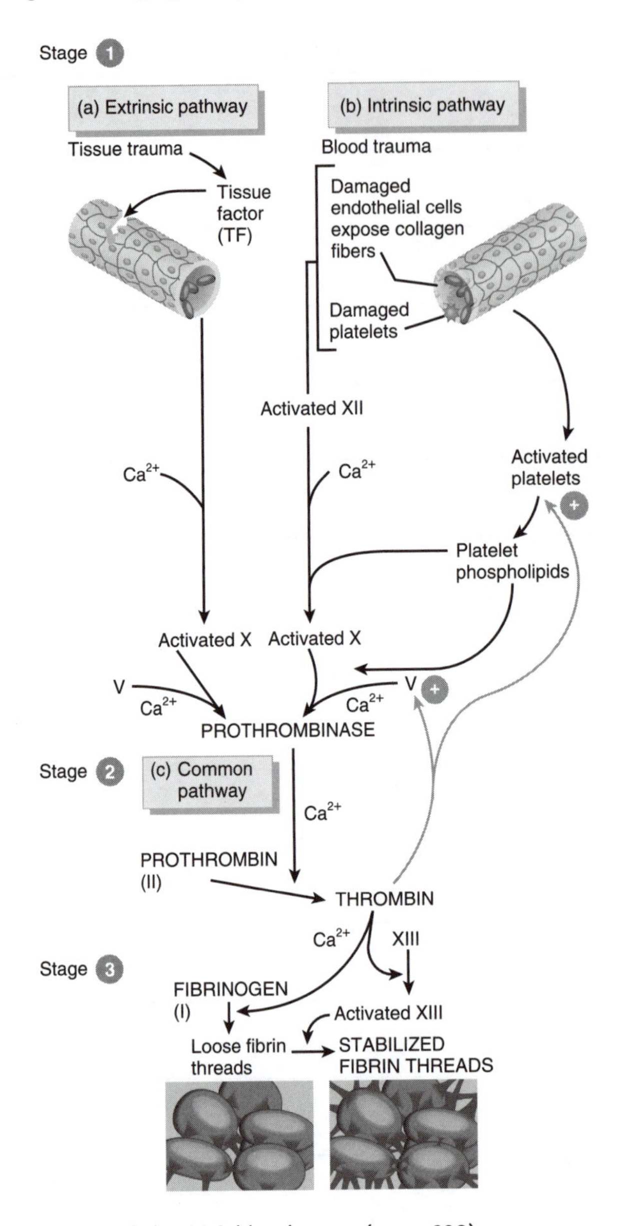

Stage 1
(a) Extrinsic pathway
(b) Intrinsic pathway
Tissue trauma
Blood trauma
Tissue factor (TF)
Damaged endothelial cells expose collagen fibers
Damaged platelets
Activated XII
Ca2+
Ca2+
Activated platelets
Platelet phospholipids
Activated X
Activated X
V
Ca2+
V
Ca2+
PROTHROMBINASE
Stage 2
(c) Common pathway
Ca2+
PROTHROMBIN (II)
THROMBIN
Ca2+
XIII
Stage 3
FIBRINOGEN (I)
Activated XIII
Loose fibrin threads
STABILIZED FIBRIN THREADS

Figure 19.12 Antigens and antibodies of the ABO blood types (page 628).

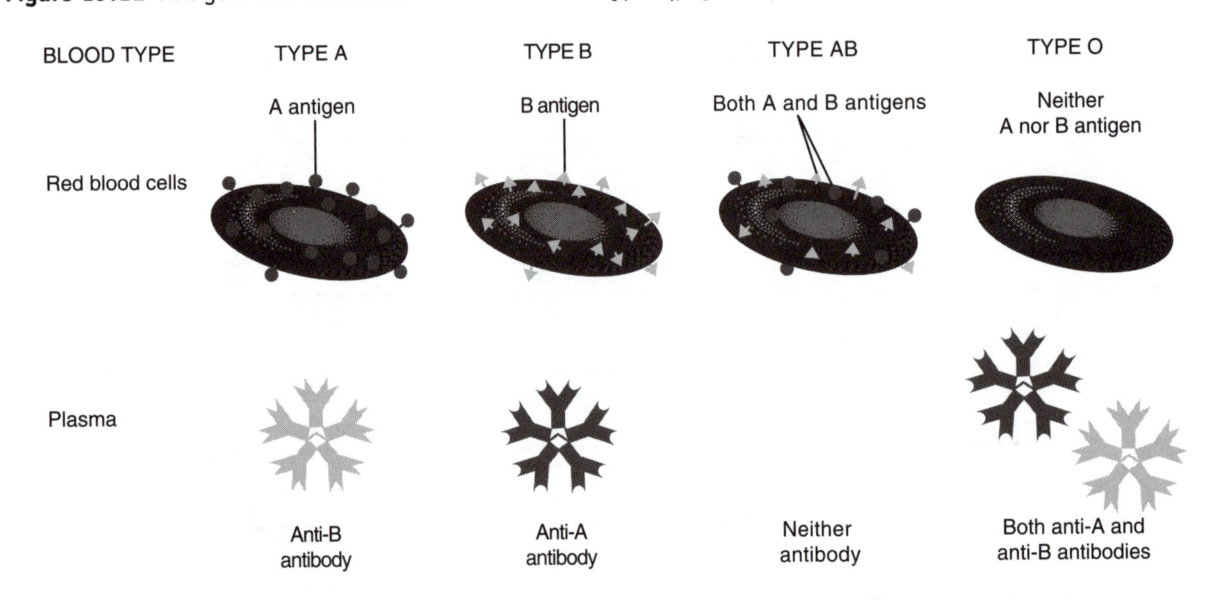

BLOOD TYPE
TYPE A
TYPE B
TYPE AB
TYPE O
A antigen
B antigen
Both A and B antigens
Neither A nor B antigen
Red blood cells
Plasma
Anti-B antibody
Anti-A antibody
Neither antibody
Both anti-A and anti-B antibodies

NOTES

19

Figure 19.13 Development of hemolytic disease of the newborn (page 629).

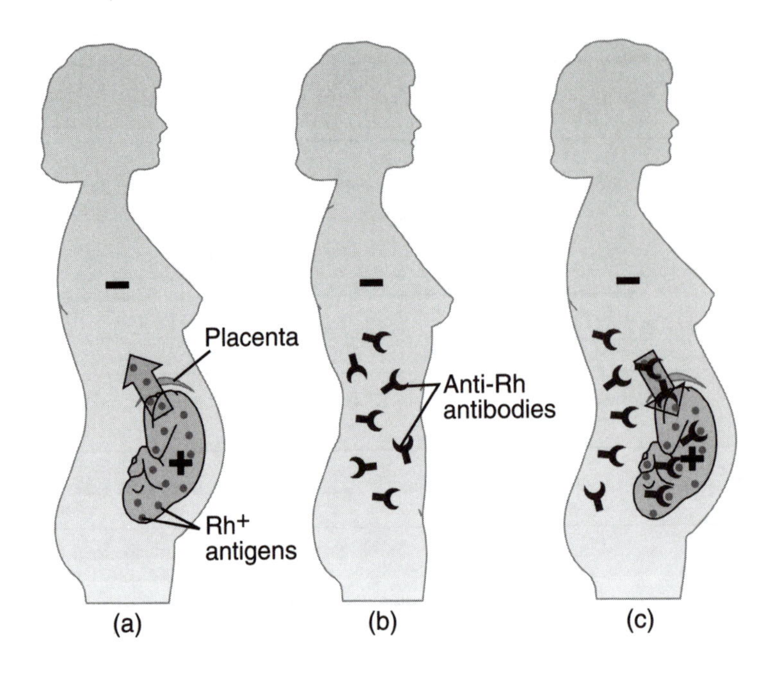
–
–
–
Placenta
Rh+
antigens
Anti-Rh
antibodies
+
+
(a)
(b)
(c)

NOTES

19

Figure 20.1 Position of the heart and associated structures in the mediastinum (page 637).

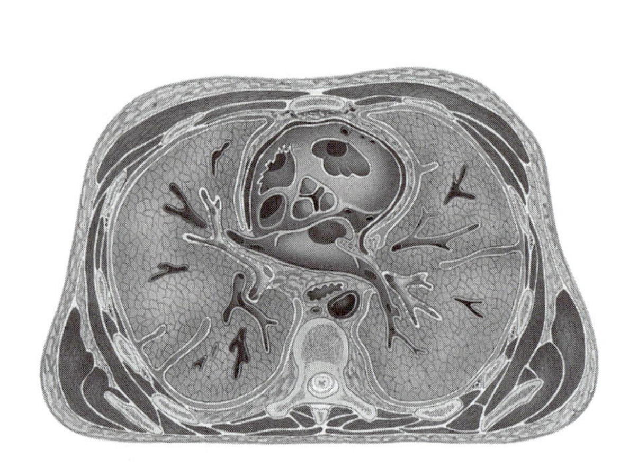

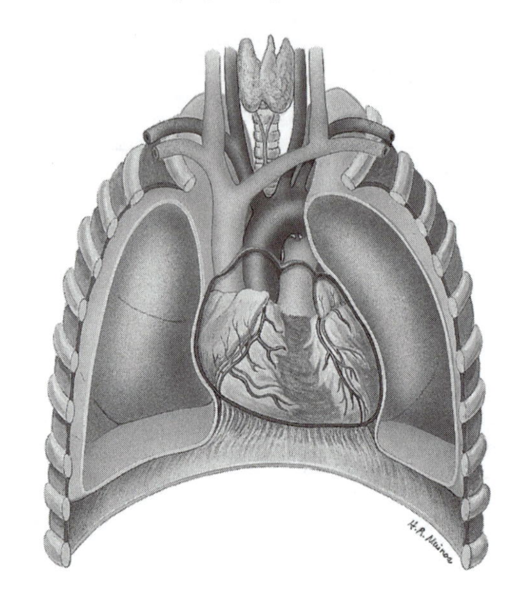

Figure 20.2 Pericardium and heart wall (page 639).

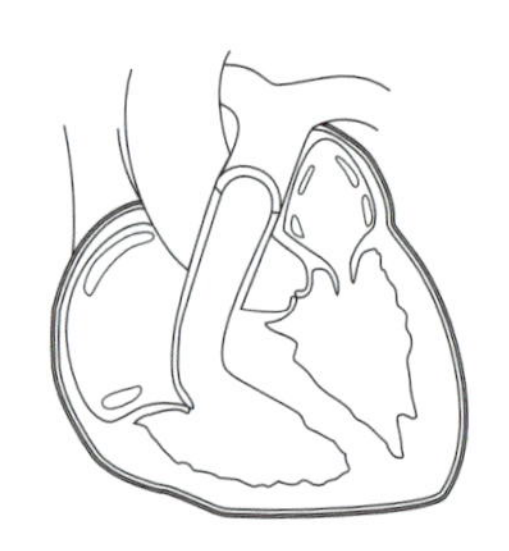

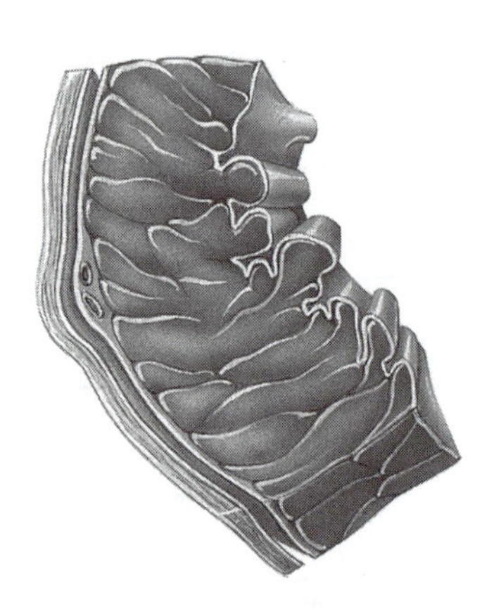

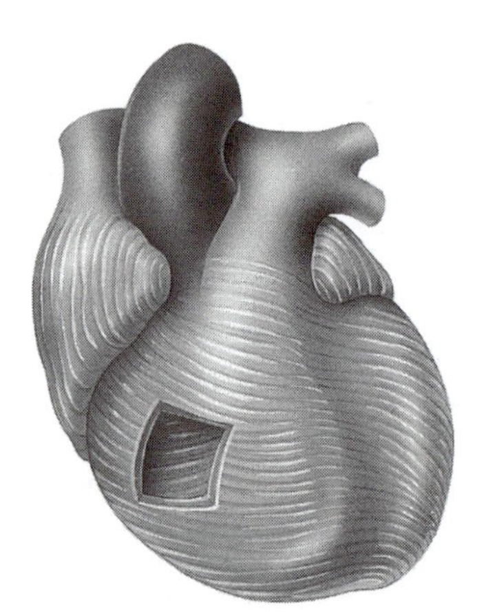

NOTES

20

Figure 20.3a-c Structure of the heart: surface features (page 640, 641).

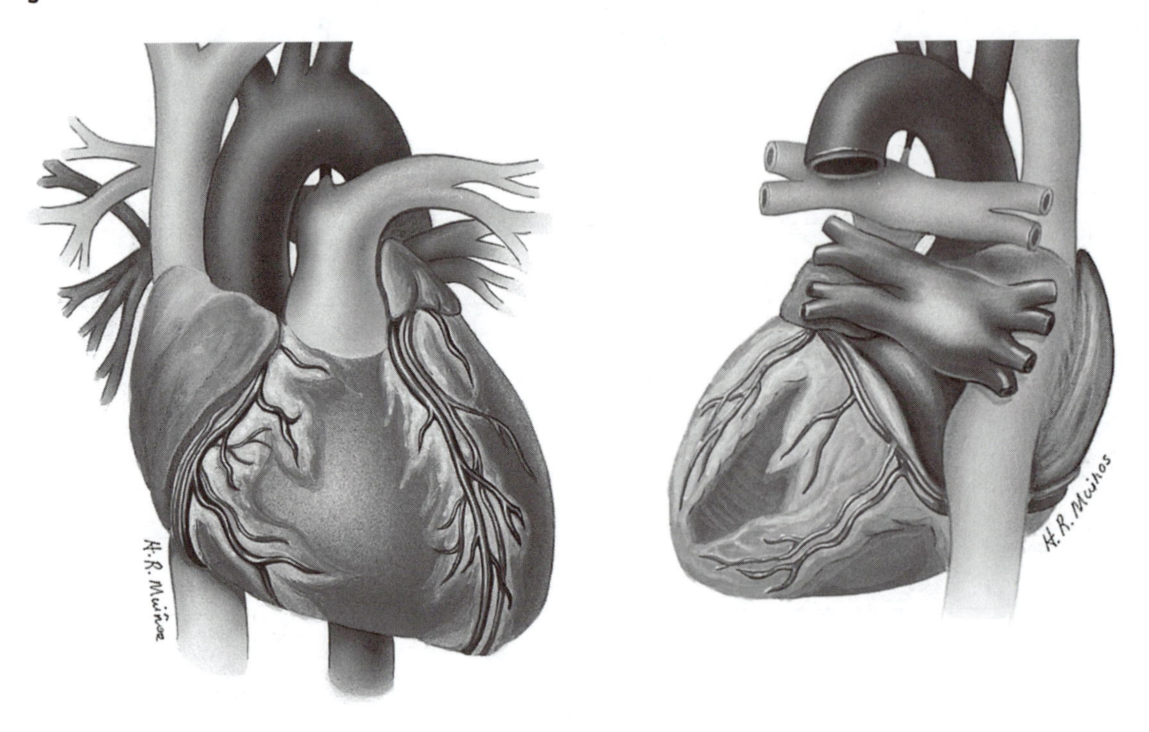

Figure 20.4a-c Structure of the heart: internal anatomy (page 642, 643).

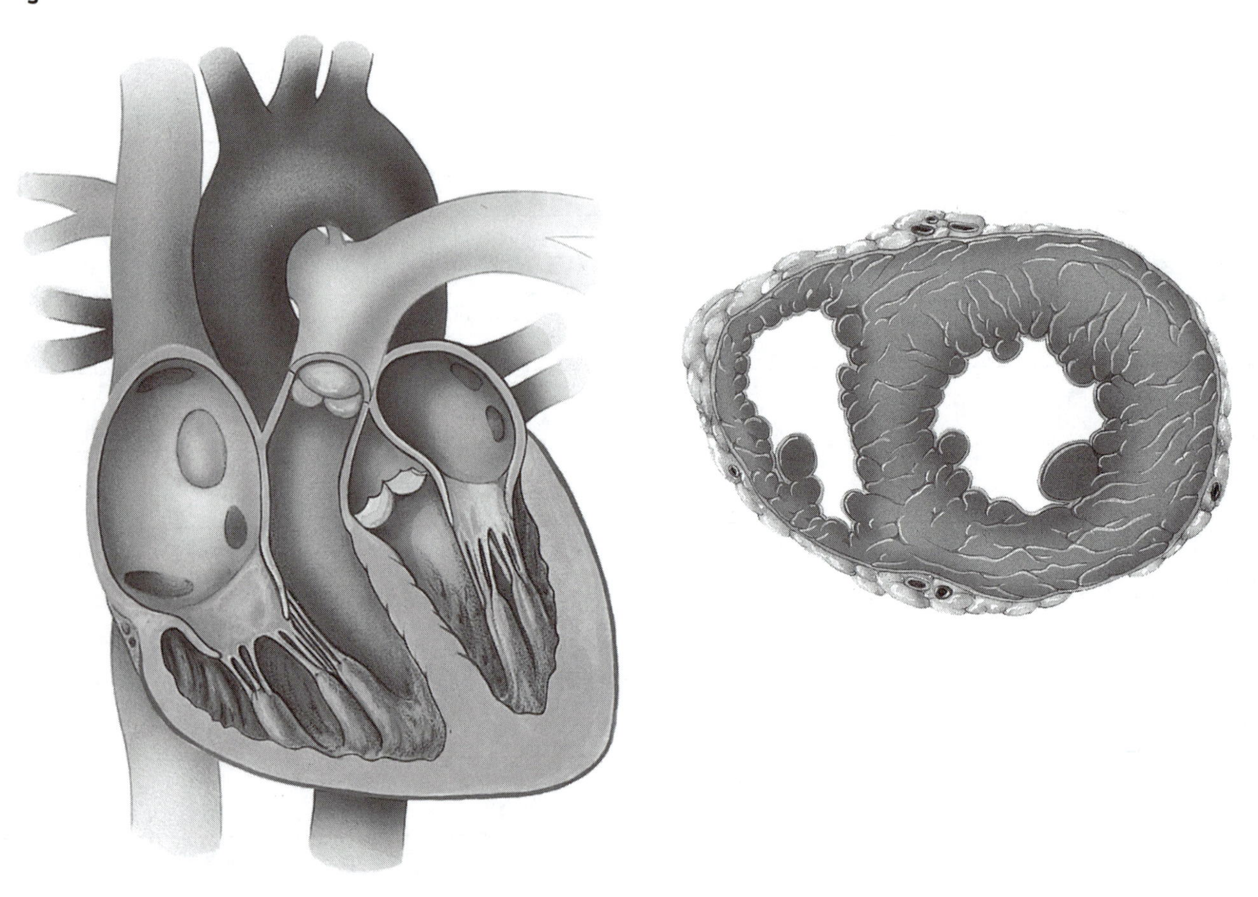

NOTES

20

Figure 20.5 Fibrous skeleton of the heart (page 644).

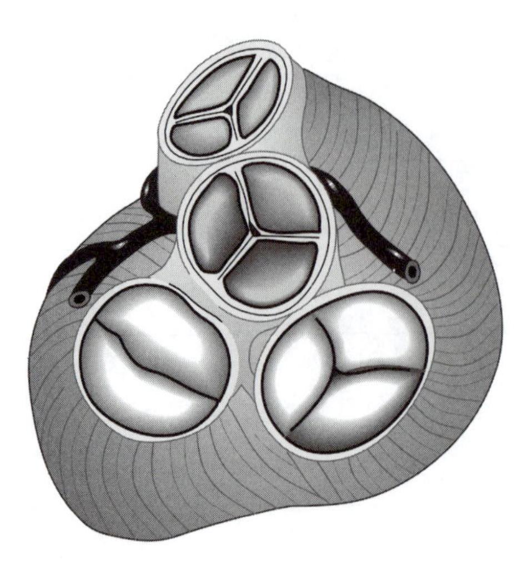

Figure 20.6 Responses of the valves to the pumping of the heart (page 645).

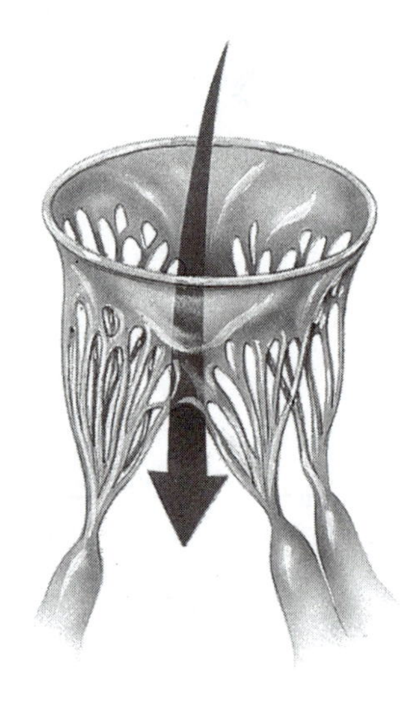

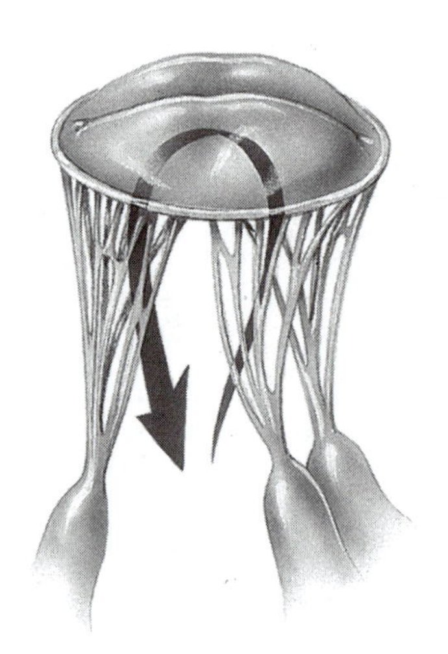

NOTES

20

Figure 20.7 Systemic and pulmonary circulations (page 646).

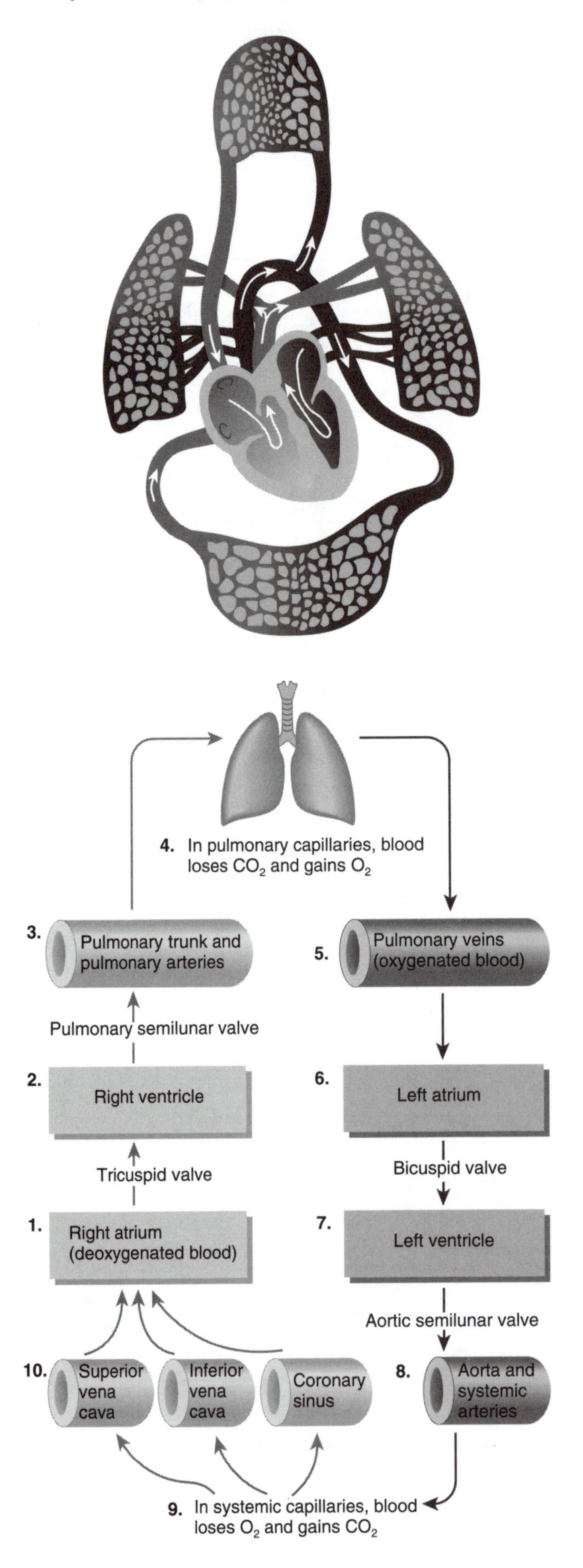

4. In pulmonary capillaries, blood loses CO_2 and gains O_2
3. Pulmonary trunk and pulmonary arteries
5. Pulmonary veins (oxygenated blood)
Pulmonary semilunar valve
2. Right ventricle
6. Left atrium
Tricuspid valve
Bicuspid valve
1. Right atrium (deoxygenated blood)
7. Left ventricle
Aortic semilunar valve
10. Superior vena cava
Inferior vena cava
Coronary sinus
8. Aorta and systemic arteries
9. In systemic capillaries, blood loses O_2 and gains CO_2

NOTES

20

Figure 20.8 Coronary circulation (page 647).

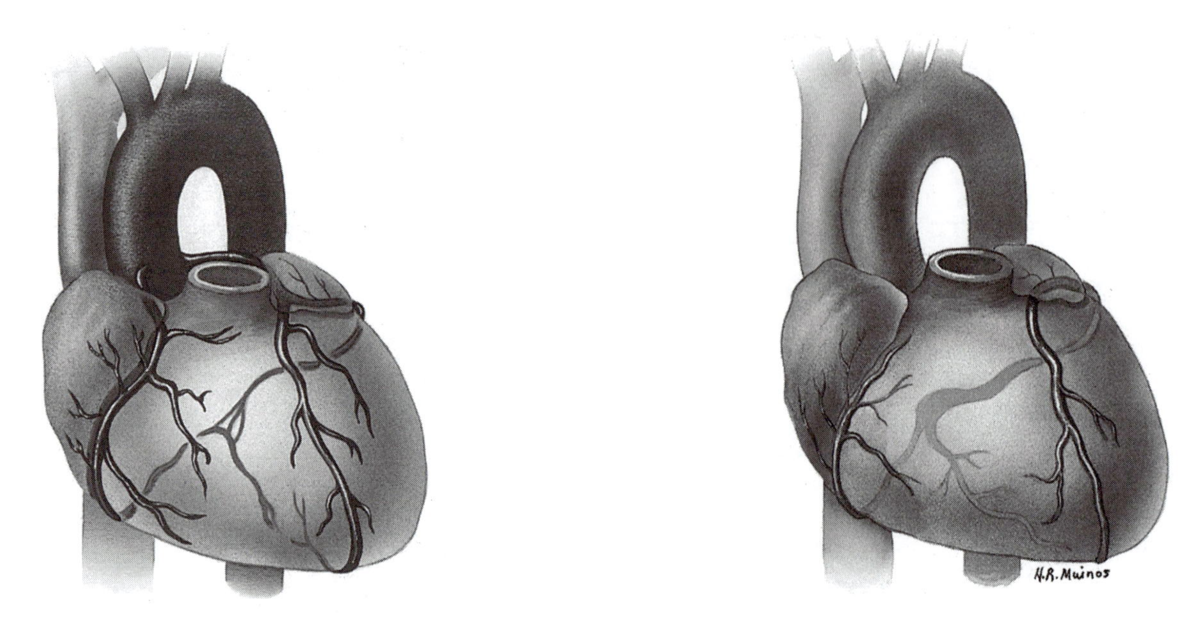
H.R.Muinos

Figure 20.9a Histology of cardiac muscle (page 649).

NOTES

20

Figure 20.9b Histology of cardiac muscle (page 649).

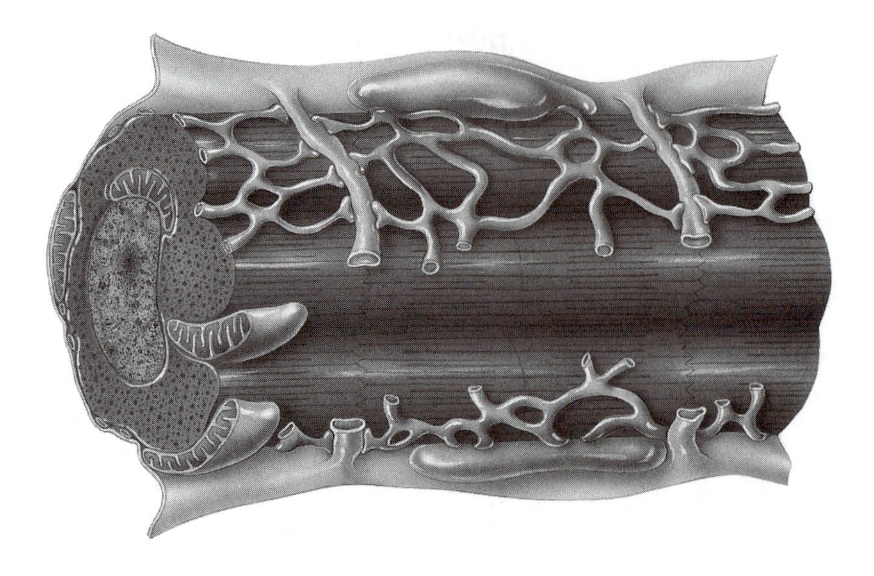

Figure 20.10 The conduction system of the heart (page 650).

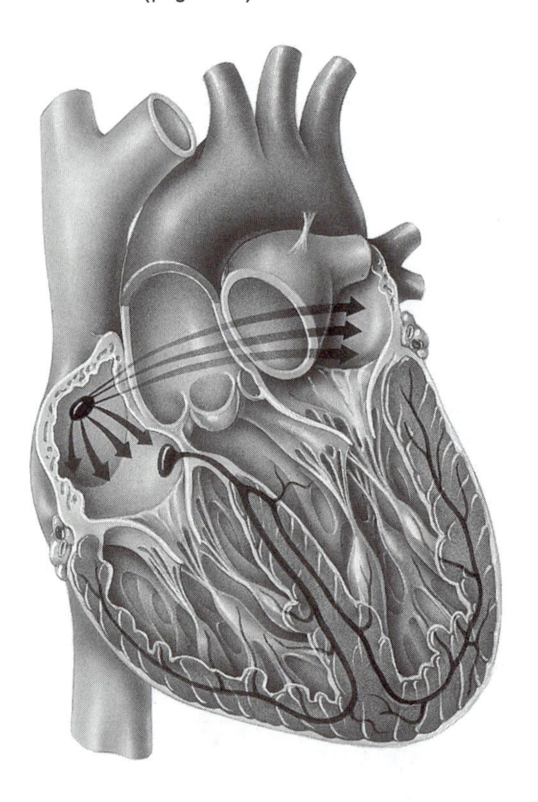

NOTES

20

Figure 20.11 Action potential in a ventricular contractile fiber (page 652).

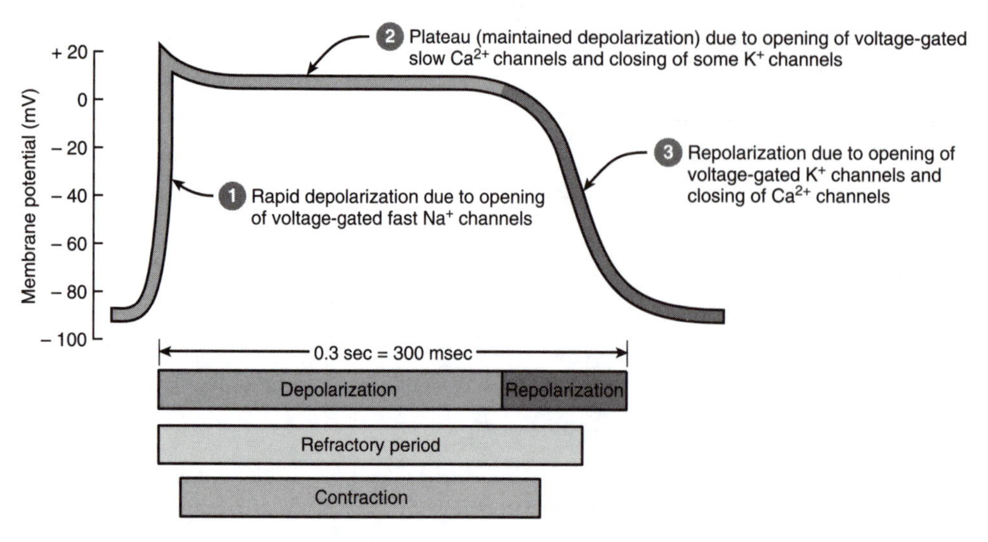

2 Plateau (maintained depolarization) due to opening of voltage-gated slow Ca^{2+} channels and closing of some K^+ channels
3 Repolarization due to opening of voltage-gated K^+ channels and closing of Ca^{2+} channels
1 Rapid depolarization due to opening of voltage-gated fast Na^+ channels
Membrane potential (mV)
+20
0
−20
−40
−60
−80
−100
0.3 sec = 300 msec
Depolarization
Repolarization
Refractory period
Contraction

(a) Action potential, refractory period, and contraction

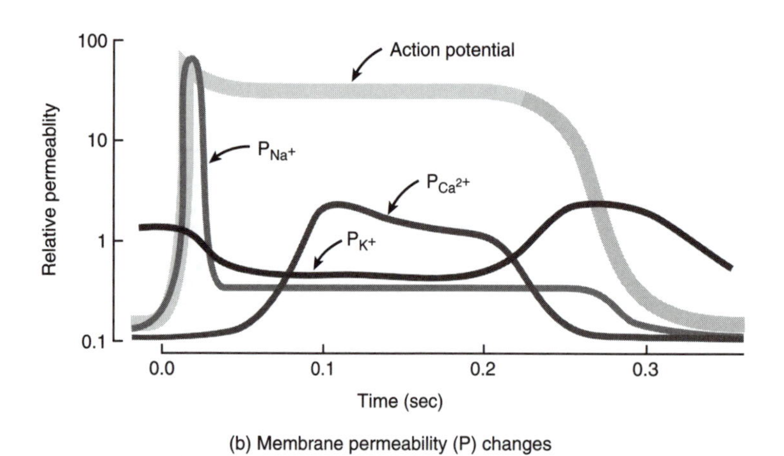

Action potential
P_{Na^+}
$P_{Ca^{2+}}$
P_{K^+}
Relative permeability
100
10
1
0.1
0.0
0.1
0.2
0.3
Time (sec)

(b) Membrane permeability (P) changes

Figure 20.12 Electrocardiogram or ECG (page 653).

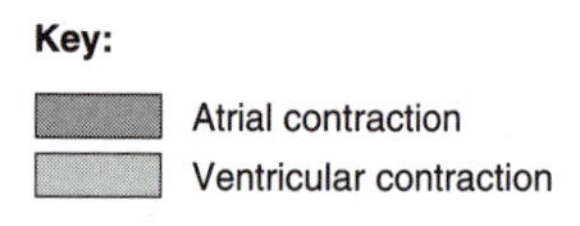

Millivolts (mV)
1.0
0.5
0
−0.5
R
P
T
Q
S
S–T segment
P–Q interval
Q–T interval
0
0.2
0.4
0.6
0.8
Seconds
Key:
Atrial contraction
Ventricular contraction

NOTES

20

Figure 20.13 Cardiac cycle (page 655).

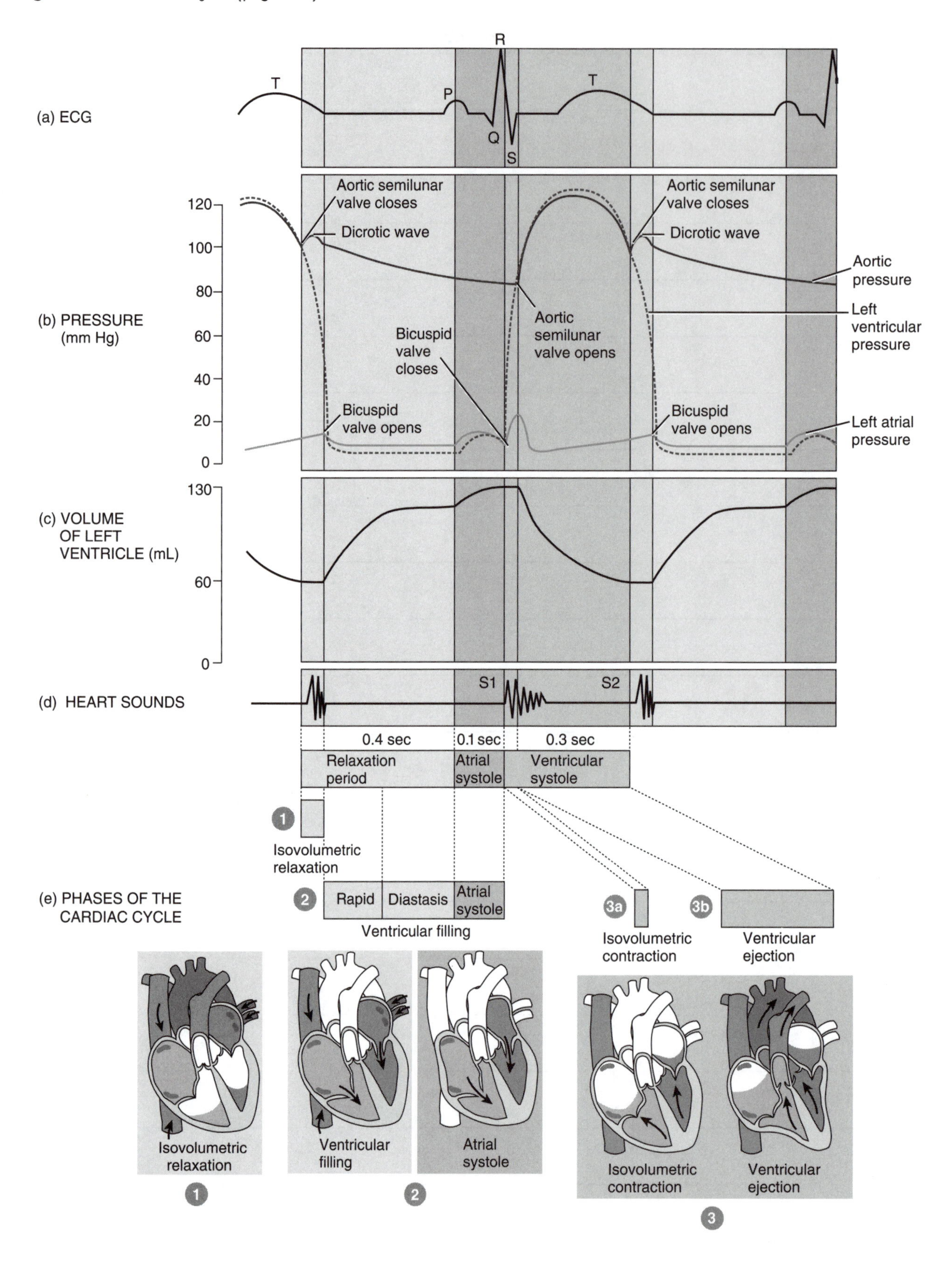

(a) ECG
T
P
R
Q
S
T
(b) PRESSURE (mm Hg)
120
100
80
60
40
20
0
Aortic semilunar valve closes
Dicrotic wave
Bicuspid valve closes
Aortic semilunar valve opens
Bicuspid valve opens
Aortic pressure
Left ventricular pressure
Left atrial pressure
(c) VOLUME OF LEFT VENTRICLE (mL)
130
60
0
(d) HEART SOUNDS
S1
S2
0.4 sec
0.1 sec
0.3 sec
Relaxation period
Atrial systole
Ventricular systole
(e) PHASES OF THE CARDIAC CYCLE
1
Isovolumetric relaxation
2
Rapid
Diastasis
Atrial systole
Ventricular filling
3a
Isovolumetric contraction
3b
Ventricular ejection
Isovolumetric relaxation
Ventricular filling
Atrial systole
Isovolumetric contraction
Ventricular ejection
3

NOTES

20

Figure 20.14 Location of valves and auscultation sites for heart sounds (page 657).

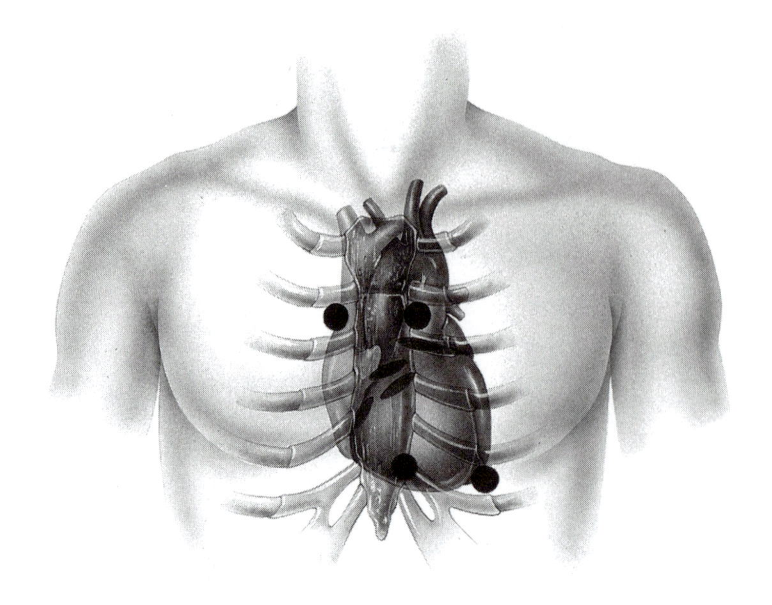

Figure 20.15 Nervous system control of the heart (page 660).

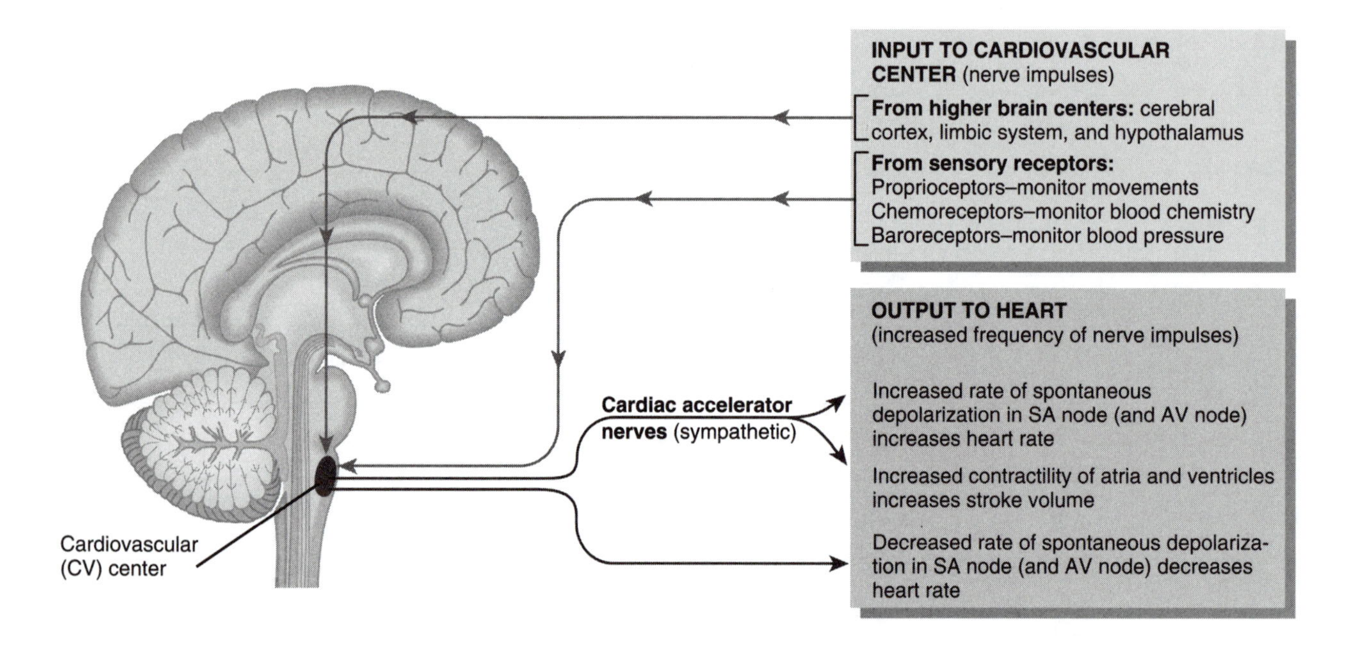

INPUT TO CARDIOVASCULAR CENTER (nerve impulses)
From higher brain centers: cerebral cortex, limbic system, and hypothalamus
From sensory receptors:
Proprioceptors–monitor movements
Chemoreceptors–monitor blood chemistry
Baroreceptors–monitor blood pressure
OUTPUT TO HEART
(increased frequency of nerve impulses)
Increased rate of spontaneous depolarization in SA node (and AV node) increases heart rate
Increased contractility of atria and ventricles increases stroke volume
Decreased rate of spontaneous depolarization in SA node (and AV node) decreases heart rate
Cardiac accelerator nerves (sympathetic)
Cardiovascular (CV) center

NOTES

20

Figure 20.16 Factors that increase cardiac output (page 661).

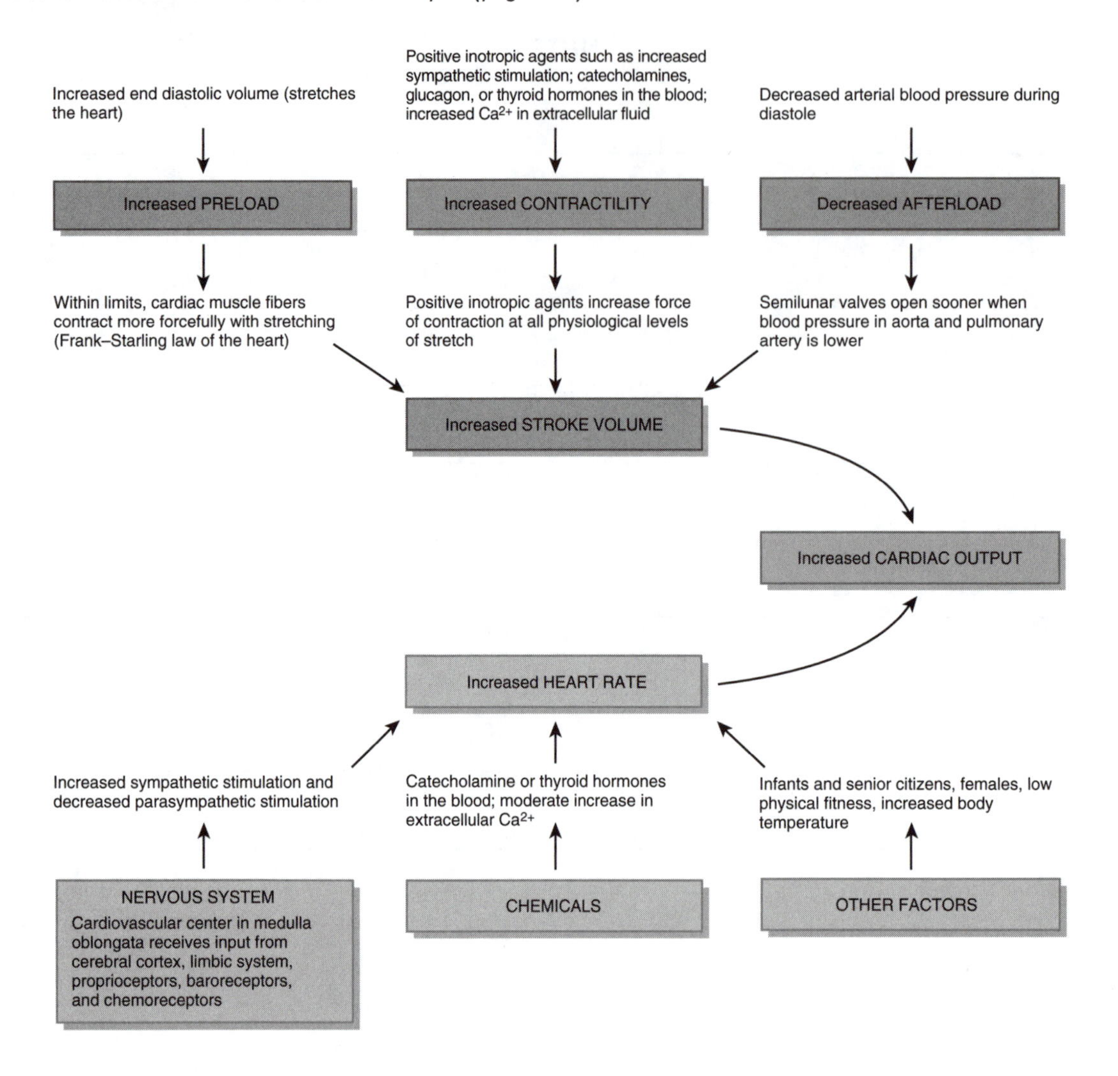

Increased end diastolic volume (stretches the heart)
Positive inotropic agents such as increased sympathetic stimulation; catecholamines, glucagon, or thyroid hormones in the blood; increased Ca^{2+} in extracellular fluid
Decreased arterial blood pressure during diastole
Increased PRELOAD
Increased CONTRACTILITY
Decreased AFTERLOAD
Within limits, cardiac muscle fibers contract more forcefully with stretching (Frank–Starling law of the heart)
Positive inotropic agents increase force of contraction at all physiological levels of stretch
Semilunar valves open sooner when blood pressure in aorta and pulmonary artery is lower
Increased STROKE VOLUME
Increased CARDIAC OUTPUT
Increased HEART RATE
Increased sympathetic stimulation and decreased parasympathetic stimulation
Catecholamine or thyroid hormones in the blood; moderate increase in extracellular Ca^{2+}
Infants and senior citizens, females, low physical fitness, increased body temperature
NERVOUS SYSTEM
Cardiovascular center in medulla oblongata receives input from cerebral cortex, limbic system, proprioceptors, baroreceptors, and chemoreceptors
CHEMICALS
OTHER FACTORS

Figure 20.17 Development of the heart (page 662).

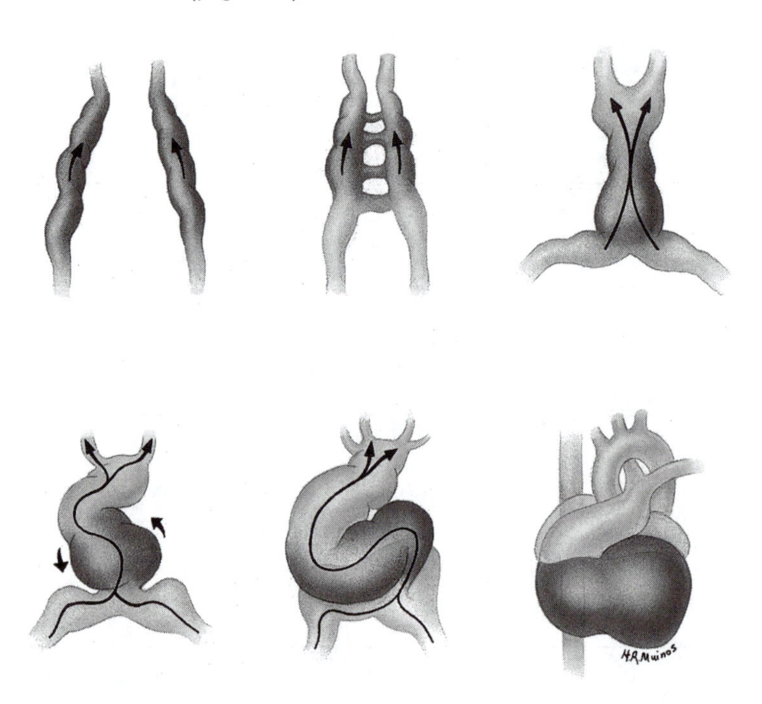

NOTES

20

Figure 21.1 Comparative structure of blood vessels (page 671).

Figure 21.2 Pressure reservoir function of elastic arteries (page 672).

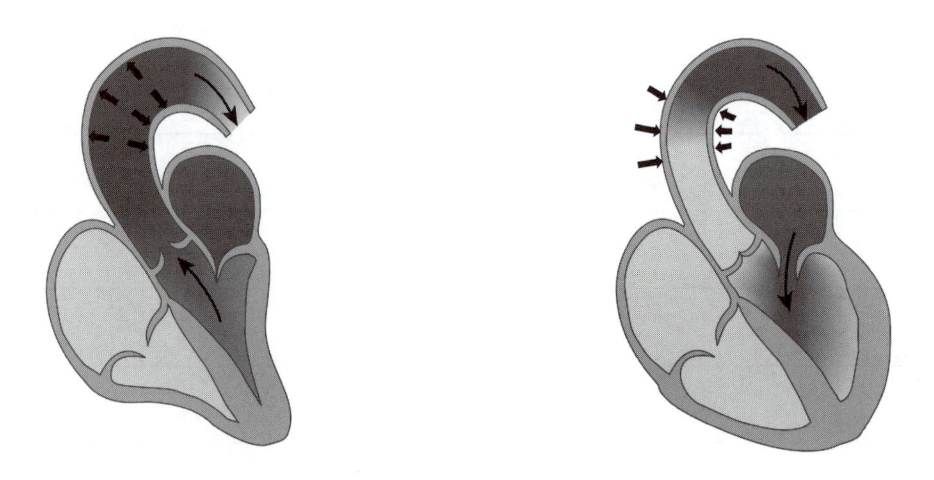

NOTES

21

Figure 21.3 Arteriole, capillaries, and venule (page 673).

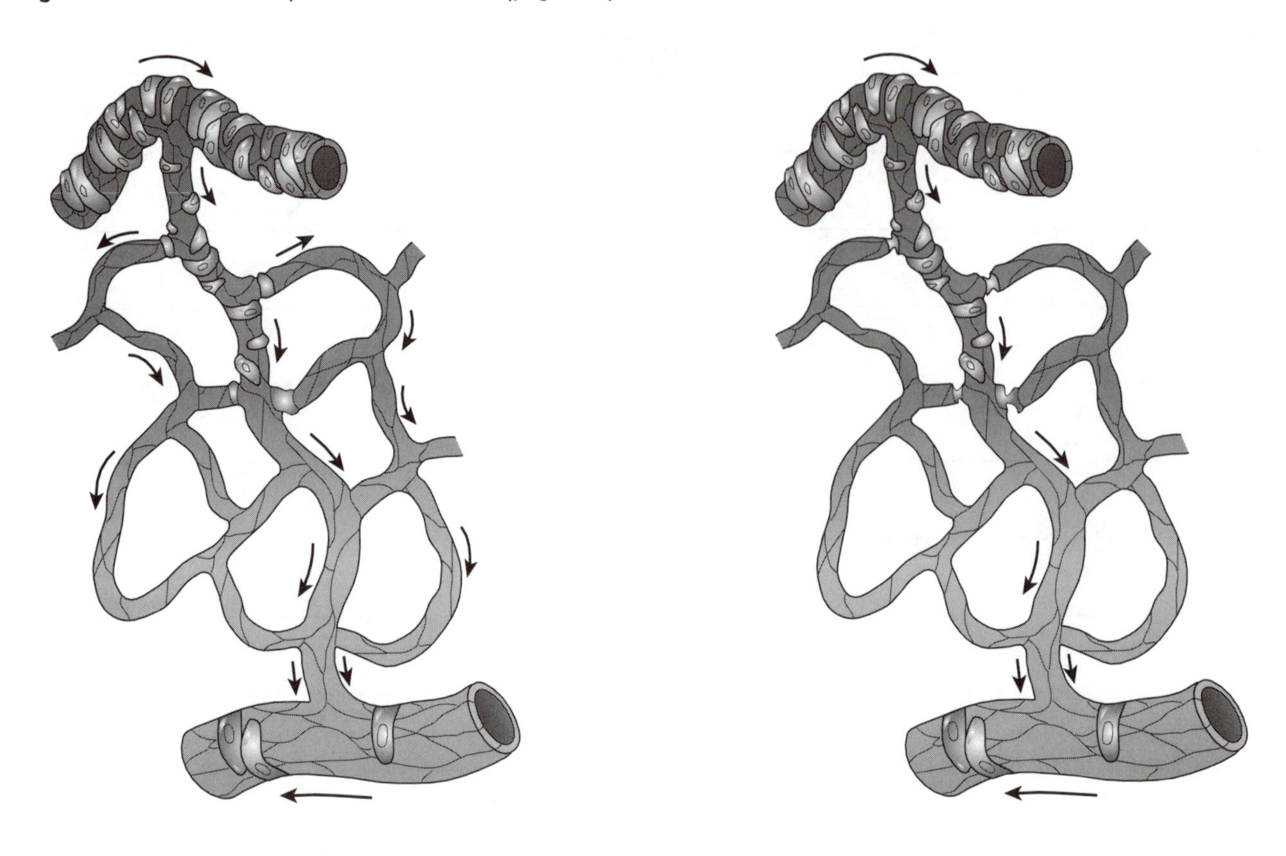

Figure 21.4 Types of capillaries (page 674).

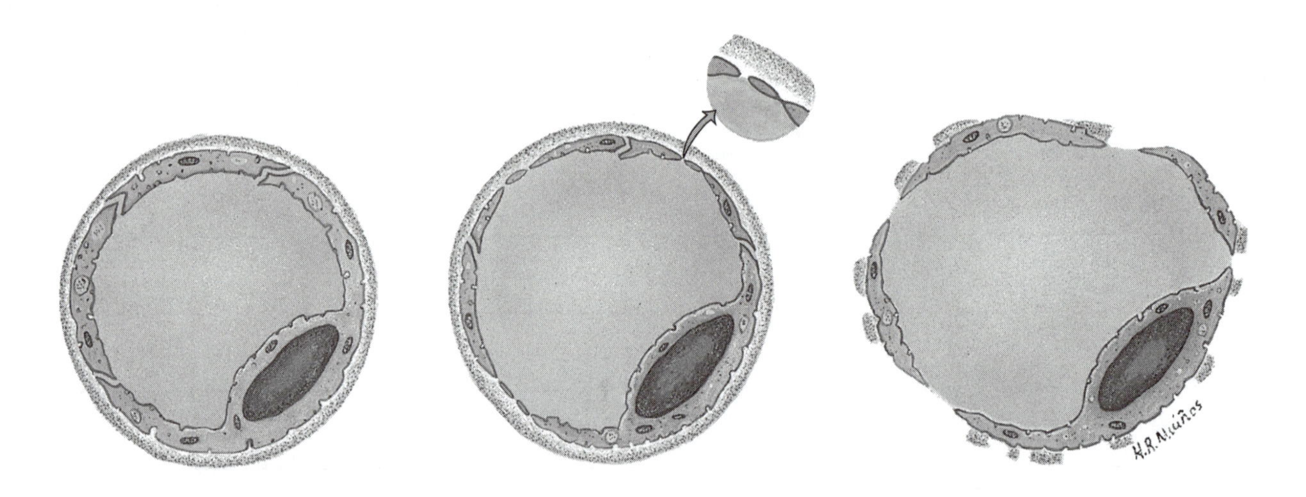

NOTES

21

Figure 21.6 Blood distribution in the cardiovascular system at rest (page 675).

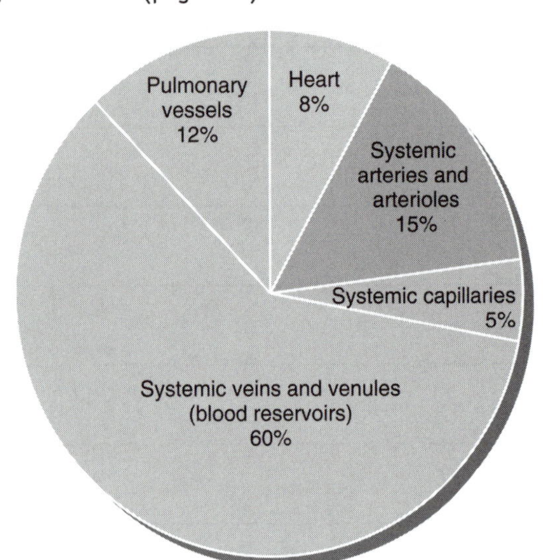

Pulmonary vessels 12%
Heart 8%
Systemic arteries and arterioles 15%
Systemic capillaries 5%
Systemic veins and venules (blood reservoirs) 60%

Figure 21.8 Relationship between velocity of blood flow and total cross-sectional area (page 679).

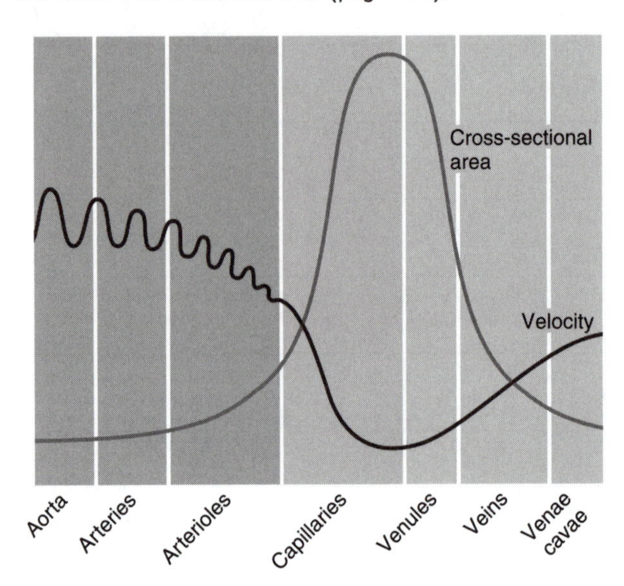

Cross-sectional area
Velocity
Aorta
Arteries
Arterioles
Capillaries
Venules
Veins
Venae cavae

Figure 21.7 Dynamics of capillary exchange (page 677).

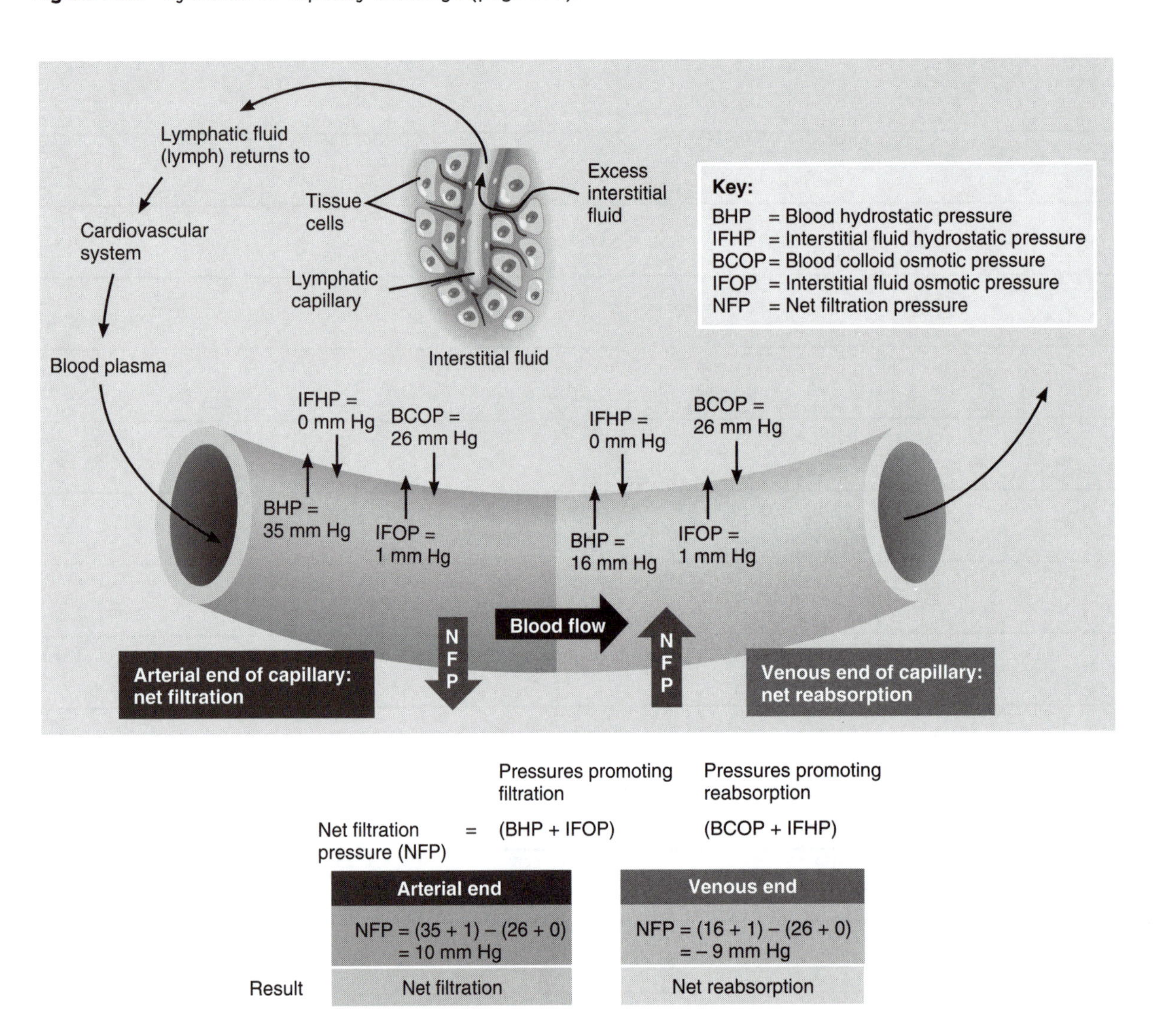

Lymphatic fluid (lymph) returns to
Cardiovascular system
Blood plasma
Tissue cells
Lymphatic capillary
Excess interstitial fluid
Interstitial fluid
Key:
BHP = Blood hydrostatic pressure
IFHP = Interstitial fluid hydrostatic pressure
BCOP = Blood colloid osmotic pressure
IFOP = Interstitial fluid osmotic pressure
NFP = Net filtration pressure
IFHP = 0 mm Hg
BCOP = 26 mm Hg
BHP = 35 mm Hg
IFOP = 1 mm Hg
IFHP = 0 mm Hg
BCOP = 26 mm Hg
BHP = 16 mm Hg
IFOP = 1 mm Hg
NFP
Blood flow
NFP
Arterial end of capillary: net filtration
Venous end of capillary: net reabsorption
Pressures promoting filtration
Pressures promoting reabsorption
Net filtration pressure (NFP) = (BHP + IFOP) (BCOP + IFHP)
Arterial end
NFP = (35 + 1) – (26 + 0) = 10 mm Hg
Venous end
NFP = (16 + 1) – (26 + 0) = – 9 mm Hg
Result
Net filtration
Net reabsorption

NOTES

21

Figure 21.9 Blood pressures in various parts of the cardiovascular system (page 679).

Figure 21.10 Action of the skeletal muscle pump returning blood to the heart (page 680).

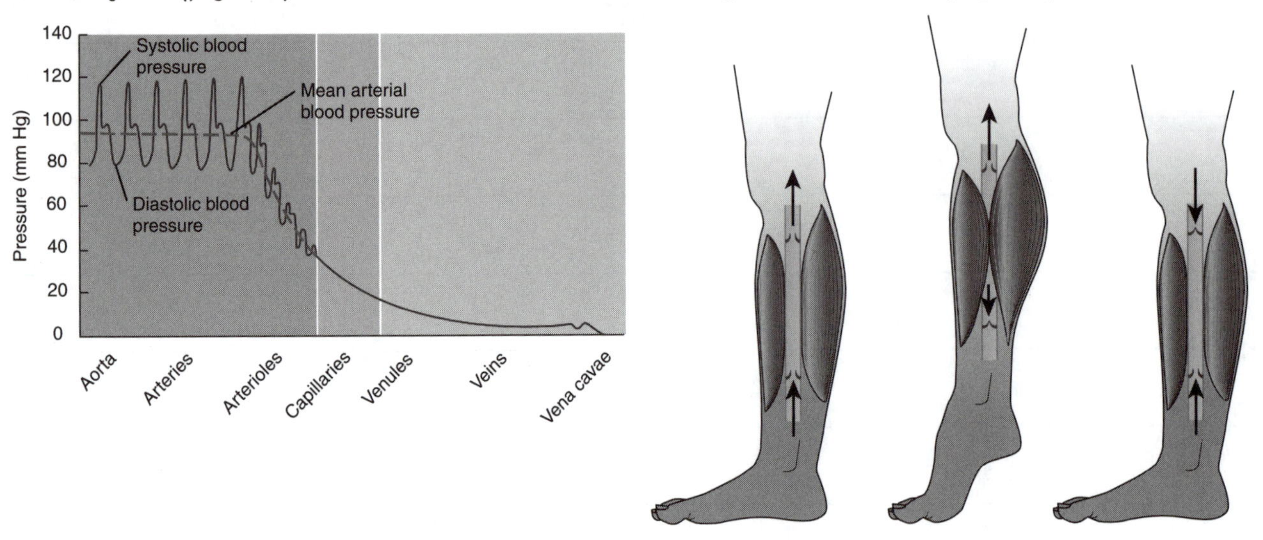

140
120
100
80
60
40
20
0
Pressure (mm Hg)
Systolic blood pressure
Mean arterial blood pressure
Diastolic blood pressure
Aorta
Arteries
Arterioles
Capillaries
Venules
Veins
Vena cavae

Figure 21.11 Summary of factors that increase blood pressure (page 681).

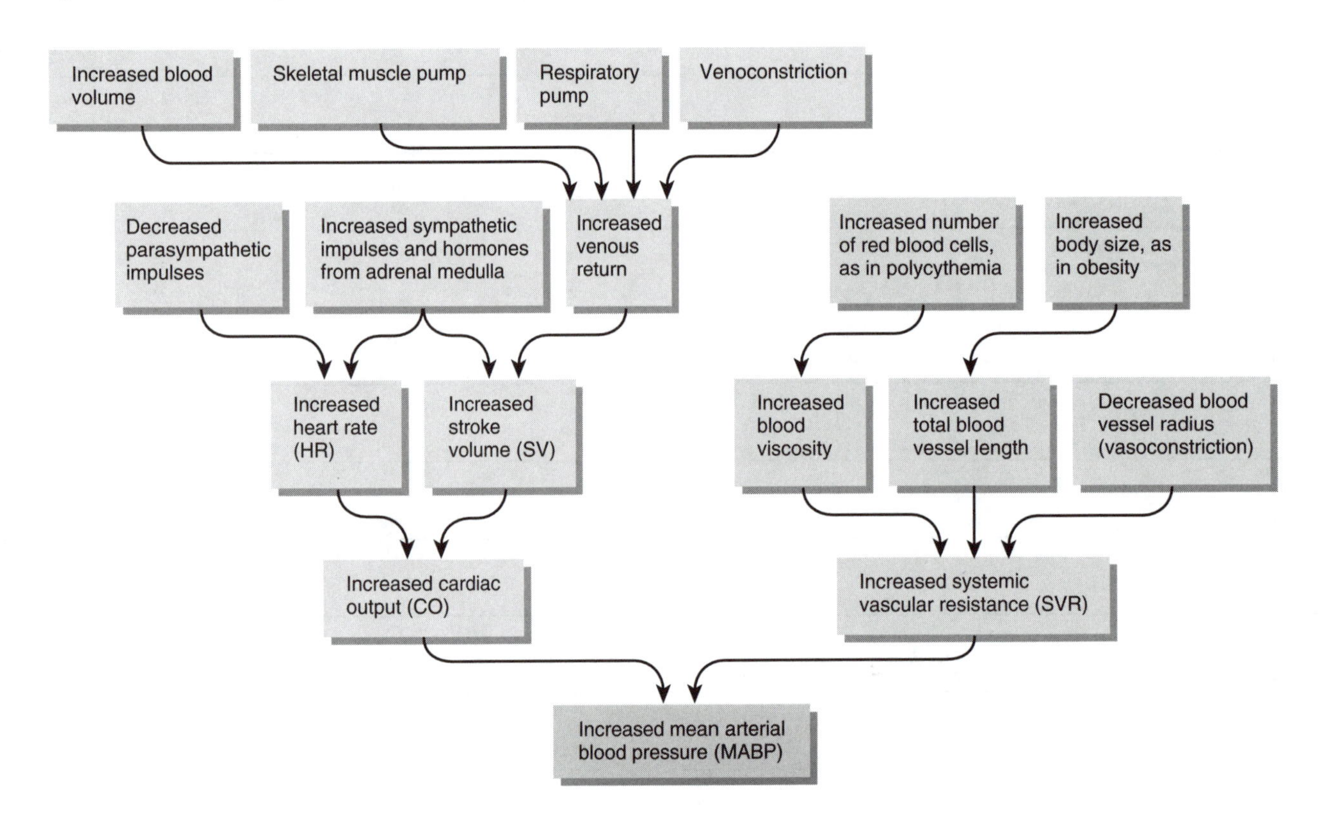

Increased blood volume
Skeletal muscle pump
Respiratory pump
Venoconstriction
Decreased parasympathetic impulses
Increased sympathetic impulses and hormones from adrenal medulla
Increased venous return
Increased number of red blood cells, as in polycythemia
Increased body size, as in obesity
Increased heart rate (HR)
Increased stroke volume (SV)
Increased blood viscosity
Increased total blood vessel length
Decreased blood vessel radius (vasoconstriction)
Increased cardiac output (CO)
Increased systemic vascular resistance (SVR)
Increased mean arterial blood pressure (MABP)

NOTES

21

Figure 21.12 The cardiovascular center (page 683).

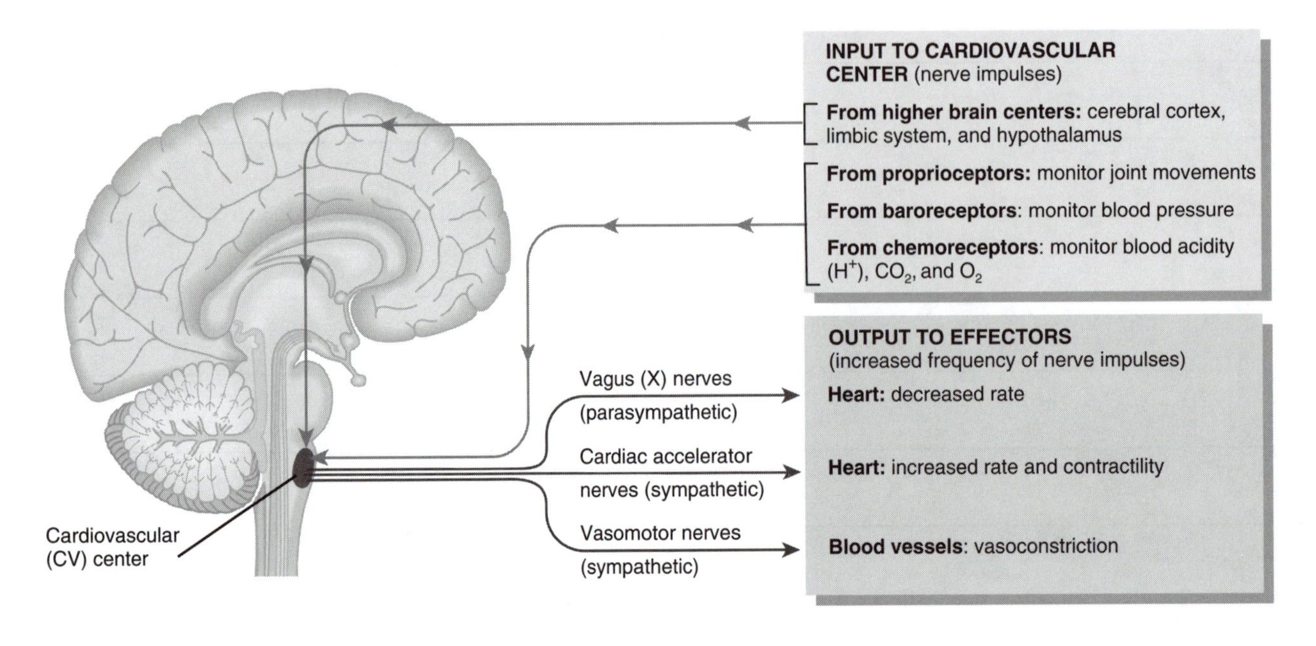

INPUT TO CARDIOVASCULAR CENTER (nerve impulses)
From higher brain centers: cerebral cortex, limbic system, and hypothalamus
From proprioceptors: monitor joint movements
From baroreceptors: monitor blood pressure
From chemoreceptors: monitor blood acidity (H^+), CO_2, and O_2
OUTPUT TO EFFECTORS (increased frequency of nerve impulses)
Heart: decreased rate
Heart: increased rate and contractility
Blood vessels: vasoconstriction
Vagus (X) nerves (parasympathetic)
Cardiac accelerator nerves (sympathetic)
Vasomotor nerves (sympathetic)
Cardiovascular (CV) center

Figure 21.13 ANS innervation of the heart and baroreceptor reflexes (page 684).

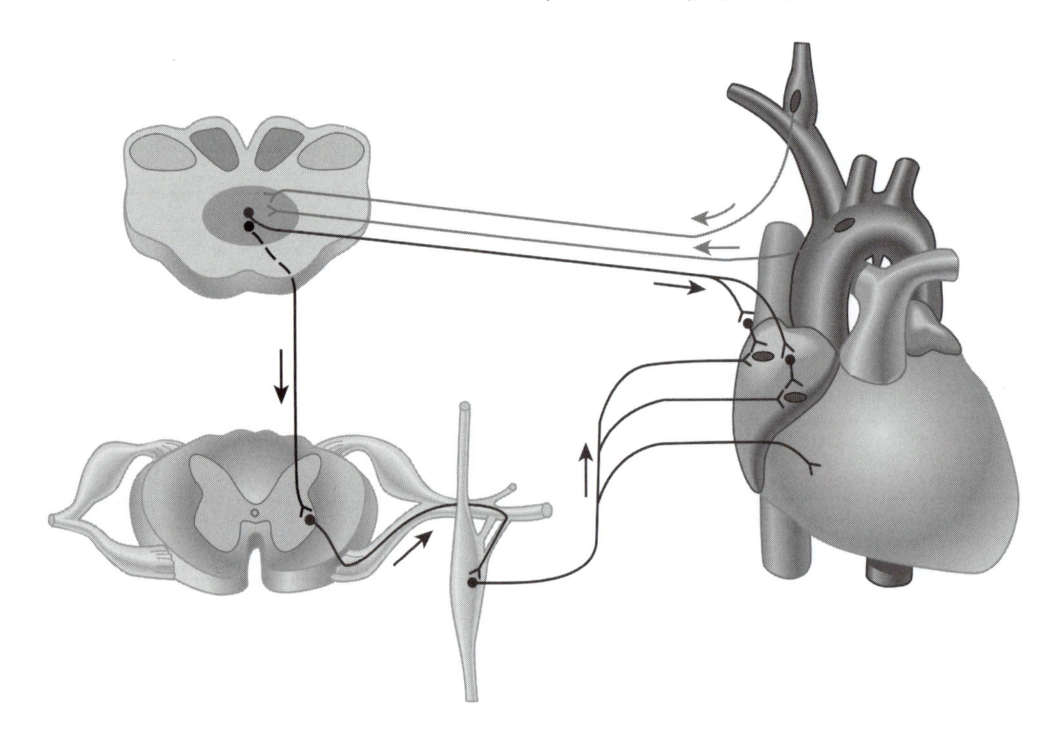

NOTES

21

Figure 21.14 Negative feedback regulation of blood pressure via baroreceptor reflexes (page 685).

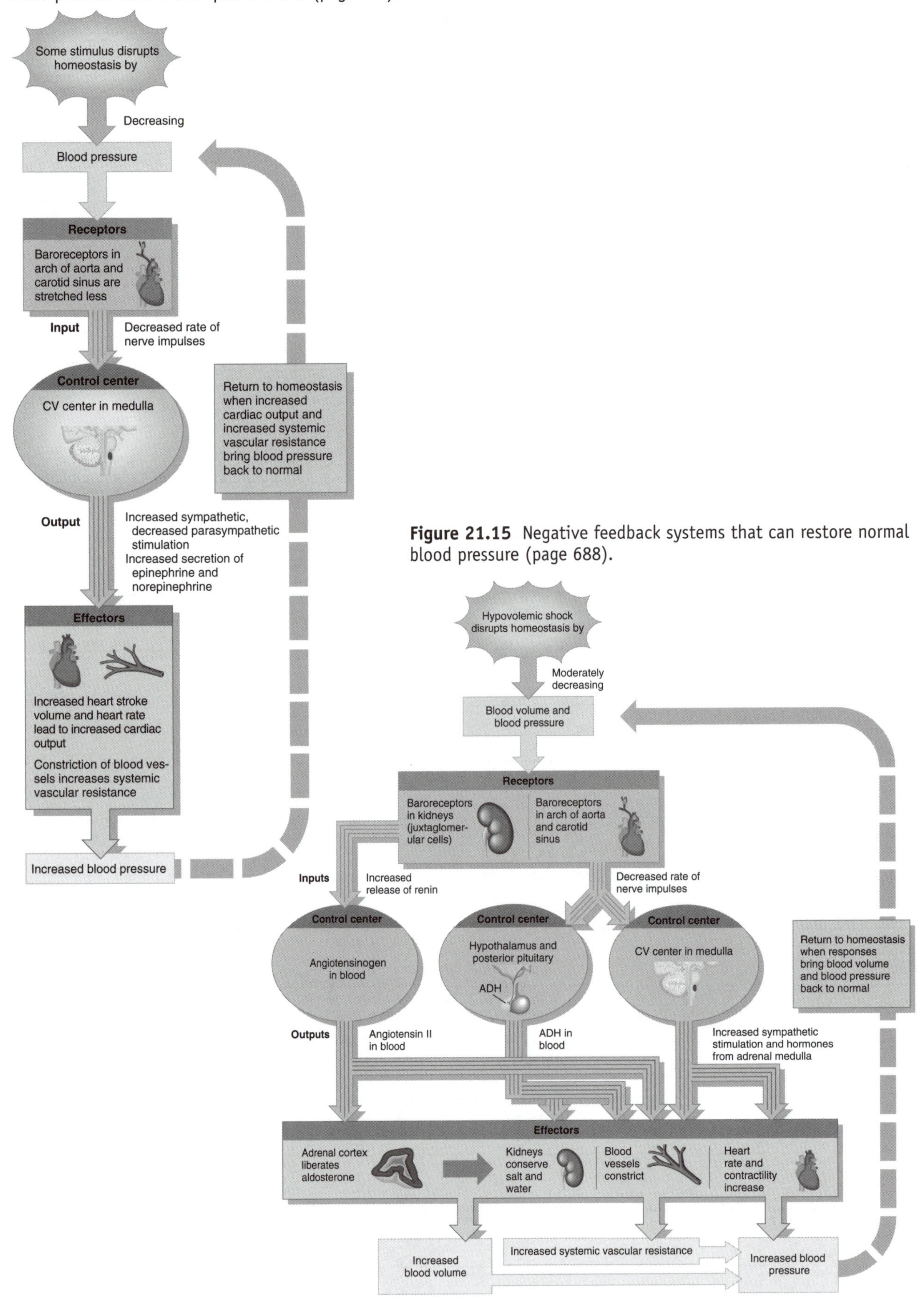
Some stimulus disrupts homeostasis by
Decreasing
Blood pressure
Receptors
Baroreceptors in arch of aorta and carotid sinus are stretched less
Input
Decreased rate of nerve impulses
Control center
CV center in medulla
Output
Increased sympathetic, decreased parasympathetic stimulation
Increased secretion of epinephrine and norepinephrine
Effectors
Increased heart stroke volume and heart rate lead to increased cardiac output
Constriction of blood vessels increases systemic vascular resistance
Increased blood pressure
Return to homeostasis when increased cardiac output and increased systemic vascular resistance bring blood pressure back to normal
Hypovolemic shock disrupts homeostasis by
Moderately decreasing
Blood volume and blood pressure
Receptors
Baroreceptors in kidneys (juxtaglomerular cells)
Baroreceptors in arch of aorta and carotid sinus
Inputs
Increased release of renin
Decreased rate of nerve impulses
Control center
Angiotensinogen in blood
Control center
Hypothalamus and posterior pituitary
ADH
Control center
CV center in medulla
Return to homeostasis when responses bring blood volume and blood pressure back to normal
Outputs
Angiotensin II in blood
ADH in blood
Increased sympathetic stimulation and hormones from adrenal medulla
Effectors
Adrenal cortex liberates aldosterone
Kidneys conserve salt and water
Blood vessels constrict
Heart rate and contractility increase
Increased blood volume
Increased systemic vascular resistance
Increased blood pressure

Figure 21.15 Negative feedback systems that can restore normal blood pressure (page 688).

NOTES

21

Figure 21.16 Relationship of blood pressure changes to cuff pressure (page 690).

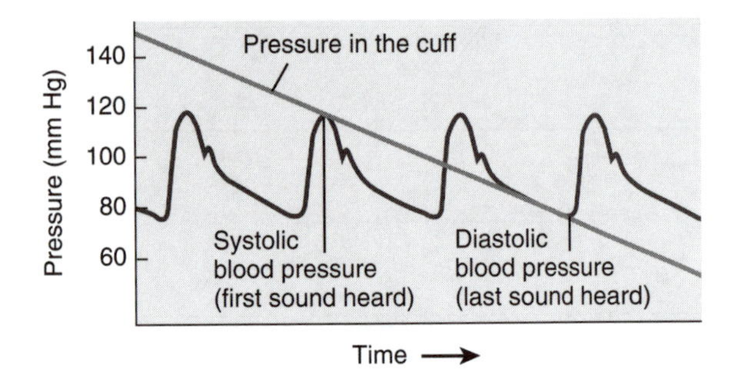
Pressure in the cuff
140
120
100
80
60
Pressure (mm Hg)
Systolic
blood pressure
(first sound heard)
Diastolic
blood pressure
(last sound heard)
Time

Figure 21.17 Circulatory routes (page 691).

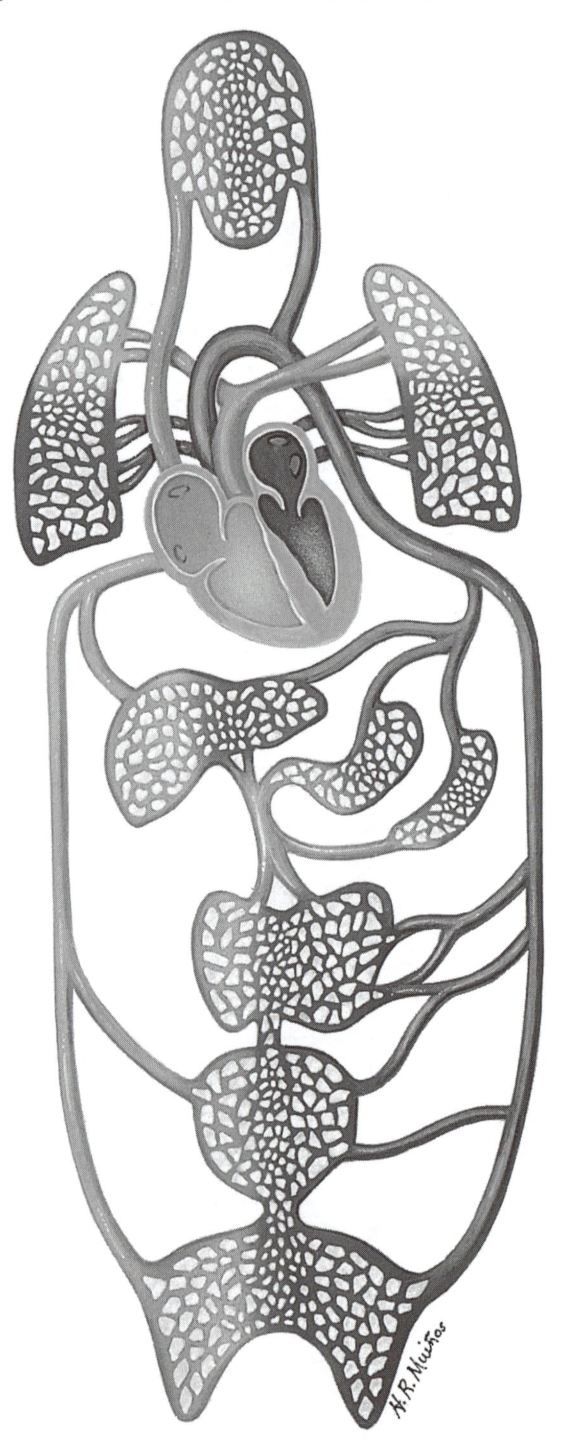

NOTES

21

Figure 21.18 Aorta and its principal branches (page 693, 694).

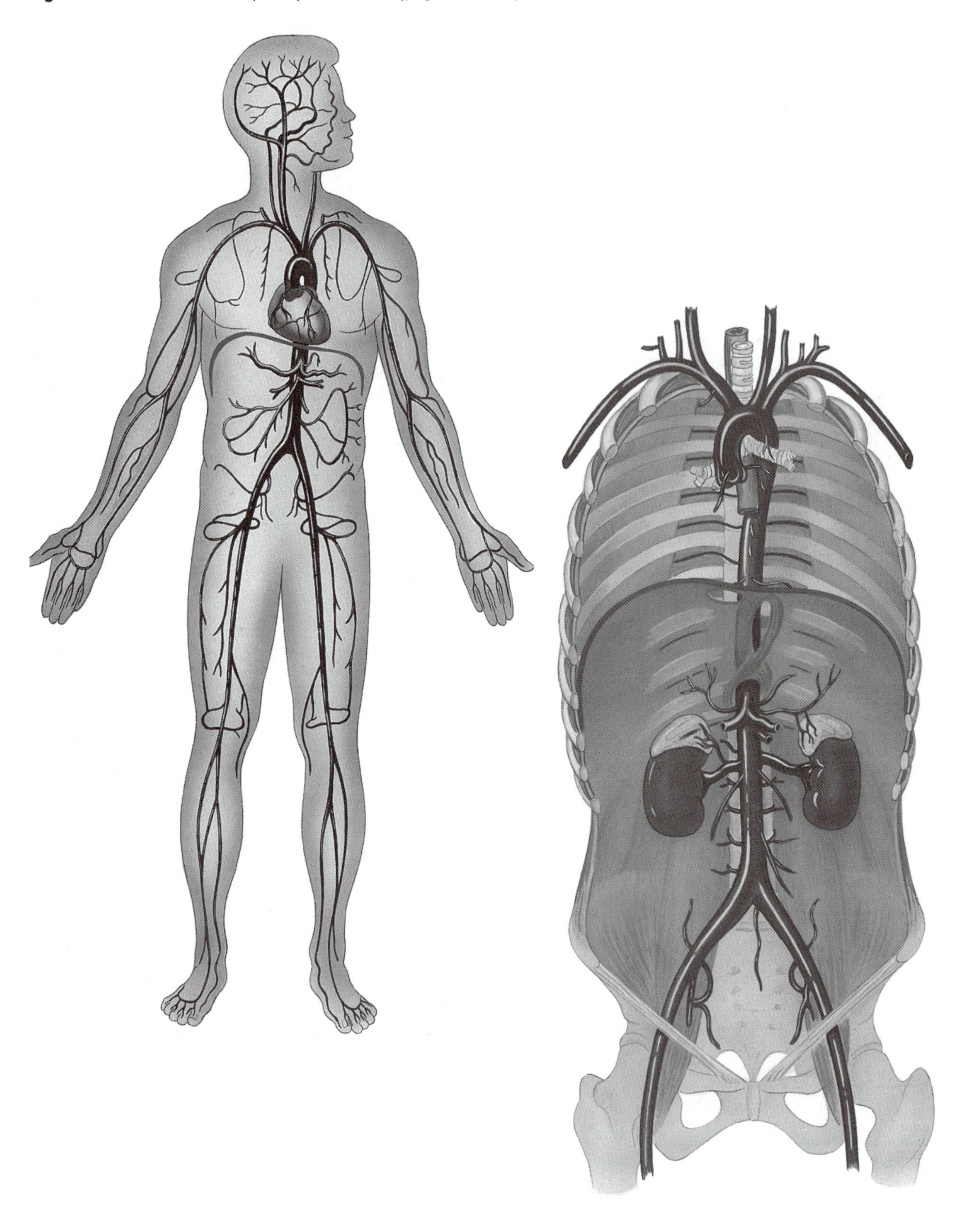

NOTES

21

Figure 21.19 Ascending aorta and its branches (page 695).

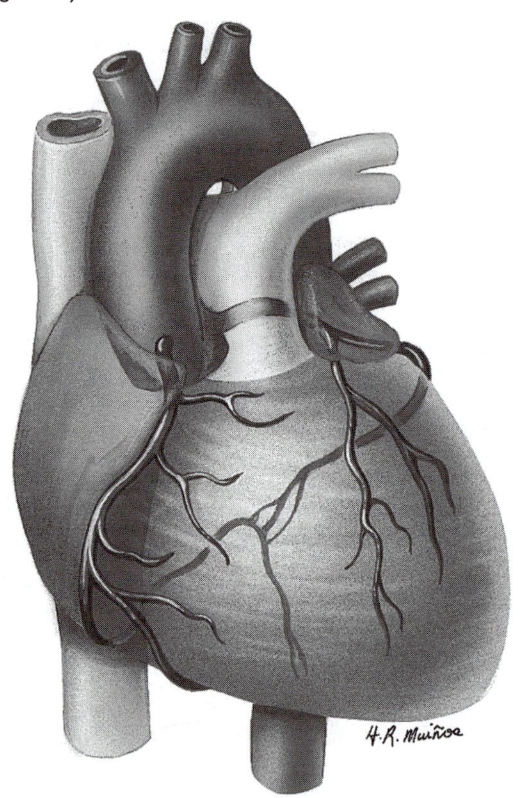
H.R. Muiños

Figure 21.20a-b Arch of the aorta and its branches (page 699).

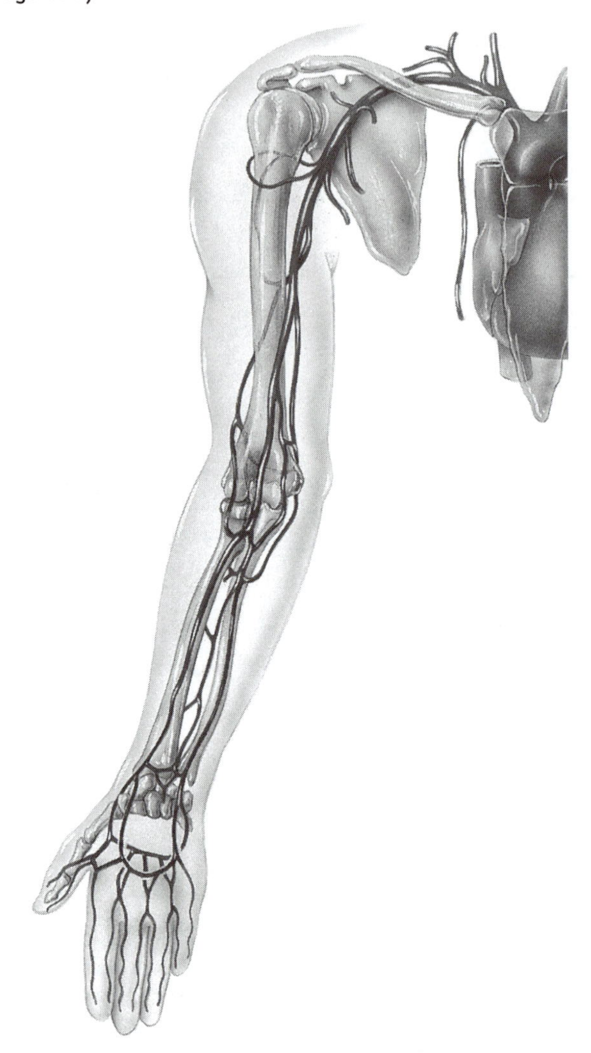

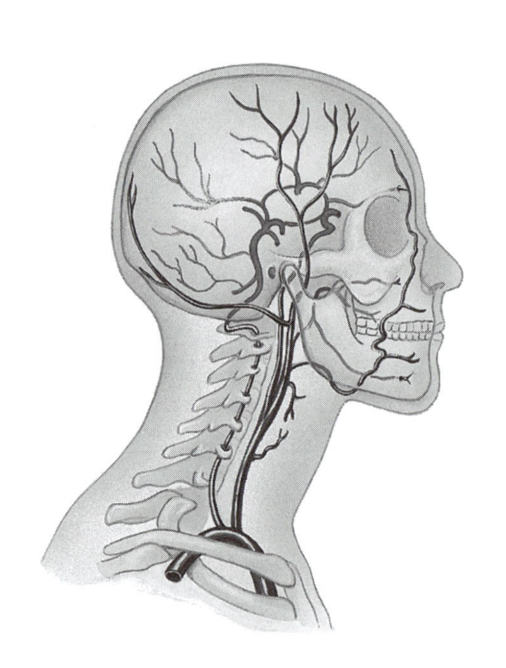

NOTES

21

Figure 21.20c Arch of the aorta and its branches (page 699).

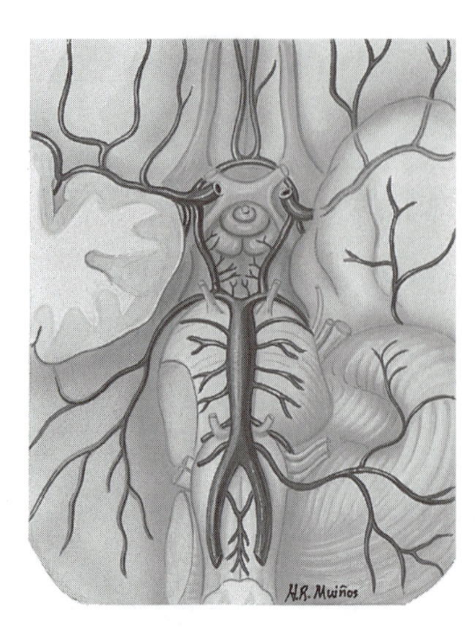
H.R. Muiños

Figure 21.21 Thoracic and abdominal aorta (page 701).

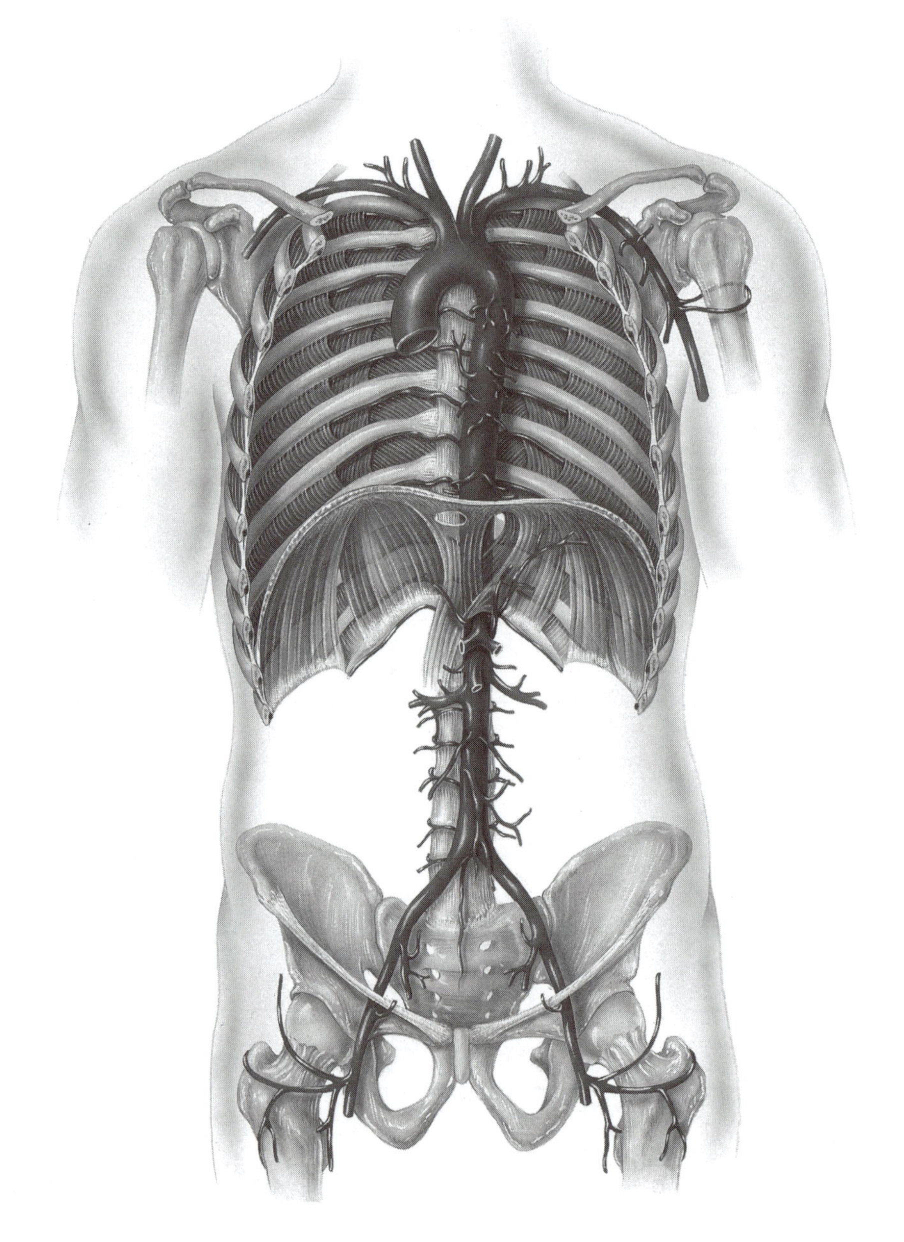

NOTES

21

Figure 21.22a-b Abdominal aorta (page 705).

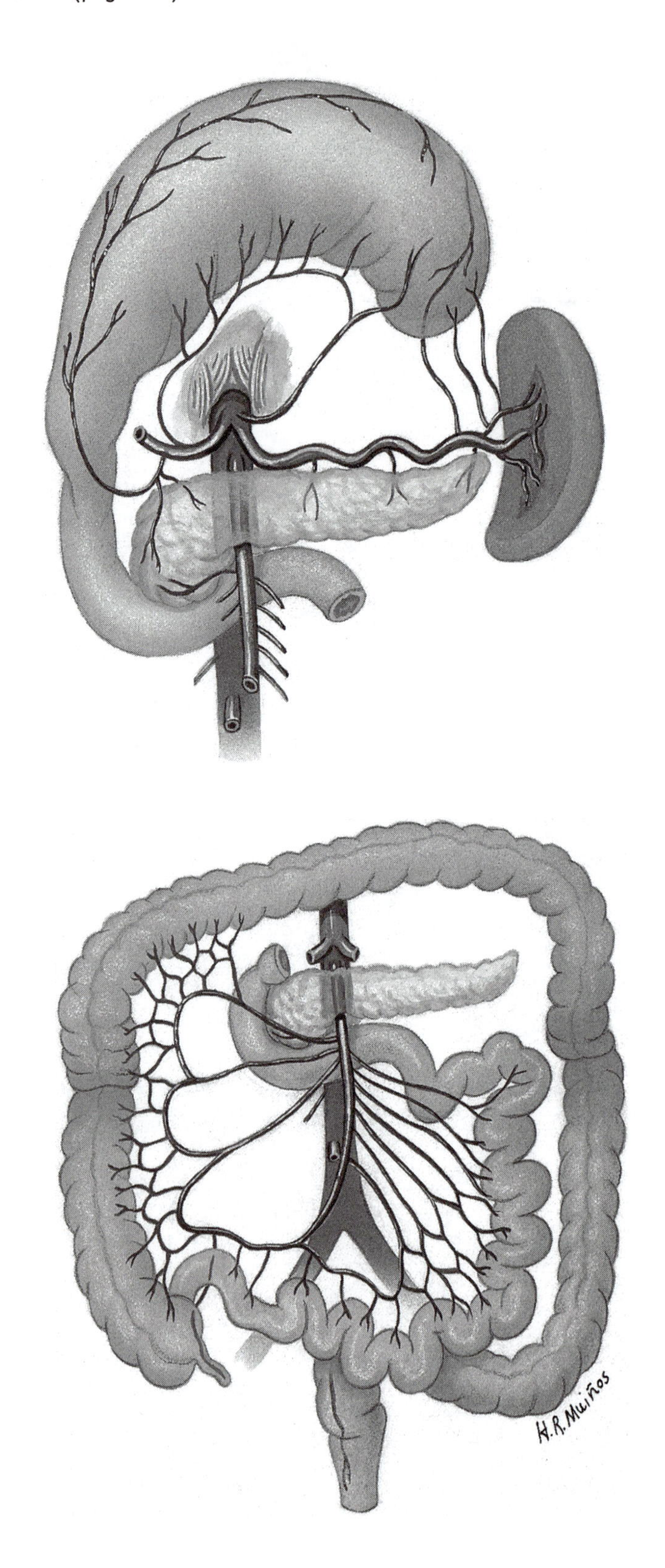
H.R.Muiños

NOTES

21

Figure 21.22c Abdominal aorta (page 706).

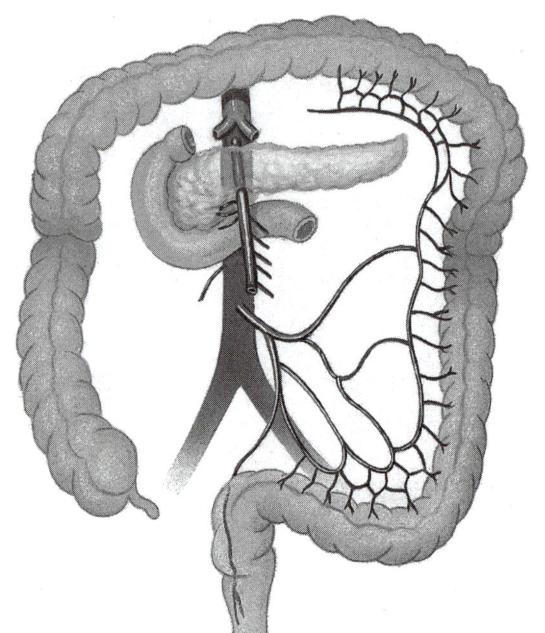

Figure 21.23 Arteries of the pelvis and right lower limb (page 709).

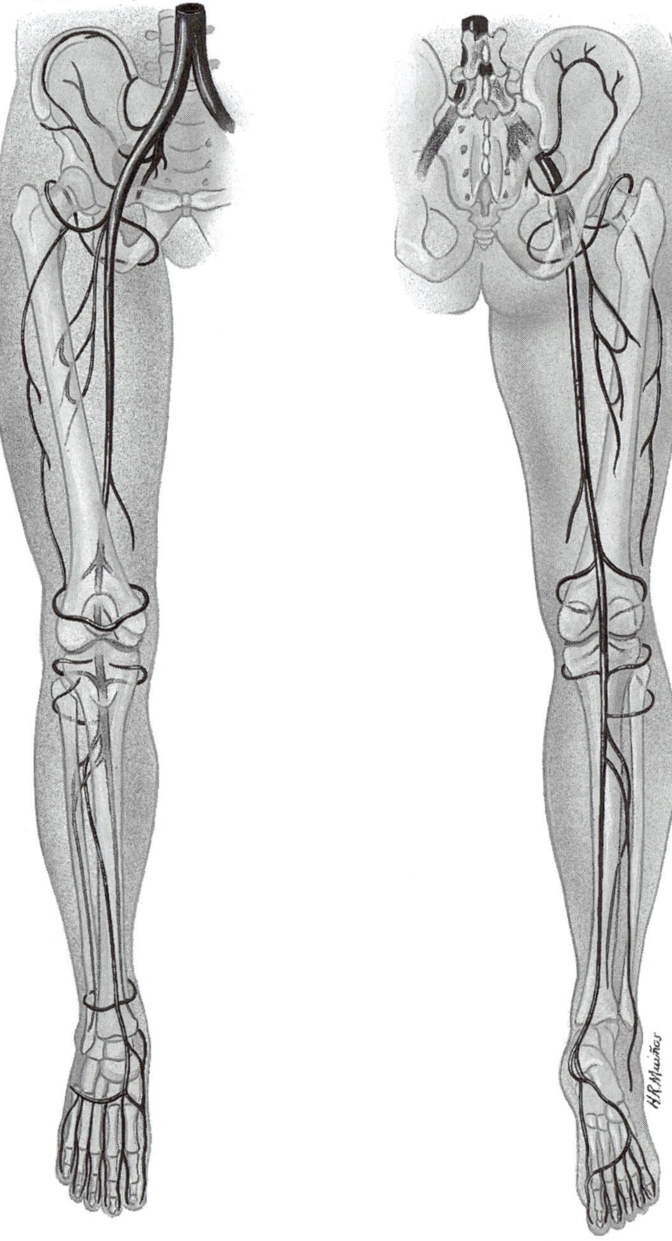

NOTES

21

Figure 21.24 Principal veins (page 711).

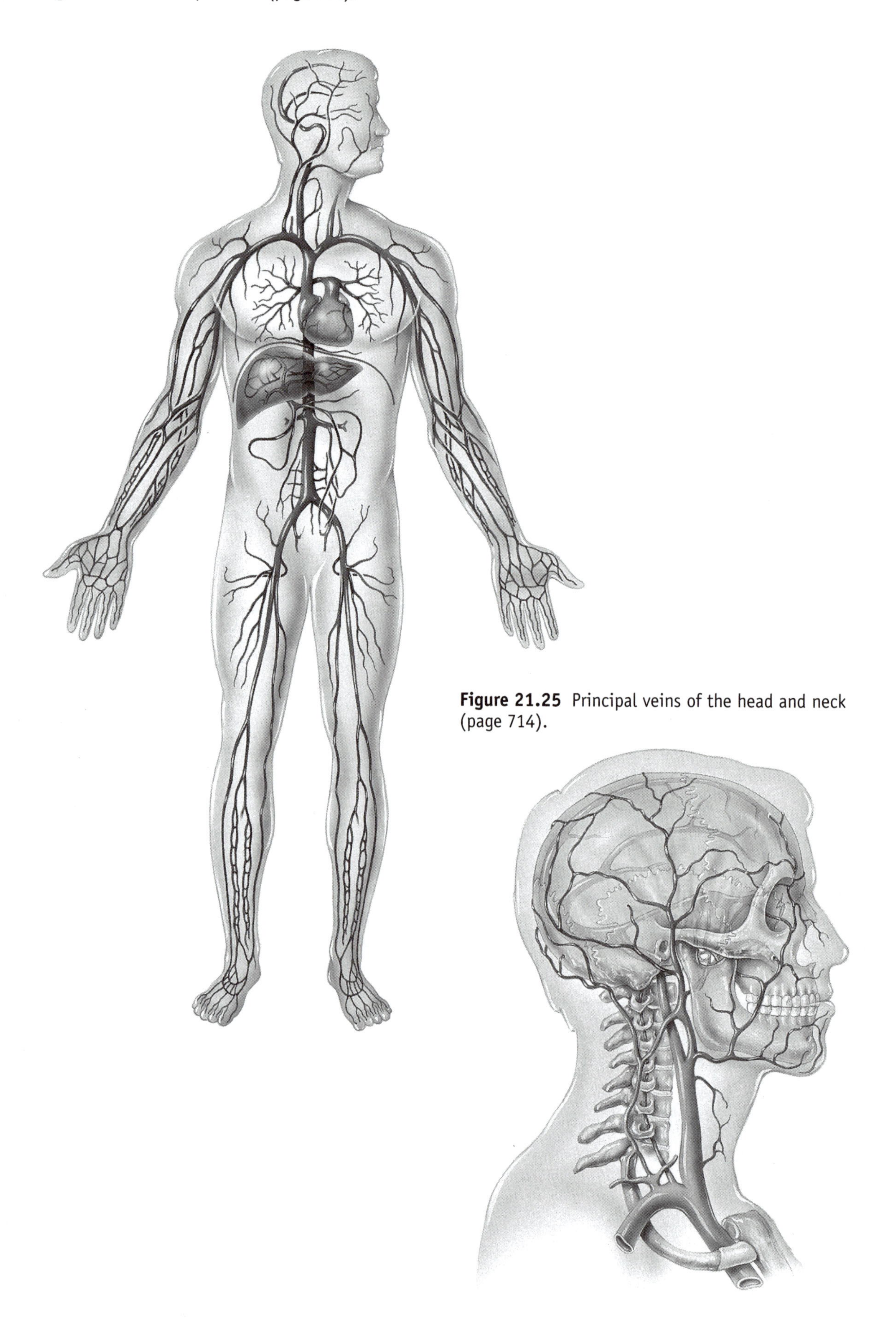

Figure 21.25 Principal veins of the head and neck (page 714).

NOTES

21

Figure 21.26 Principal veins of the right upper limb (page 717).

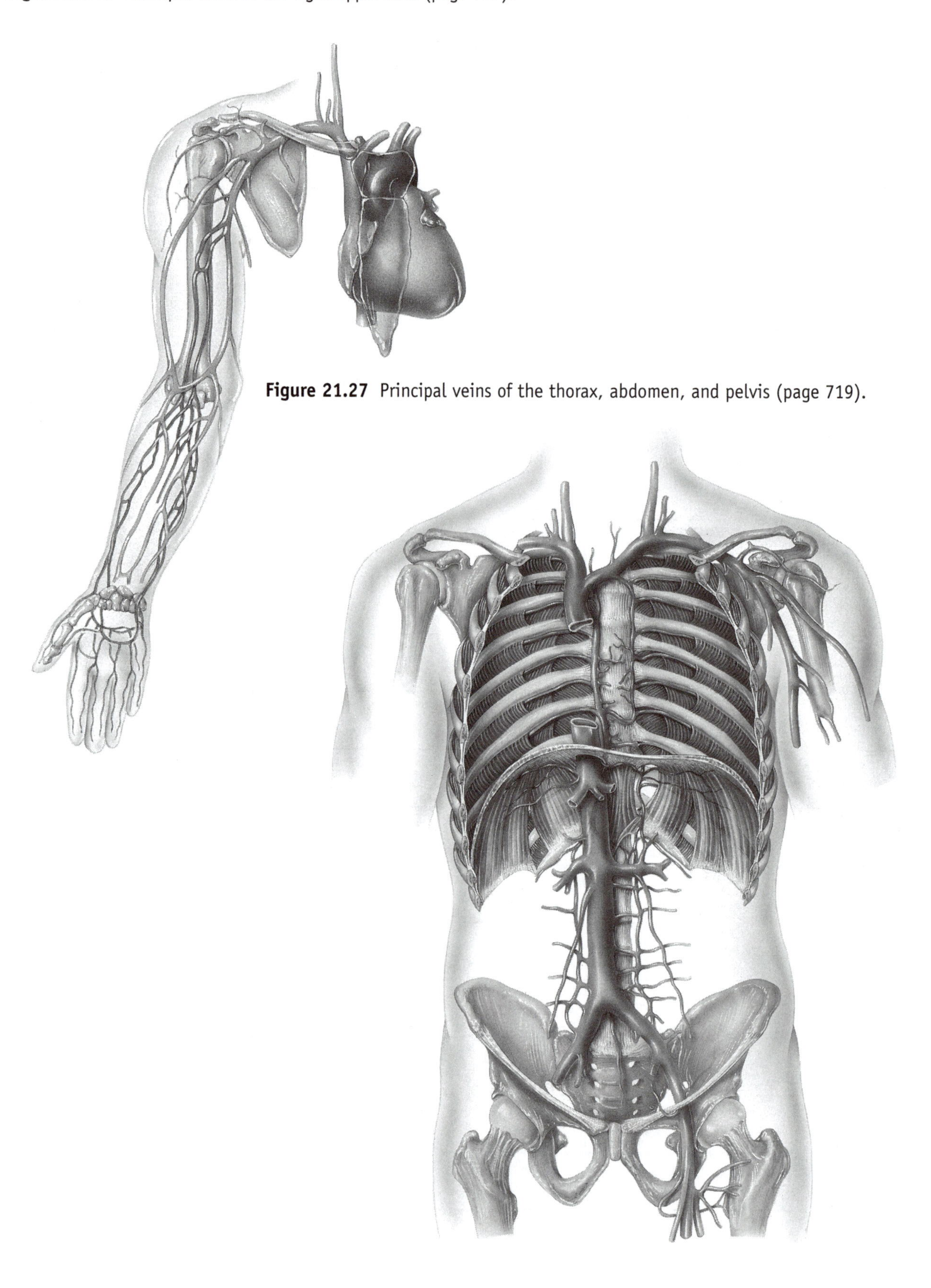

Figure 21.27 Principal veins of the thorax, abdomen, and pelvis (page 719).

NOTES

Figure 21.28 Principal veins of the pelvis and lower limbs (page 725).

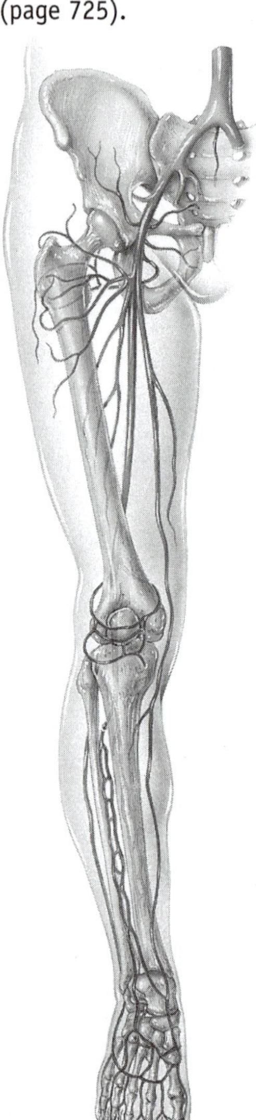

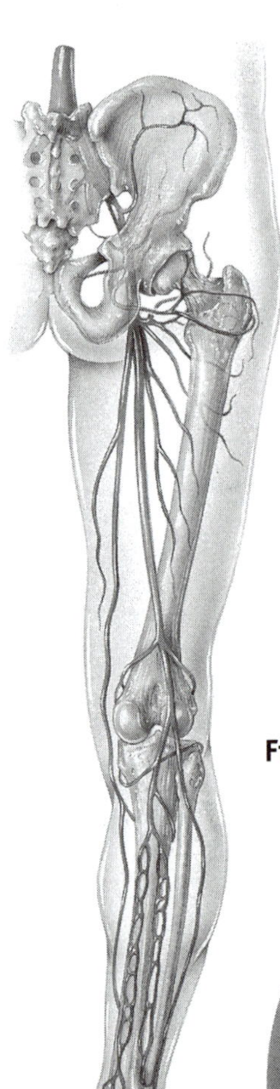

Figure 21.29 Hepatic portal circulation (page 727).

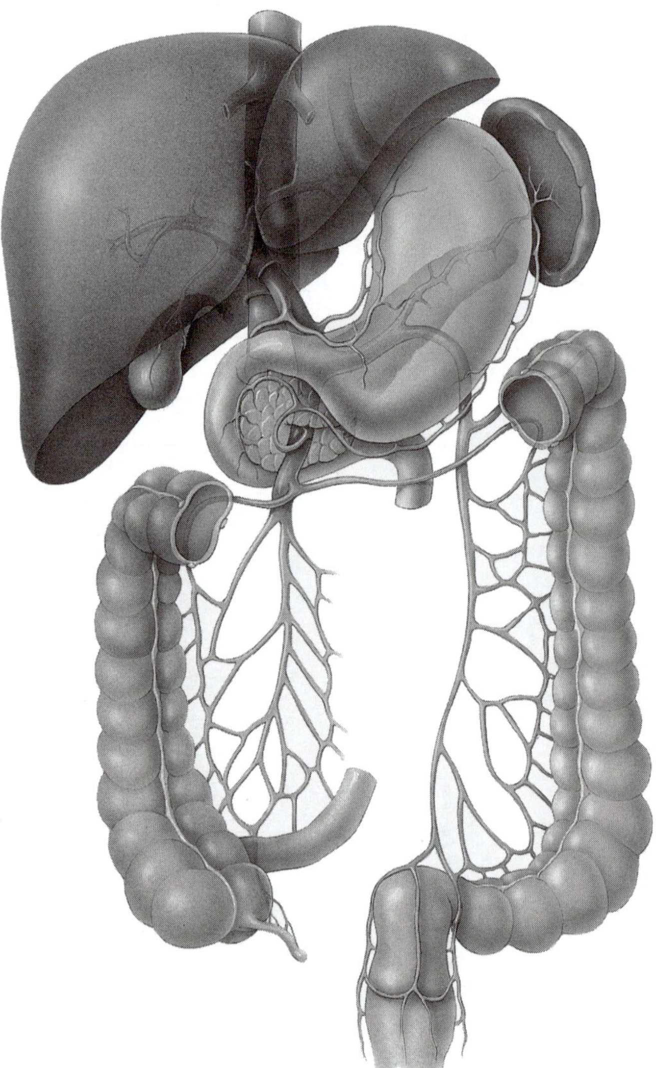

NOTES

21

Figure 21.30 Pulmonary circulation (page 729).

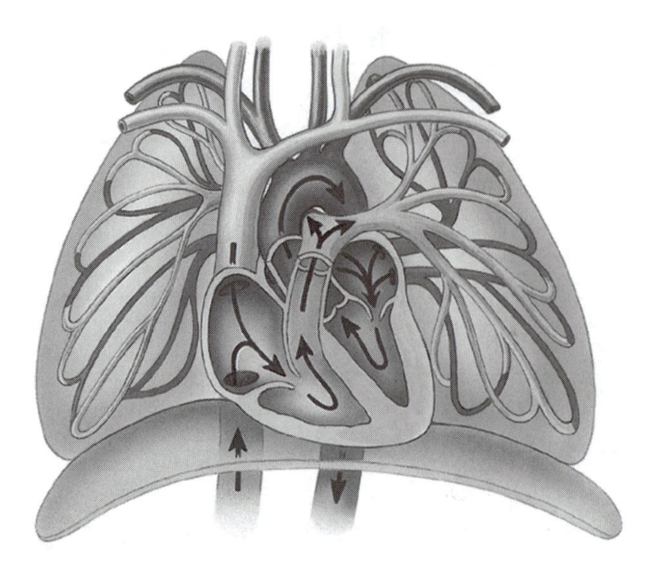

Figure 21.31 Fetal circulation and changes at birth (page 730).

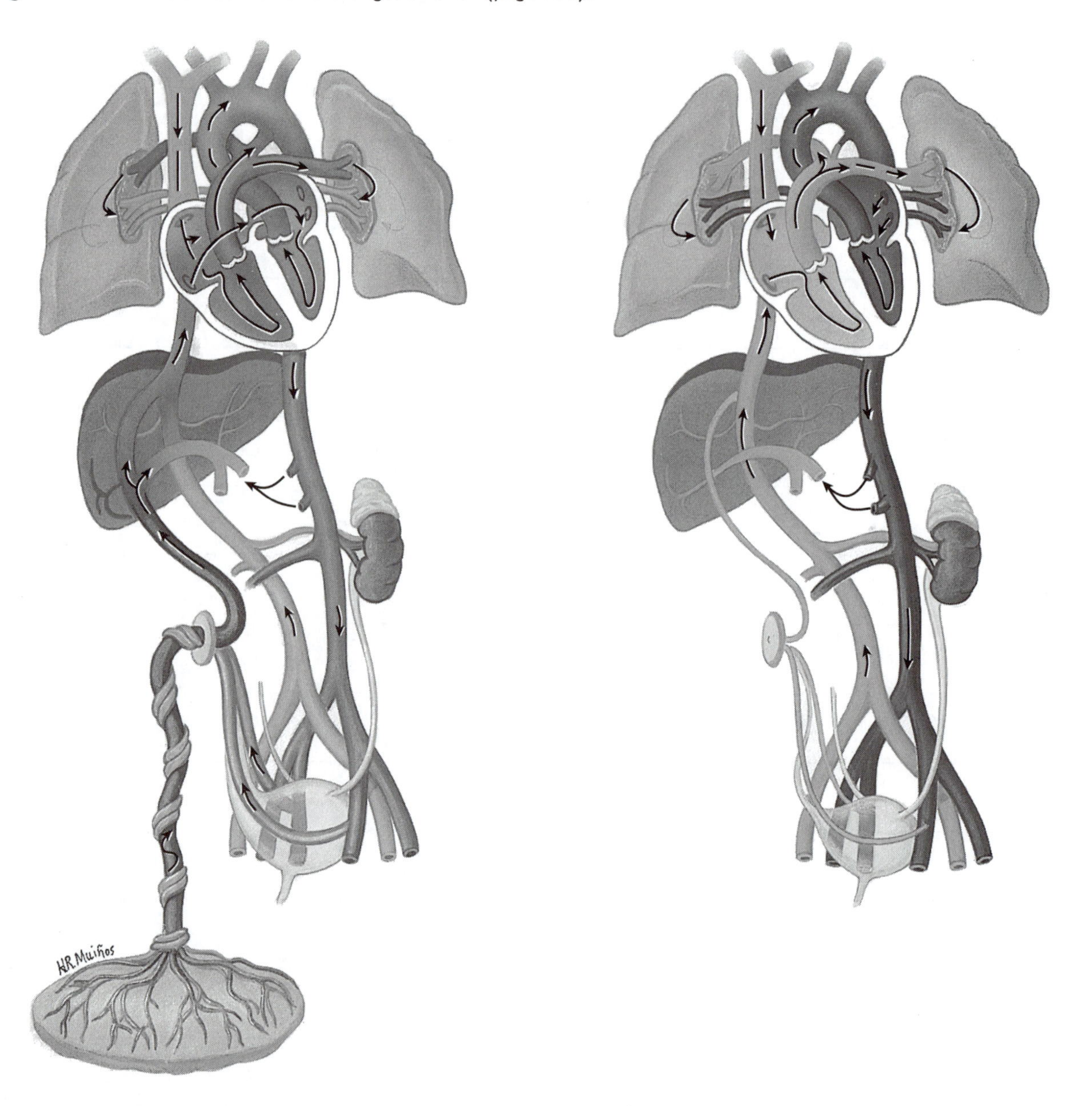

NOTES

Figure 21.32 Development of blood vessels and blood cells from blood islands (page 732).

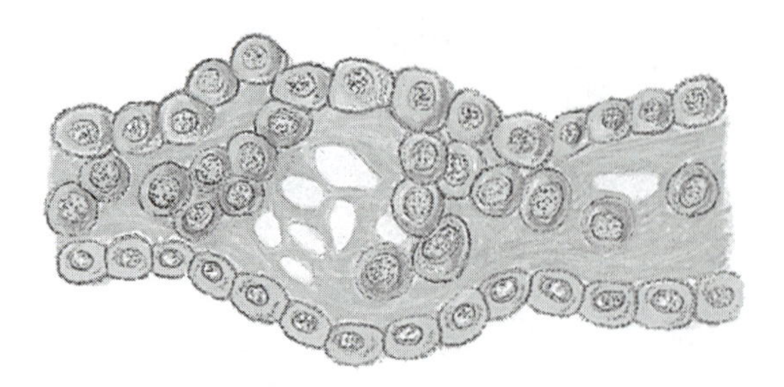

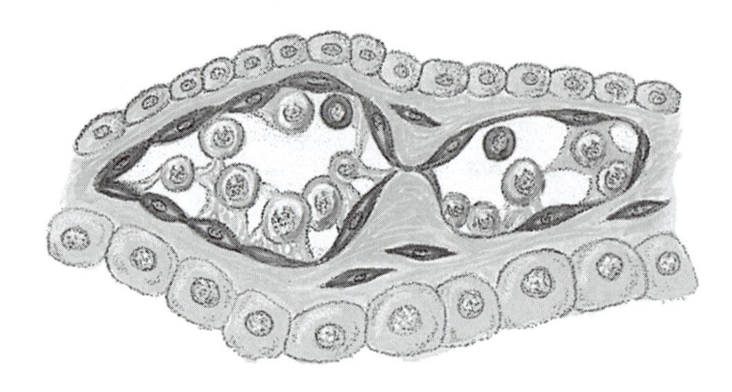

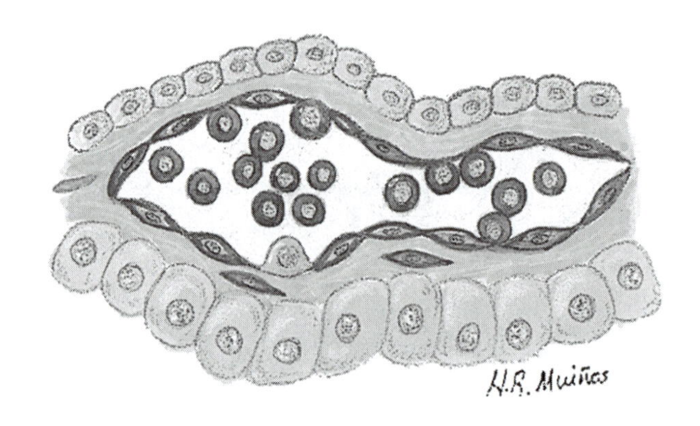
H.R. Muiños

NOTES

21

Figure 22.1 Components of the lymphatic system (page 739).

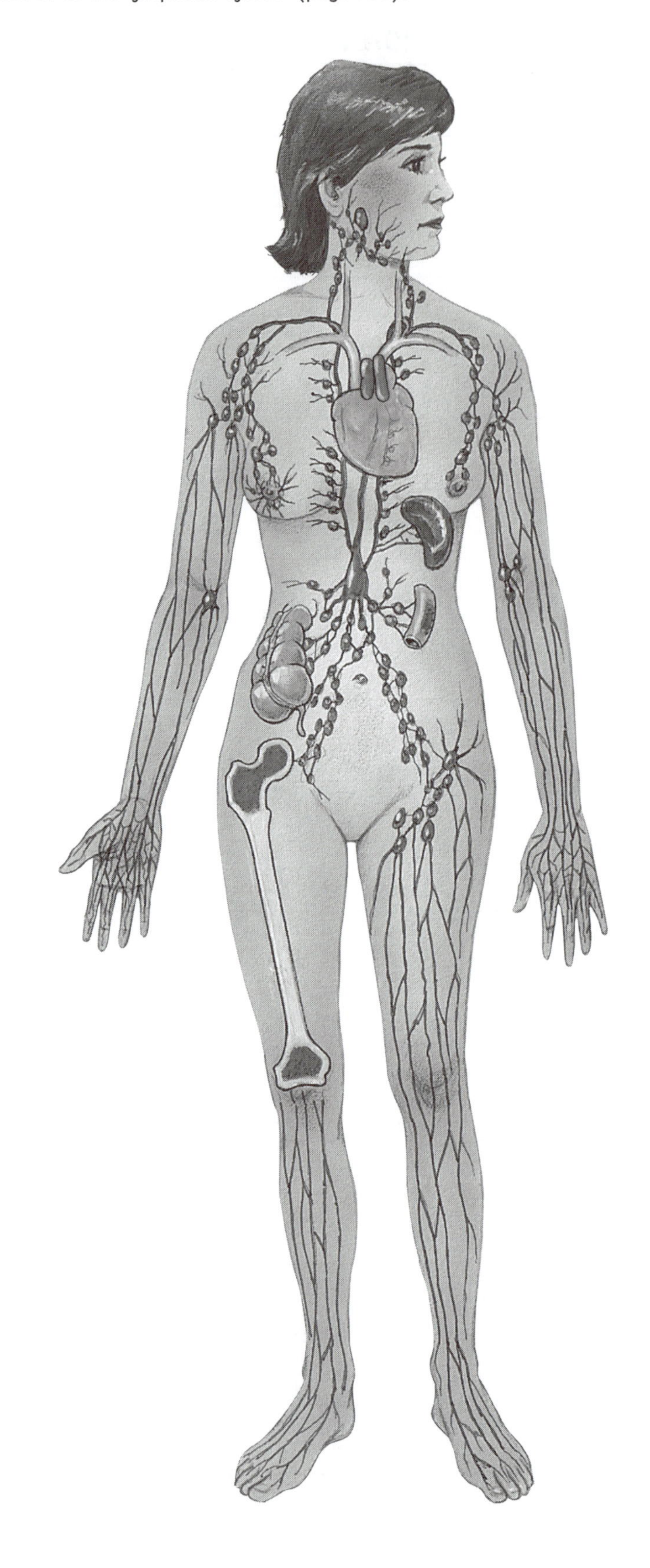

NOTES

Figure 22.2 Lymphatic capillaries (page 740).

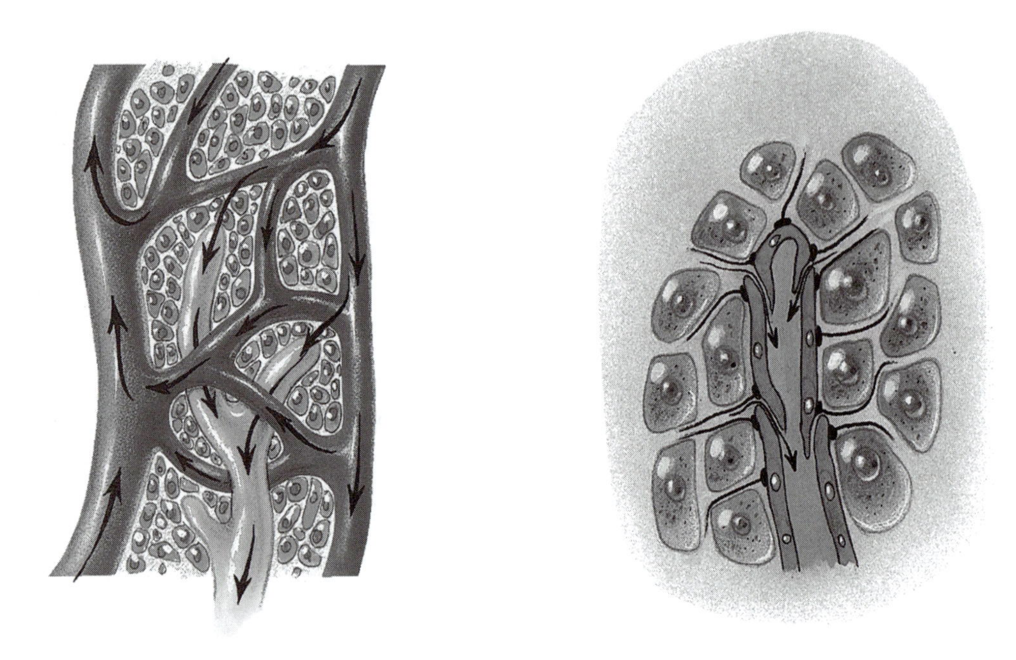

Figure 22.3 Routes for drainage of lymph (page 741).

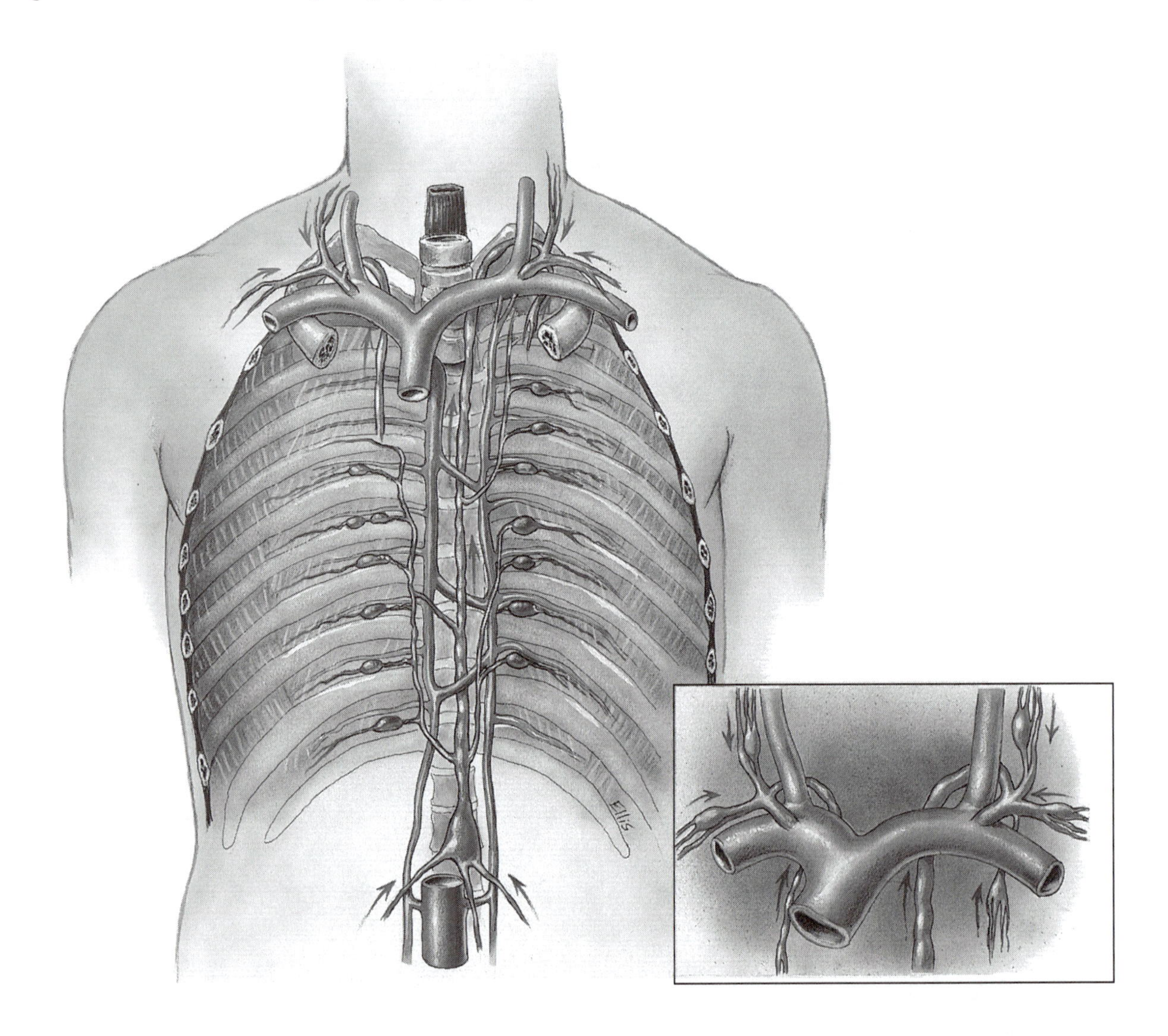
Ellis

NOTES

22

Figure 22.4 Relationship of the lymphatic system to the cardiovascular system (page 743).

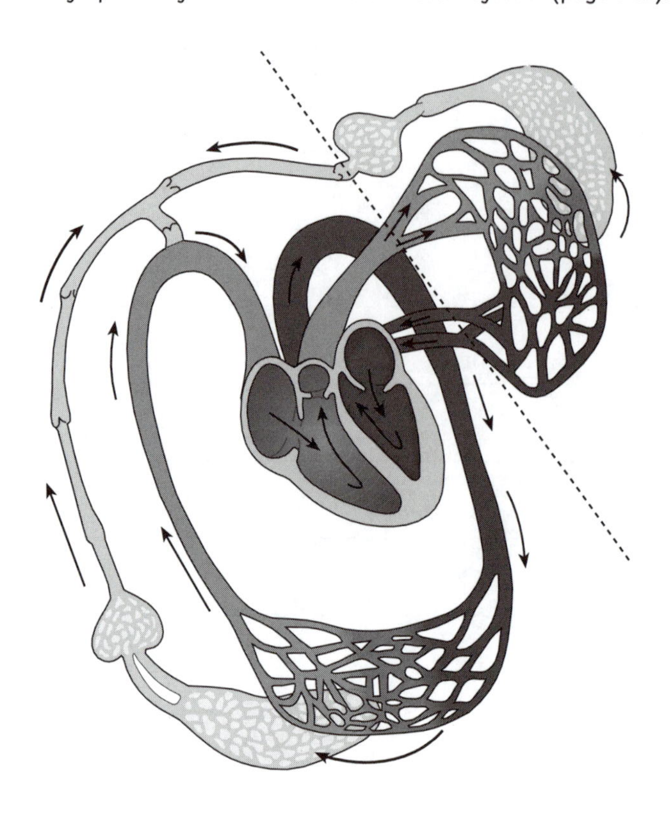

Figure 22.5 Thymus gland (page 744).

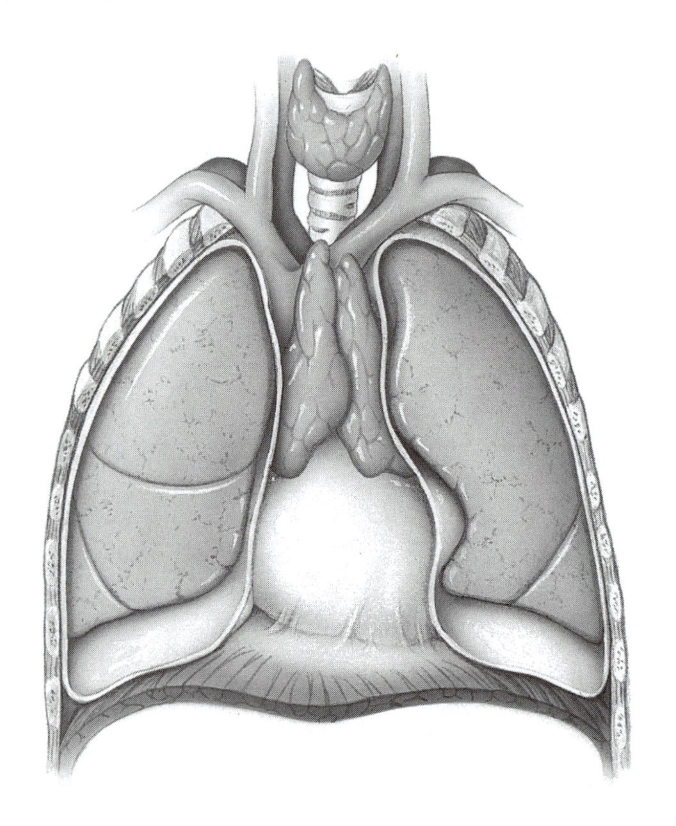

NOTES

22

Figure 22.6 Structure of a lymph node (page 745).

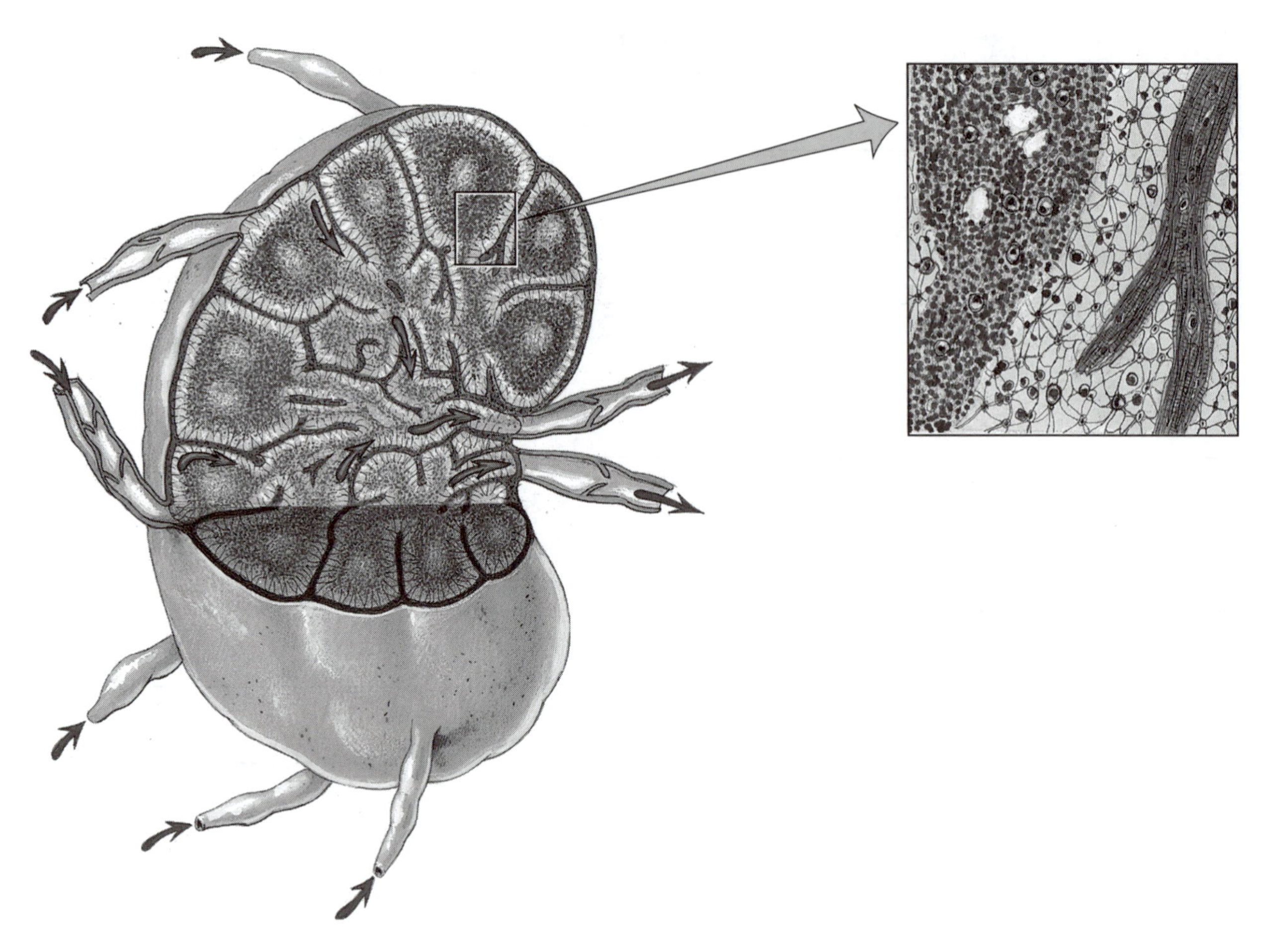

Figure 22.7 Structure of the spleen (page 746).

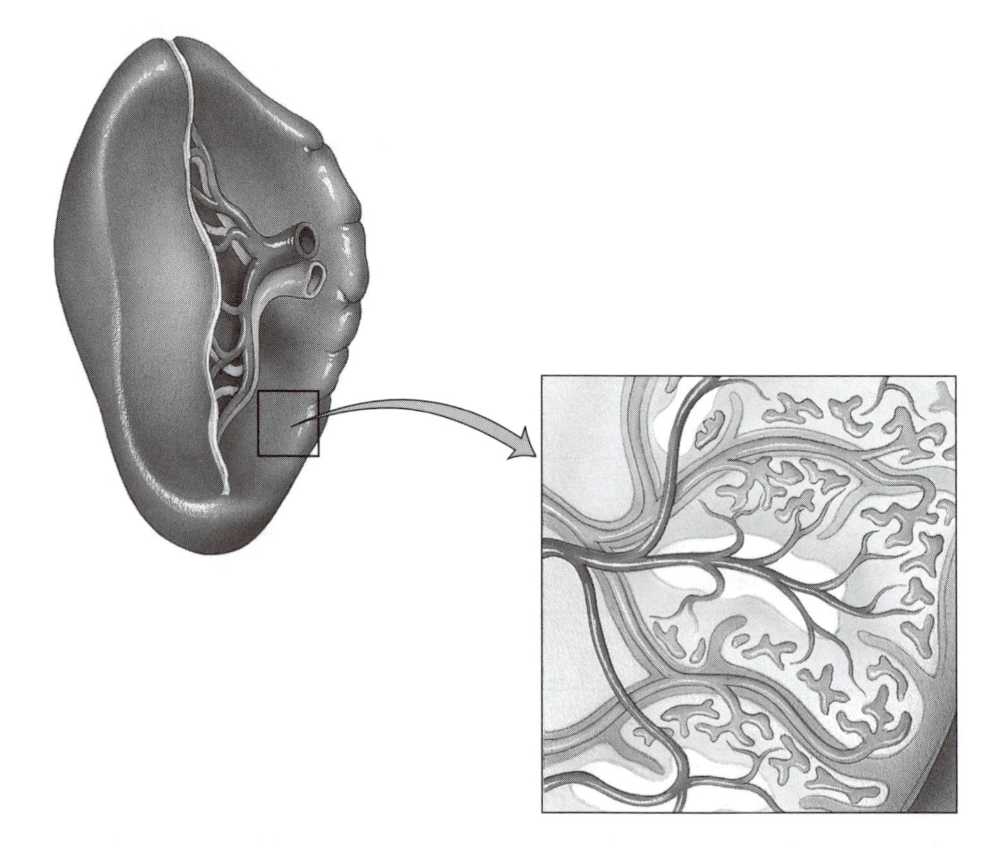

NOTES

22

Figure 22.8 Development of the lymphatic system (page 747).

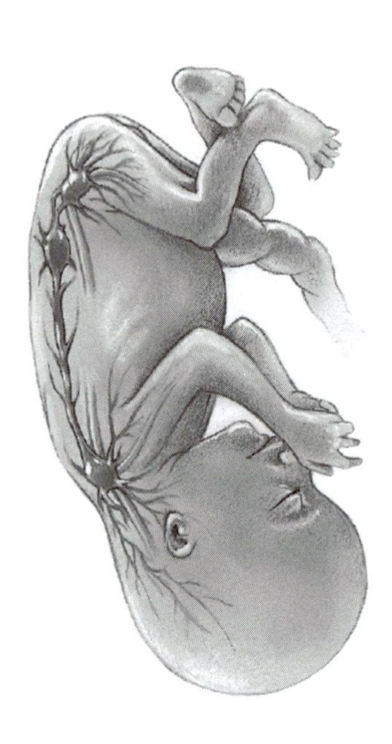

Figure 22.9 Phagocytosis of a microbe (page 749).

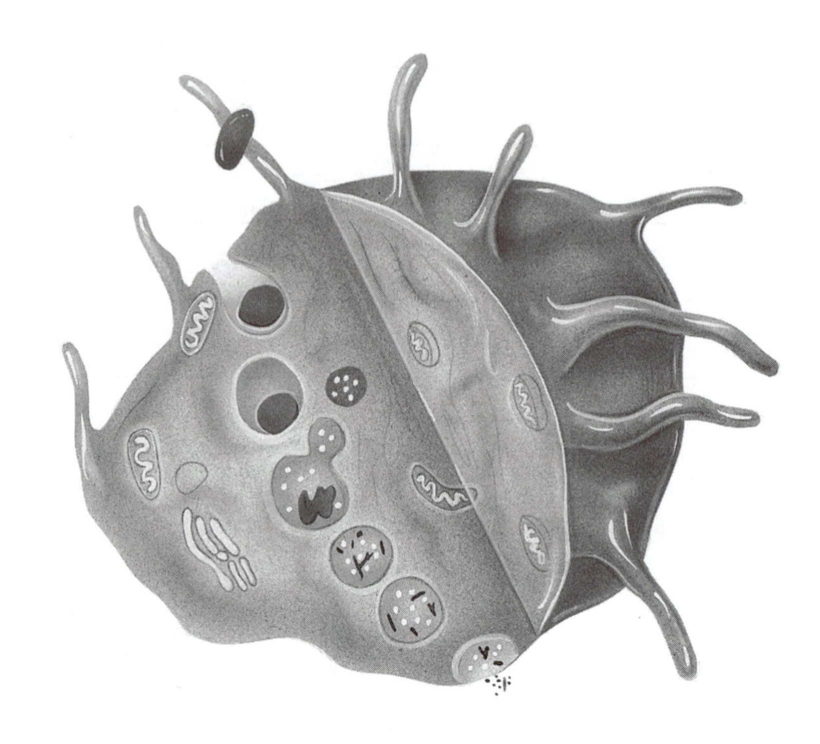

Figure 22.10 Inflammation (page 750).

NOTES

22

Figure 22.11 Maturation of lymphocytes (page 753).

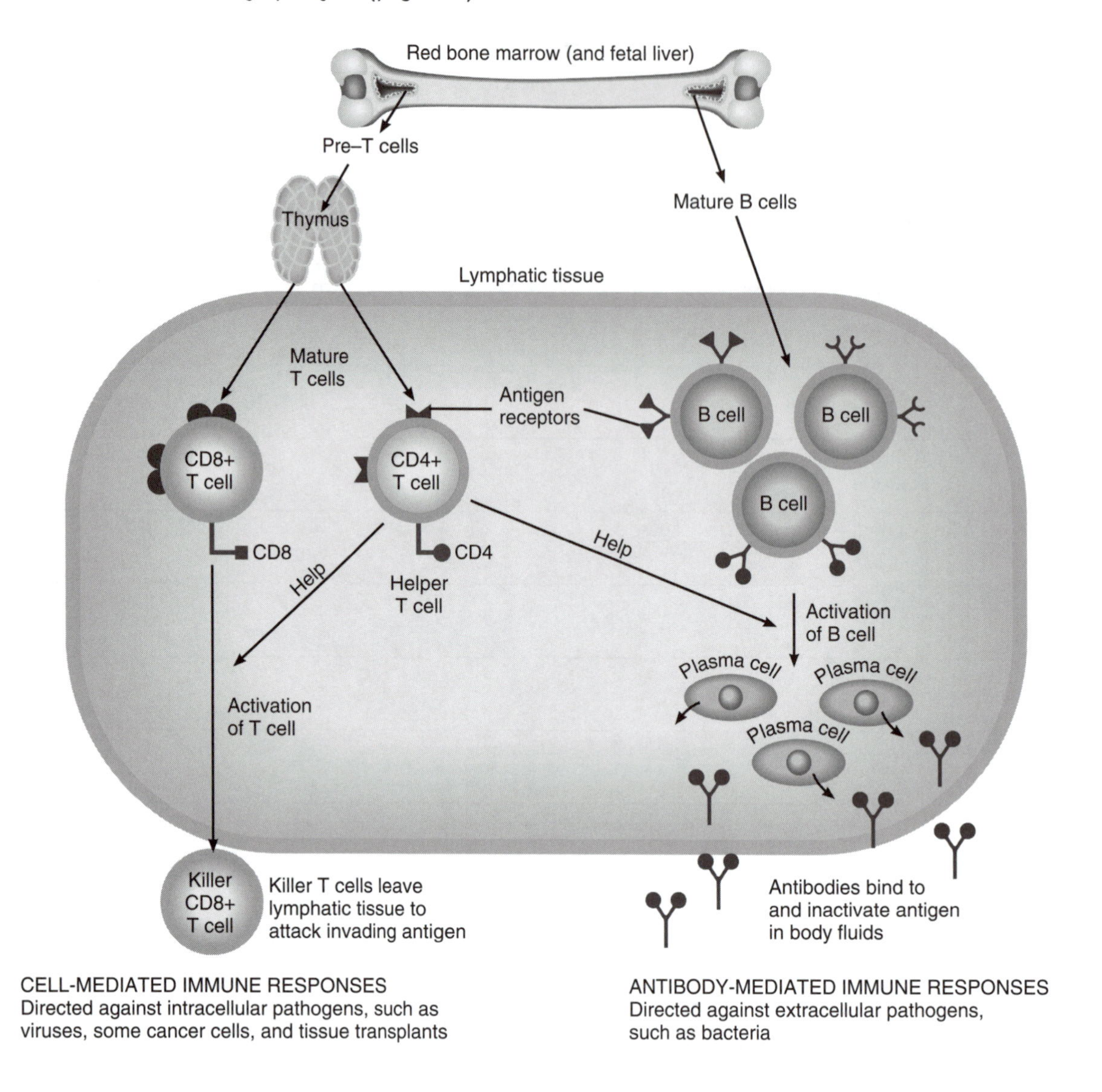
Red bone marrow (and fetal liver)
Pre–T cells
Thymus
Mature B cells
Lymphatic tissue
Mature T cells
Antigen receptors
B cell
B cell
B cell
CD8+ T cell
CD4+ T cell
CD8
CD4
Help
Help
Helper T cell
Activation of B cell
Plasma cell
Plasma cell
Plasma cell
Activation of T cell
Killer CD8+ T cell
Killer T cells leave lymphatic tissue to attack invading antigen
Antibodies bind to and inactivate antigen in body fluids
CELL-MEDIATED IMMUNE RESPONSES
Directed against intracellular pathogens, such as viruses, some cancer cells, and tissue transplants
ANTIBODY-MEDIATED IMMUNE RESPONSES
Directed against extracellular pathogens, such as bacteria

Figure 22.12 Epitopes (page 754).

NOTES

22

Figure 22.13 Processing and presenting of exogenous antigen (page 756).

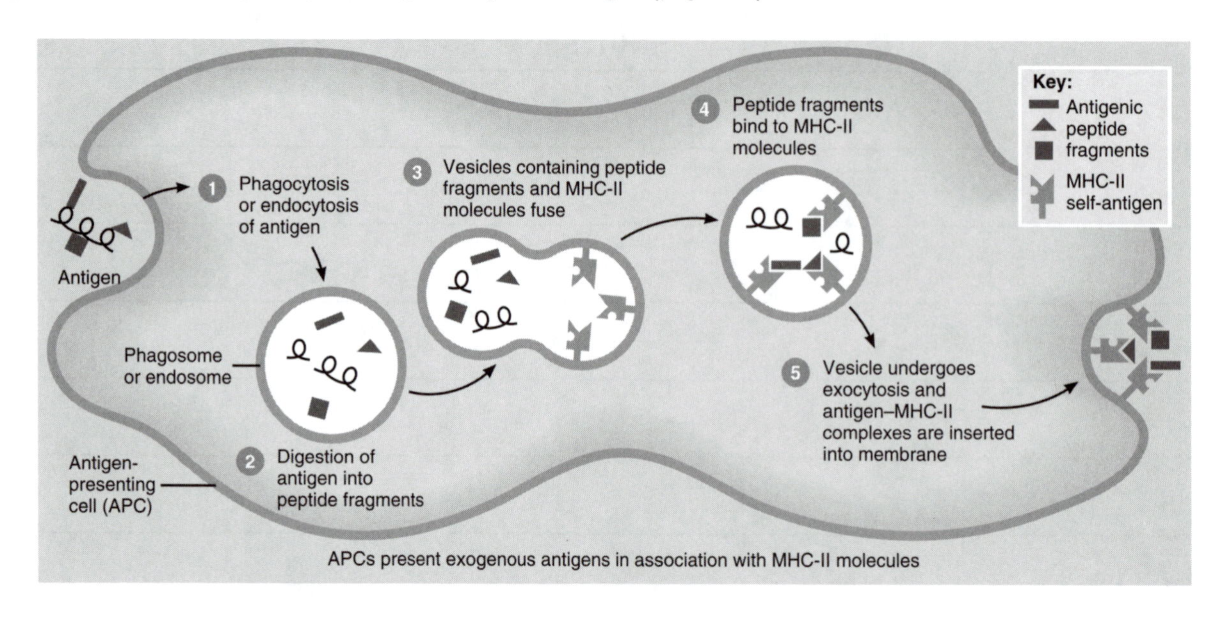

Key:
Antigenic peptide fragments
MHC-II self-antigen
Antigen
1 Phagocytosis or endocytosis of antigen
Phagosome or endosome
2 Digestion of antigen into peptide fragments
Antigen-presenting cell (APC)
3 Vesicles containing peptide fragments and MHC-II molecules fuse
4 Peptide fragments bind to MHC-II molecules
5 Vesicle undergoes exocytosis and antigen–MHC-II complexes are inserted into membrane
APCs present exogenous antigens in association with MHC-II molecules

Figure 22.14 Activation, proliferation, and differentiation of T cells (page 758).

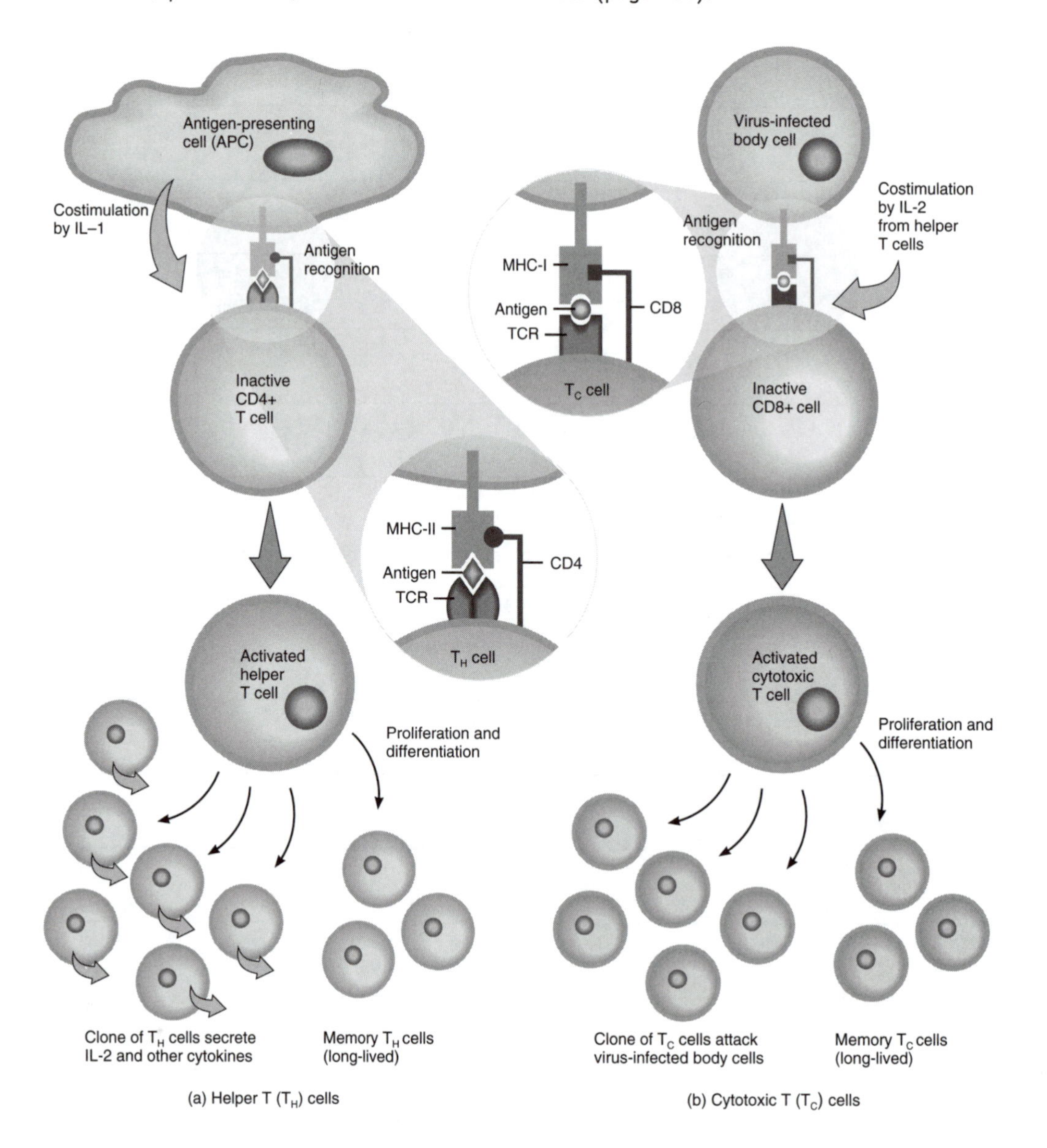

Antigen-presenting cell (APC)
Costimulation by IL–1
Antigen recognition
Inactive CD4+ T cell
MHC-II
Antigen
TCR
CD4
T$_H$ cell
Activated helper T cell
Proliferation and differentiation
Clone of T$_H$ cells secrete IL-2 and other cytokines
Memory T$_H$ cells (long-lived)
(a) Helper T (T$_H$) cells
Virus-infected body cell
Costimulation by IL-2 from helper T cells
Antigen recognition
MHC-I
Antigen
TCR
CD8
T$_C$ cell
Inactive CD8+ cell
Activated cytotoxic T cell
Proliferation and differentiation
Clone of T$_C$ cells attack virus-infected body cells
Memory T$_C$ cells (long-lived)
(b) Cytotoxic T (T$_C$) cells

NOTES

22

Figure 22.15 Activity of cytotoxic T cells (page 759).

Figure 22.16 Activation, proliferation, and differentiation of B cells (page 761).

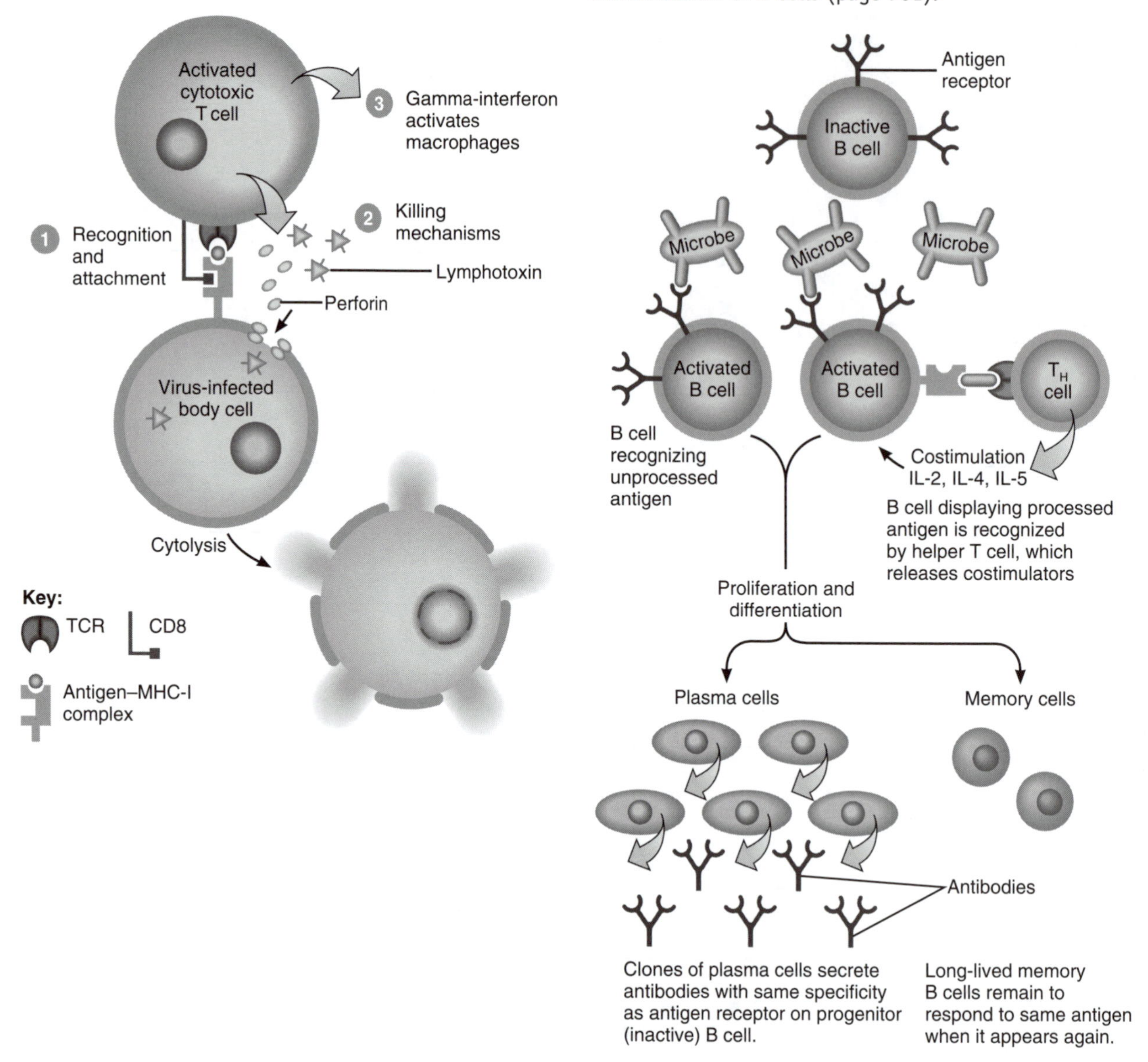

Activated cytotoxic T cell
3 Gamma-interferon activates macrophages
2 Killing mechanisms
1 Recognition and attachment
Lymphotoxin
Perforin
Virus-infected body cell
Cytolysis
Key:
TCR
CD8
Antigen–MHC-I complex
Antigen receptor
Inactive B cell
Microbe
Microbe
Microbe
Activated B cell
Activated B cell
T_H cell
B cell recognizing unprocessed antigen
Costimulation IL-2, IL-4, IL-5
B cell displaying processed antigen is recognized by helper T cell, which releases costimulators
Proliferation and differentiation
Plasma cells
Memory cells
Antibodies
Clones of plasma cells secrete antibodies with same specificity as antigen receptor on progenitor (inactive) B cell.
Long-lived memory B cells remain to respond to same antigen when it appears again.

Figure 22.17 Chemical structure of the immunoglobulin G class of antibody (page 762).

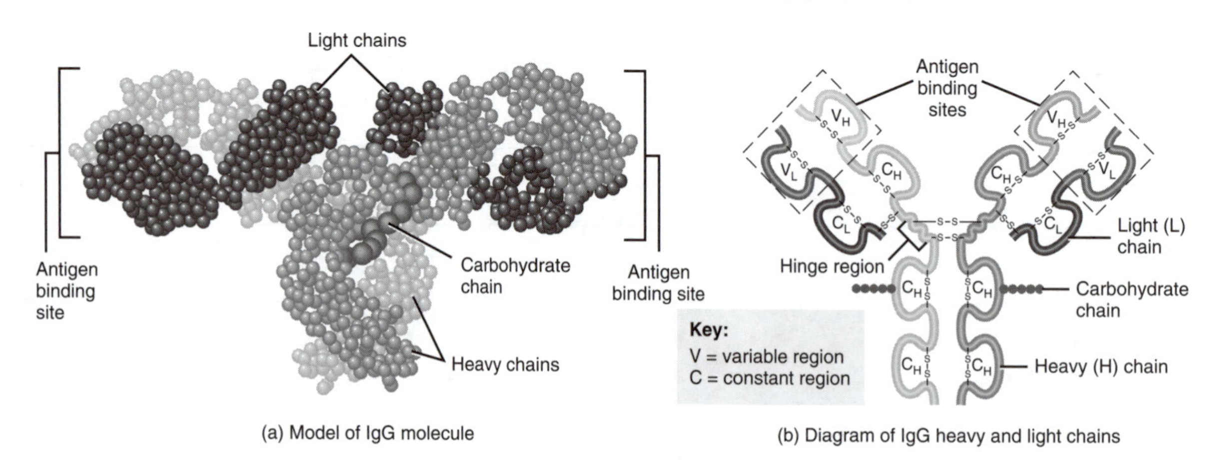

Light chains
Antigen binding site
Carbohydrate chain
Heavy chains
Antigen binding site
Antigen binding sites
V_H
V_L
C_H
C_L
S–S
Hinge region
Light (L) chain
Carbohydrate chain
Heavy (H) chain
Key:
V = variable region
C = constant region

(a) Model of IgG molecule

(b) Diagram of IgG heavy and light chains

NOTES

22

Figure 22.18 The classical and alternative pathways of the complement system (page 764).

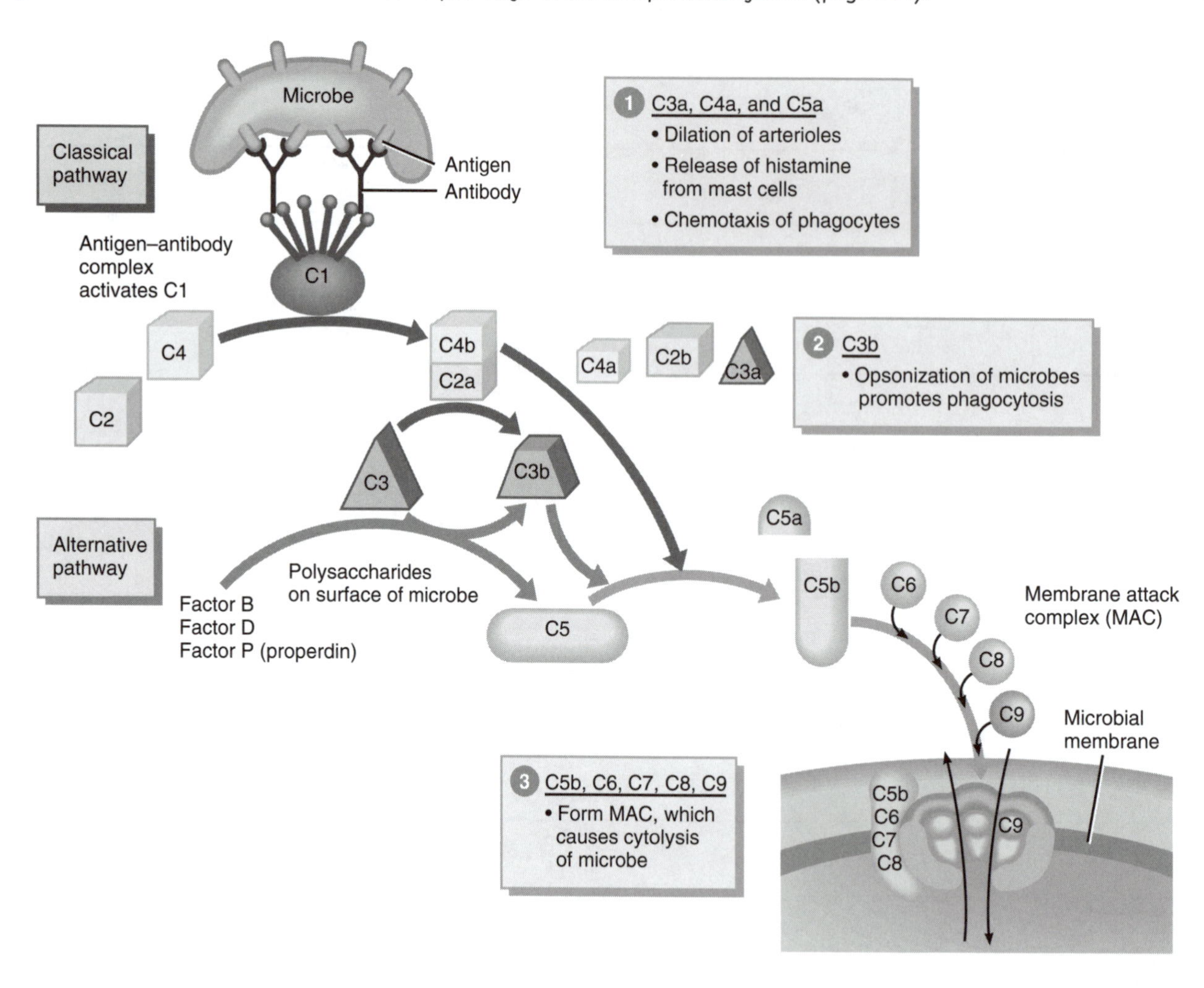

Microbe
Classical pathway
Antigen
Antibody
Antigen–antibody complex activates C1
C1
C4
C2
C4b
C2a
C4a
C2b
C3a
C3
C3b
1 C3a, C4a, and C5a
• Dilation of arterioles
• Release of histamine from mast cells
• Chemotaxis of phagocytes
2 C3b
• Opsonization of microbes promotes phagocytosis
Alternative pathway
Factor B
Factor D
Factor P (properdin)
Polysaccharides on surface of microbe
C5
C5a
C5b
C6
C7
C8
C9
Membrane attack complex (MAC)
Microbial membrane
3 C5b, C6, C7, C8, C9
• Form MAC, which causes cytolysis of microbe

Figure 22.19 Production of antibodies in the primary and secondary responses to a given antigen (page 765).

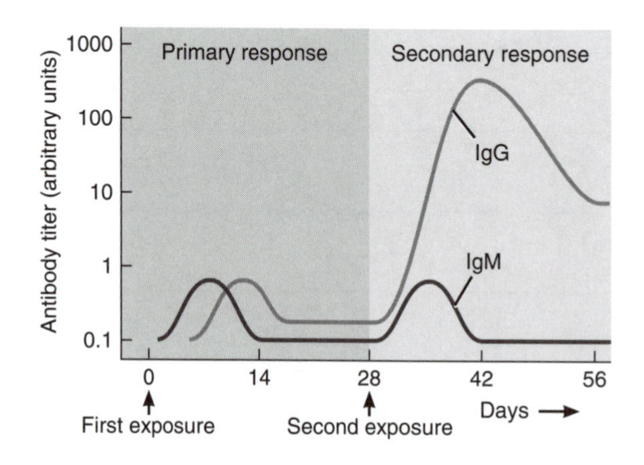

Primary response
Secondary response
Antibody titer (arbitrary units)
1000
100
10
1
0.1
IgG
IgM
0
14
28
42
56
First exposure
Second exposure
Days

NOTES

22

Figure 22.20 Development of self-recognition and immunological tolerance (page 766).

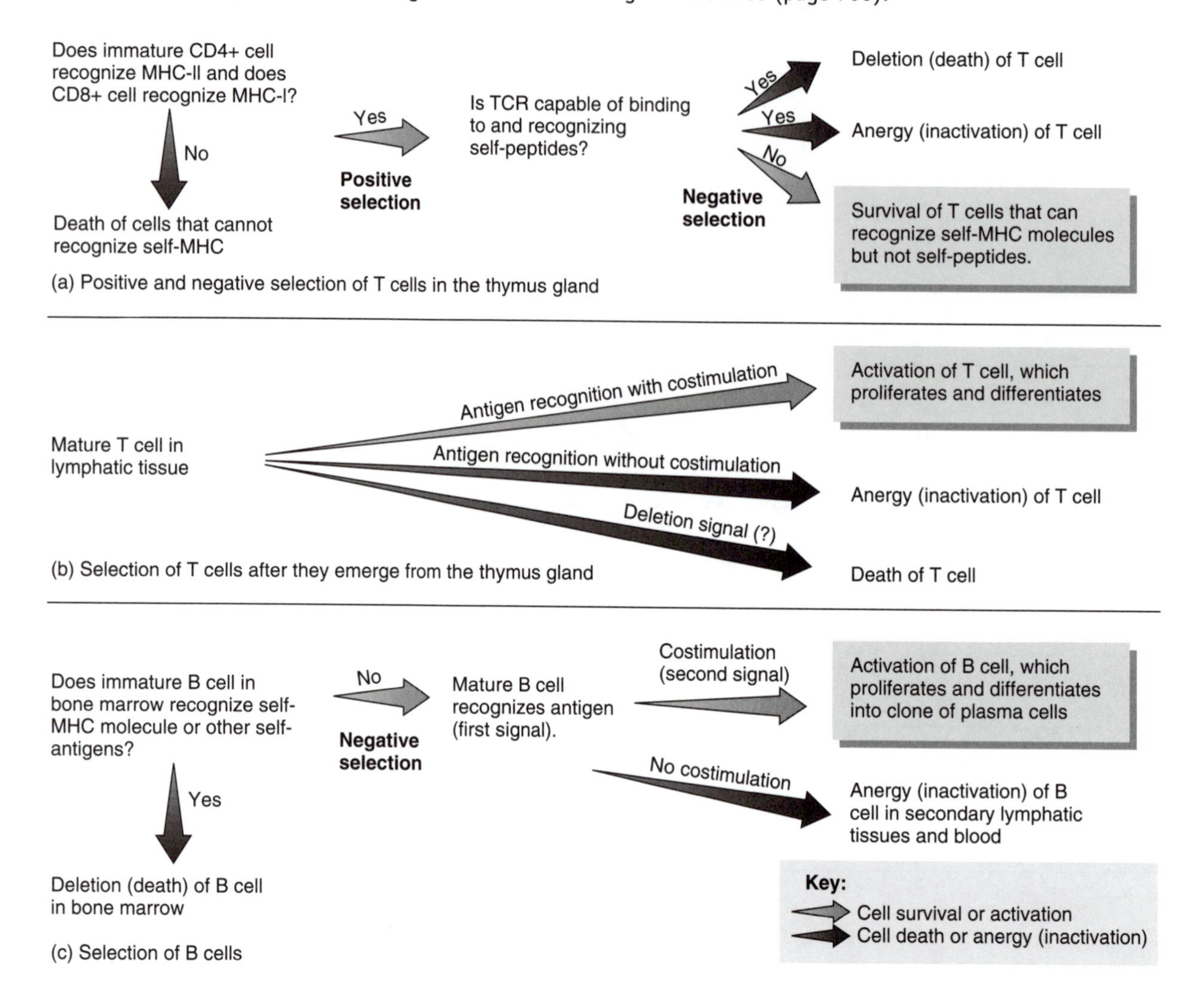
Does immature CD4+ cell recognize MHC-II and does CD8+ cell recognize MHC-I?
No
Death of cells that cannot recognize self-MHC
Yes
Positive selection
Is TCR capable of binding to and recognizing self-peptides?
Yes
Yes
No
Negative selection
Deletion (death) of T cell
Anergy (inactivation) of T cell
Survival of T cells that can recognize self-MHC molecules but not self-peptides.
(a) Positive and negative selection of T cells in the thymus gland
Mature T cell in lymphatic tissue
Antigen recognition with costimulation
Antigen recognition without costimulation
Deletion signal (?)
Activation of T cell, which proliferates and differentiates
Anergy (inactivation) of T cell
Death of T cell
(b) Selection of T cells after they emerge from the thymus gland
Does immature B cell in bone marrow recognize self-MHC molecule or other self-antigens?
Yes
Deletion (death) of B cell in bone marrow
No
Negative selection
Mature B cell recognizes antigen (first signal).
Costimulation (second signal)
No costimulation
Activation of B cell, which proliferates and differentiates into clone of plasma cells
Anergy (inactivation) of B cell in secondary lymphatic tissues and blood
Key:
Cell survival or activation
Cell death or anergy (inactivation)
(c) Selection of B cells

Figure 22.21 Human immunodeficiency virus (page 768).

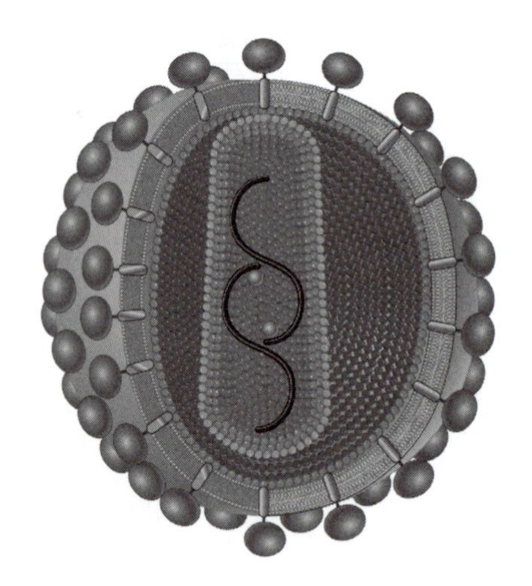

NOTES

22

Figure 23.1 Structures of the respiratory system (page 776).

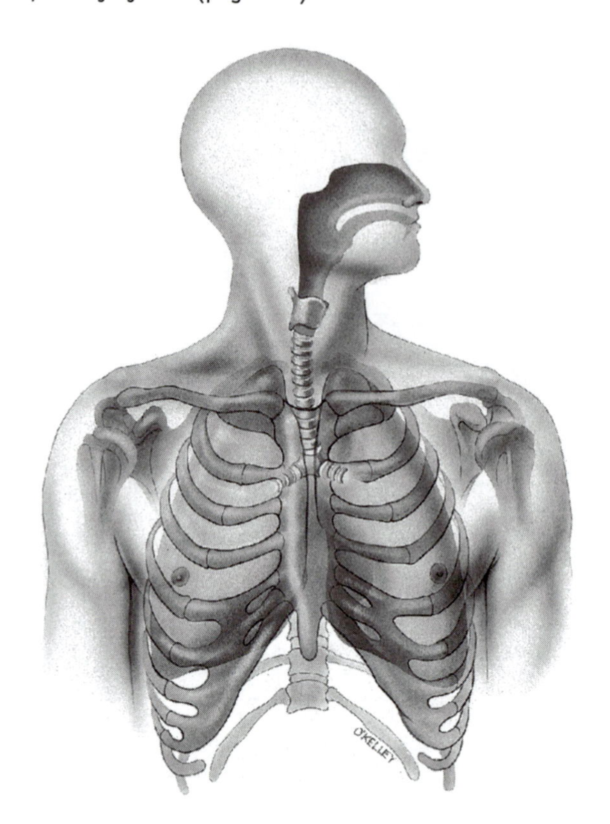

Figure 23.2a Respiratory structures in the head and neck (page 778).

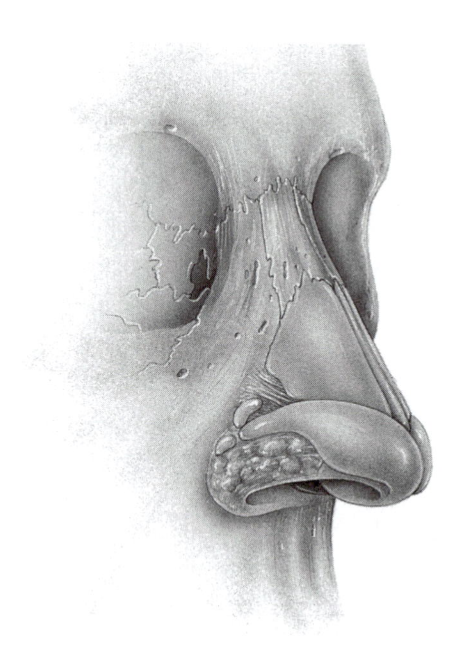

NOTES

23

Figure 23.2b Respiratory structures in the head and neck (page 778).

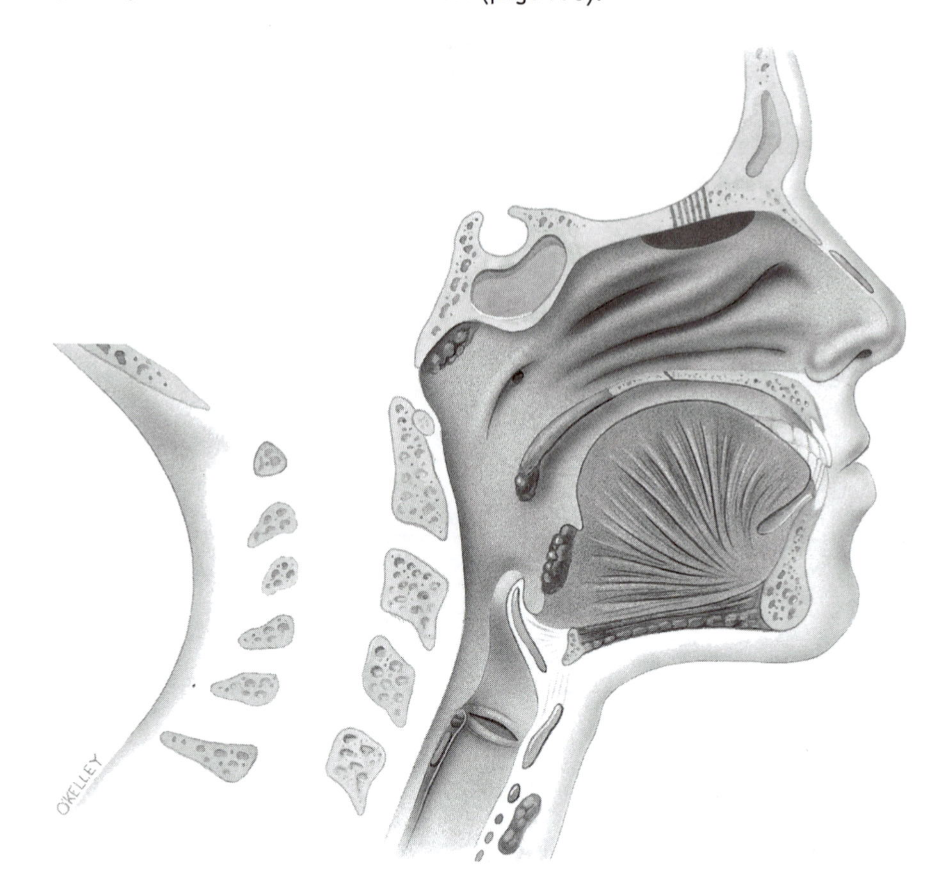
O'KELLEY

Figure 23.4 Pharynx (page 779).

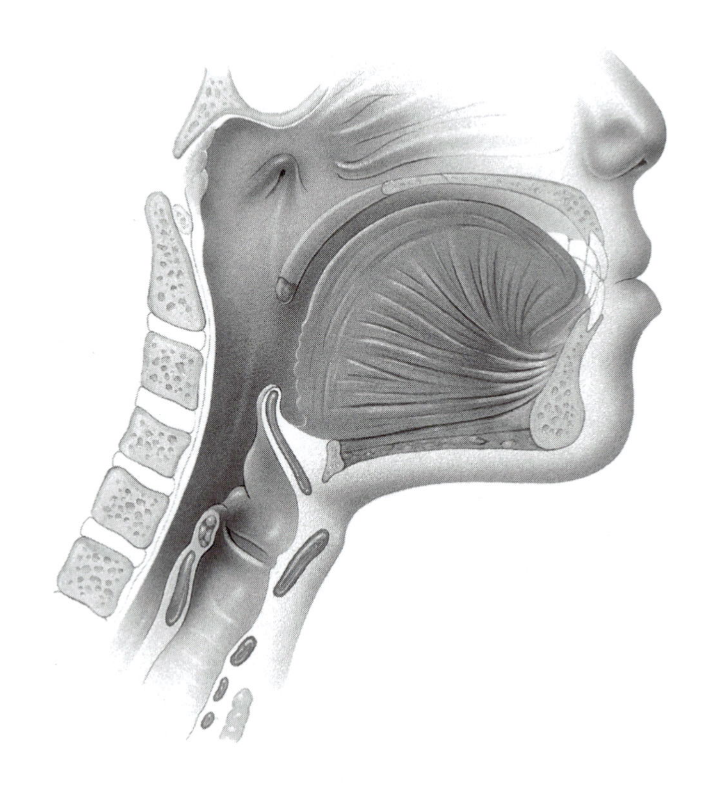

NOTES

23

Figure 23.5 Larynx (page 781).

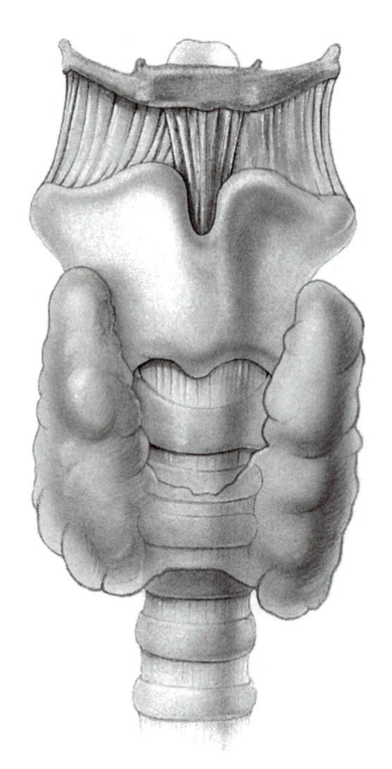

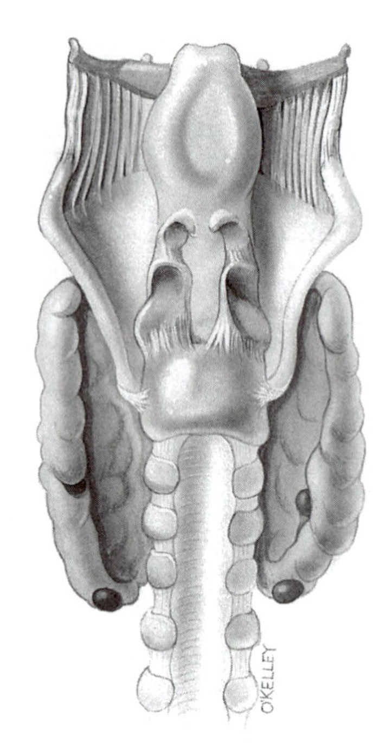
O'KELLEY

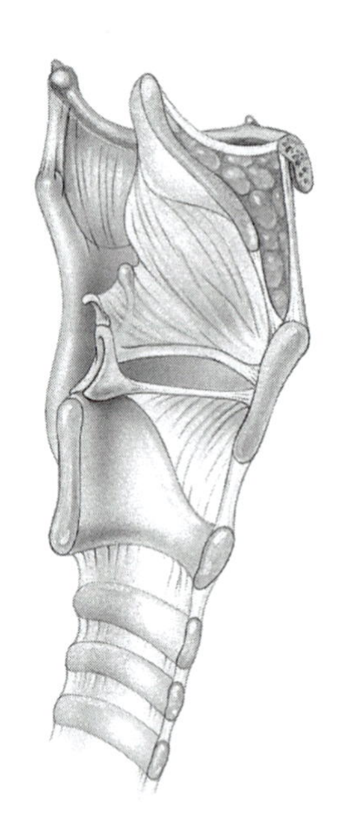

NOTES

23

Figure 23.6 Movement of the vocal folds (page 782).

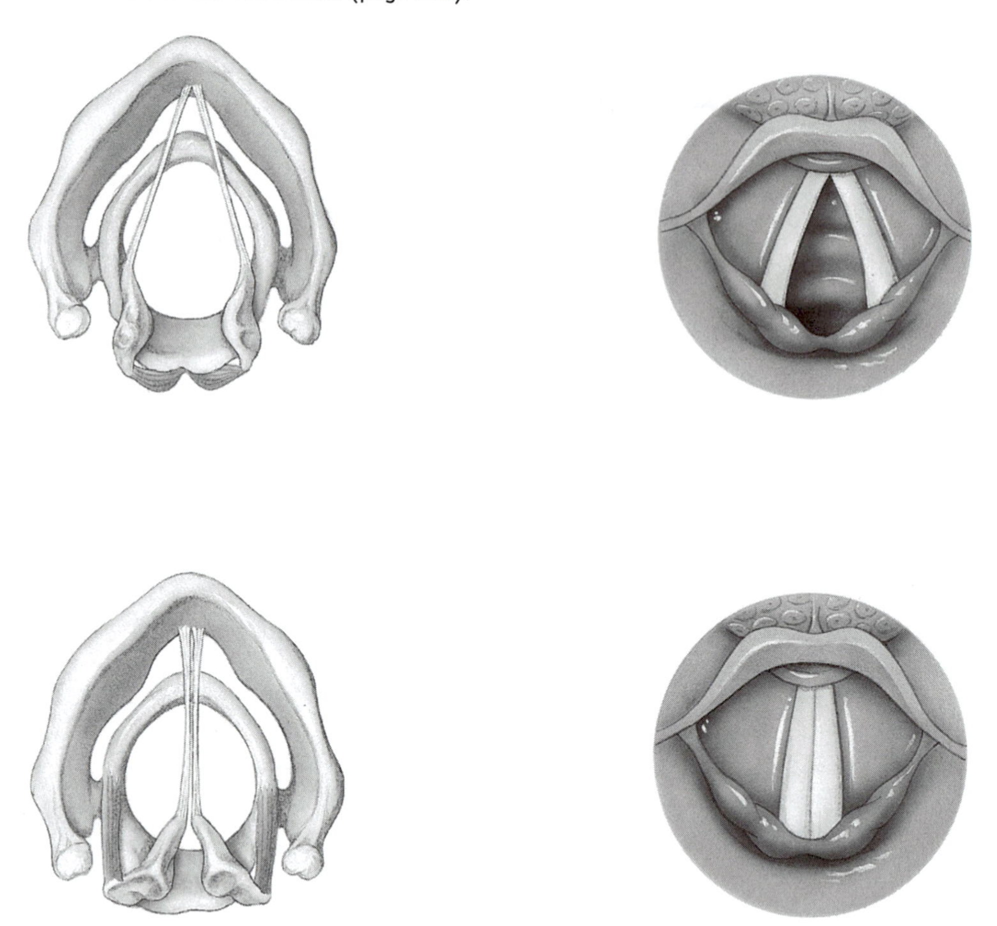

Figure 23.8 Branching of airways from the trachea: the bronchial tree (page 784).

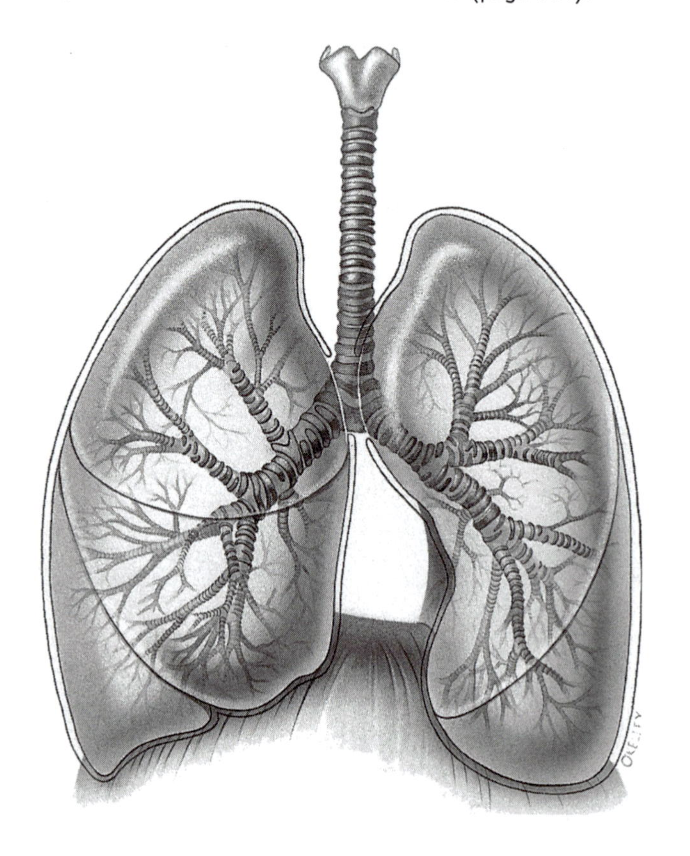

NOTES

23

Figure 23.10 Surface anatomy of the lungs (page 787).

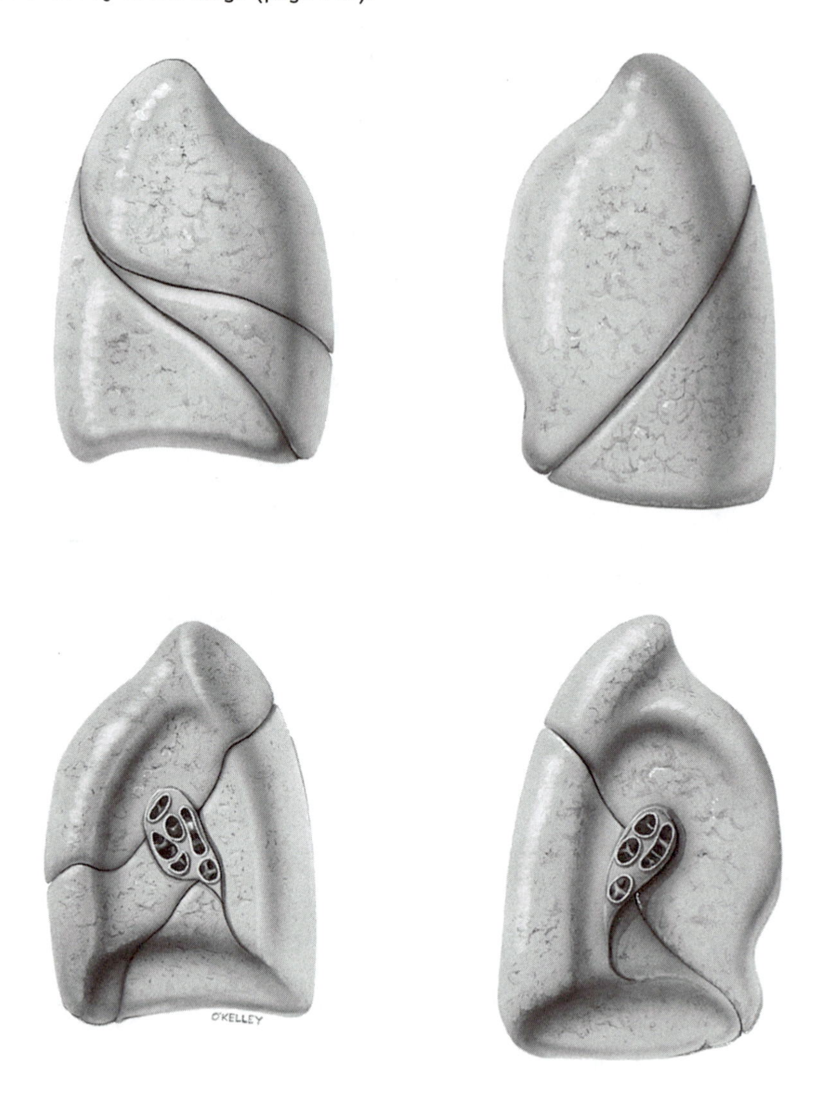
O'KELLEY

Figure 23.11 Microscopic anatomy of a lobule of the lungs (page 788).

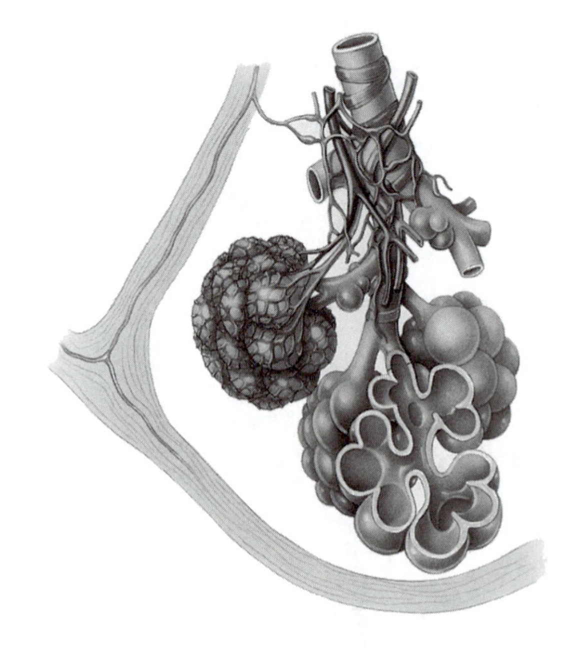

NOTES

23

Figure 23.12 Structural components and function of an alveolus (page 789).

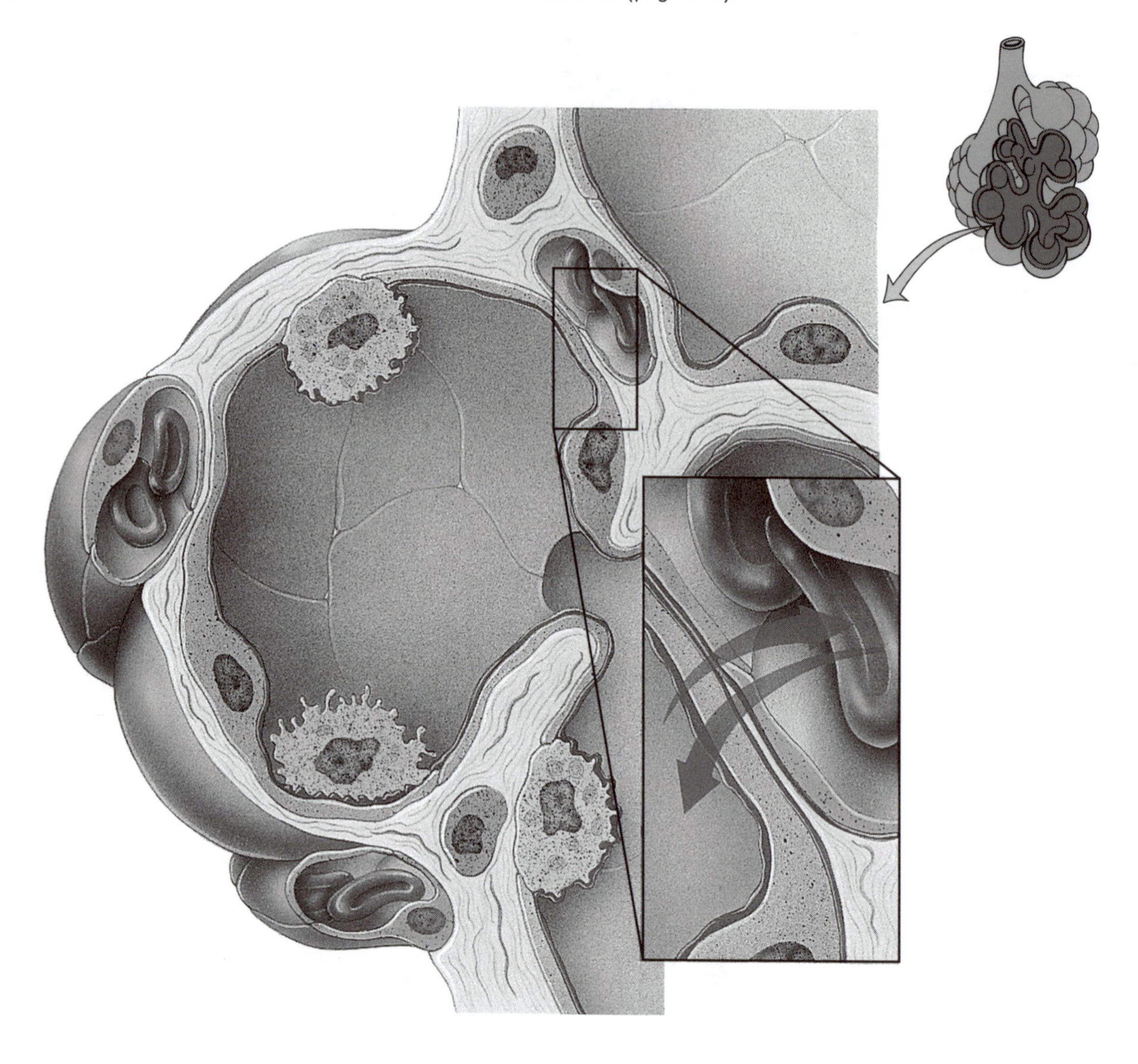

Figure 23.13 Boyle's law (page 790).

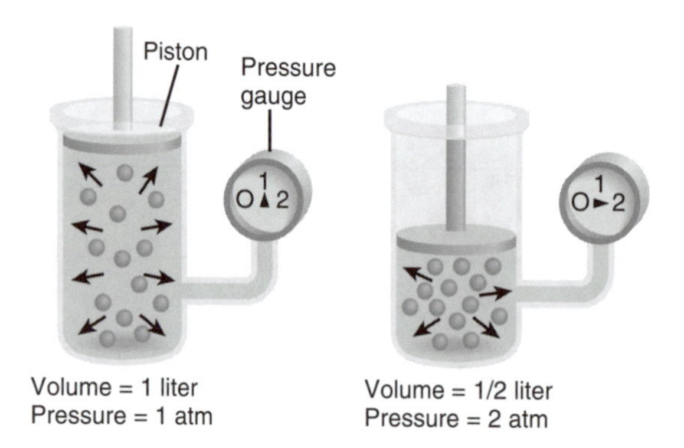
Piston
Pressure gauge
Volume = 1 liter
Pressure = 1 atm
Volume = 1/2 liter
Pressure = 2 atm

NOTES

23

Figure 23.14 Muscles of inspiration and expiration (page 791).

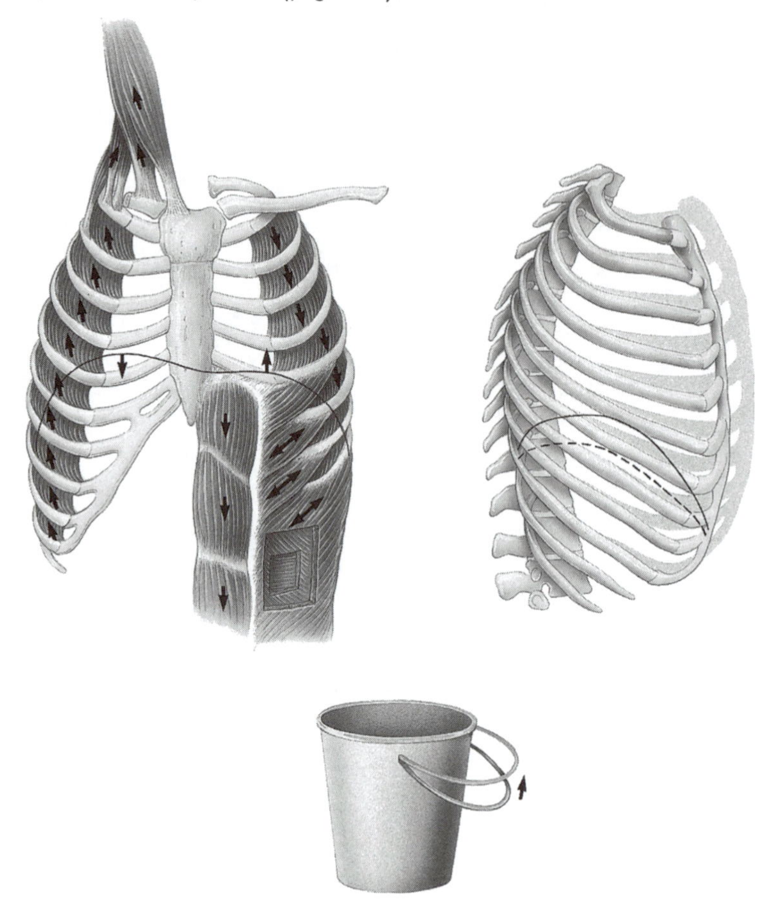

Figure 23.15 Pressure changes in pulmonary ventilation (page 792).

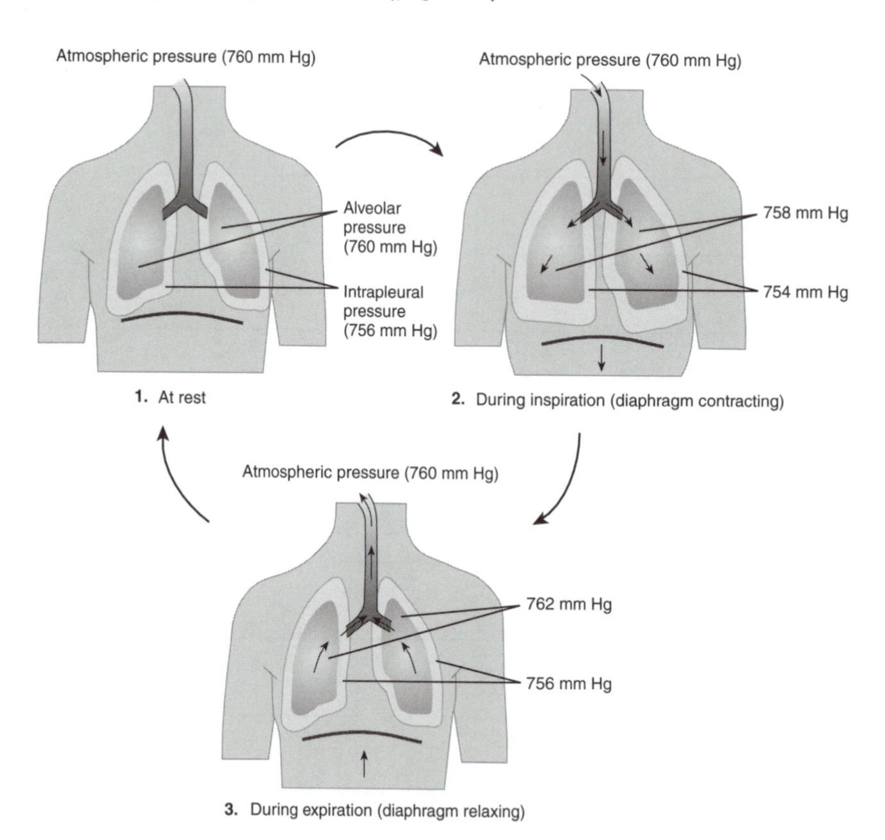
Atmospheric pressure (760 mm Hg)
Alveolar pressure (760 mm Hg)
Intrapleural pressure (756 mm Hg)
1. At rest
Atmospheric pressure (760 mm Hg)
758 mm Hg
754 mm Hg
2. During inspiration (diaphragm contracting)
Atmospheric pressure (760 mm Hg)
762 mm Hg
756 mm Hg
3. During expiration (diaphragm relaxing)

NOTES

23

Figure 23.16 Summary of events of inspiration and expiration (page 793).

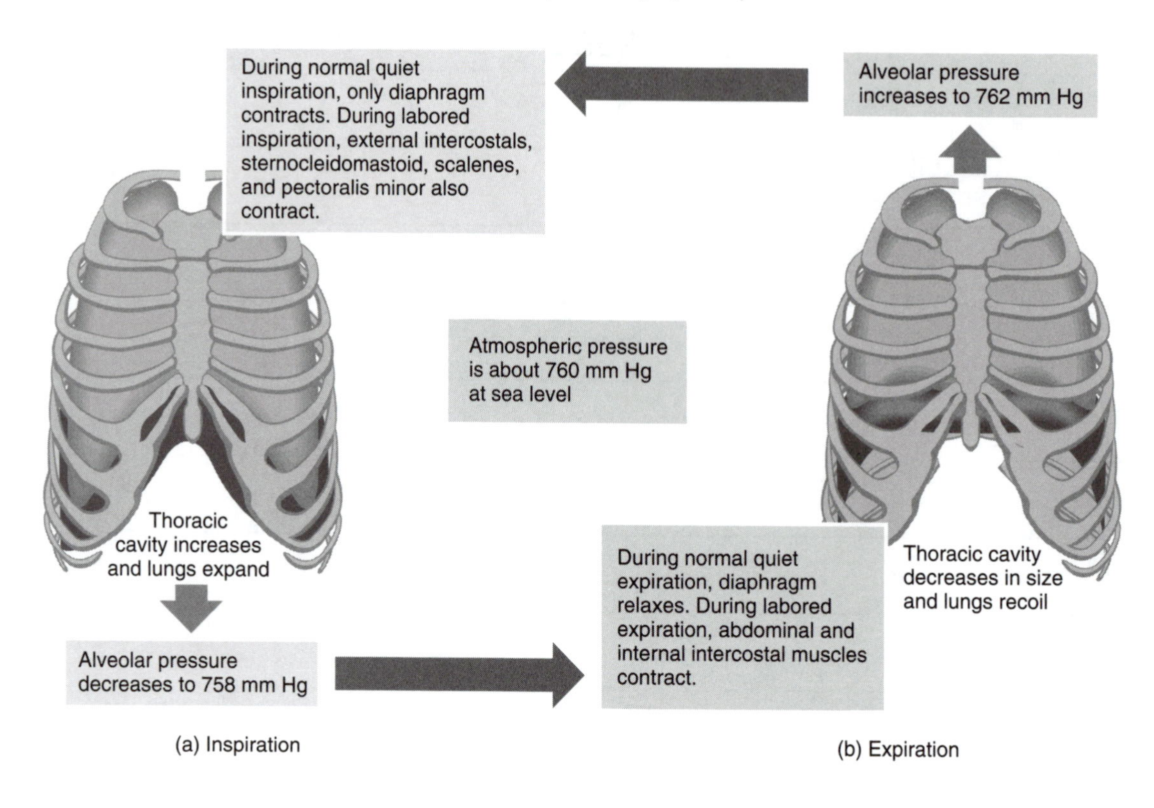
During normal quiet inspiration, only diaphragm contracts. During labored inspiration, external intercostals, sternocleidomastoid, scalenes, and pectoralis minor also contract.
Alveolar pressure increases to 762 mm Hg
Atmospheric pressure is about 760 mm Hg at sea level
Thoracic cavity increases and lungs expand
Alveolar pressure decreases to 758 mm Hg
During normal quiet expiration, diaphragm relaxes. During labored expiration, abdominal and internal intercostal muscles contract.
Thoracic cavity decreases in size and lungs recoil
(a) Inspiration
(b) Expiration

Figure 23.17 Spirogram of lung volumes and capacities (page 795).

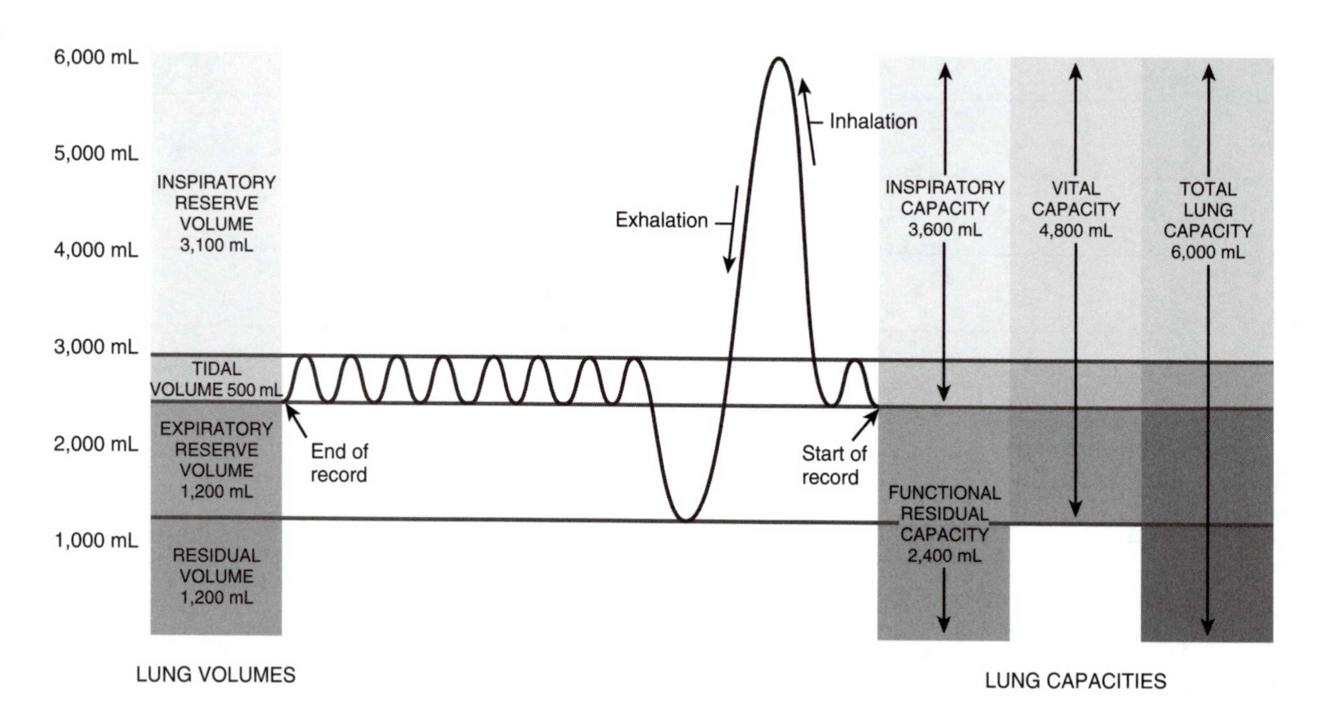
6,000 mL
5,000 mL
4,000 mL
3,000 mL
2,000 mL
1,000 mL
INSPIRATORY RESERVE VOLUME 3,100 mL
TIDAL VOLUME 500 mL
EXPIRATORY RESERVE VOLUME 1,200 mL
RESIDUAL VOLUME 1,200 mL
Inhalation
Exhalation
End of record
Start of record
INSPIRATORY CAPACITY 3,600 mL
VITAL CAPACITY 4,800 mL
TOTAL LUNG CAPACITY 6,000 mL
FUNCTIONAL RESIDUAL CAPACITY 2,400 mL
LUNG VOLUMES
LUNG CAPACITIES

NOTES

23

Figure 23.18 Changes in partial pressures of oxygen and carbon dioxide during external and internal respiration (page 797).

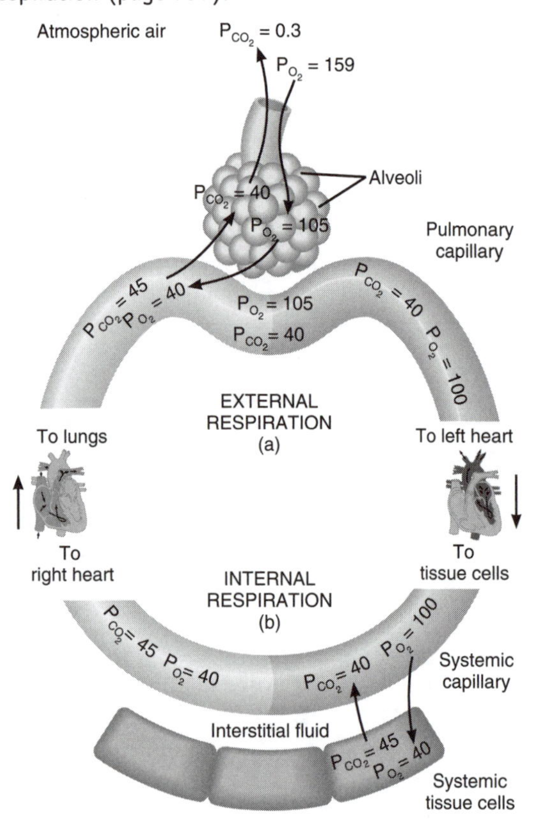

Atmospheric air
P_{CO_2} = 0.3
P_{O_2} = 159
Alveoli
P_{CO_2} = 40
P_{O_2} = 105
Pulmonary capillary
P_{CO_2} = 45
P_{O_2} = 40
P_{O_2} = 105
P_{CO_2} = 40
P_{CO_2} = 40
P_{O_2} = 100
EXTERNAL RESPIRATION
(a)
To lungs
To left heart
To right heart
To tissue cells
INTERNAL RESPIRATION
(b)
P_{CO_2} = 45 P_{O_2} = 40
P_{CO_2} = 40
P_{O_2} = 100
Systemic capillary
Interstitial fluid
P_{CO_2} = 45
P_{O_2} = 40
Systemic tissue cells

Figure 23.19 Transport of oxygen and carbon dioxide in the blood (page 799).

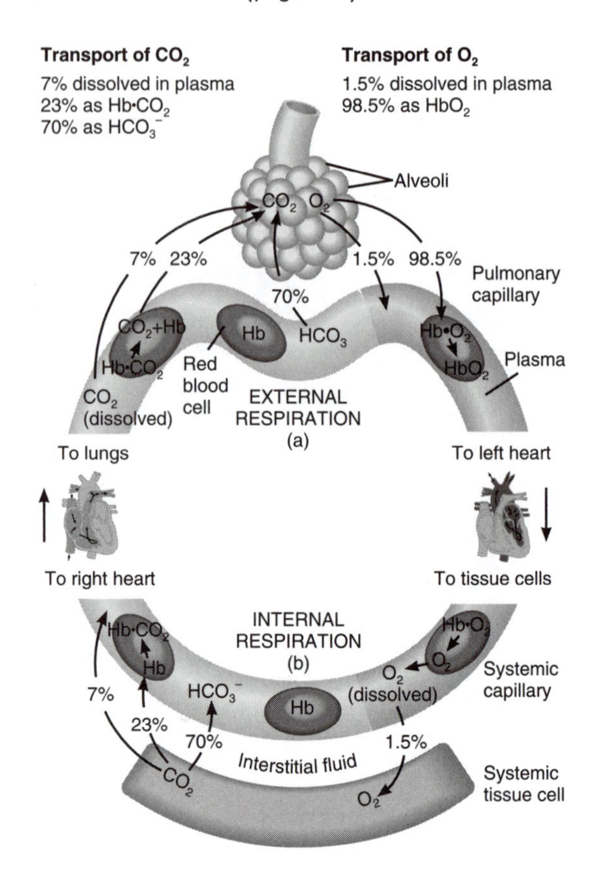

Transport of CO_2
7% dissolved in plasma
23% as Hb•CO_2
70% as HCO_3^-
Transport of O_2
1.5% dissolved in plasma
98.5% as HbO_2
Alveoli
CO_2
O_2
7% 23%
70%
1.5% 98.5%
Pulmonary capillary
CO_2+Hb
Hb•CO_2
Hb
HCO_3
Red blood cell
Hb•O_2
HbO_2
Plasma
CO_2 (dissolved)
EXTERNAL RESPIRATION
(a)
To lungs
To left heart
To right heart
To tissue cells
INTERNAL RESPIRATION
(b)
Hb•CO_2
Hb
Hb•O_2
O_2
O_2 (dissolved)
Systemic capillary
7%
23%
HCO_3^-
70%
Hb
Interstitial fluid
1.5%
CO_2
O_2
Systemic tissue cell

Figure 23.20 Oxygen–hemoglobin dissociation curve (page 800).

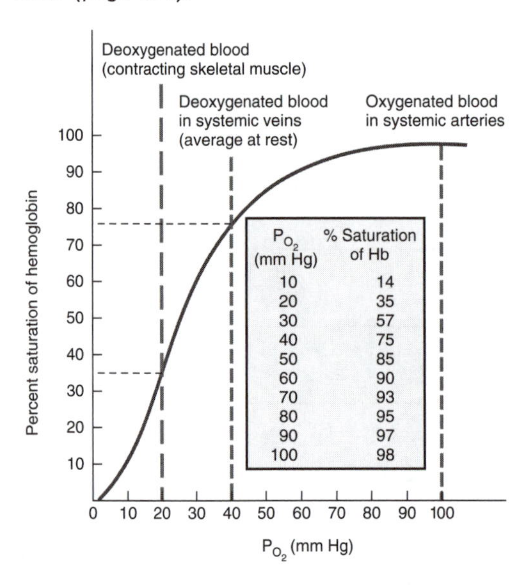

Deoxygenated blood (contracting skeletal muscle)
Deoxygenated blood in systemic veins (average at rest)
Oxygenated blood in systemic arteries
Percent saturation of hemoglobin
100
90
80
70
60
50
40
30
20
10
0 10 20 30 40 50 60 70 80 90 100
P_{O_2} (mm Hg)
P_{O_2} (mm Hg) | % Saturation of Hb
10 | 14
20 | 35
30 | 57
40 | 75
50 | 85
60 | 90
70 | 93
80 | 95
90 | 97
100 | 98

NOTES

23

Figure 23.21 Oxygen–hemoglobin dissociation curves showing the relationship of (a) pH and (b) P_{CO_2} to hemoglobin saturation at normal body temperature (page 801).

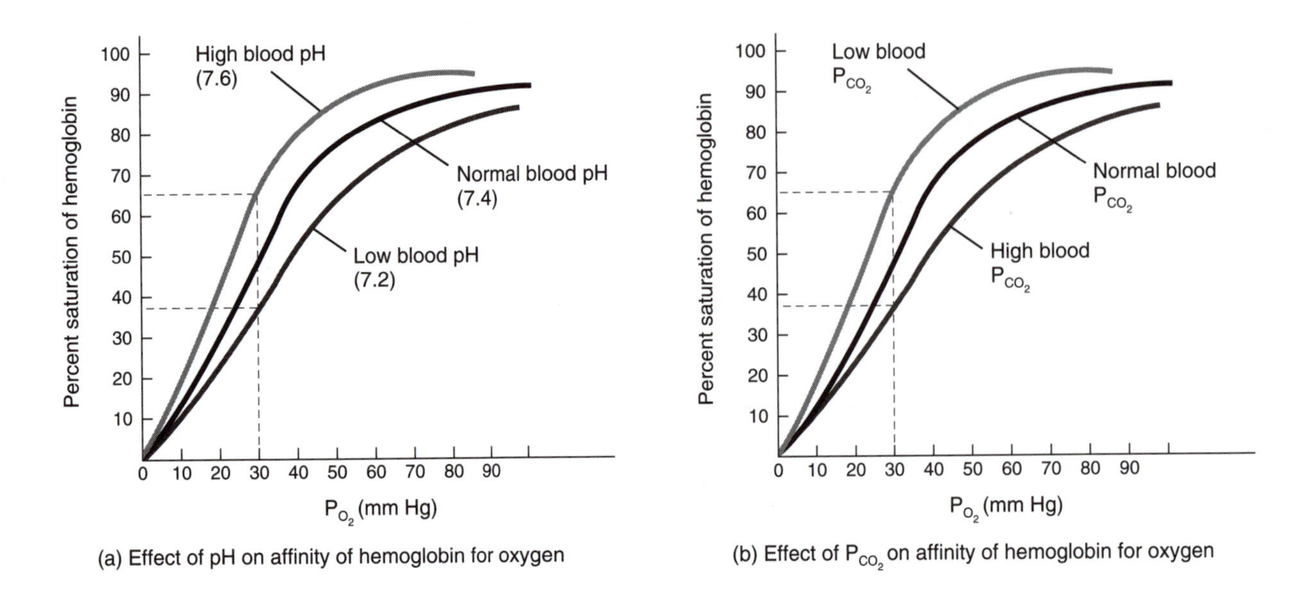

High blood pH
(7.6)
Normal blood pH
(7.4)
Low blood pH
(7.2)
Percent saturation of hemoglobin
100
90
80
70
60
50
40
30
20
10
0 10 20 30 40 50 60 70 80 90
P_{O_2} (mm Hg)
(a) Effect of pH on affinity of hemoglobin for oxygen
Low blood
P_{CO_2}
Normal blood
P_{CO_2}
High blood
P_{CO_2}
Percent saturation of hemoglobin
100
90
80
70
60
50
40
30
20
10
0 10 20 30 40 50 60 70 80 90
P_{O_2} (mm Hg)
(b) Effect of P_{CO_2} on affinity of hemoglobin for oxygen

Figure 23.22 Oxygen–hemoglobin dissociation curves showing the relationship between temperature and hemoglobin saturation with O_2 (page 801).

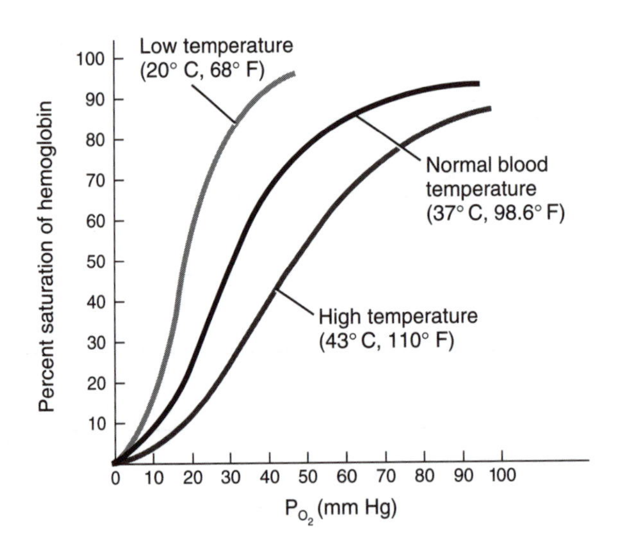

Low temperature
(20° C, 68° F)
Normal blood
temperature
(37° C, 98.6° F)
High temperature
(43° C, 110° F)
Percent saturation of hemoglobin
100
90
80
70
60
50
40
30
20
10
0 10 20 30 40 50 60 70 80 90 100
P_{O_2} (mm Hg)

NOTES

23

Figure 23.23 Oxygen-hemoglobin dissociation curves comparing fetal and maternal hemoglobin (page 802).

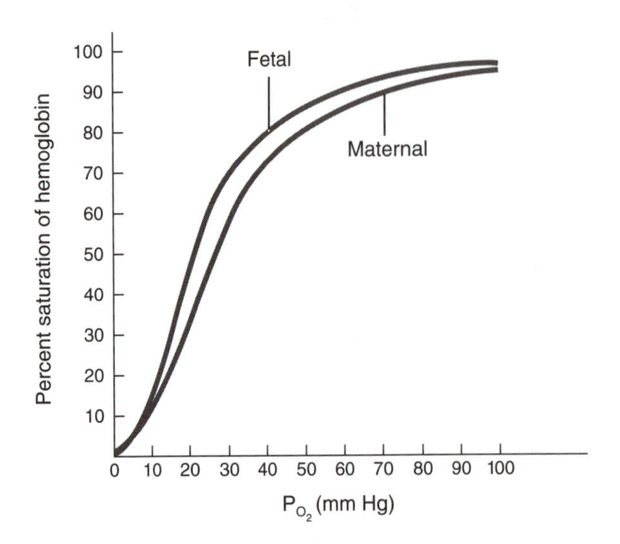

Fetal
Maternal
Percent saturation of hemoglobin
100
90
80
70
60
50
40
30
20
10
0
10 20 30 40 50 60 70 80 90 100
P_{O_2} (mm Hg)

Figure 23.24 Summary of gas exchange and transport (page 803).

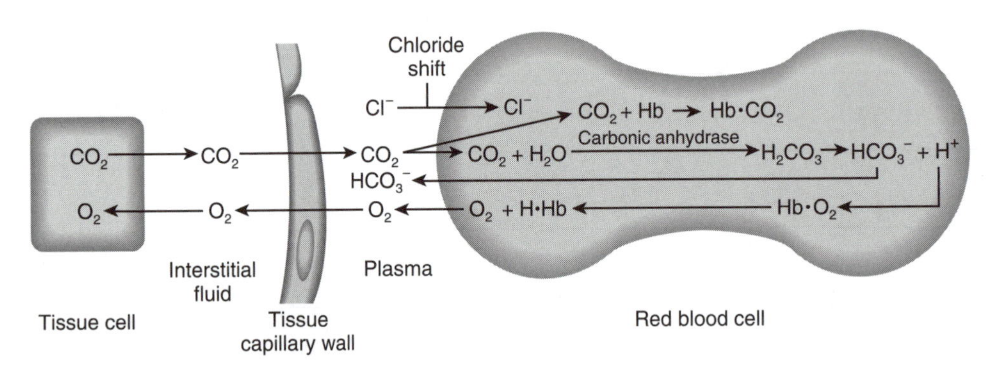

Chloride shift
Cl^-
Cl^-
$CO_2 + Hb \rightarrow Hb \cdot CO_2$
Carbonic anhydrase
CO_2
CO_2
CO_2
$CO_2 + H_2O \rightarrow H_2CO_3 \rightarrow HCO_3^- + H^+$
HCO_3^-
O_2
O_2
O_2
$O_2 + H \cdot Hb$
$Hb \cdot O_2$
Interstitial fluid
Plasma
Tissue cell
Tissue capillary wall
Red blood cell

(a) Exchange of O_2 and CO_2 in the tissues (internal respiration)

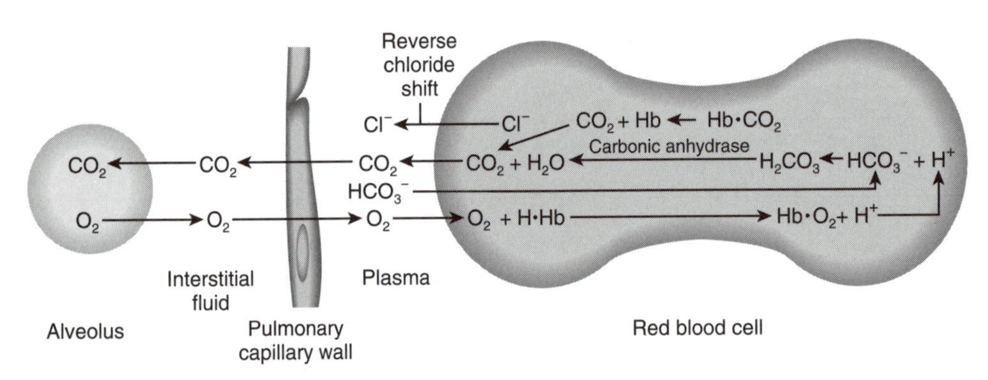

Reverse chloride shift
Cl^-
Cl^-
$CO_2 + Hb \leftarrow Hb \cdot CO_2$
Carbonic anhydrase
CO_2
CO_2
CO_2
$CO_2 + H_2O \leftarrow H_2CO_3 \leftarrow HCO_3^- + H^+$
HCO_3^-
O_2
O_2
O_2
$O_2 + H \cdot Hb$
$Hb \cdot O_2 + H^+$
Interstitial fluid
Plasma
Alveolus
Pulmonary capillary wall
Red blood cell

(b) Exchange of O_2 and CO_2 in the lungs (external respiration)

NOTES

23

Figure 23.25 Locations of areas of the respiratory center (page 804).

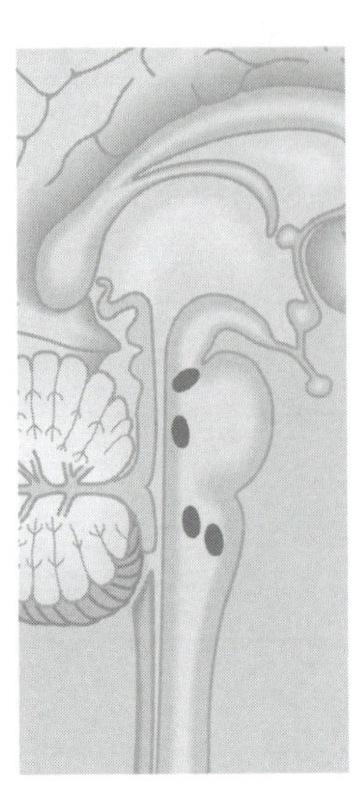

Figure 23.26 Proposed roles of the medullary rhythmicity area in controlling the basic rhythm of respiration and labored breathing (page 805).

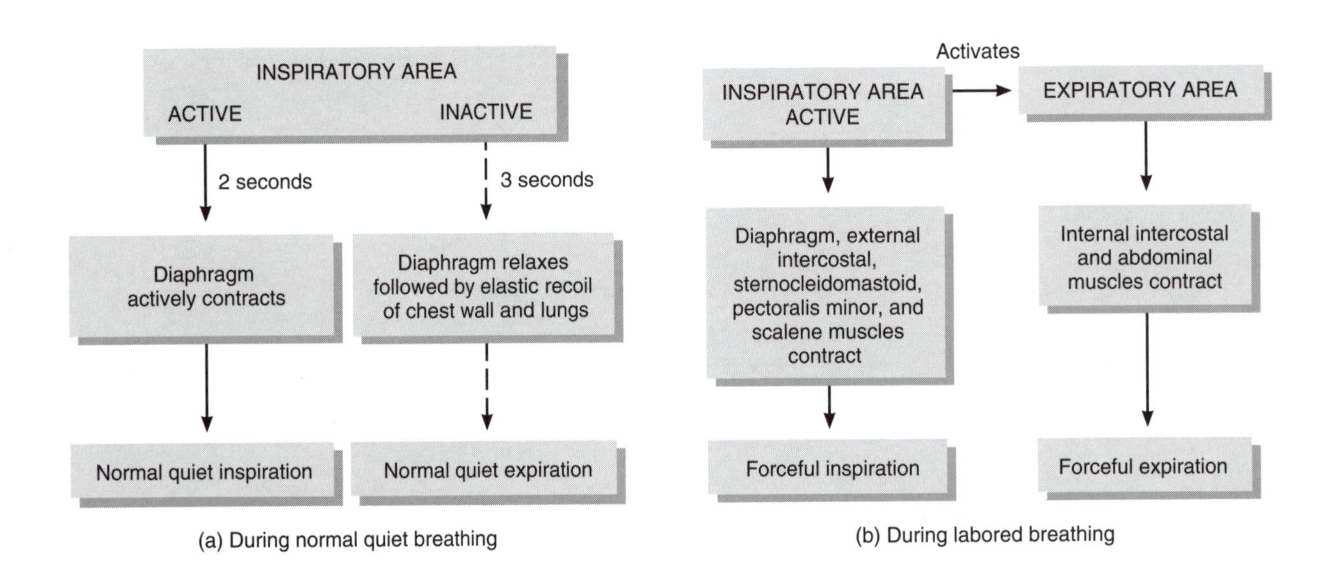

INSPIRATORY AREA
ACTIVE
INACTIVE
2 seconds
3 seconds
Diaphragm actively contracts
Diaphragm relaxes followed by elastic recoil of chest wall and lungs
Normal quiet inspiration
Normal quiet expiration
(a) During normal quiet breathing
Activates
INSPIRATORY AREA ACTIVE
EXPIRATORY AREA
Diaphragm, external intercostal, sternocleidomastoid, pectoralis minor, and scalene muscles contract
Internal intercostal and abdominal muscles contract
Forceful inspiration
Forceful expiration
(b) During labored breathing

NOTES

23

Figure 23.27 Regulation of breathing in response to changes in blood P_{CO_2}, P_{O_2}, and pH via negative feedback control (page 806).

Some stimulus disrupts homeostasis by

Increasing

Arterial blood P_{CO_2} (or decreasing pH or P_{O_2})

Receptors

Central chemoreceptors in medulla

Peripheral chemoreceptors in aortic and carotid bodies

Input Nerve impulses

Control center

Inspiratory area in medulla oblongata

Output Nerve impulses

Effectors

Diaphragm and other muscles of respiration contract more forcefully and more frequently (hyperventilation)

Decrease in arterial blood P_{CO_2}, increase in pH, and increase in P_{O_2}

Return to homeostasis when response brings arterial blood P_{CO_2}, pH, and P_{O_2} back to normal

Figure 23.28 Reduction of P_{O_2} in blood via positive feedback (page 807).

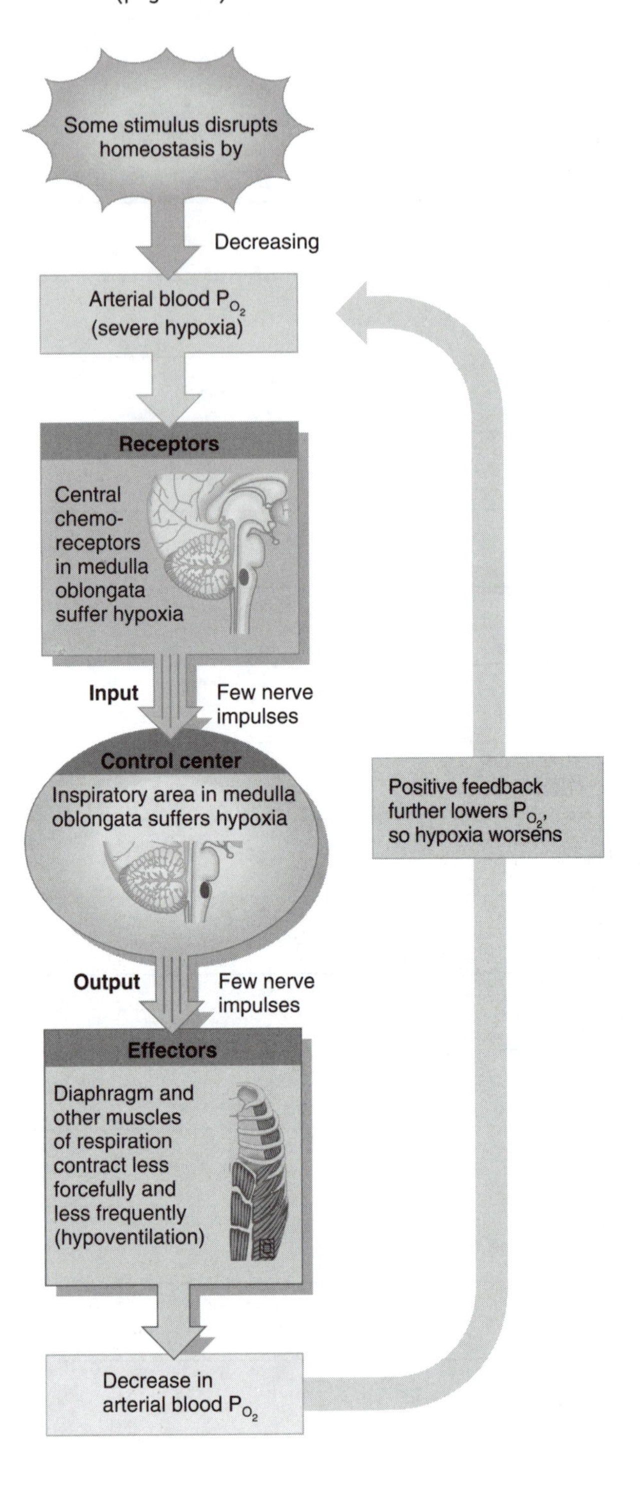

Some stimulus disrupts homeostasis by
Decreasing
Arterial blood P_{O_2} (severe hypoxia)
Receptors
Central chemoreceptors in medulla oblongata suffer hypoxia
Input
Few nerve impulses
Control center
Inspiratory area in medulla oblongata suffers hypoxia
Output
Few nerve impulses
Effectors
Diaphragm and other muscles of respiration contract less forcefully and less frequently (hypoventilation)
Decrease in arterial blood P_{O_2}
Positive feedback further lowers P_{O_2}, so hypoxia worsens

NOTES

23

Figure 23.29 Development of the bronchial tubes and lungs (page 810).

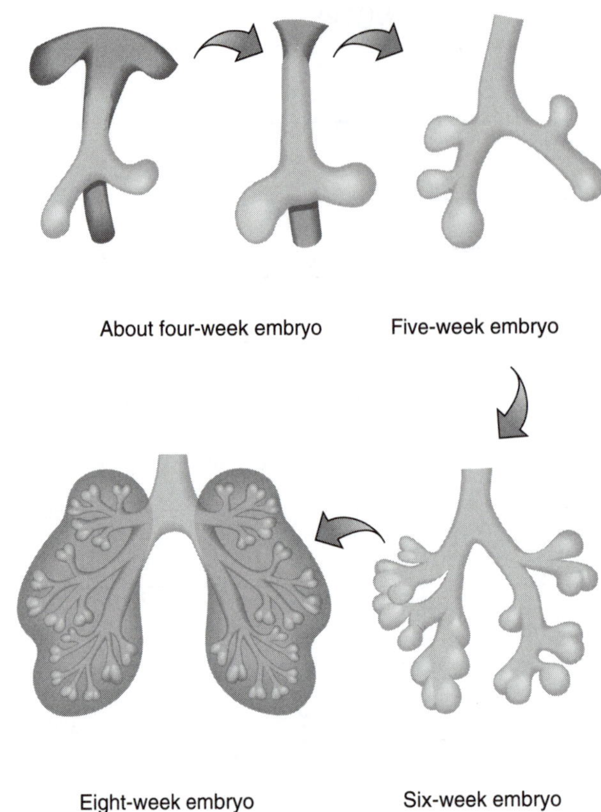
About four-week embryo
Five-week embryo
Eight-week embryo
Six-week embryo

NOTES

23

Figure 24.1 Organs of the digestive system (page 819).

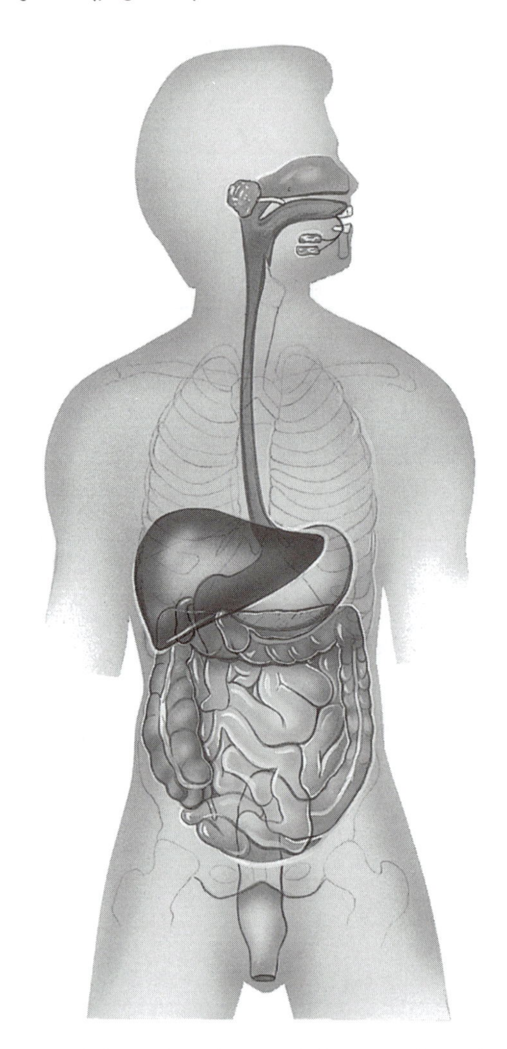

Figure 24.2 Three-dimensional depiction of the various layers of the gastrointestinal tract (page 820).

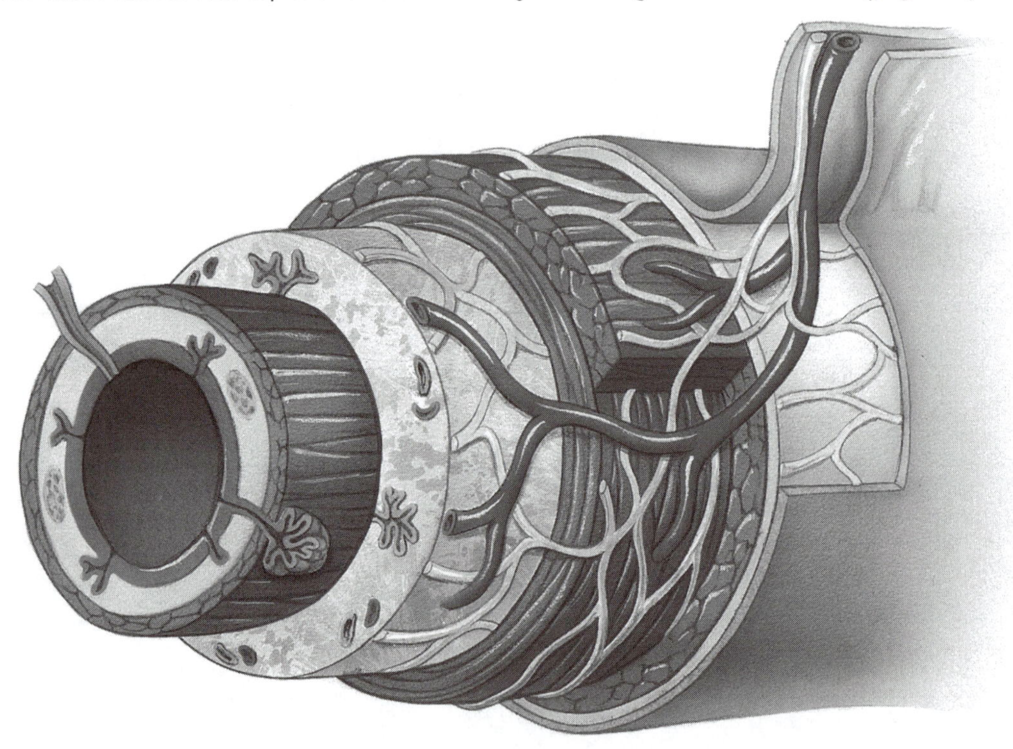

NOTES

24

Figure 24.3 Relationship of the peritoneal folds to each other and to organs of the digestive system (page 822).

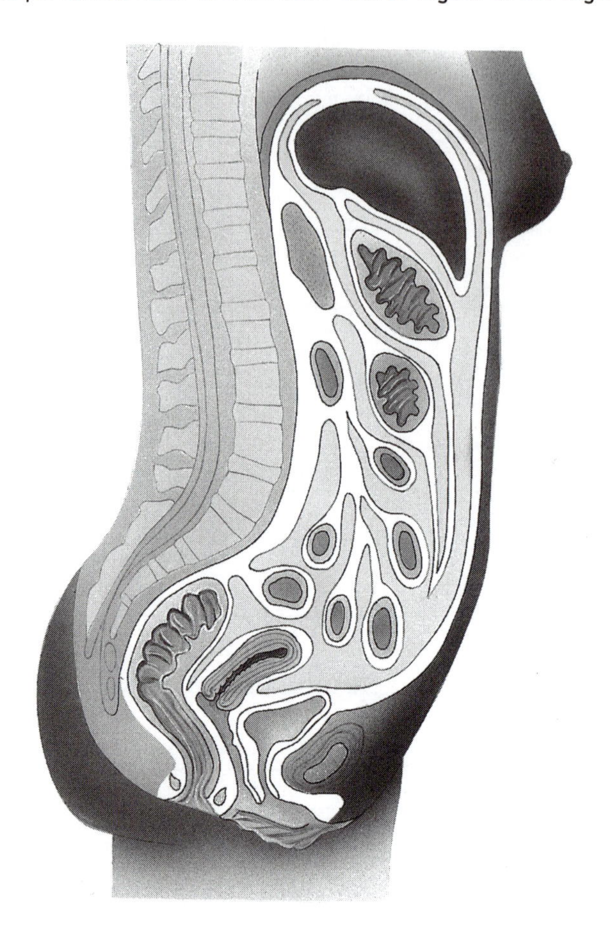

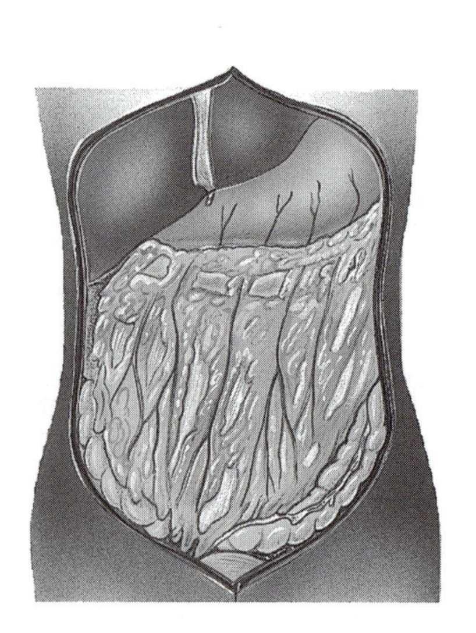

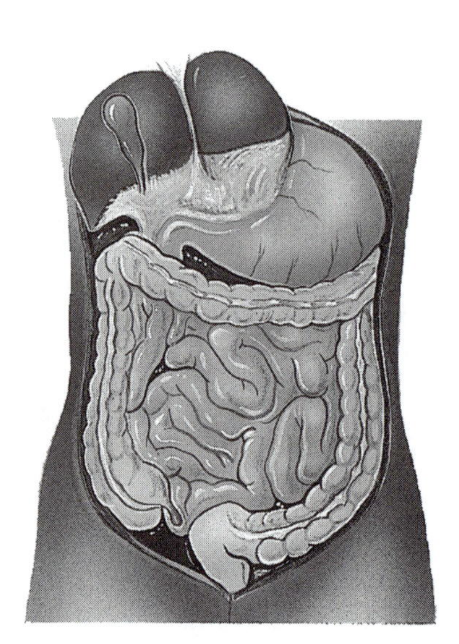

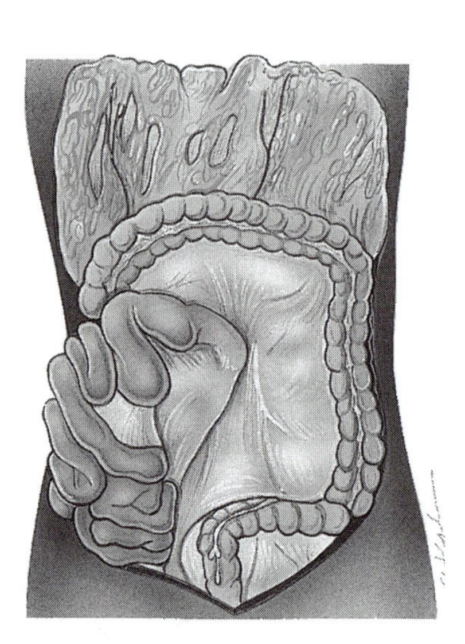

NOTES

24

Figure 24.4 Structures of the mouth (page 824).

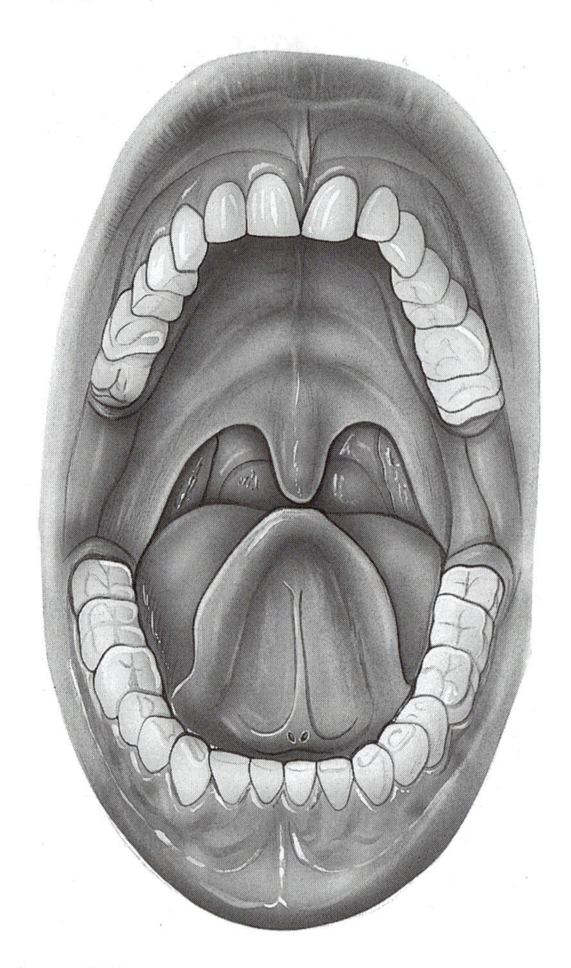

Figure 24.5 Major salivary glands (page 826).

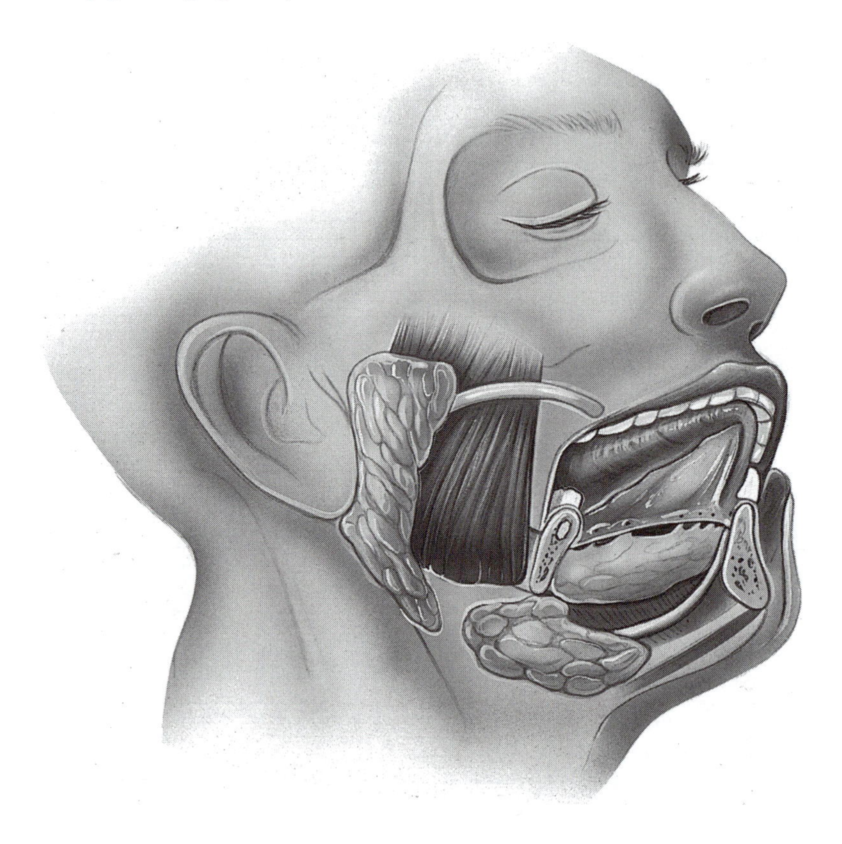

NOTES

24

Figure 24.6 A typical tooth and surrounding structures (page 828).

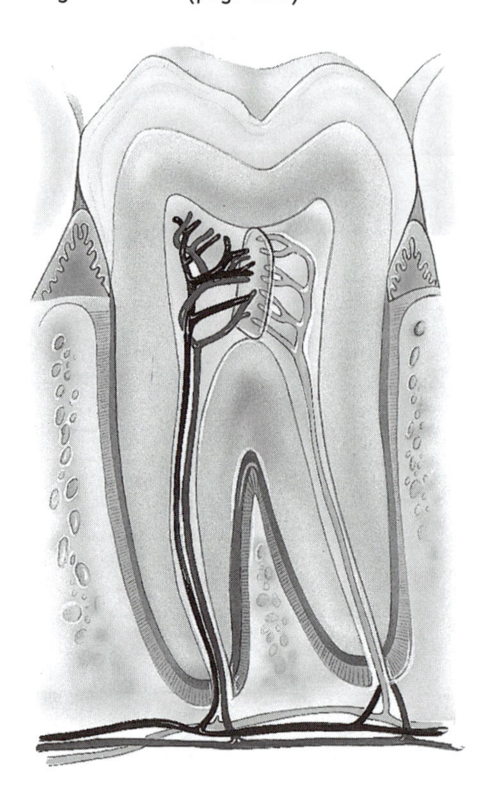

Figure 24.7 Dentitions and times of eruptions (page 829).

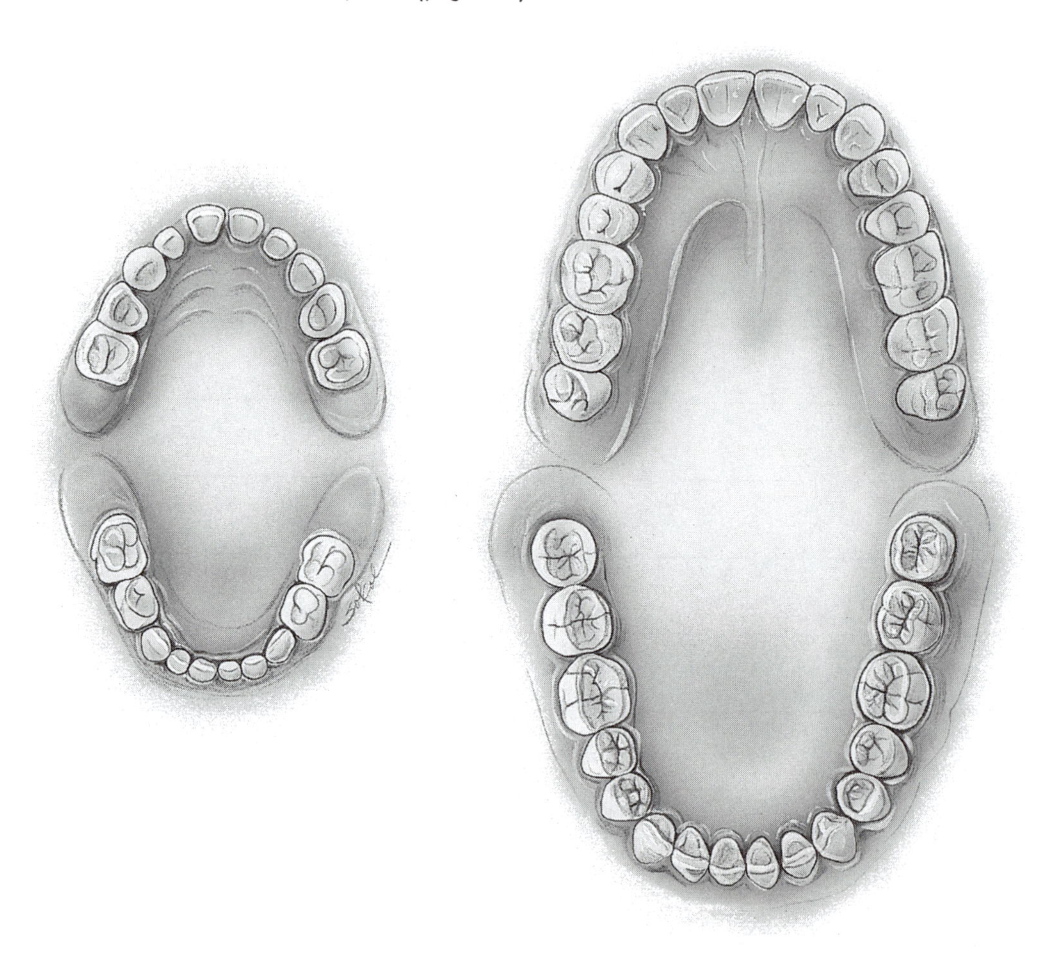

NOTES

24

Figure 24.8 Deglutition (page 831).

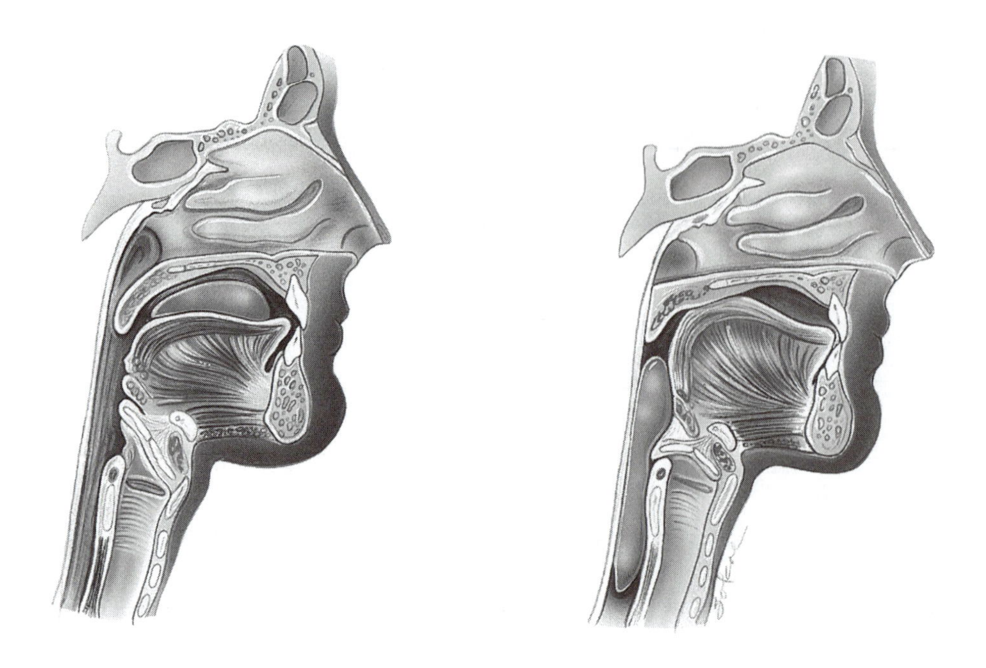

Figure 24.10 Peristalsis during the esophageal stage of deglutition (page 832).

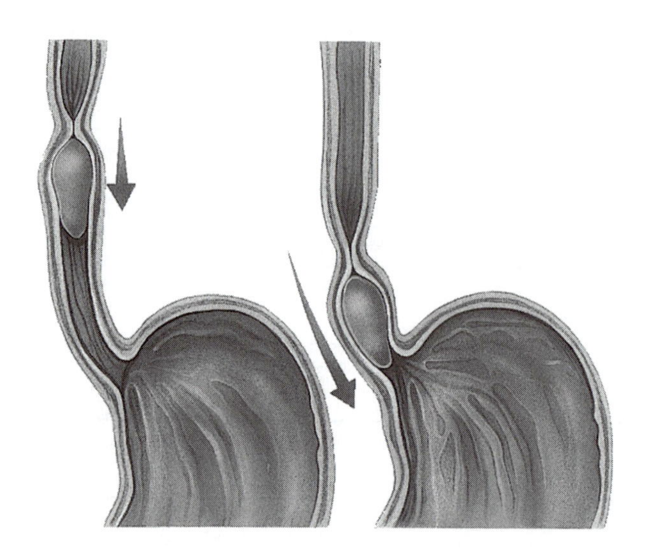

NOTES

24

Figure 24.11 External and internal anatomy of the stomach (page 834).

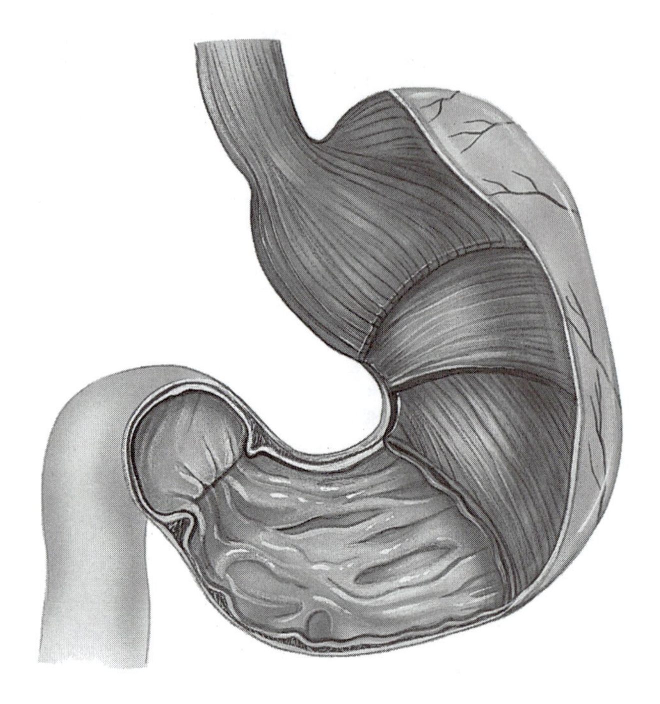

Figure 24.12 Histology of the stomach (page 835).

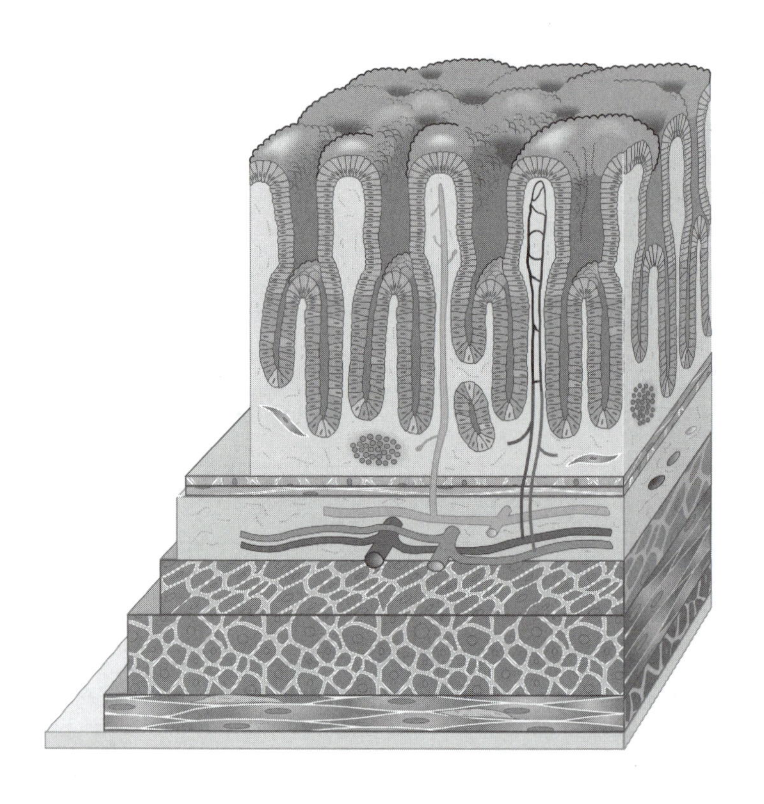

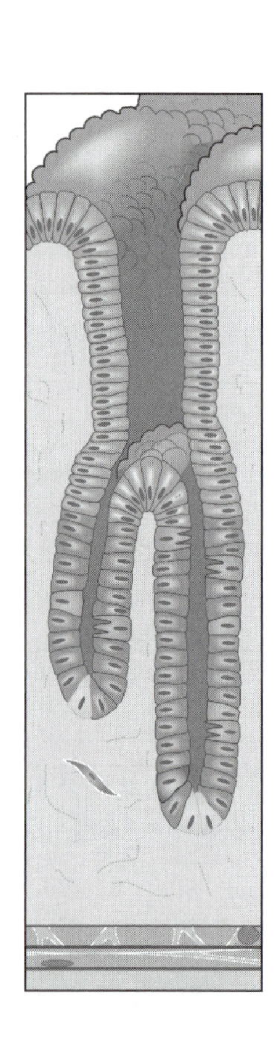

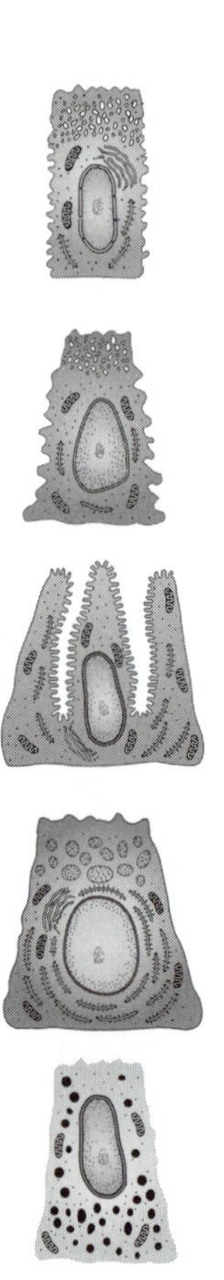

NOTES

24

Figure 24.13 The cephalic, gastric, and intestinal phases of gastric digestion (page 838).

Figure 24.14 Neural negative feedback regulation of pH of gastric juice and gastric motility during the gastric phase of digestion (page 838).

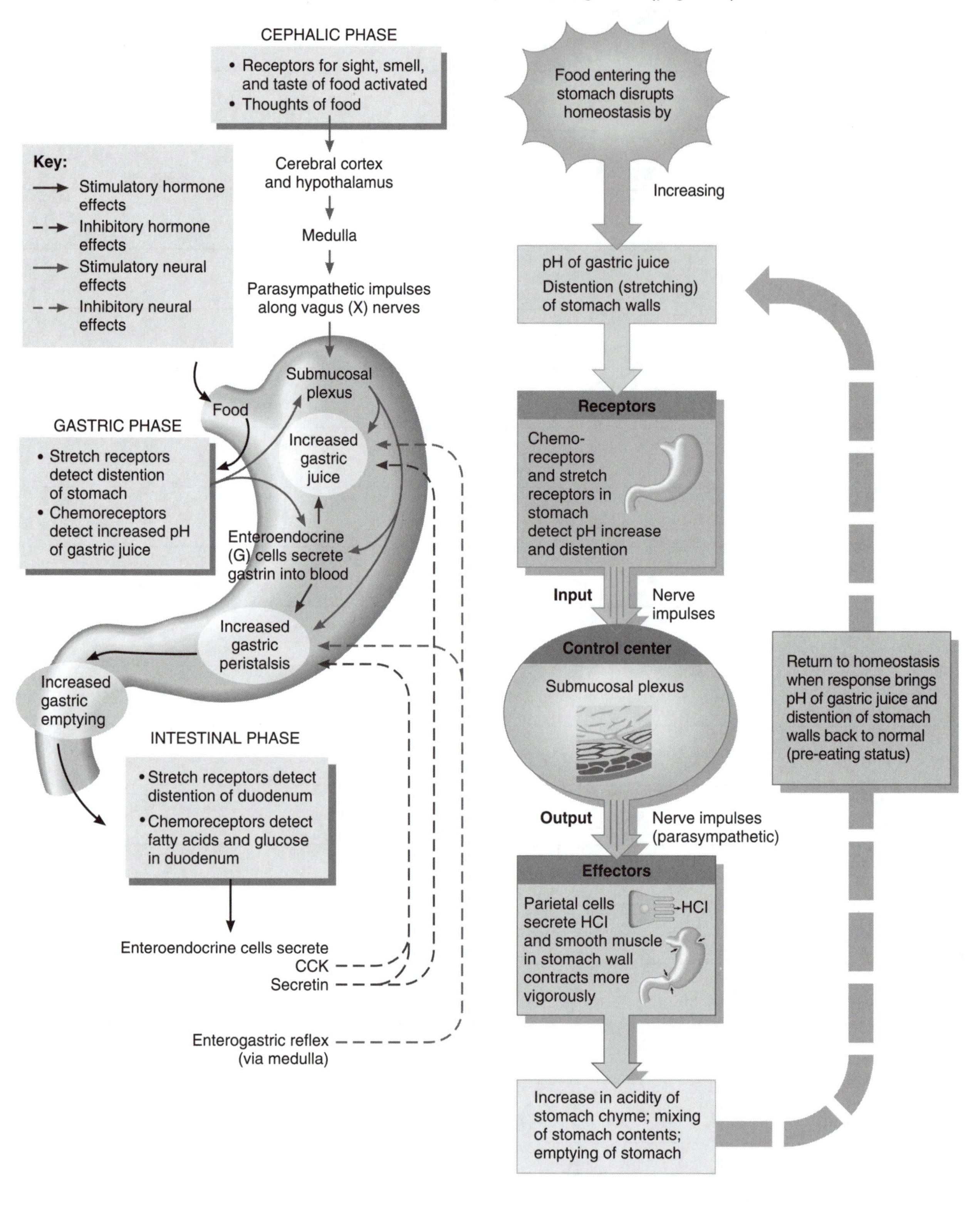

CEPHALIC PHASE
• Receptors for sight, smell, and taste of food activated
• Thoughts of food
Cerebral cortex and hypothalamus
Medulla
Parasympathetic impulses along vagus (X) nerves
Key:
Stimulatory hormone effects
Inhibitory hormone effects
Stimulatory neural effects
Inhibitory neural effects
Food
Submucosal plexus
Increased gastric juice
GASTRIC PHASE
• Stretch receptors detect distention of stomach
• Chemoreceptors detect increased pH of gastric juice
Enteroendocrine (G) cells secrete gastrin into blood
Increased gastric peristalsis
Increased gastric emptying
INTESTINAL PHASE
• Stretch receptors detect distention of duodenum
• Chemoreceptors detect fatty acids and glucose in duodenum
Enteroendocrine cells secrete
CCK
Secretin
Enterogastric reflex (via medulla)
Food entering the stomach disrupts homeostasis by
Increasing
pH of gastric juice
Distention (stretching) of stomach walls
Receptors
Chemo-receptors and stretch receptors in stomach detect pH increase and distention
Input
Nerve impulses
Control center
Submucosal plexus
Output
Nerve impulses (parasympathetic)
Effectors
Parietal cells secrete HCl and smooth muscle in stomach wall contracts more vigorously
HCl
Increase in acidity of stomach chyme; mixing of stomach contents; emptying of stomach
Return to homeostasis when response brings pH of gastric juice and distention of stomach walls back to normal (pre-eating status)

NOTES

24

Figure 24.15 Neural and hormonal regulation of gastric emptying (page 839).

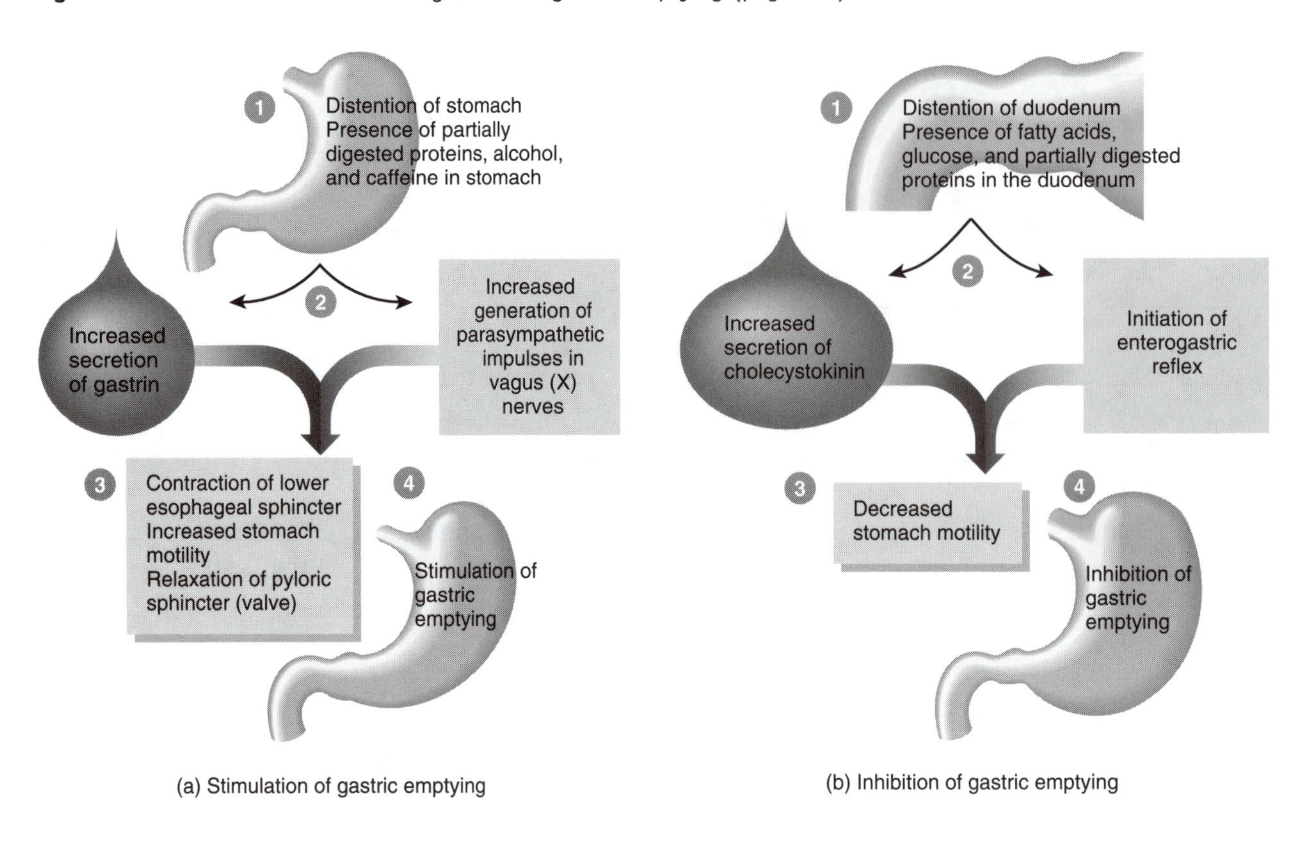

1
Distention of stomach
Presence of partially digested proteins, alcohol, and caffeine in stomach
2
Increased secretion of gastrin
Increased generation of parasympathetic impulses in vagus (X) nerves
3
Contraction of lower esophageal sphincter
Increased stomach motility
Relaxation of pyloric sphincter (valve)
4
Stimulation of gastric emptying
(a) Stimulation of gastric emptying
1
Distention of duodenum
Presence of fatty acids, glucose, and partially digested proteins in the duodenum
2
Increased secretion of cholecystokinin
Initiation of enterogastric reflex
3
Decreased stomach motility
4
Inhibition of gastric emptying
(b) Inhibition of gastric emptying

Figure 24.16 Relation of the pancreas to the liver, gallbladder, and duodenum (page 841).

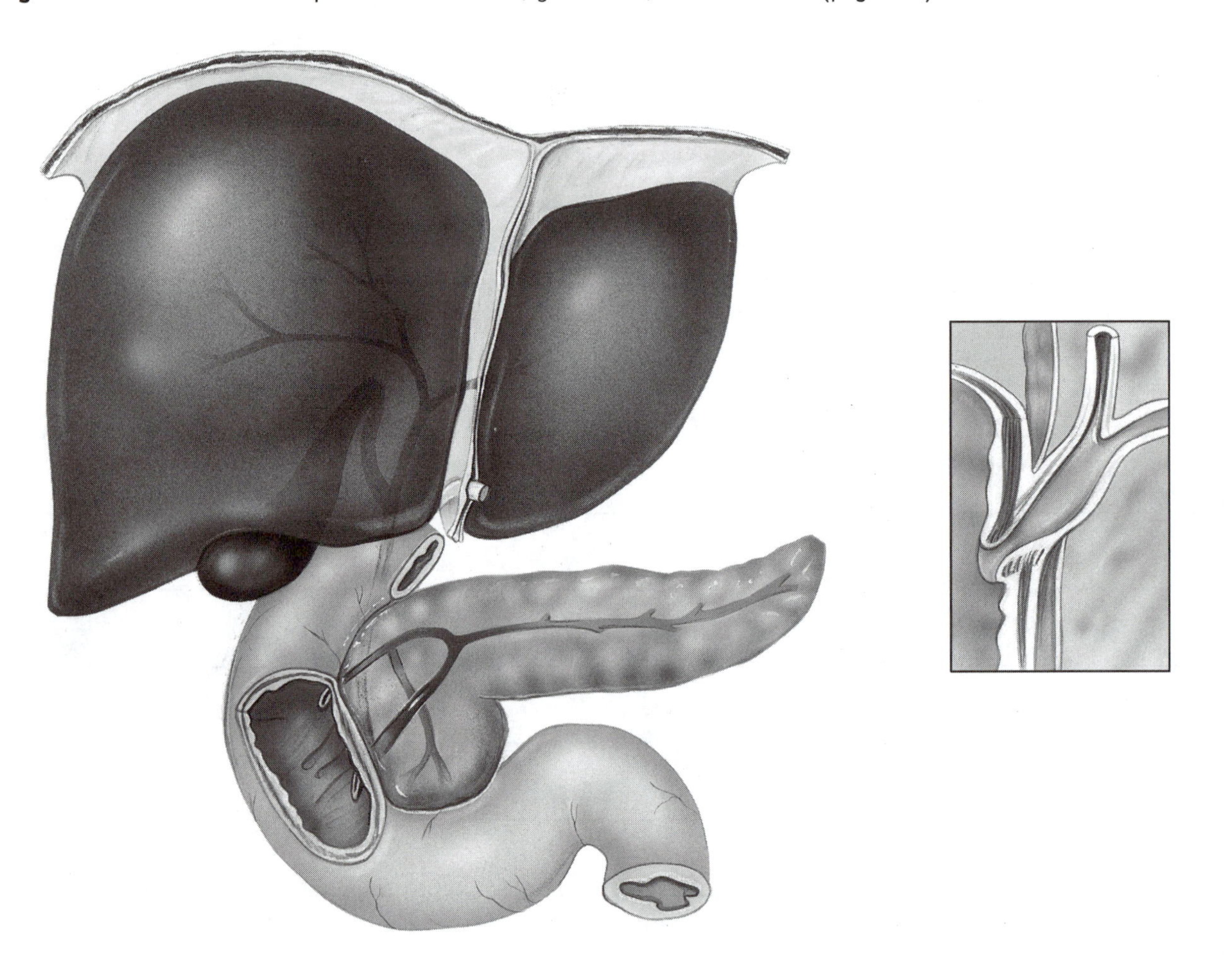

NOTES

24

Figure 24.17 Neural and hormonal enhancement of the secretion of pancreatic juice (page 842).

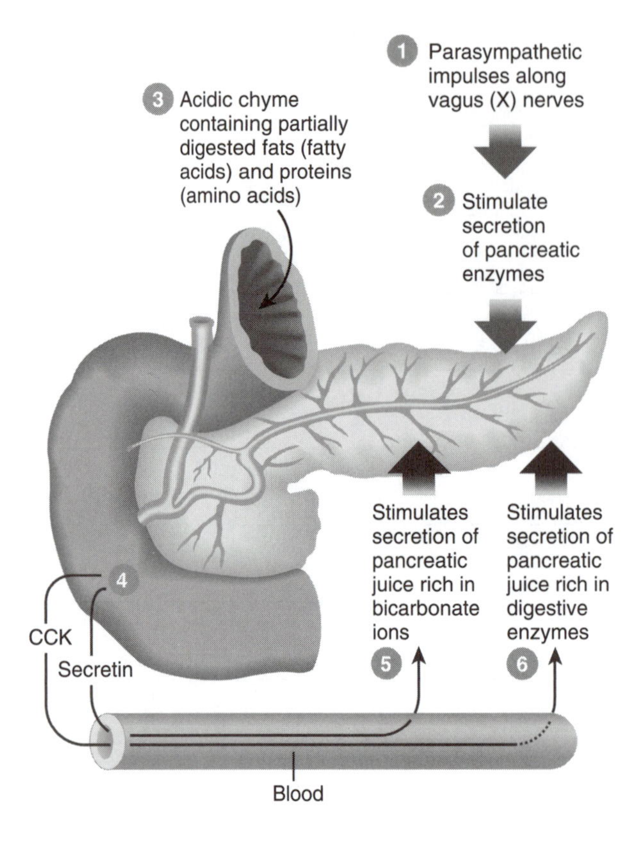
1 Parasympathetic impulses along vagus (X) nerves
2 Stimulate secretion of pancreatic enzymes
3 Acidic chyme containing partially digested fats (fatty acids) and proteins (amino acids)
4
CCK
Secretin
Stimulates secretion of pancreatic juice rich in bicarbonate ions
5
Stimulates secretion of pancreatic juice rich in digestive enzymes
6
Blood

Figure 24.18 Histology of a lobule, the functional unit of the liver (page 844).

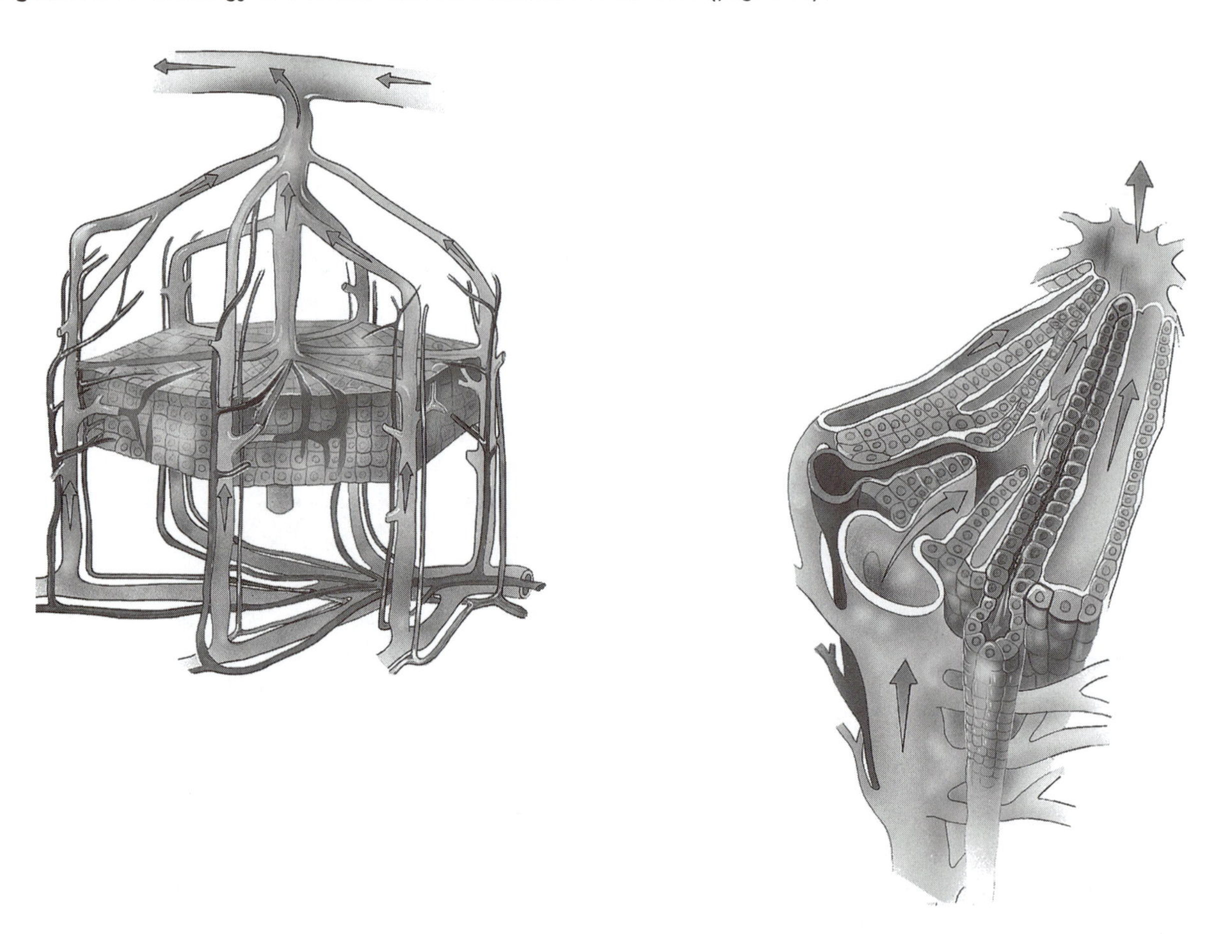

NOTES

24

Figure 24.19 Hepatic blood flow (page 845).

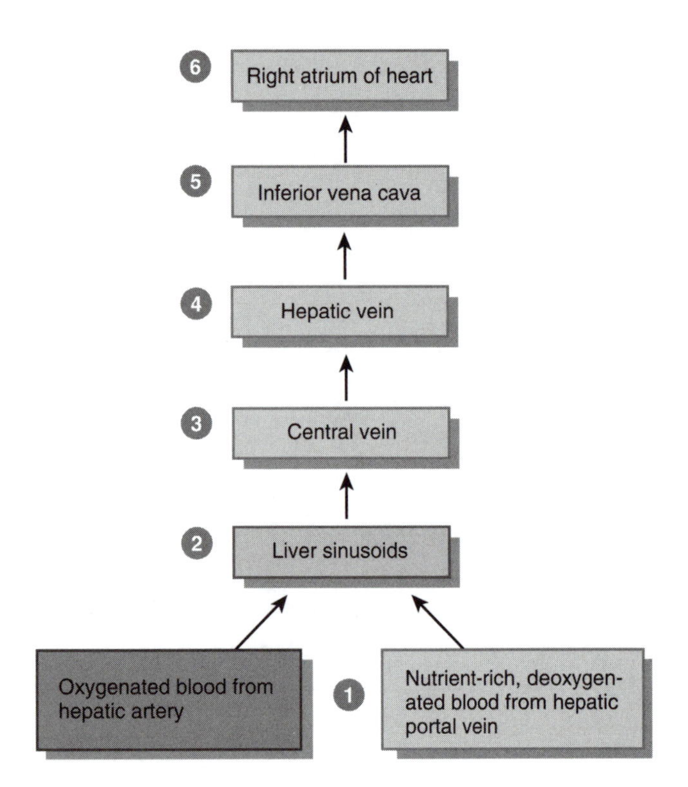

6 Right atrium of heart
5 Inferior vena cava
4 Hepatic vein
3 Central vein
2 Liver sinusoids
Oxygenated blood from hepatic artery
1
Nutrient-rich, deoxygen-ated blood from hepatic portal vein

Figure 24.21 Regions of the small intestine (page 848).

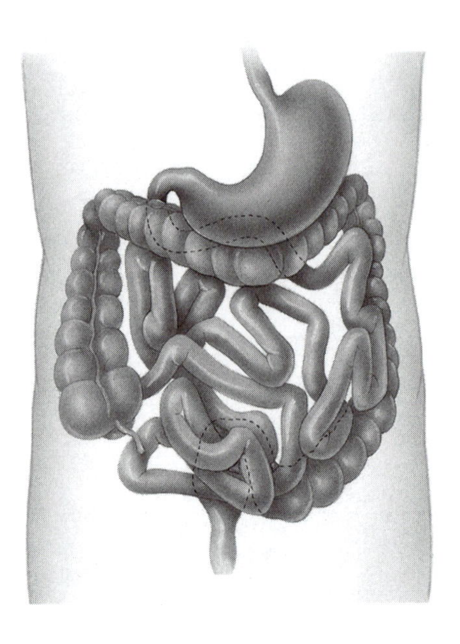

Figure 24.20 Neural and hormonal stimuli that promote production and release of bile (page 846).

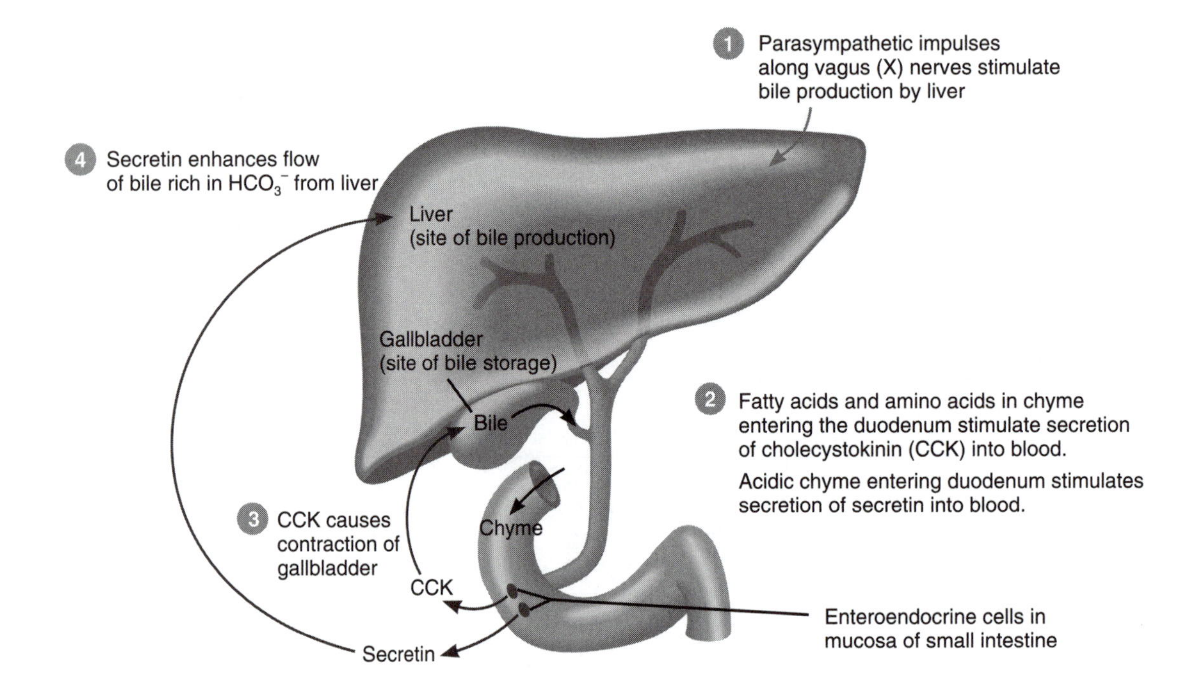

1 Parasympathetic impulses along vagus (X) nerves stimulate bile production by liver
4 Secretin enhances flow of bile rich in HCO_3^- from liver
Liver (site of bile production)
Gallbladder (site of bile storage)
Bile
2 Fatty acids and amino acids in chyme entering the duodenum stimulate secretion of cholecystokinin (CCK) into blood.
Acidic chyme entering duodenum stimulates secretion of secretin into blood.
3 CCK causes contraction of gallbladder
Chyme
CCK
Secretin
Enteroendocrine cells in mucosa of small intestine

NOTES

24

Figure 24.22 Anatomy of the small intestine (page 849).

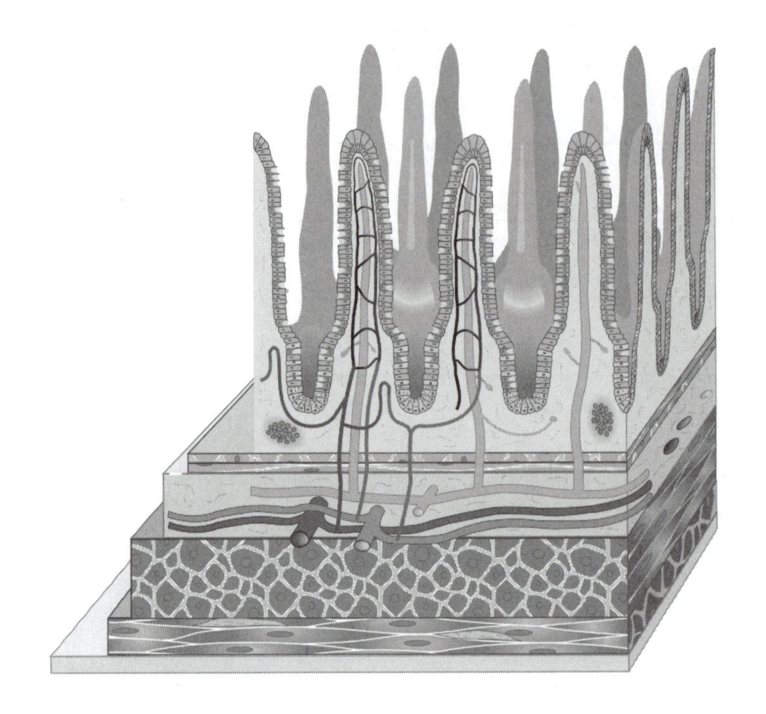

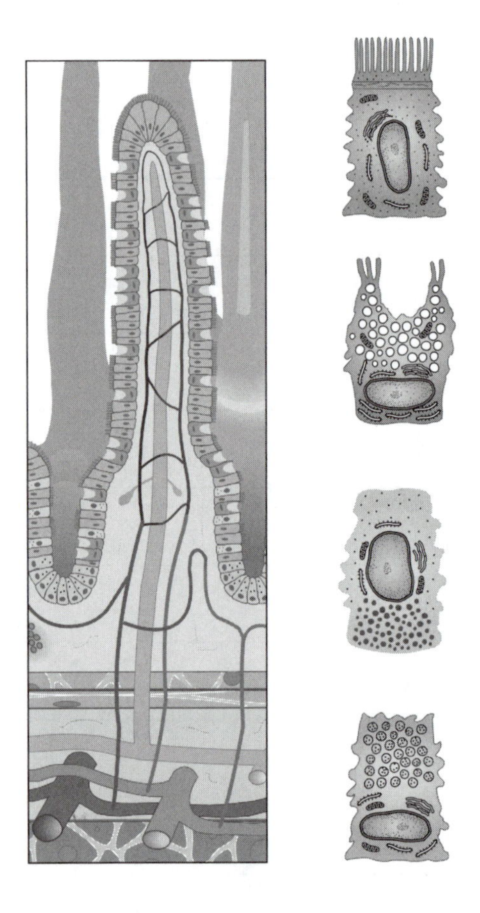

NOTES

24

Figure 24.24 Absorption of digested nutrients in the small intestine (page 854).

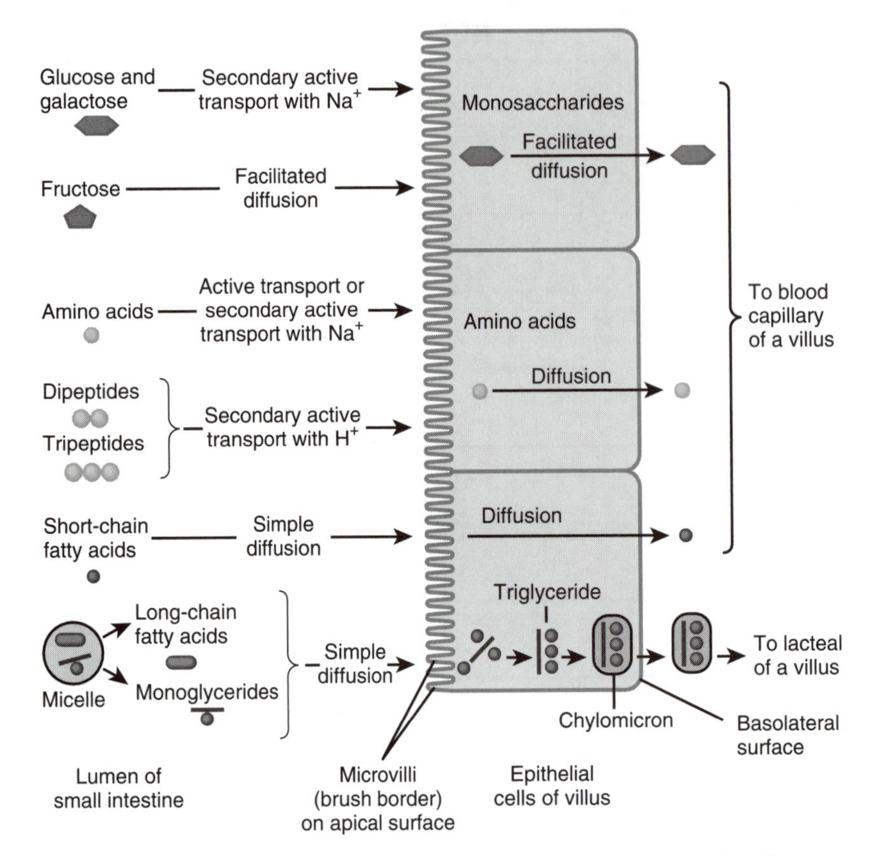

Glucose and galactose
Secondary active transport with Na^+
Monosaccharides
Facilitated diffusion
Fructose
Facilitated diffusion
Amino acids
Active transport or secondary active transport with Na^+
Amino acids
Diffusion
To blood capillary of a villus
Dipeptides
Tripeptides
Secondary active transport with H^+
Short-chain fatty acids
Simple diffusion
Diffusion
Triglyceride
Long-chain fatty acids
Micelle
Monoglycerides
Simple diffusion
To lacteal of a villus
Chylomicron
Basolateral surface
Lumen of small intestine
Microvilli (brush border) on apical surface
Epithelial cells of villus

(a) Mechanisms for movement of nutrients through epithelial cells of the villi

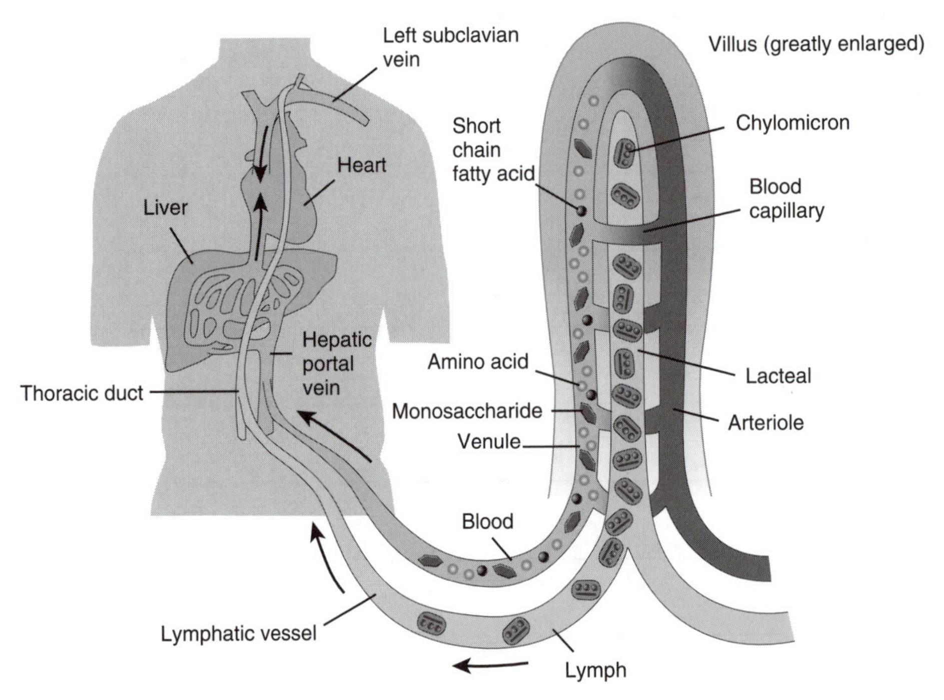

Left subclavian vein
Villus (greatly enlarged)
Short chain fatty acid
Chylomicron
Heart
Blood capillary
Liver
Hepatic portal vein
Amino acid
Lacteal
Thoracic duct
Monosaccharide
Arteriole
Venule
Blood
Lymphatic vessel
Lymph

(b) Movement of absorbed nutrients into the blood lymph

NOTES

24

Figure 24.25 Daily volumes of fluid ingested, secreted, absorbed, and excreted from the GI tract (page 857).

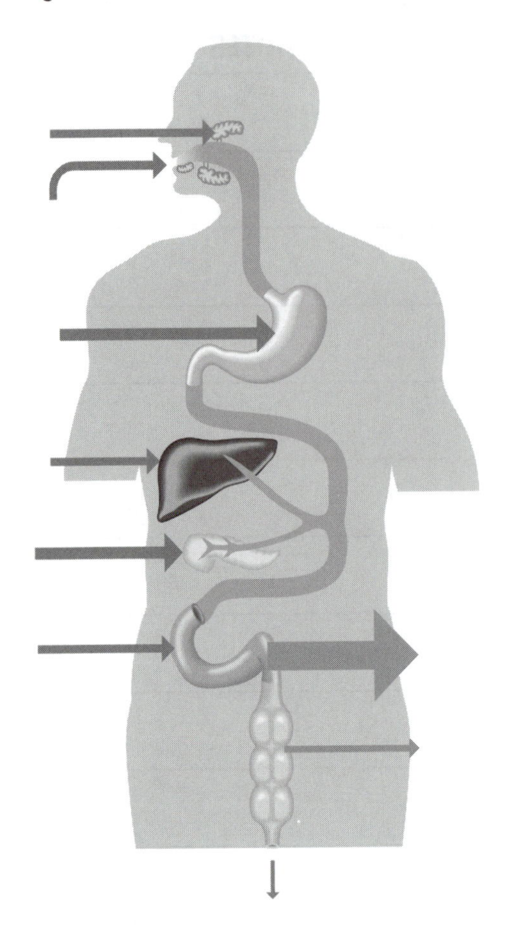

Figure 24.26 Anatomy of large intestine (page 858).

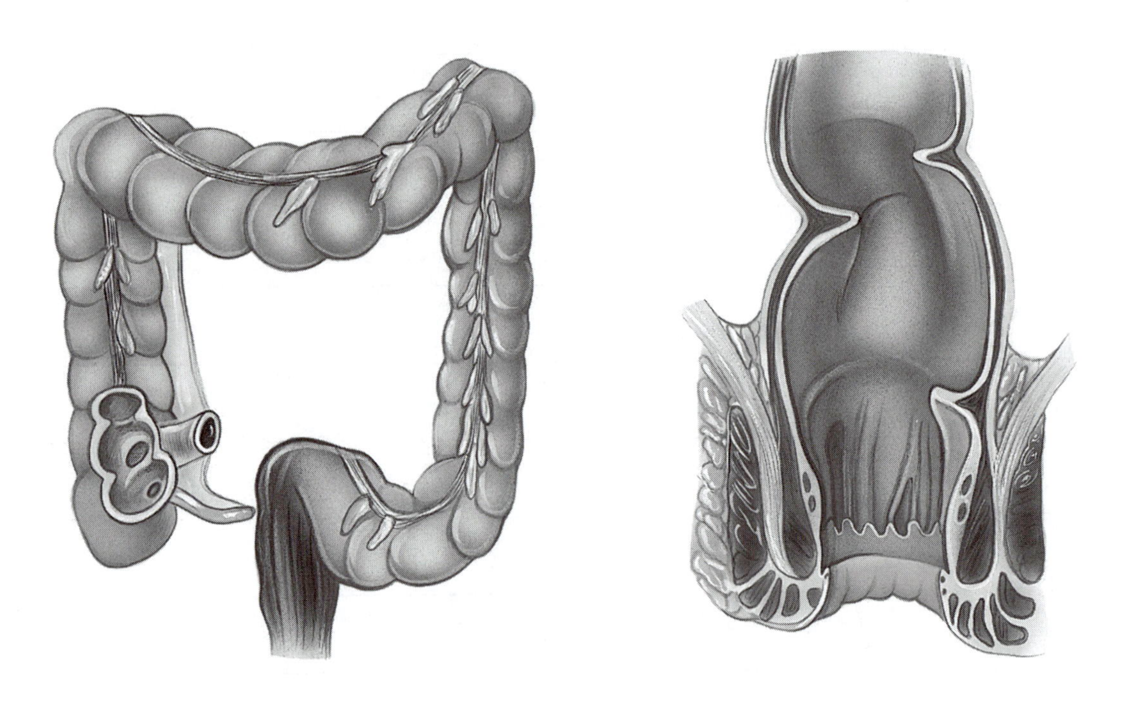

NOTES

24

Figure 24.27 Histology of the large intestine (page 860).

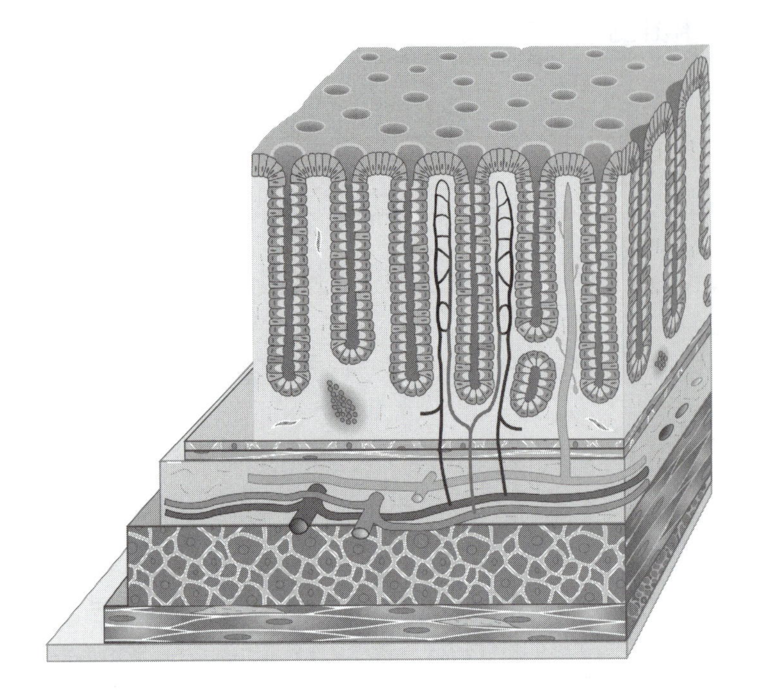

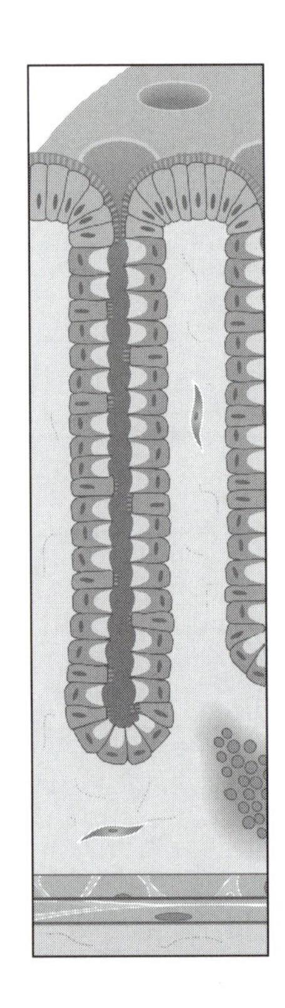

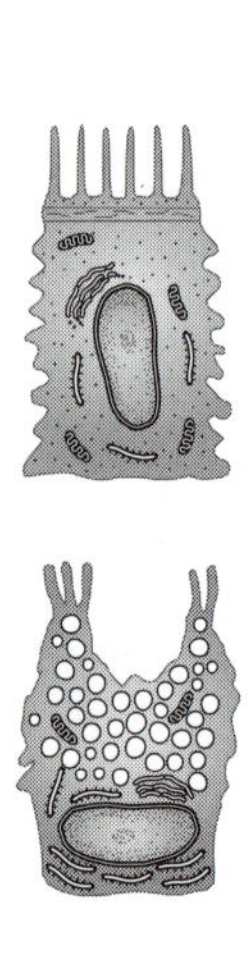

NOTES

Figure 24.28 Development of the digestive system (page 863).

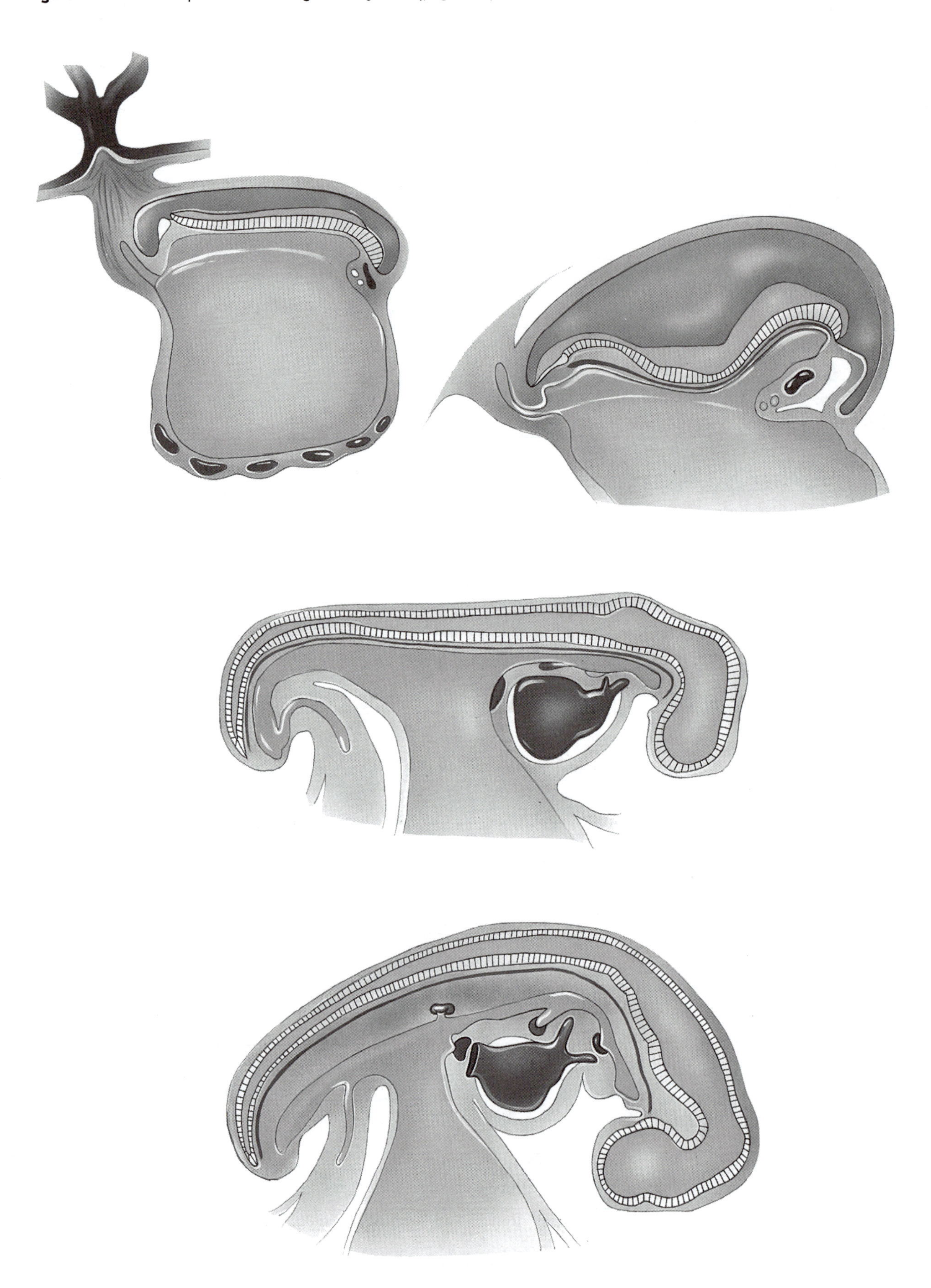

NOTES

24

Figure 25.1 Role of ATP in linking anabolic and catabolic reactions (page 872).

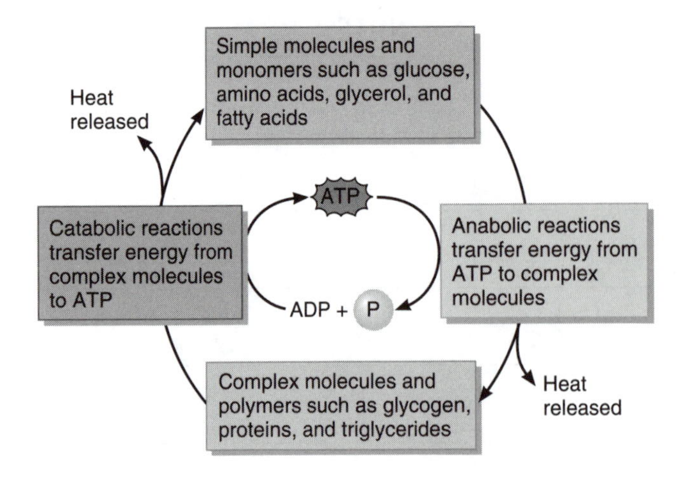
Simple molecules and monomers such as glucose, amino acids, glycerol, and fatty acids
Heat released
ATP
Catabolic reactions transfer energy from complex molecules to ATP
ADP + P
Anabolic reactions transfer energy from ATP to complex molecules
Complex molecules and polymers such as glycogen, proteins, and triglycerides
Heat released

Figure 25.2 Overview of cellular respiration (page 875).

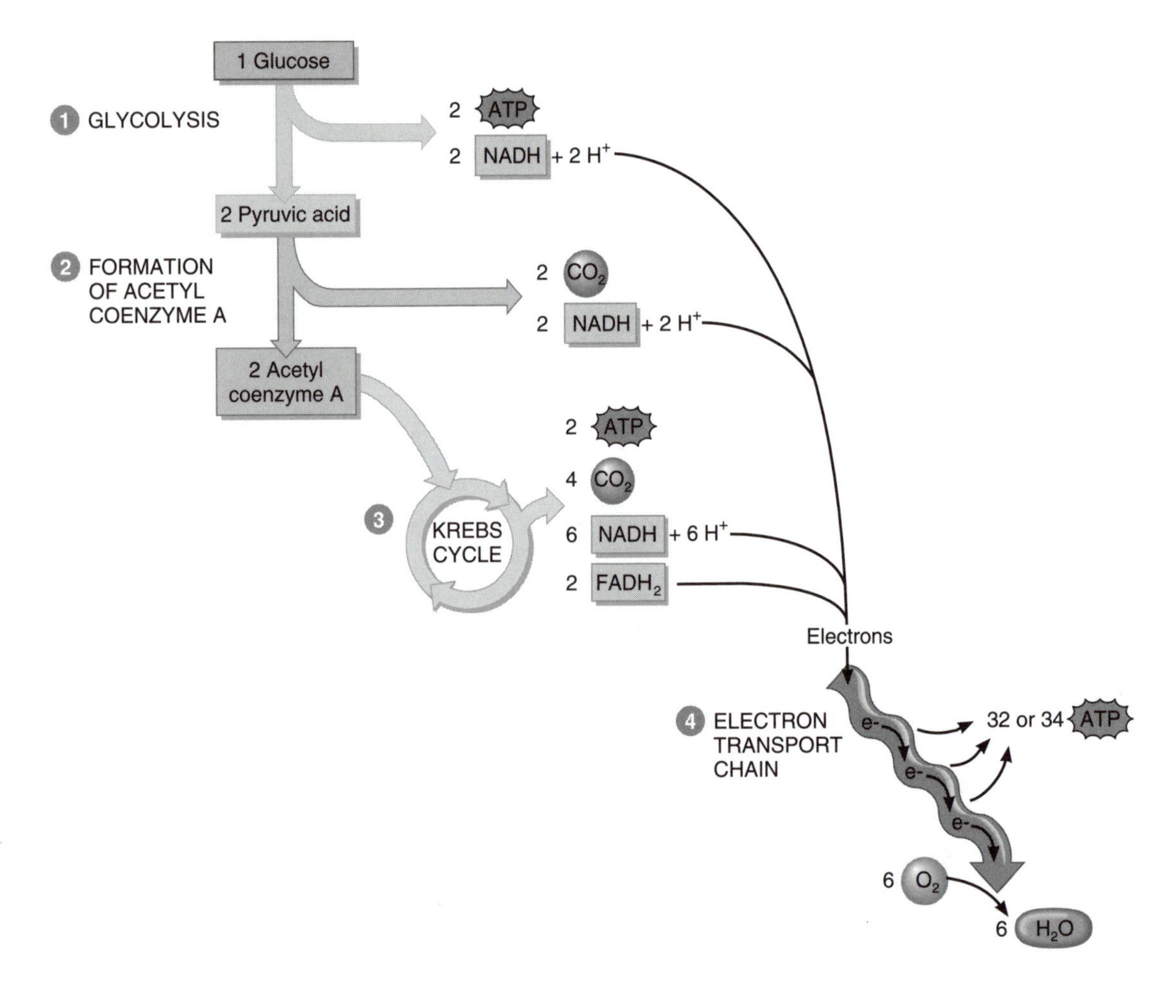
1 Glucose
1 GLYCOLYSIS
2 ATP
2 NADH + 2 H^+
2 Pyruvic acid
2 FORMATION OF ACETYL COENZYME A
2 CO_2
2 NADH + 2 H^+
2 Acetyl coenzyme A
2 ATP
4 CO_2
3 KREBS CYCLE
6 NADH + 6 H^+
2 $FADH_2$
Electrons
4 ELECTRON TRANSPORT CHAIN
e-
e-
e-
32 or 34 ATP
6 O_2
6 H_2O

NOTES

25

Figure 25.3 Glycolysis (page 876, 877).

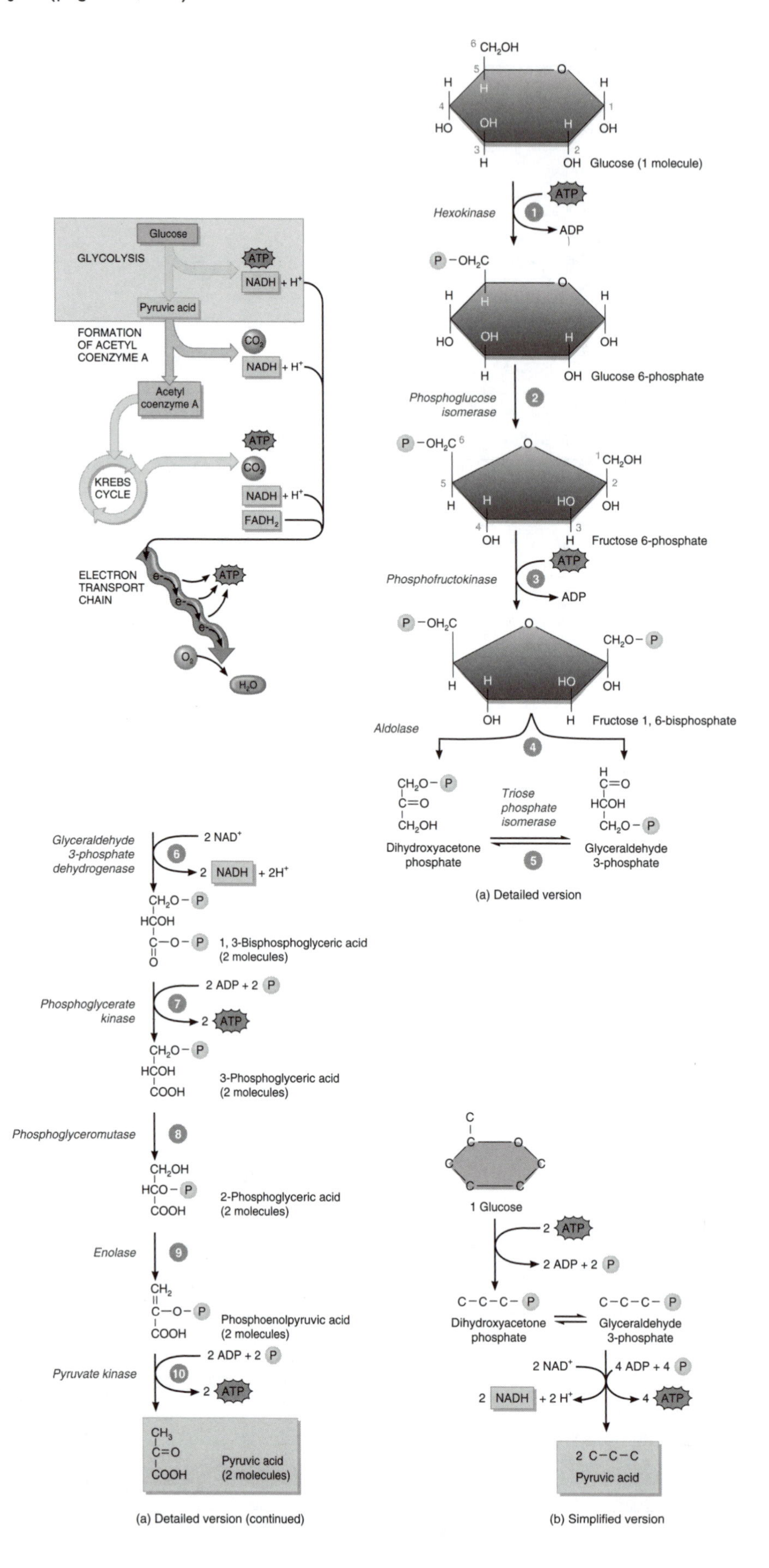
GLYCOLYSIS
Glucose
ATP
NADH + H+
Pyruvic acid
FORMATION OF ACETYL COENZYME A
CO2
Acetyl coenzyme A
KREBS CYCLE
FADH2
ELECTRON TRANSPORT CHAIN
O2
H2O
Glucose (1 molecule)
Hexokinase
ADP
Glucose 6-phosphate
Phosphoglucose isomerase
Fructose 6-phosphate
Phosphofructokinase
Fructose 1, 6-bisphosphate
Aldolase
Triose phosphate isomerase
Dihydroxyacetone phosphate
Glyceraldehyde 3-phosphate
(a) Detailed version
Glyceraldehyde 3-phosphate dehydrogenase
2 NAD+
2 NADH + 2H+
1, 3-Bisphosphoglyceric acid (2 molecules)
Phosphoglycerate kinase
2 ADP + 2 P
2 ATP
3-Phosphoglyceric acid (2 molecules)
Phosphoglyceromutase
2-Phosphoglyceric acid (2 molecules)
Enolase
Phosphoenolpyruvic acid (2 molecules)
Pyruvate kinase
Pyruvic acid (2 molecules)
(a) Detailed version (continued)
1 Glucose
4 ADP + 4 P
4 ATP
2 C-C-C
Pyruvic acid
(b) Simplified version

NOTES

25

Figure 25.4 Fate of pyruvic acid (page 878).

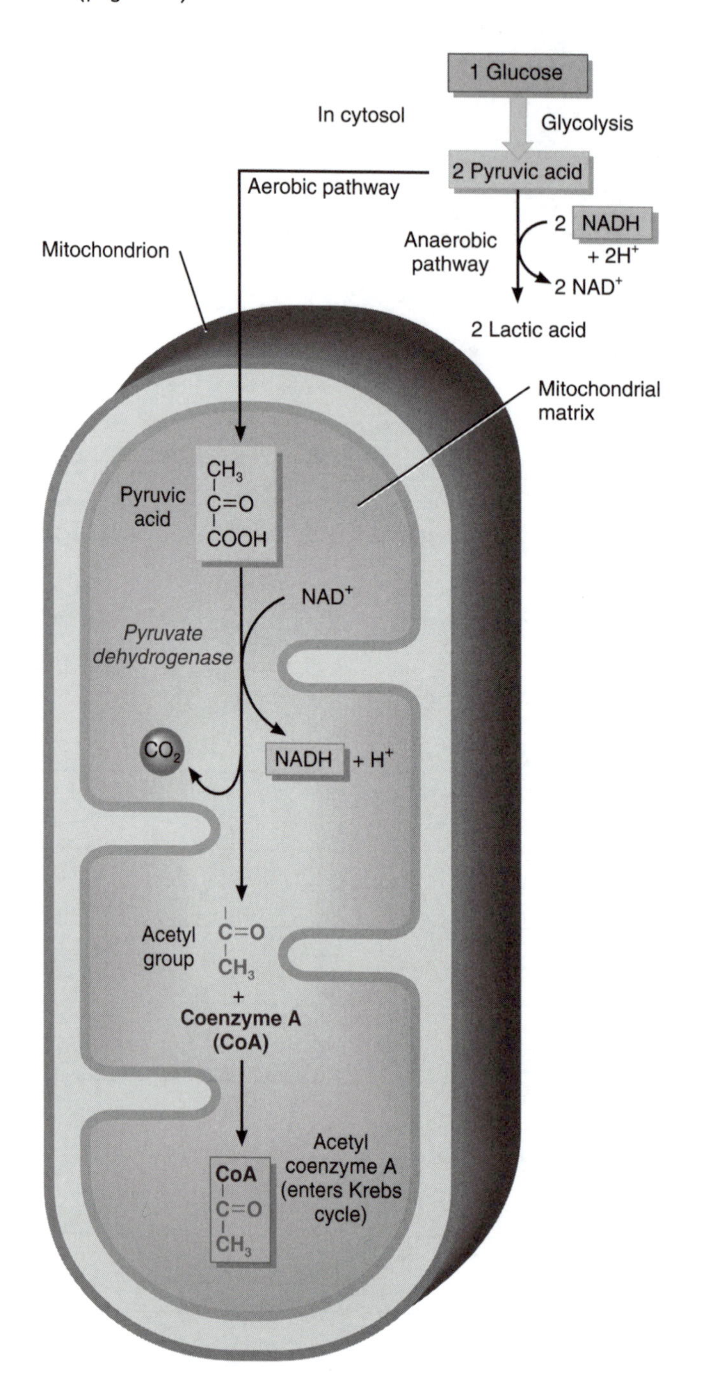
1 Glucose
In cytosol
Glycolysis
2 Pyruvic acid
Aerobic pathway
Mitochondrion
Anaerobic pathway
2 NADH
+ 2H+
2 NAD+
2 Lactic acid
Mitochondrial matrix
Pyruvic acid
CH3
C=O
COOH
NAD+
Pyruvate dehydrogenase
CO2
NADH + H+
Acetyl group
C=O
CH3
+
Coenzyme A (CoA)
CoA
C=O
CH3
Acetyl coenzyme A (enters Krebs cycle)

NOTES

25

Figure 25.5 The Krebs cycle (page 880, 881).

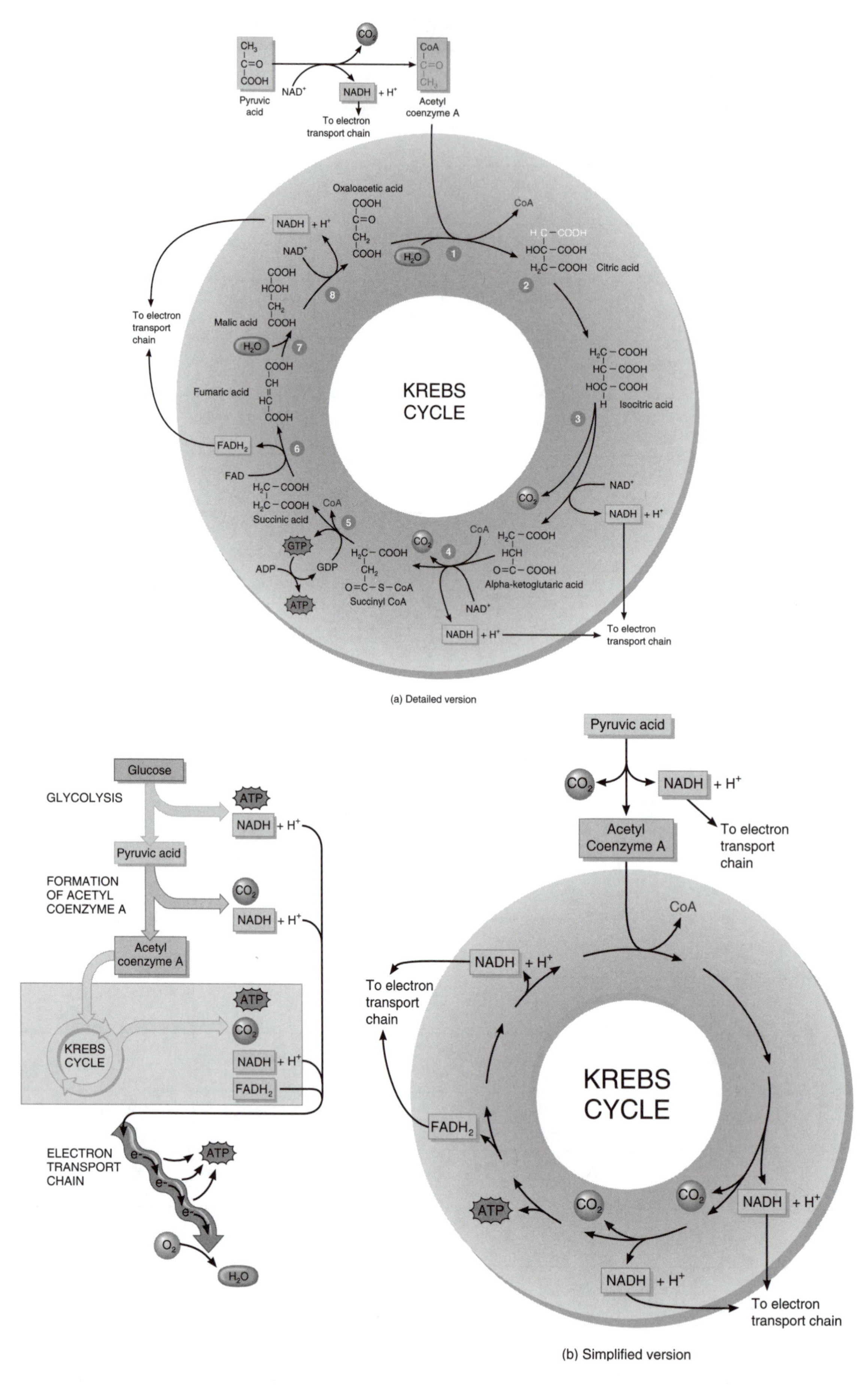

Pyruvic acid
NAD+
NADH + H+
To electron transport chain
Acetyl coenzyme A
Oxaloacetic acid
CoA
Citric acid
Isocitric acid
Alpha-ketoglutaric acid
Succinyl CoA
Succinic acid
Fumaric acid
Malic acid
FAD
FADH2
GTP
GDP
ADP
ATP
KREBS CYCLE
(a) Detailed version
Glucose
GLYCOLYSIS
FORMATION OF ACETYL COENZYME A
ELECTRON TRANSPORT CHAIN
(b) Simplified version

NOTES

25

Figure 25.6 Chemiosmosis (page 882).

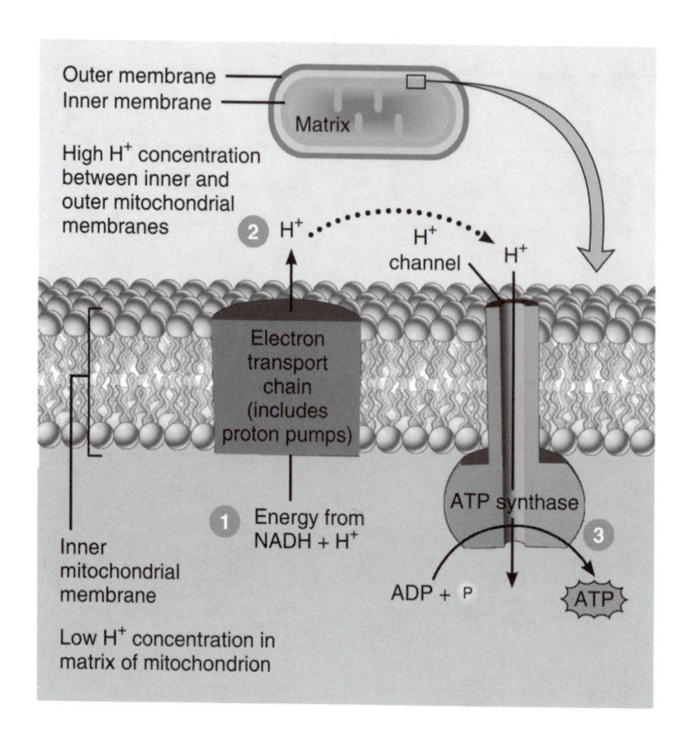
Outer membrane
Inner membrane
Matrix
High H+ concentration between inner and outer mitochondrial membranes
2 H+
H+ channel
H+
Electron transport chain (includes proton pumps)
1 Energy from NADH + H+
ATP synthase
3
Inner mitochondrial membrane
ADP + P
ATP
Low H+ concentration in matrix of mitochondrion

Figure 25.7 Proton pumps (page 883).

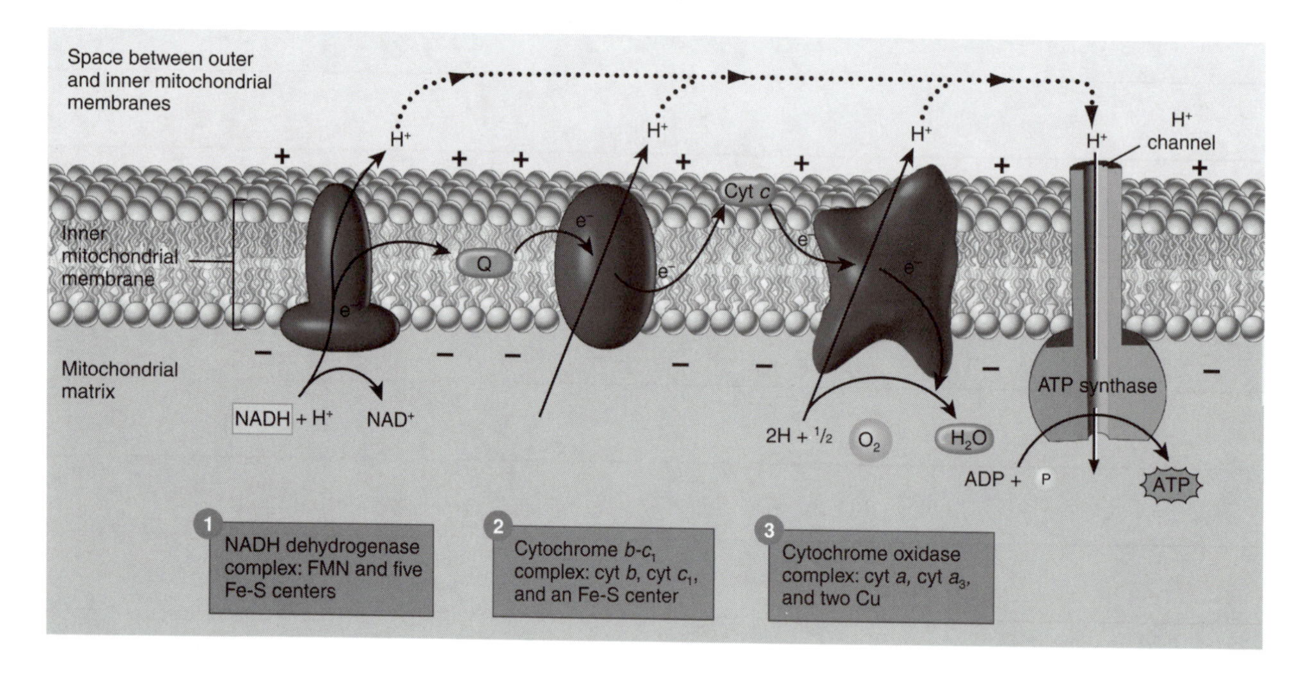
Space between outer and inner mitochondrial membranes
H+
H+
H+
H+
H+ channel
Cyt c
Inner mitochondrial membrane
Q
e-
Mitochondrial matrix
NADH + H+
NAD+
2H + 1/2 O2
H2O
ATP synthase
ADP + P
ATP
1 NADH dehydrogenase complex: FMN and five Fe-S centers
2 Cytochrome b-c1 complex: cyt b, cyt c1, and an Fe-S center
3 Cytochrome oxidase complex: cyt a, cyt a3, and two Cu

NOTES

25

Figure 25.8 Principal reactions of cellular respiration (page 884).

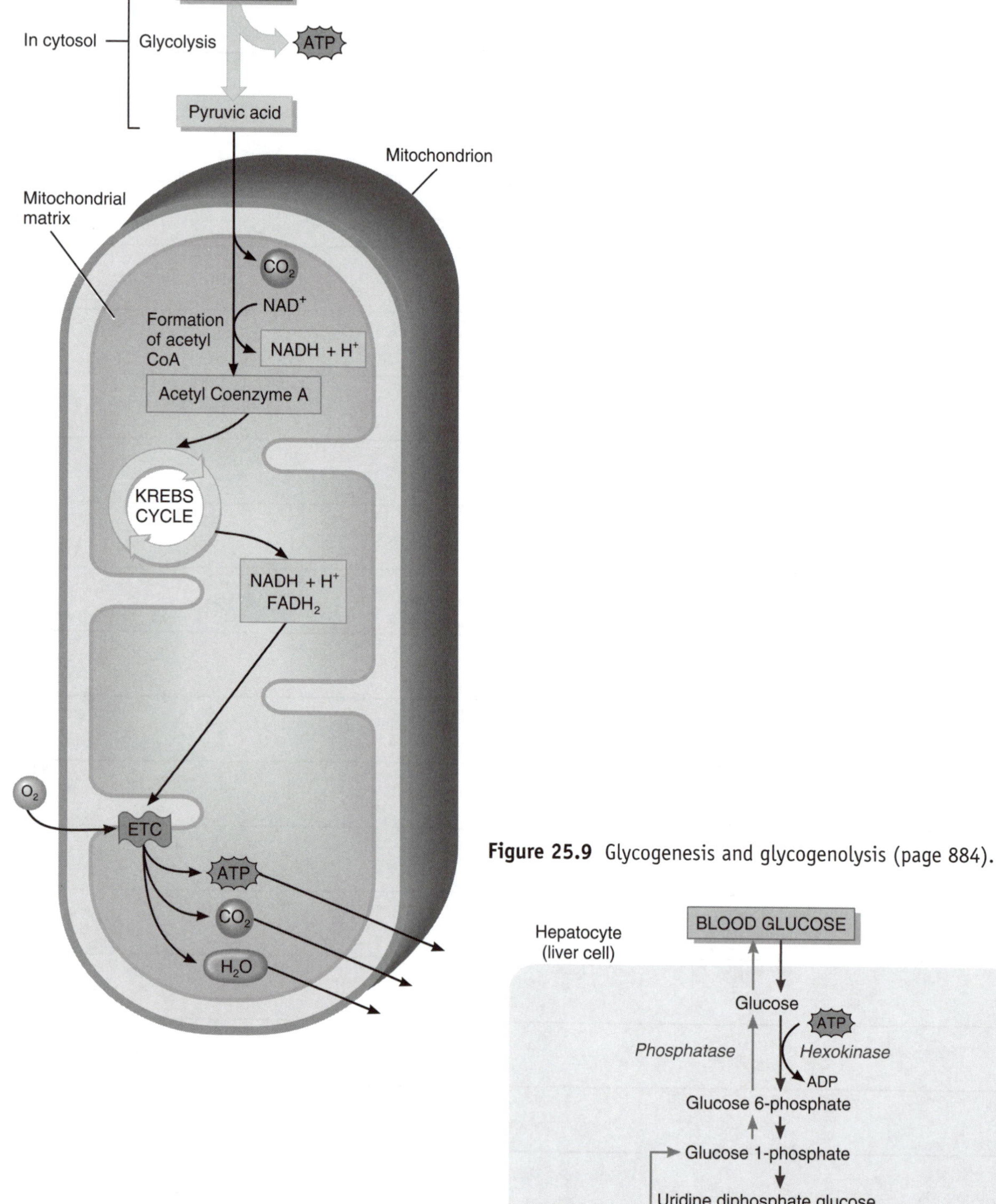

Glucose
In cytosol
Glycolysis
ATP
Pyruvic acid
Mitochondrion
Mitochondrial matrix
CO_2
NAD^+
Formation of acetyl CoA
$NADH + H^+$
Acetyl Coenzyme A
KREBS CYCLE
$NADH + H^+$
$FADH_2$
O_2
ETC
ATP
CO_2
H_2O

Figure 25.9 Glycogenesis and glycogenolysis (page 884).

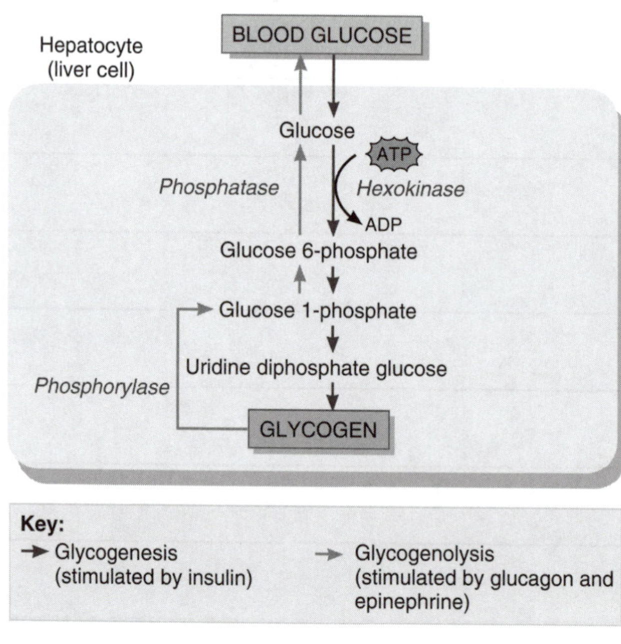

Hepatocyte (liver cell)
BLOOD GLUCOSE
Glucose
ATP
Phosphatase
Hexokinase
ADP
Glucose 6-phosphate
Glucose 1-phosphate
Uridine diphosphate glucose
Phosphorylase
GLYCOGEN

Key:
Glycogenesis (stimulated by insulin)
Glycogenolysis (stimulated by glucagon and epinephrine)

NOTES

25

Figure 25.10 Gluconeogenesis (page 885).

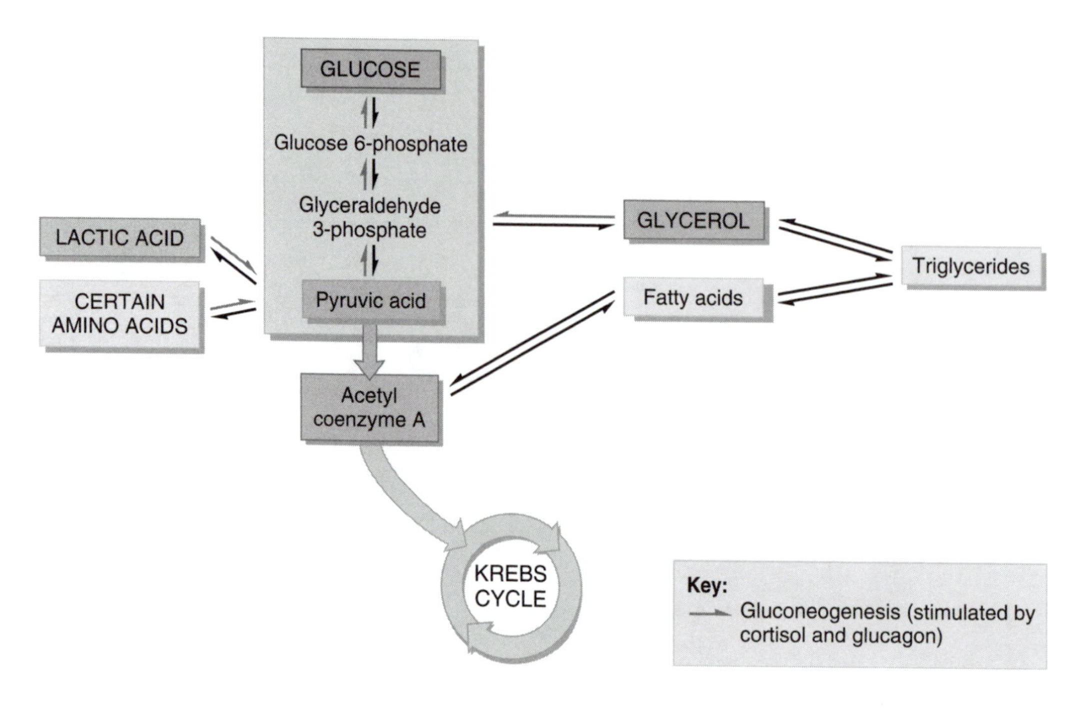
GLUCOSE
Glucose 6-phosphate
Glyceraldehyde
3-phosphate
Pyruvic acid
LACTIC ACID
CERTAIN
AMINO ACIDS
GLYCEROL
Fatty acids
Triglycerides
Acetyl
coenzyme A
KREBS
CYCLE
Key:
Gluconeogenesis (stimulated by
cortisol and glucagon)

Figure 25.11 A lipoprotein (page 887).

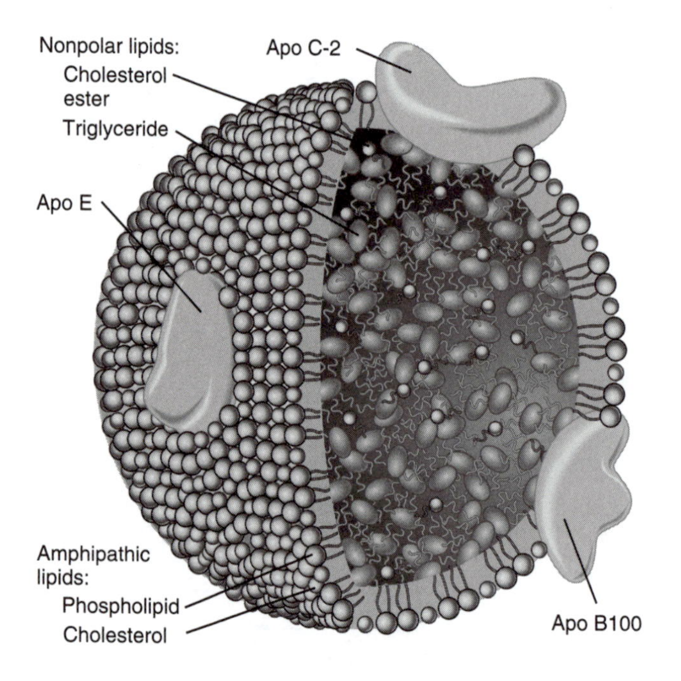
Nonpolar lipids:
Cholesterol
ester
Triglyceride
Apo C-2
Apo E
Amphipathic
lipids:
Phospholipid
Cholesterol
Apo B100

NOTES

25

Figure 25.12 Pathways of lipid metabolism (page 888).

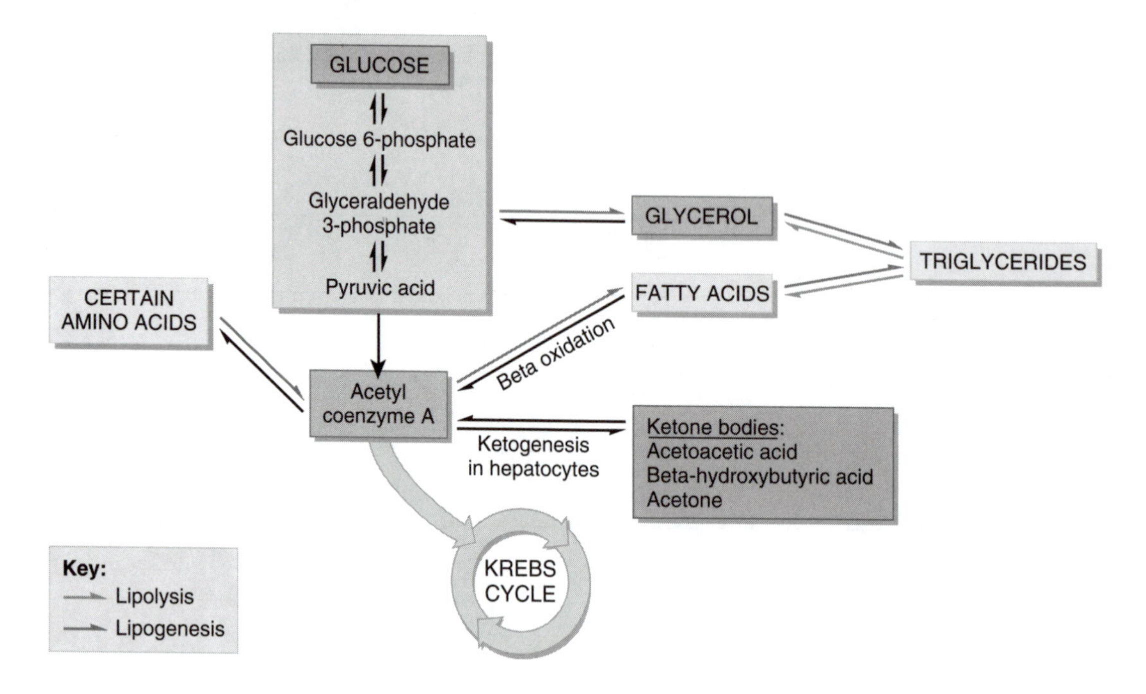
GLUCOSE
Glucose 6-phosphate
Glyceraldehyde 3-phosphate
Pyruvic acid
CERTAIN AMINO ACIDS
Acetyl coenzyme A
GLYCEROL
FATTY ACIDS
TRIGLYCERIDES
Beta oxidation
Ketogenesis in hepatocytes
Ketone bodies:
Acetoacetic acid
Beta-hydroxybutyric acid
Acetone
KREBS CYCLE
Key:
Lipolysis
Lipogenesis

Figure 25.13 Various points at which amino acids enter the Krebs cycle (page 890).

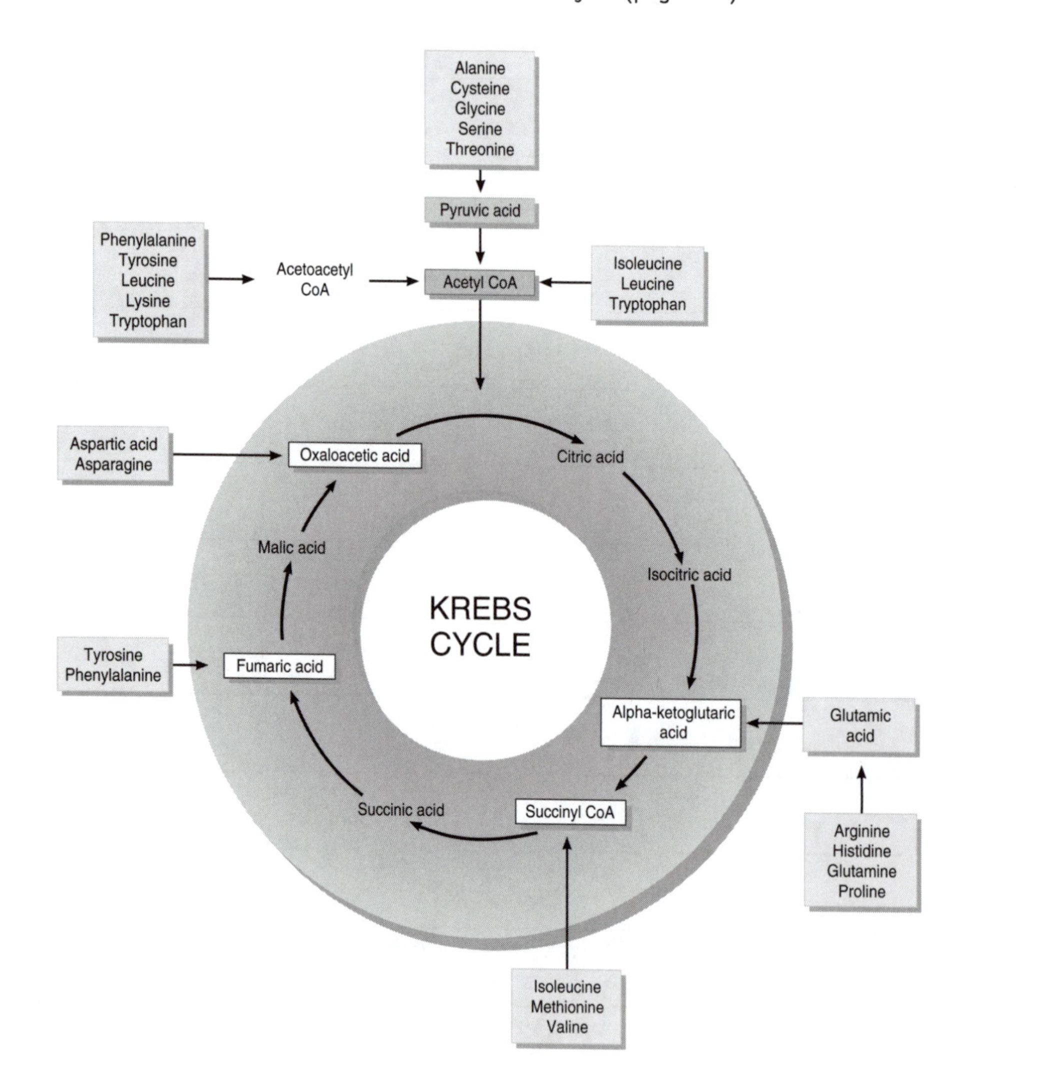
Alanine
Cysteine
Glycine
Serine
Threonine
Pyruvic acid
Phenylalanine
Tyrosine
Leucine
Lysine
Tryptophan
Acetoacetyl CoA
Acetyl CoA
Isoleucine
Leucine
Tryptophan
Aspartic acid
Asparagine
Oxaloacetic acid
Citric acid
Malic acid
Isocitric acid
KREBS CYCLE
Tyrosine
Phenylalanine
Fumaric acid
Alpha-ketoglutaric acid
Glutamic acid
Succinic acid
Succinyl CoA
Arginine
Histidine
Glutamine
Proline
Isoleucine
Methionine
Valine

NOTES

25

Figure 25.14 Summary of the roles of the key molecules in metabolic pathways (page 892).

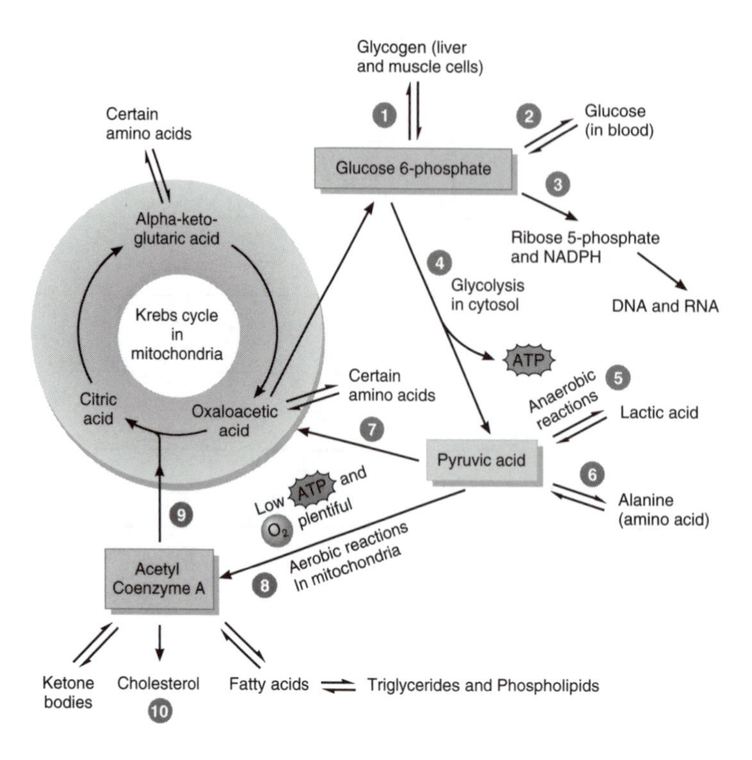

Glycogen (liver and muscle cells)
Certain amino acids
Glucose (in blood)
Glucose 6-phosphate
Alpha-keto-glutaric acid
Ribose 5-phosphate and NADPH
Glycolysis in cytosol
DNA and RNA
Krebs cycle in mitochondria
ATP
Certain amino acids
Anaerobic reactions
Citric acid
Oxaloacetic acid
Lactic acid
Pyruvic acid
Low ATP and plentiful O2
Alanine (amino acid)
Aerobic reactions in mitochondria
Acetyl Coenzyme A
Ketone bodies
Cholesterol
Fatty acids
Triglycerides and Phospholipids

Figure 25.17 Negative feedback mechanisms that conserve heat and increase heat production (page 900).

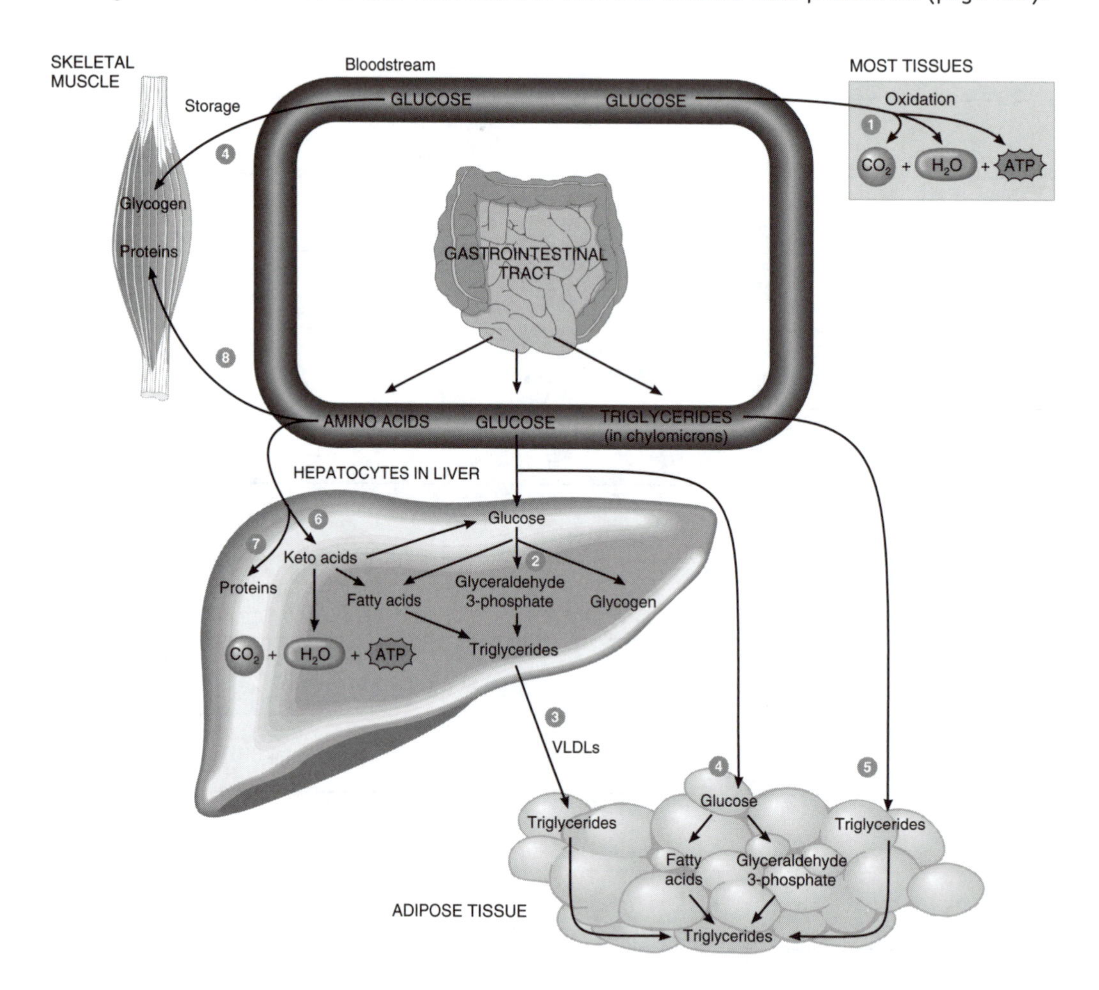

SKELETAL MUSCLE
Bloodstream
MOST TISSUES
Storage
GLUCOSE
GLUCOSE
Oxidation
CO2 + H2O + ATP
Glycogen
Proteins
GASTROINTESTINAL TRACT
AMINO ACIDS
GLUCOSE
TRIGLYCERIDES (in chylomicrons)
HEPATOCYTES IN LIVER
Glucose
Keto acids
Proteins
Fatty acids
Glyceraldehyde 3-phosphate
Glycogen
CO2 + H2O + ATP
Triglycerides
VLDLs
Glucose
Triglycerides
Triglycerides
Fatty acids
Glyceraldehyde 3-phosphate
ADIPOSE TISSUE
Triglycerides

NOTES

25

Figure 25.16 Principal metabolic pathways during the postabsorptive state (page 896).

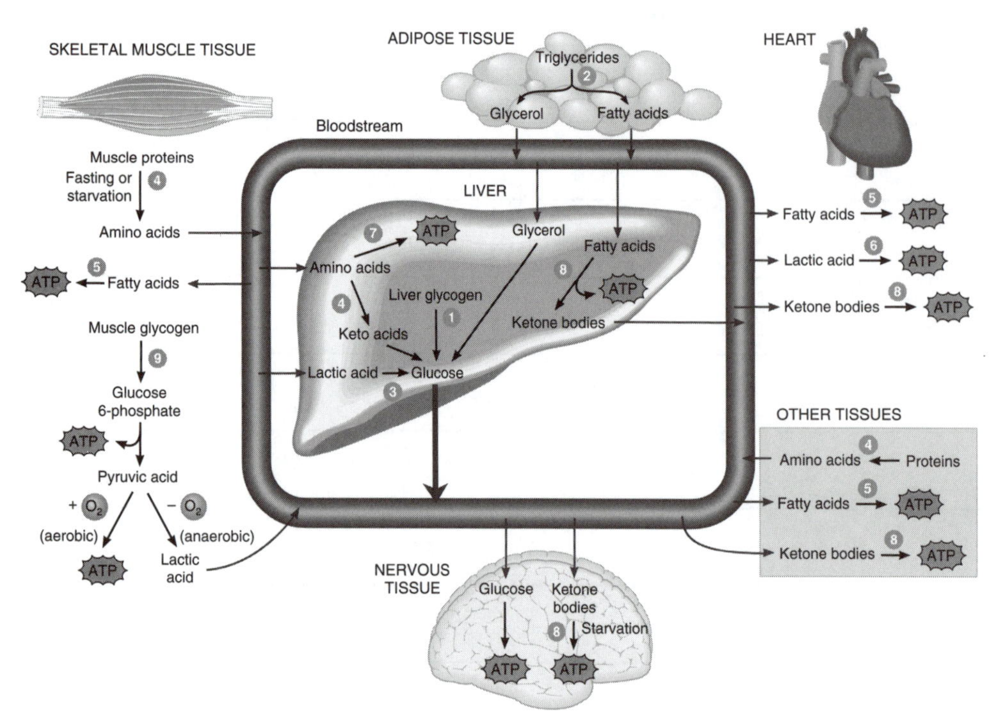

SKELETAL MUSCLE TISSUE
ADIPOSE TISSUE
Triglycerides
2
Glycerol
Fatty acids
HEART
Bloodstream
Muscle proteins
Fasting or starvation
4
Amino acids
LIVER
7
ATP
Glycerol
Fatty acids
Amino acids
5
ATP
Fatty acids
8
ATP
Liver glycogen
1
Ketone bodies
4
Keto acids
Muscle glycogen
9
Lactic acid
Glucose
3
Glucose 6-phosphate
ATP
Pyruvic acid
+ O_2
- O_2
(aerobic)
(anaerobic)
ATP
Lactic acid
5 Fatty acids ATP
6 Lactic acid ATP
8 Ketone bodies ATP
OTHER TISSUES
4
Amino acids
Proteins
5 Fatty acids ATP
8 Ketone bodies ATP
NERVOUS TISSUE
Glucose
Ketone bodies
8
Starvation
ATP
ATP

Figure 25.17 Negative feedback mechanisms that conserve heat and increase heat production (page 900).

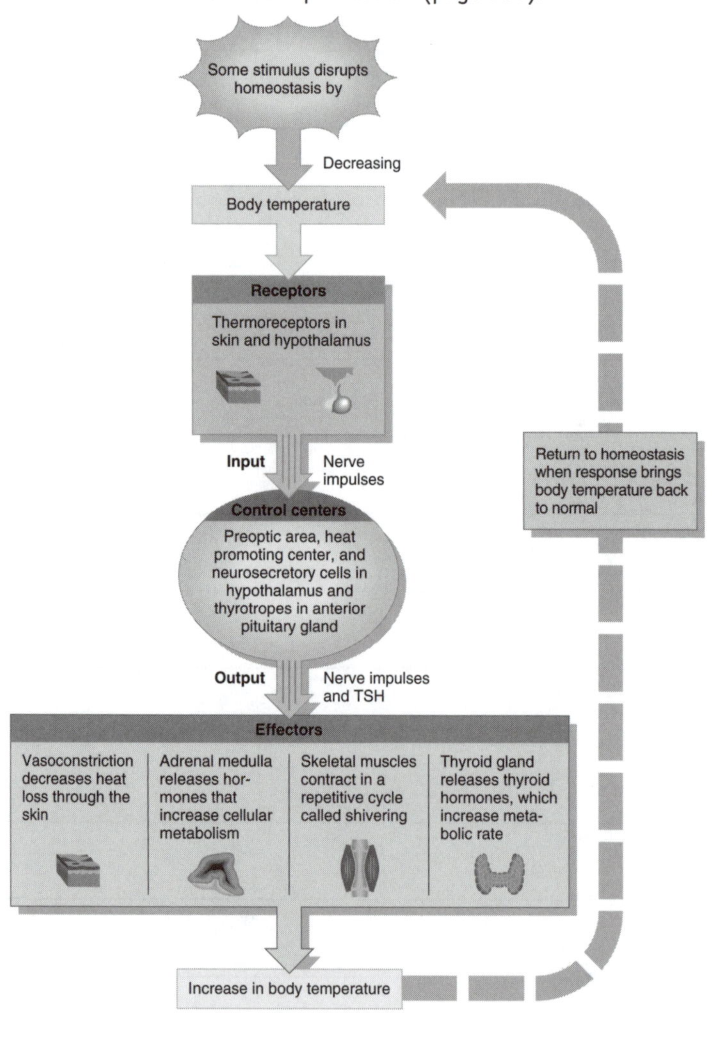

Some stimulus disrupts homeostasis by
Decreasing
Body temperature
Receptors
Thermoreceptors in skin and hypothalamus
Input
Nerve impulses
Control centers
Preoptic area, heat promoting center, and neurosecretory cells in hypothalamus and thyrotropes in anterior pituitary gland
Output
Nerve impulses and TSH
Effectors
Vasoconstriction decreases heat loss through the skin
Adrenal medulla releases hormones that increase cellular metabolism
Skeletal muscles contract in a repetitive cycle called shivering
Thyroid gland releases thyroid hormones, which increase metabolic rate
Increase in body temperature
Return to homeostasis when response brings body temperature back to normal

NOTES

25

Figure 25.18 The food guide pyramid (page 903).

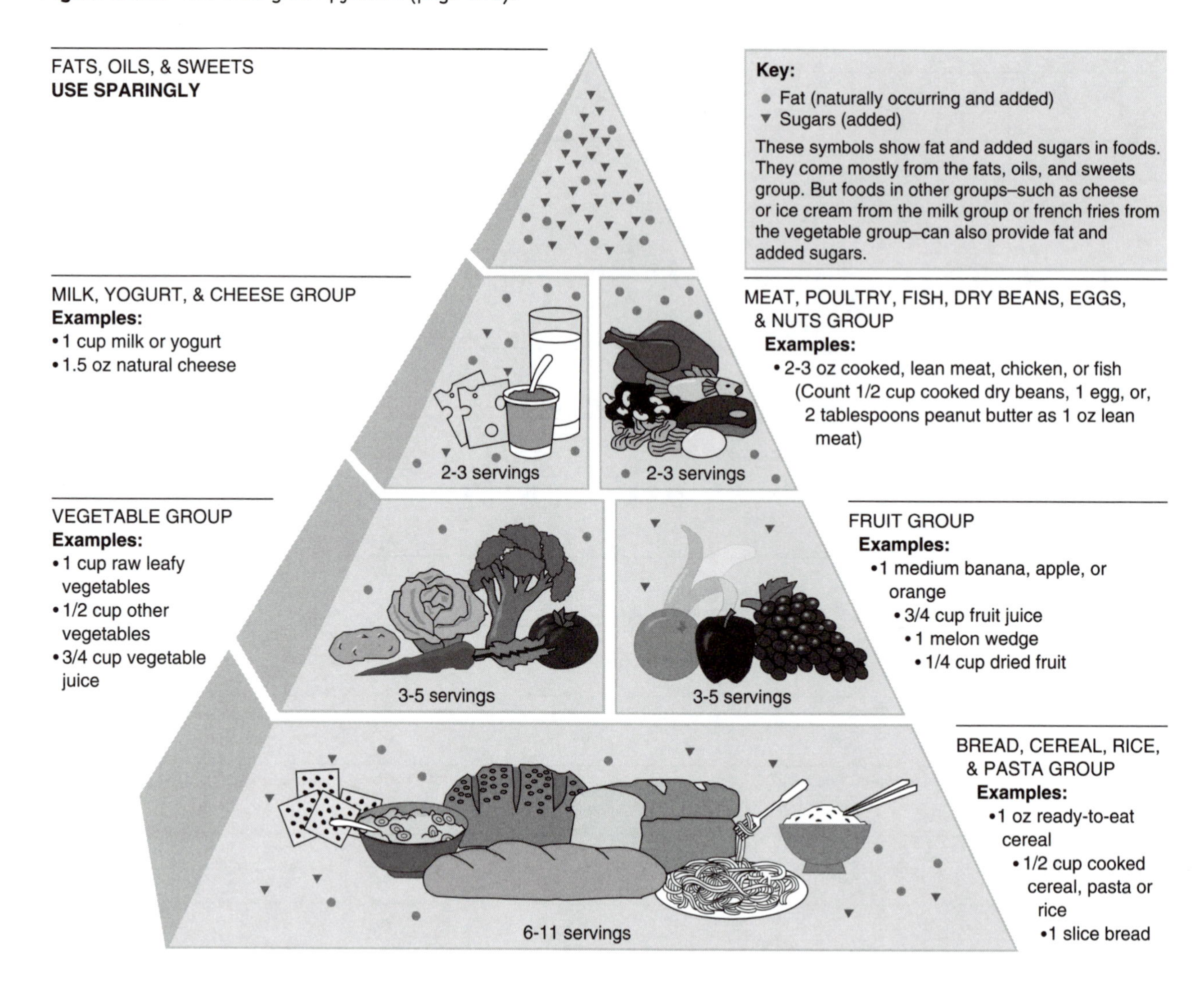

FATS, OILS, & SWEETS
USE SPARINGLY
Key:
Fat (naturally occurring and added)
Sugars (added)
These symbols show fat and added sugars in foods. They come mostly from the fats, oils, and sweets group. But foods in other groups–such as cheese or ice cream from the milk group or french fries from the vegetable group–can also provide fat and added sugars.
MILK, YOGURT, & CHEESE GROUP
Examples:
• 1 cup milk or yogurt
• 1.5 oz natural cheese
2-3 servings
MEAT, POULTRY, FISH, DRY BEANS, EGGS, & NUTS GROUP
Examples:
• 2-3 oz cooked, lean meat, chicken, or fish (Count 1/2 cup cooked dry beans, 1 egg, or, 2 tablespoons peanut butter as 1 oz lean meat)
2-3 servings
VEGETABLE GROUP
Examples:
• 1 cup raw leafy vegetables
• 1/2 cup other vegetables
• 3/4 cup vegetable juice
3-5 servings
FRUIT GROUP
Examples:
• 1 medium banana, apple, or orange
• 3/4 cup fruit juice
• 1 melon wedge
• 1/4 cup dried fruit
3-5 servings
BREAD, CEREAL, RICE, & PASTA GROUP
Examples:
• 1 oz ready-to-eat cereal
• 1/2 cup cooked cereal, pasta or rice
• 1 slice bread
6-11 servings

NOTES

25

Figure 26.1 Organs of the urinary system (page 915).

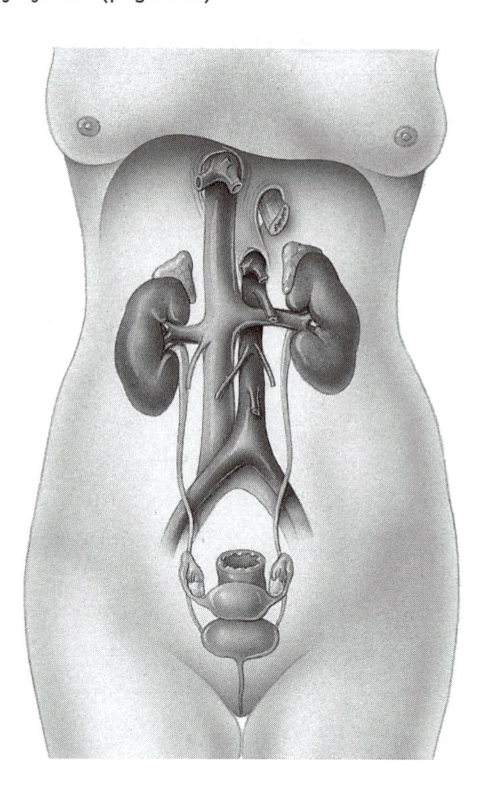

Figure 26.2 Position and coverings of the kidneys (page 916).

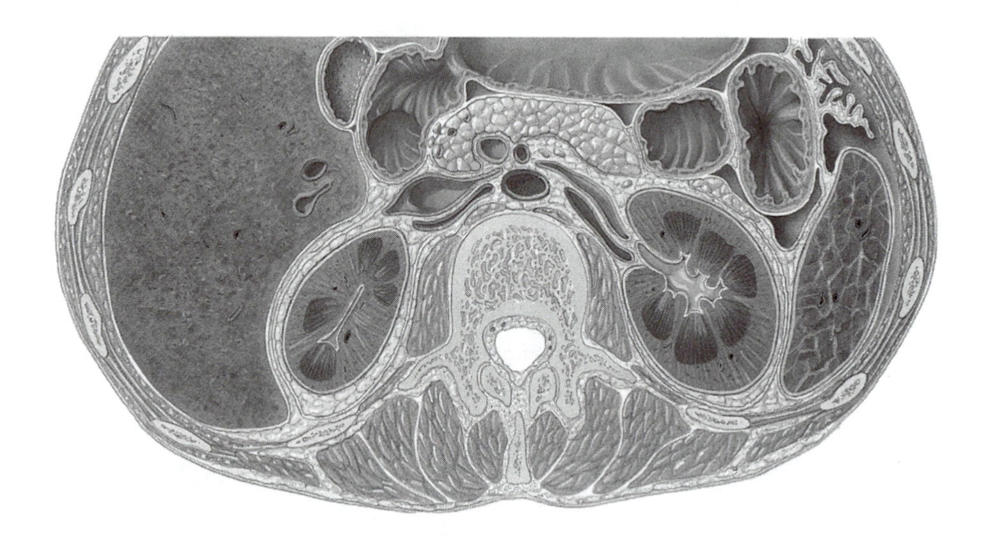

NOTES

26

Figure 26.3 Internal anatomy of the kidneys (page 917).

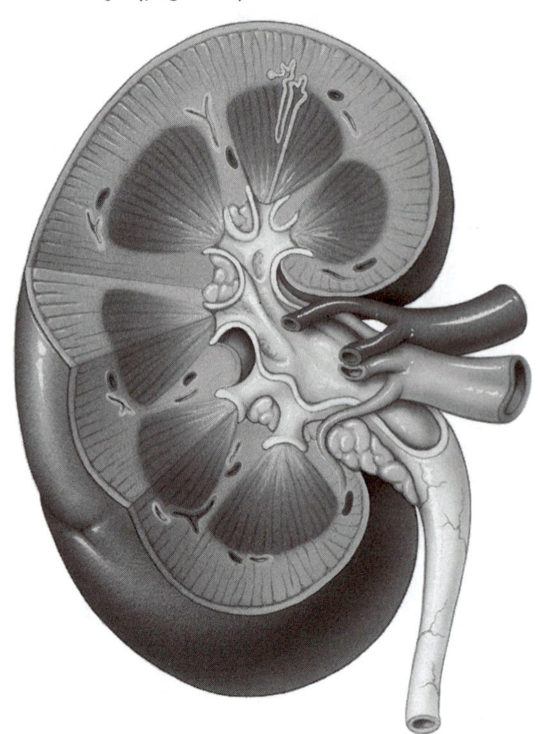

Figure 26.4a Blood supply of the kidneys (page 919).

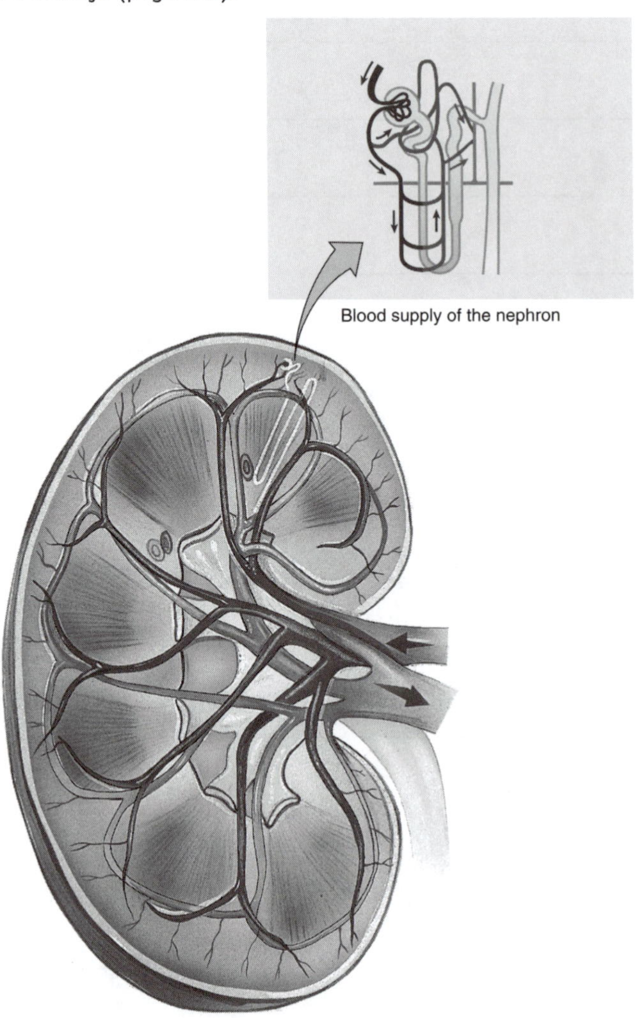
Blood supply of the nephron

NOTES

26

Figure 26.4b Blood supply of the kidneys (page 919).

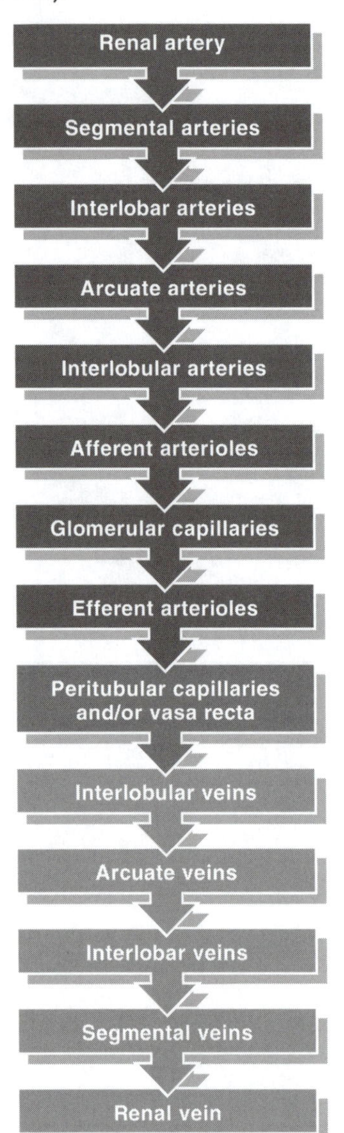
Renal artery
Segmental arteries
Interlobar arteries
Arcuate arteries
Interlobular arteries
Afferent arterioles
Glomerular capillaries
Efferent arterioles
Peritubular capillaries and/or vasa recta
Interlobular veins
Arcuate veins
Interlobar veins
Segmental veins
Renal vein

Figure 26.5a The structure of nephrons (page 920).

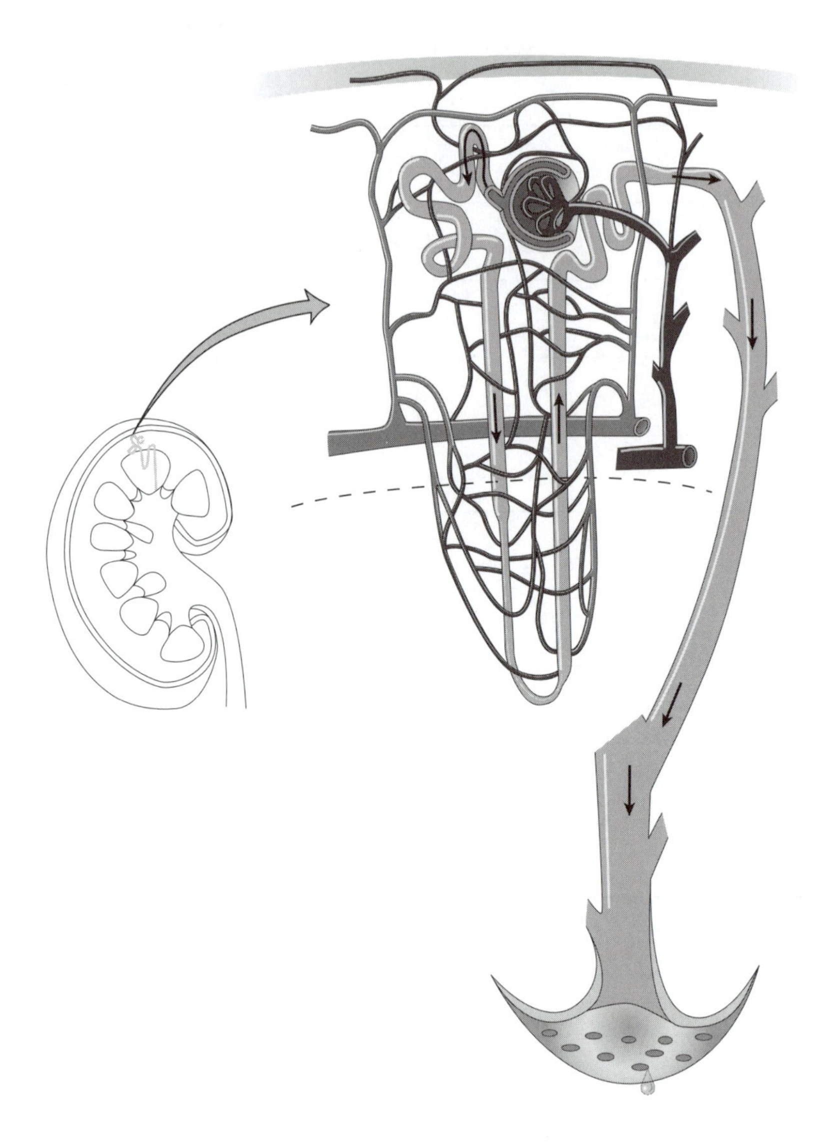

NOTES

26

Figure 26.5b The structure of nephrons (page 920).

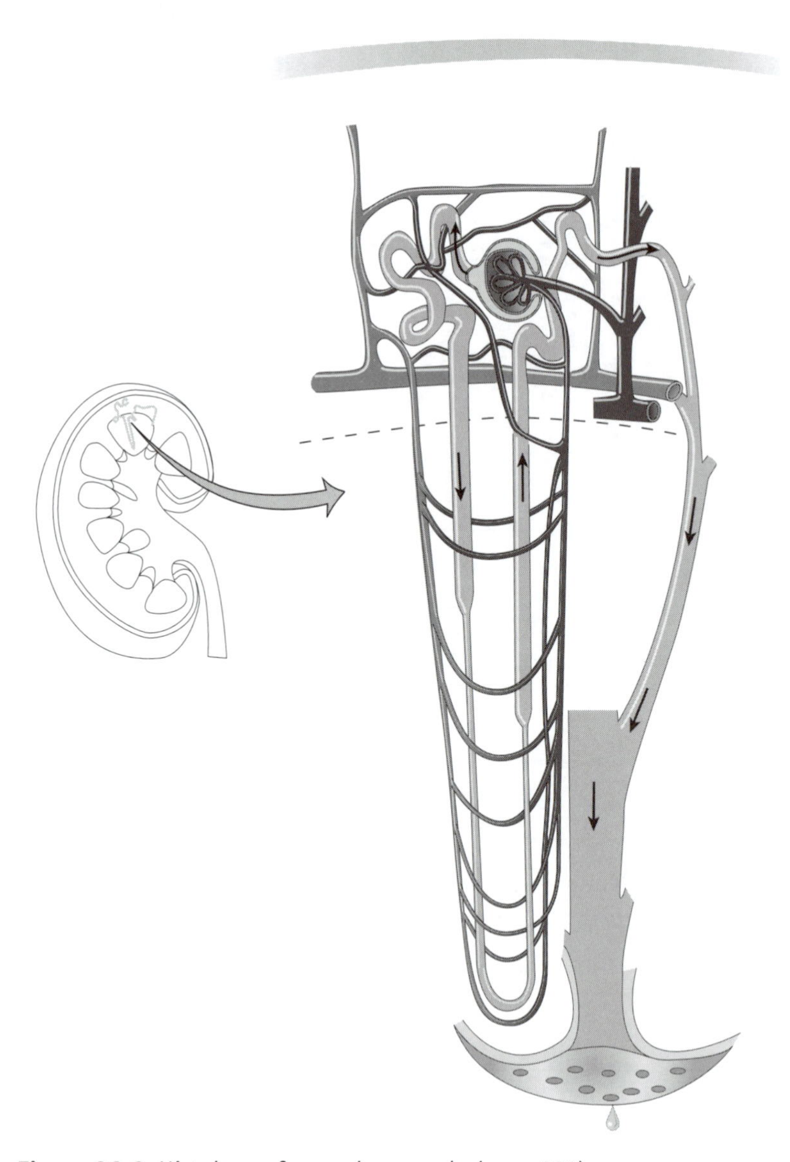

Figure 26.6 Histology of a renal corpuscle (page 922).

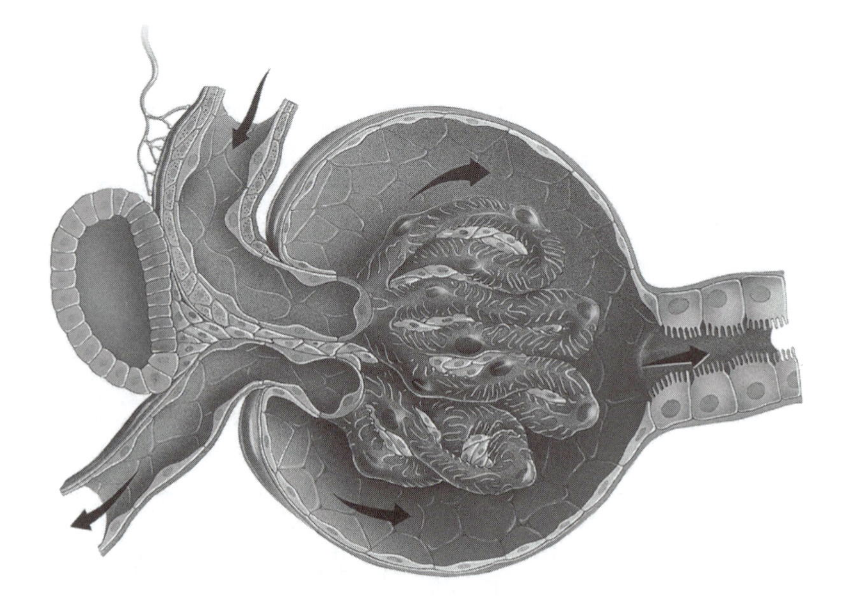

NOTES

26

Figure 26.7 Relation of a nephron's structure to its three basic functions (page 924).

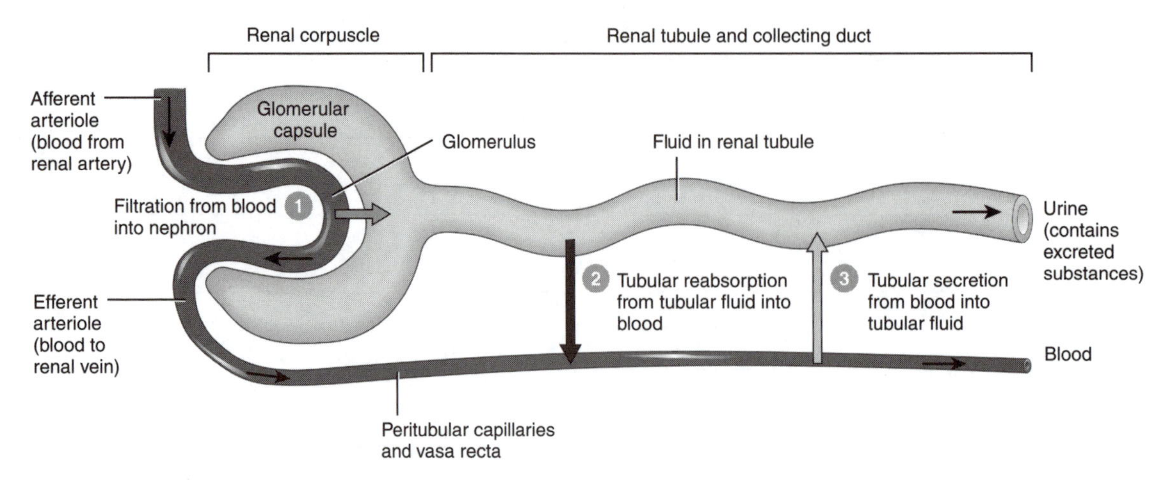

Renal corpuscle
Renal tubule and collecting duct
Afferent arteriole (blood from renal artery)
Glomerular capsule
Glomerulus
Fluid in renal tubule
1 Filtration from blood into nephron
Urine (contains excreted substances)
2 Tubular reabsorption from tubular fluid into blood
3 Tubular secretion from blood into tubular fluid
Efferent arteriole (blood to renal vein)
Blood
Peritubular capillaries and vasa recta

Figure 26.8 The filtration membrane (page 926).

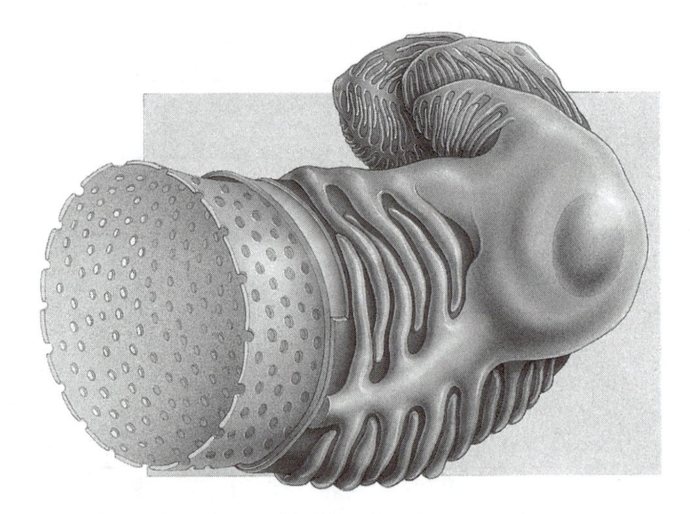

Figure 26.9 The pressures that drive glomeruler filtration (page 927).

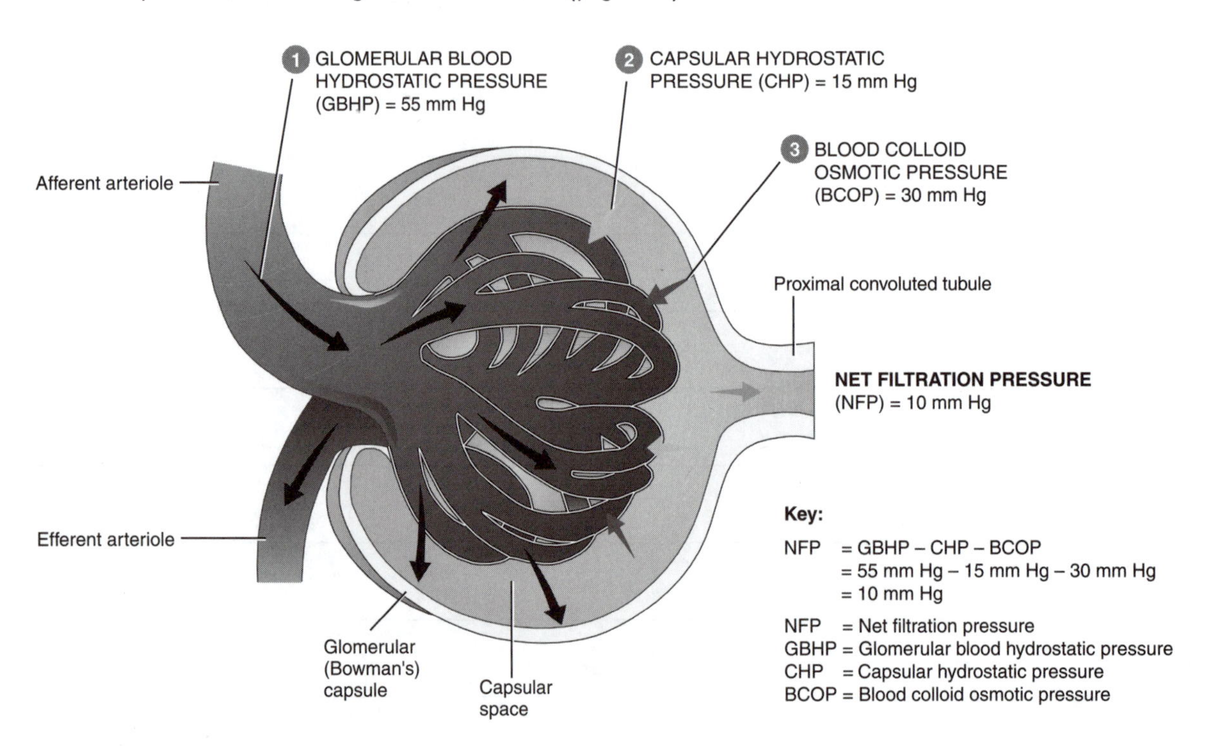

1 GLOMERULAR BLOOD HYDROSTATIC PRESSURE (GBHP) = 55 mm Hg
2 CAPSULAR HYDROSTATIC PRESSURE (CHP) = 15 mm Hg
3 BLOOD COLLOID OSMOTIC PRESSURE (BCOP) = 30 mm Hg
Afferent arteriole
Proximal convoluted tubule
NET FILTRATION PRESSURE (NFP) = 10 mm Hg
Key:
NFP = GBHP – CHP – BCOP
= 55 mm Hg – 15 mm Hg – 30 mm Hg
= 10 mm Hg
NFP = Net filtration pressure
GBHP = Glomerular blood hydrostatic pressure
CHP = Capsular hydrostatic pressure
BCOP = Blood colloid osmotic pressure
Efferent arteriole
Glomerular (Bowman's) capsule
Capsular space

NOTES

26

Figure 26.10 Tubulogomerular feedback (page 928).

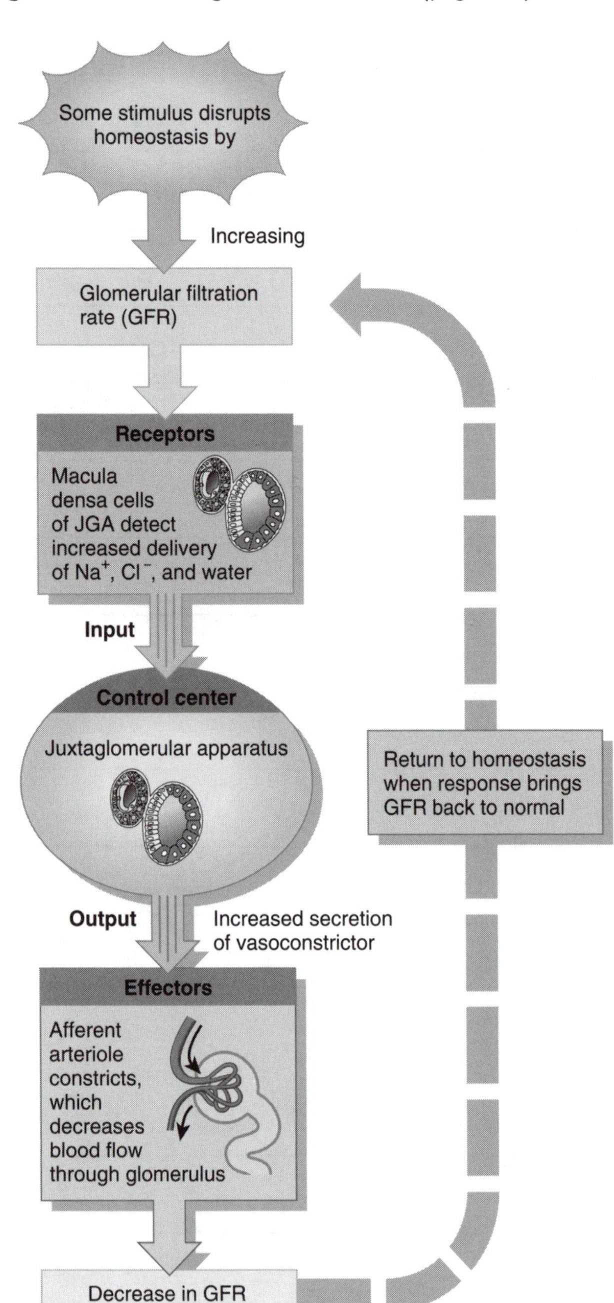
Some stimulus disrupts homeostasis by
Increasing
Glomerular filtration rate (GFR)
Receptors
Macula densa cells of JGA detect increased delivery of Na^+, Cl^-, and water
Input
Control center
Juxtaglomerular apparatus
Output
Increased secretion of vasoconstrictor
Effectors
Afferent arteriole constricts, which decreases blood flow through glomerulus
Decrease in GFR
Return to homeostasis when response brings GFR back to normal

Figure 26.11 Reabsorption routes (page 930).

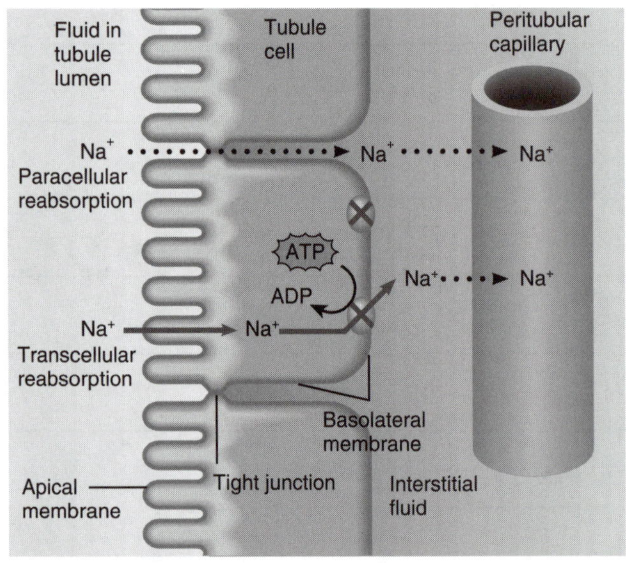
Fluid in tubule lumen
Tubule cell
Peritubular capillary
Na^+
Na^+
Na^+
Paracellular reabsorption
ATP
ADP
Na^+
Na^+
Na^+
Na^+
Transcellular reabsorption
Basolateral membrane
Apical membrane
Tight junction
Interstitial fluid

Key:

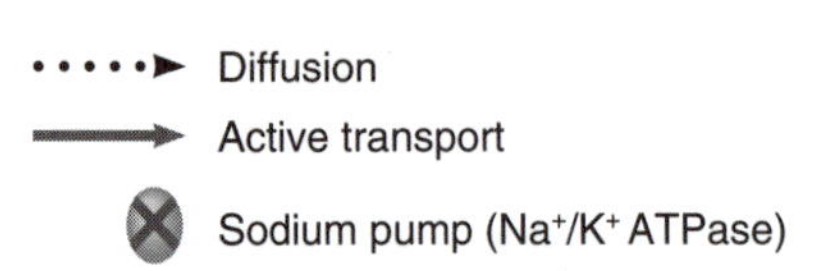
Diffusion
Active transport
Sodium pump (Na^+/K^+ ATPase)

NOTES

26

Figure 26.12 Reabsorption of glucose by Na^+ (page 932).

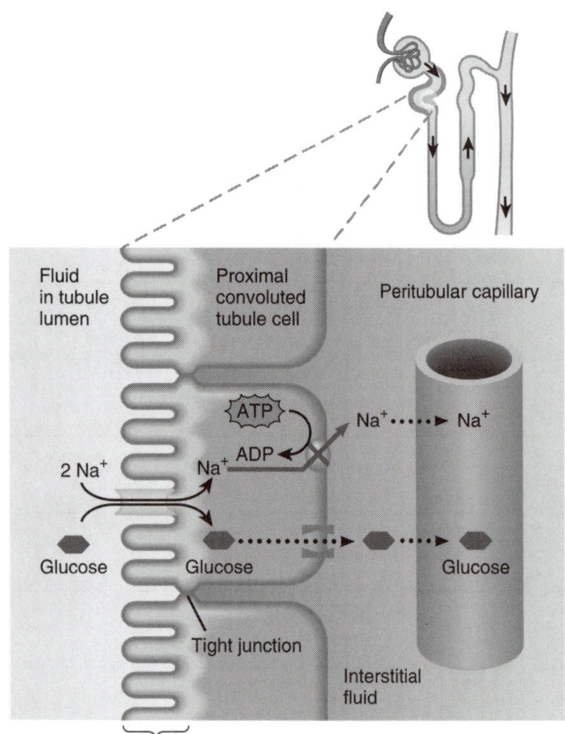

Fluid in tubule lumen
Proximal convoluted tubule cell
Peritubular capillary
ATP
ADP
Na+
Na+ Na+
2 Na+
Glucose
Glucose
Glucose
Tight junction
Interstitial fluid
Brush border (microvilli)
Key:
Na+– glucose symporter
Secondary active transport
Glucose facilitated diffusion transporter
Diffusion
Sodium pump

Figure 26.13 Actions of Na^+/H^+ antiporters in proximal convoluted tubule cells (page 933).

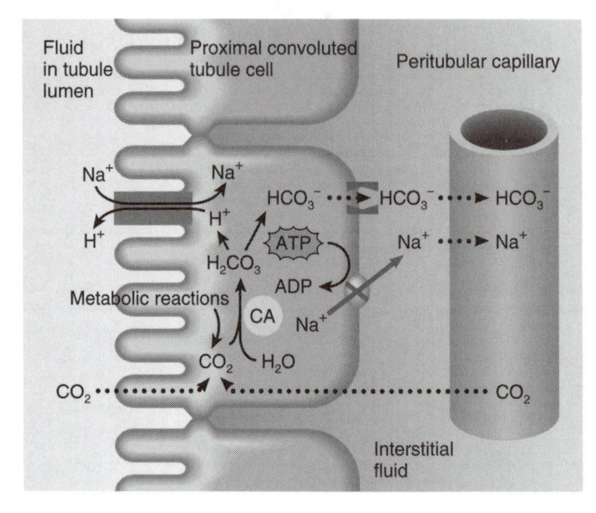

Fluid in tubule lumen
Proximal convoluted tubule cell
Peritubular capillary
Na+
Na+
H+
H+
HCO3-
HCO3-
HCO3-
H2CO3
ATP
ADP
Na+
Na+
Na+
Metabolic reactions
CA
CO2
H2O
CO2
CO2
Interstitial fluid

(a) Na^+ reabsorption and H^+ secretion

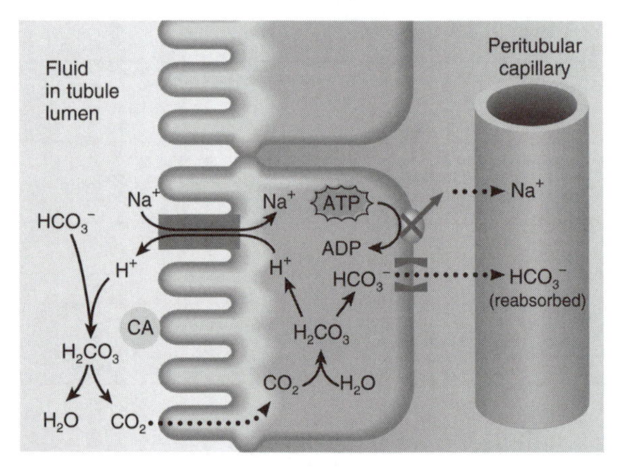

Fluid in tubule lumen
Peritubular capillary
HCO3-
Na+
Na+
ATP
Na+
ADP
H+
H+
HCO3-
HCO3-
(reabsorbed)
CA
H2CO3
H2CO3
CO2
H2O
H2O
CO2

(b) HCO_3^- reabsorption

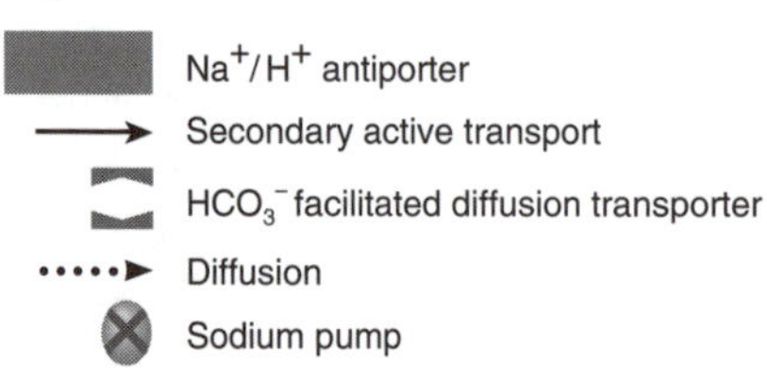

Key:
Na+/H+ antiporter
Secondary active transport
HCO3- facilitated diffusion transporter
Diffusion
Sodium pump

NOTES

26

Figure 26.14 Passive reabsorption of Cl^-, K^+, Ca^{2+}, Mg^{2+}, urea, and water (page 934).

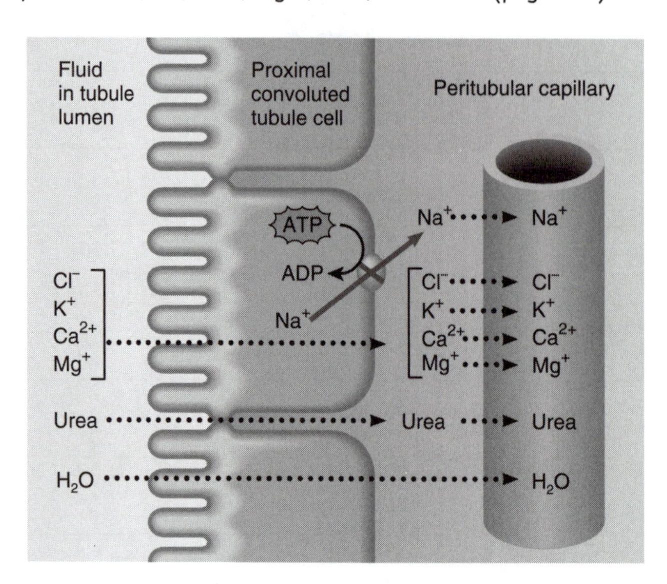

Fluid in tubule lumen
Proximal convoluted tubule cell
Peritubular capillary
ATP
ADP
Na+
Na+ Na+
Cl–
K+
Ca2+
Mg+
Cl– Cl–
K+ K+
Ca2+ Ca2+
Mg+ Mg+
Urea
Urea
Urea
H2O
H2O

Figure 26.15 Na^+ – K^+ – $2Cl^-$ symporter in the thick ascending limb of the loop of Henle (page 935).

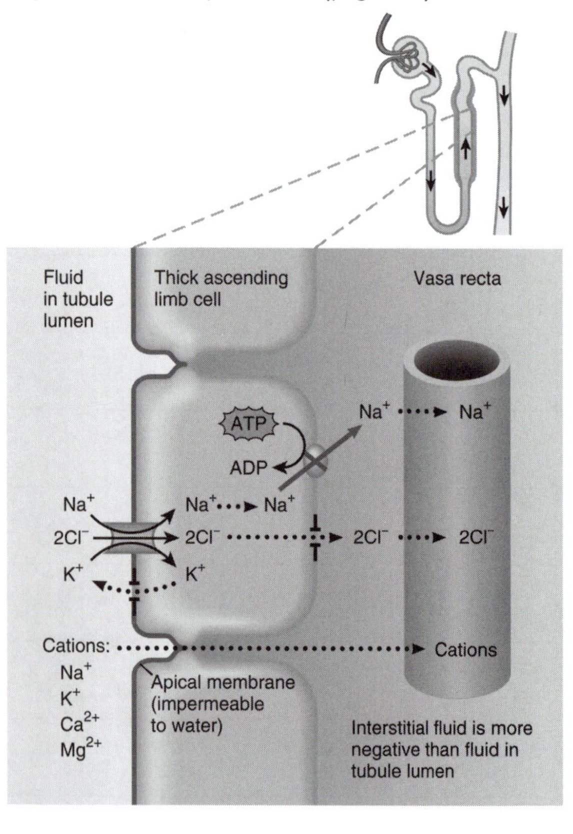

Fluid in tubule lumen
Thick ascending limb cell
Vasa recta
ATP
ADP
Na+ Na+
Na+
2Cl–
K+
Na+ Na+
2Cl–
2Cl– 2Cl–
K+
Cations:
Na+
K+
Ca2+
Mg2+
Cations
Apical membrane (impermeable to water)
Interstitial fluid is more negative than fluid in tubule lumen

Key:

Na+–K+–2Cl– symporter

Leakage channels

Sodium pump

Diffusion

Figure 26.16 Reabsorption of Na^+ and secretion of K^+ by principal cells (page 935).

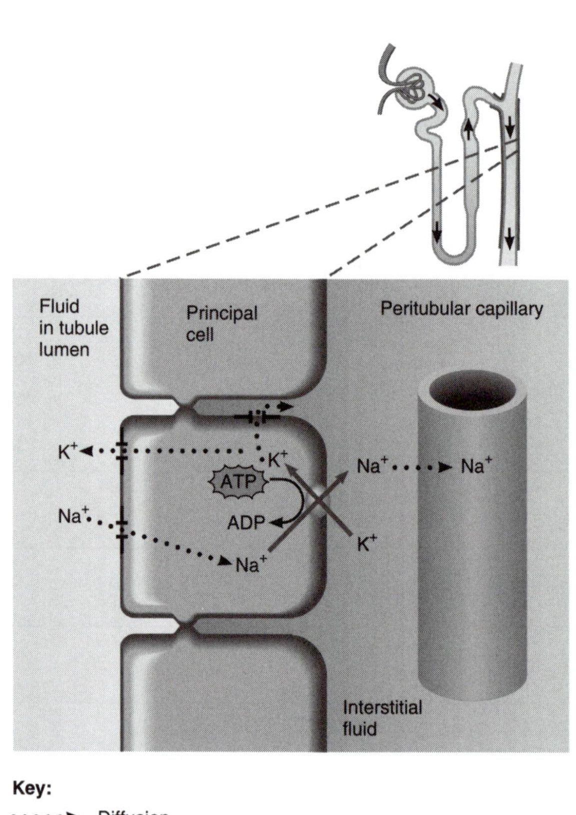

Fluid in tubule lumen
Principal cell
Peritubular capillary
K+
K+
ATP
ADP
Na+
Na+
Na+ Na+
K+
Interstitial fluid

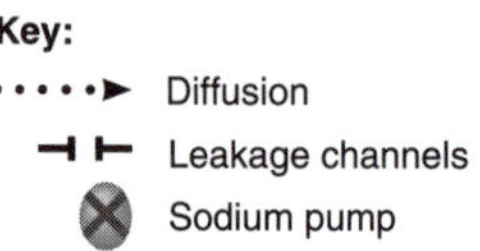

Key:
Diffusion
Leakage channels
Sodium pump

NOTES

26

Figure 26.17 Secretion of H^+ by intercalated cells in the collecting duct (page 936).

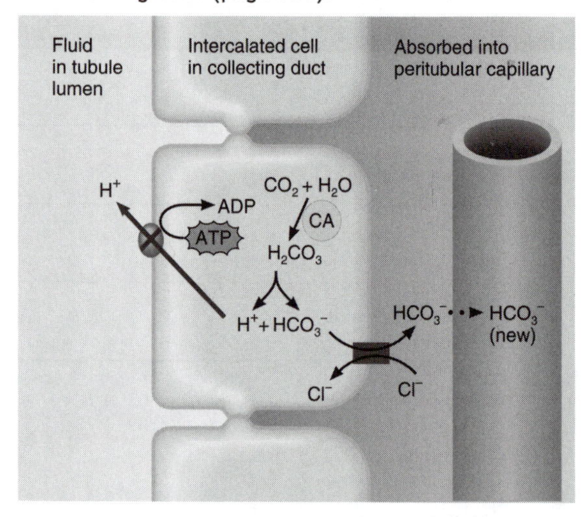

Fluid in tubule lumen
Intercalated cell in collecting duct
Absorbed into peritubular capillary
H^+
ADP
ATP
$CO_2 + H_2O$
CA
H_2CO_3
$H^+ + HCO_3^-$
HCO_3^-
HCO_3^- (new)
Cl^-
Cl^-

(a) Secretion of H^+

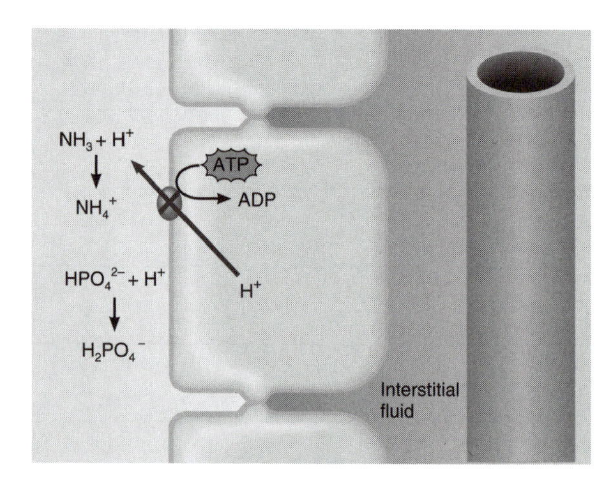

$NH_3 + H^+$
NH_4^+
ATP
ADP
$HPO_4^{2-} + H^+$
H^+
$H_2PO_4^-$
Interstitial fluid

(b) Buffering of H^+ in urine

Key:

- Proton pump (H^+ ATPase)
- HCO_3^-/Cl^- antiporter
- Diffusion

Figure 26.18 Negative feedback regulation of facultative water reabsorption by ADH (page 937).

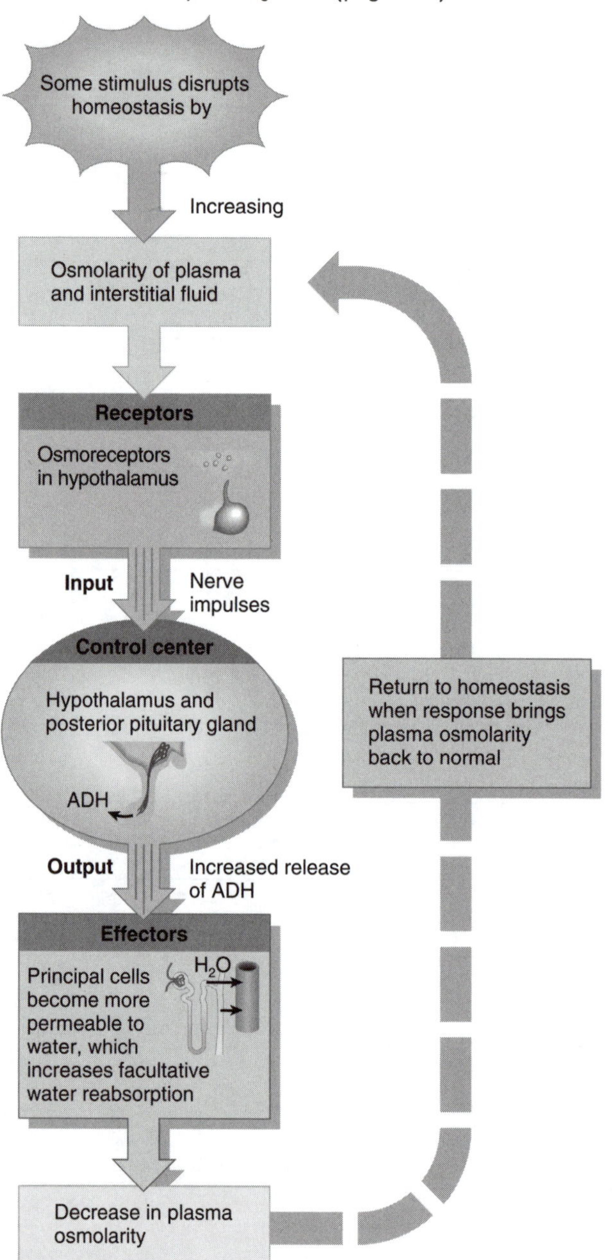

Some stimulus disrupts homeostasis by
Increasing
Osmolarity of plasma and interstitial fluid
Receptors
Osmoreceptors in hypothalamus
Input
Nerve impulses
Control center
Hypothalamus and posterior pituitary gland
ADH
Output
Increased release of ADH
Effectors
Principal cells become more permeable to water, which increases facultative water reabsorption
H_2O
Decrease in plasma osmolarity
Return to homeostasis when response brings plasma osmolarity back to normal

NOTES

26

Figure 26.19 Formation of dilute urine (page 939).

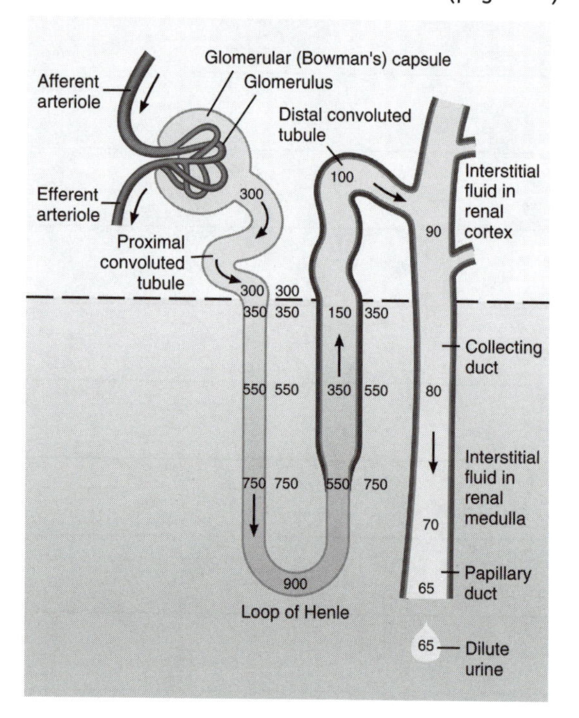

Glomerular (Bowman's) capsule
Afferent arteriole
Glomerulus
Distal convoluted tubule
Efferent arteriole
Proximal convoluted tubule
Interstitial fluid in renal cortex
Collecting duct
Interstitial fluid in renal medulla
Papillary duct
Loop of Henle
Dilute urine

Figure 26.20 Mechanism of urine concentration in long-loop juxtamedullary nephrons (page 940).

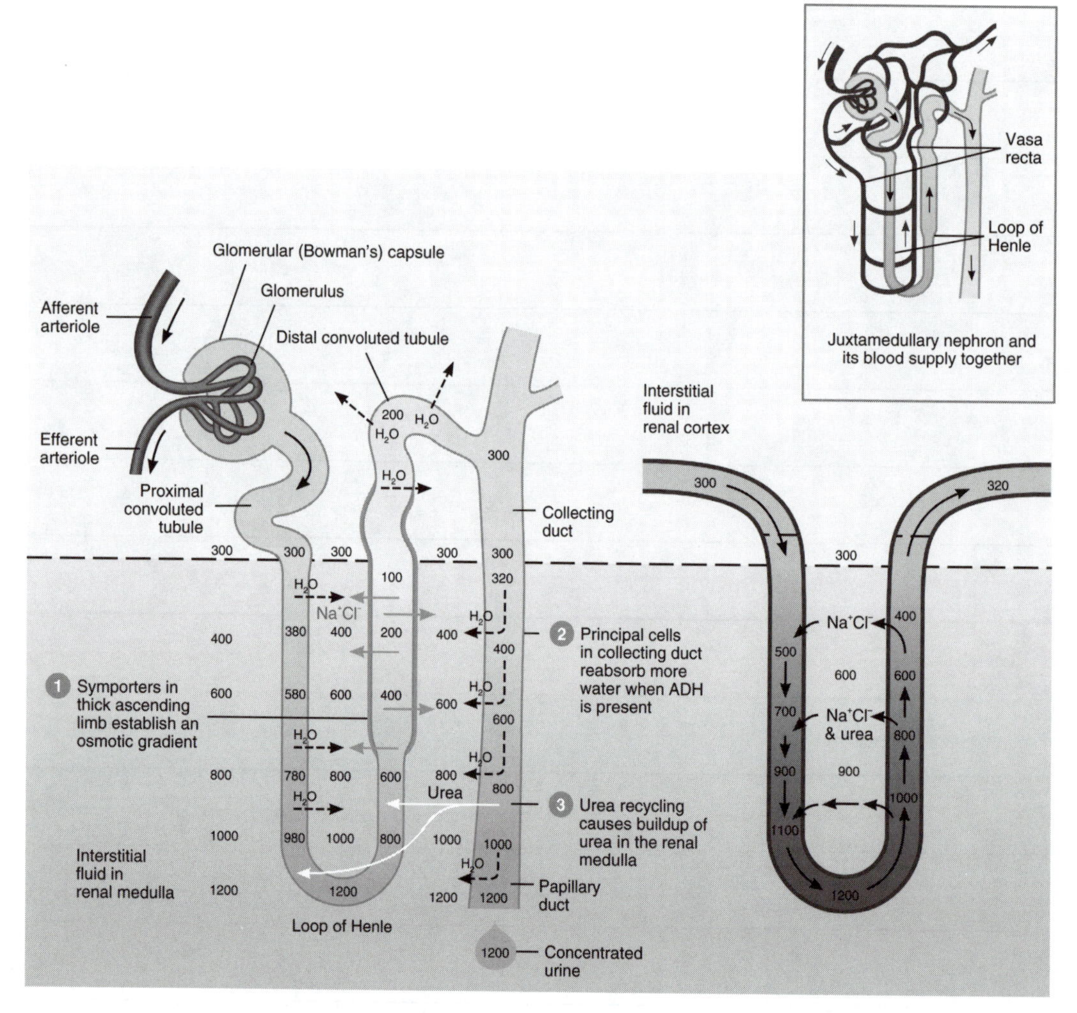

Vasa recta
Loop of Henle
Juxtamedullary nephron and its blood supply together
Glomerular (Bowman's) capsule
Glomerulus
Afferent arteriole
Distal convoluted tubule
Efferent arteriole
Proximal convoluted tubule
Collecting duct
Interstitial fluid in renal cortex
1 Symporters in thick ascending limb establish an osmotic gradient
2 Principal cells in collecting duct reabsorb more water when ADH is present
3 Urea recycling causes buildup of urea in the renal medulla
Interstitial fluid in renal medulla
Papillary duct
Loop of Henle
Concentrated urine
Urea
Na+Cl−
Na+Cl− & urea

(a) Reabsorption of Na^+, Cl^- and water in a long-loop juxtamedullary nephron

(b) Recycling of salts and urea in the vasa recta

NOTES

26

Figure 26.21 Summary of filtration, reabsorption, and secretion in the nephron and collecting duct (page 942).

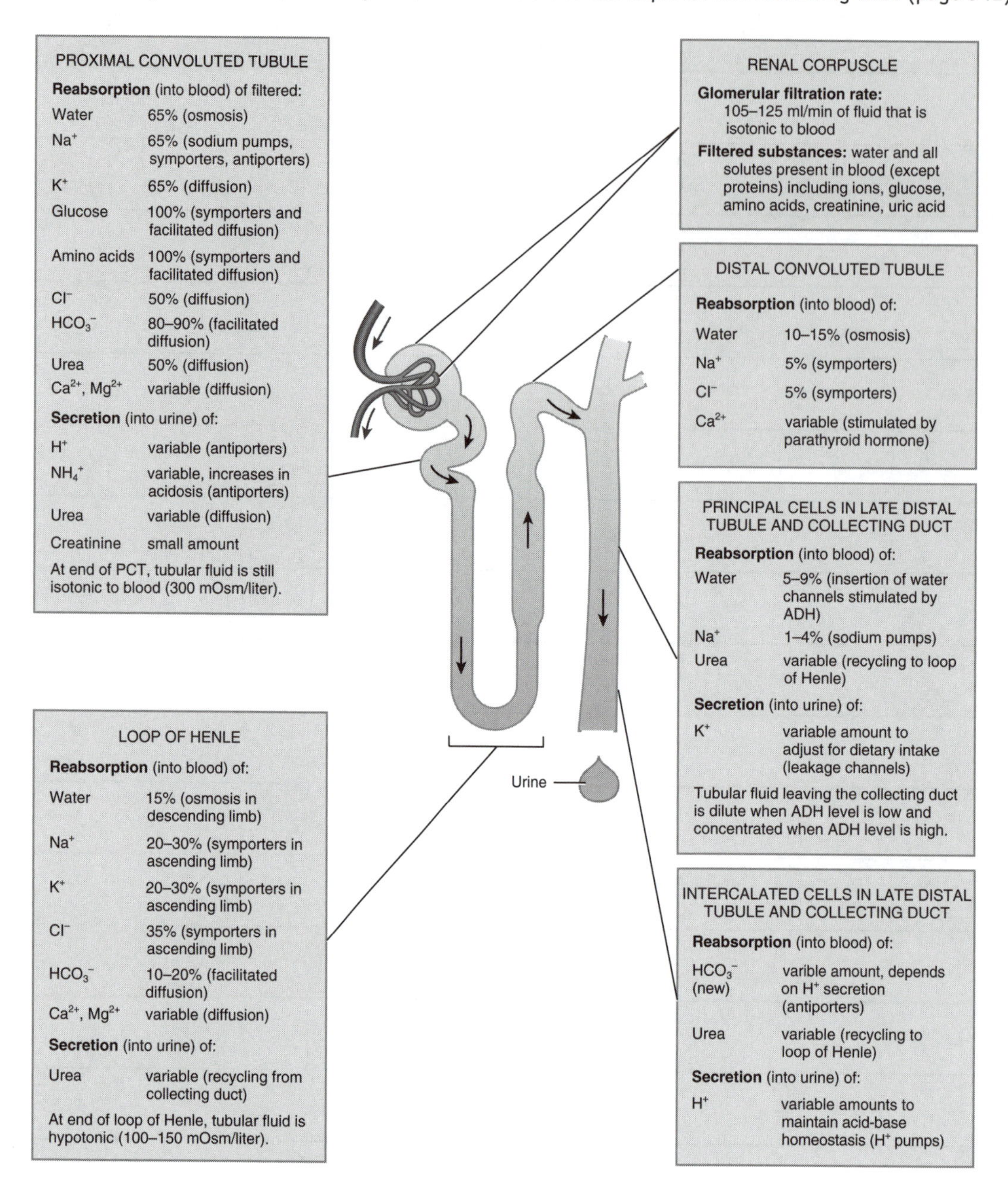

PROXIMAL CONVOLUTED TUBULE
Reabsorption (into blood) of filtered:
Water 65% (osmosis)
Na+ 65% (sodium pumps, symporters, antiporters)
K+ 65% (diffusion)
Glucose 100% (symporters and facilitated diffusion)
Amino acids 100% (symporters and facilitated diffusion)
Cl− 50% (diffusion)
HCO3− 80–90% (facilitated diffusion)
Urea 50% (diffusion)
Ca2+, Mg2+ variable (diffusion)
Secretion (into urine) of:
H+ variable (antiporters)
NH4+ variable, increases in acidosis (antiporters)
Urea variable (diffusion)
Creatinine small amount
At end of PCT, tubular fluid is still isotonic to blood (300 mOsm/liter).
RENAL CORPUSCLE
Glomerular filtration rate:
105–125 ml/min of fluid that is isotonic to blood
Filtered substances: water and all solutes present in blood (except proteins) including ions, glucose, amino acids, creatinine, uric acid
DISTAL CONVOLUTED TUBULE
Reabsorption (into blood) of:
Water 10–15% (osmosis)
Na+ 5% (symporters)
Cl− 5% (symporters)
Ca2+ variable (stimulated by parathyroid hormone)
PRINCIPAL CELLS IN LATE DISTAL TUBULE AND COLLECTING DUCT
Reabsorption (into blood) of:
Water 5–9% (insertion of water channels stimulated by ADH)
Na+ 1–4% (sodium pumps)
Urea variable (recycling to loop of Henle)
Secretion (into urine) of:
K+ variable amount to adjust for dietary intake (leakage channels)
Tubular fluid leaving the collecting duct is dilute when ADH level is low and concentrated when ADH level is high.
LOOP OF HENLE
Reabsorption (into blood) of:
Water 15% (osmosis in descending limb)
Na+ 20–30% (symporters in ascending limb)
K+ 20–30% (symporters in ascending limb)
Cl− 35% (symporters in ascending limb)
HCO3− 10–20% (facilitated diffusion)
Ca2+, Mg2+ variable (diffusion)
Secretion (into urine) of:
Urea variable (recycling from collecting duct)
At end of loop of Henle, tubular fluid is hypotonic (100–150 mOsm/liter).
Urine
INTERCALATED CELLS IN LATE DISTAL TUBULE AND COLLECTING DUCT
Reabsorption (into blood) of:
HCO3− (new) varible amount, depends on H+ secretion (antiporters)
Urea variable (recycling to loop of Henle)
Secretion (into urine) of:
H+ variable amounts to maintain acid-base homeostasis (H+ pumps)

Figure 26.22 Ureters, urinary bladder, and urethra (shown in a female) (page 946).

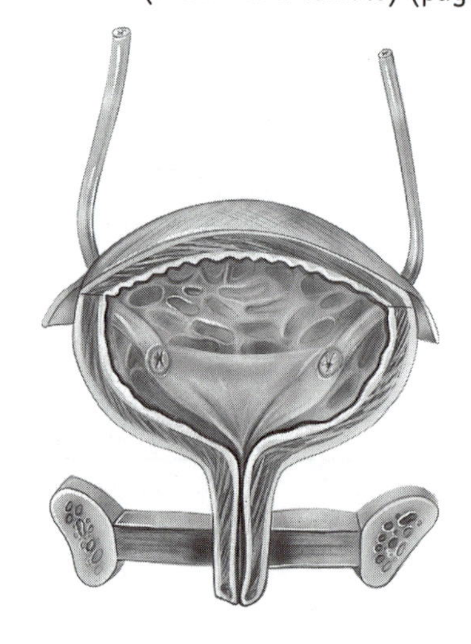

NOTES

26

Figure 26.23 Development of the urinary system (page 948).

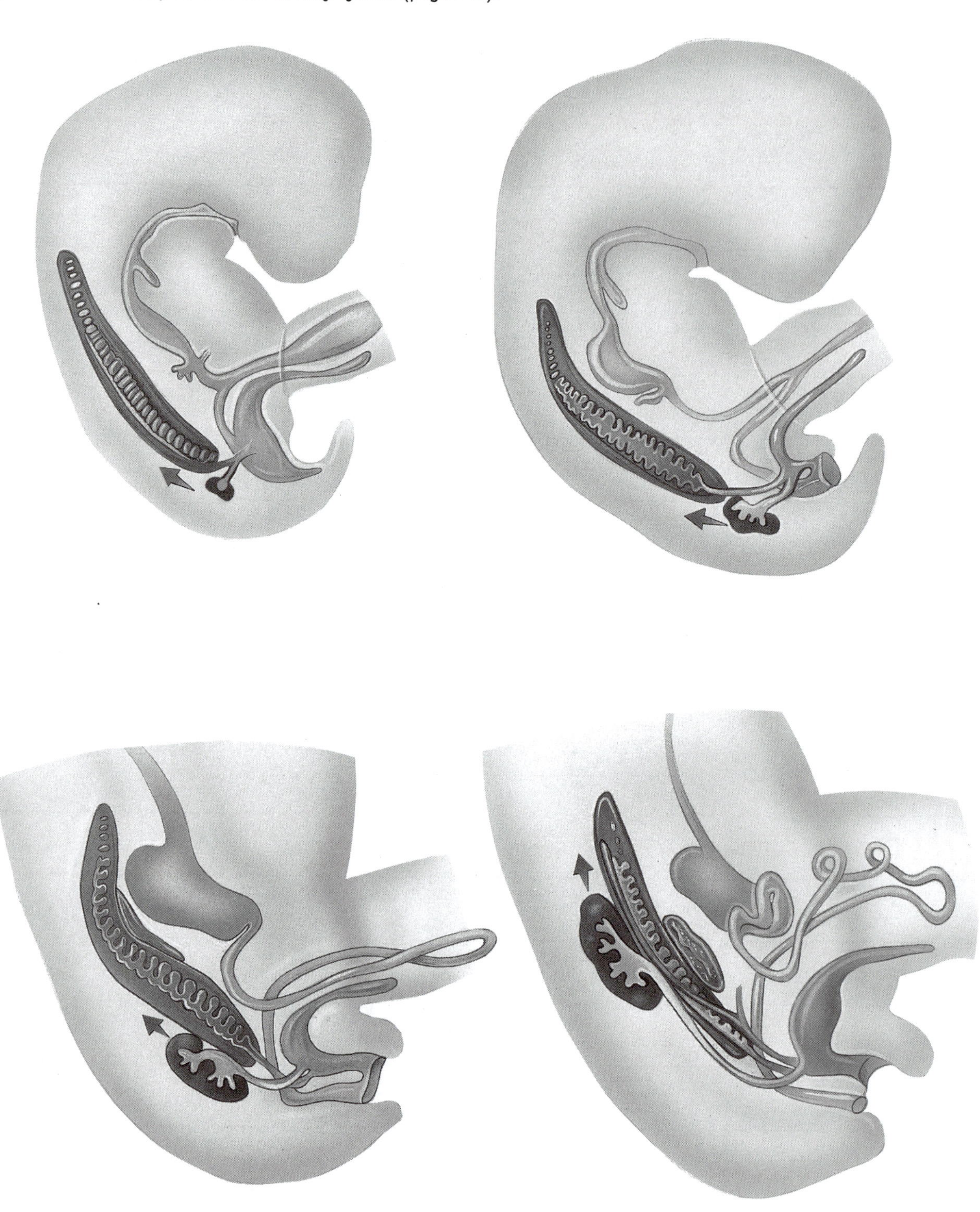

NOTES

26

Figure 27.1 Body fluid compartments (page 957).

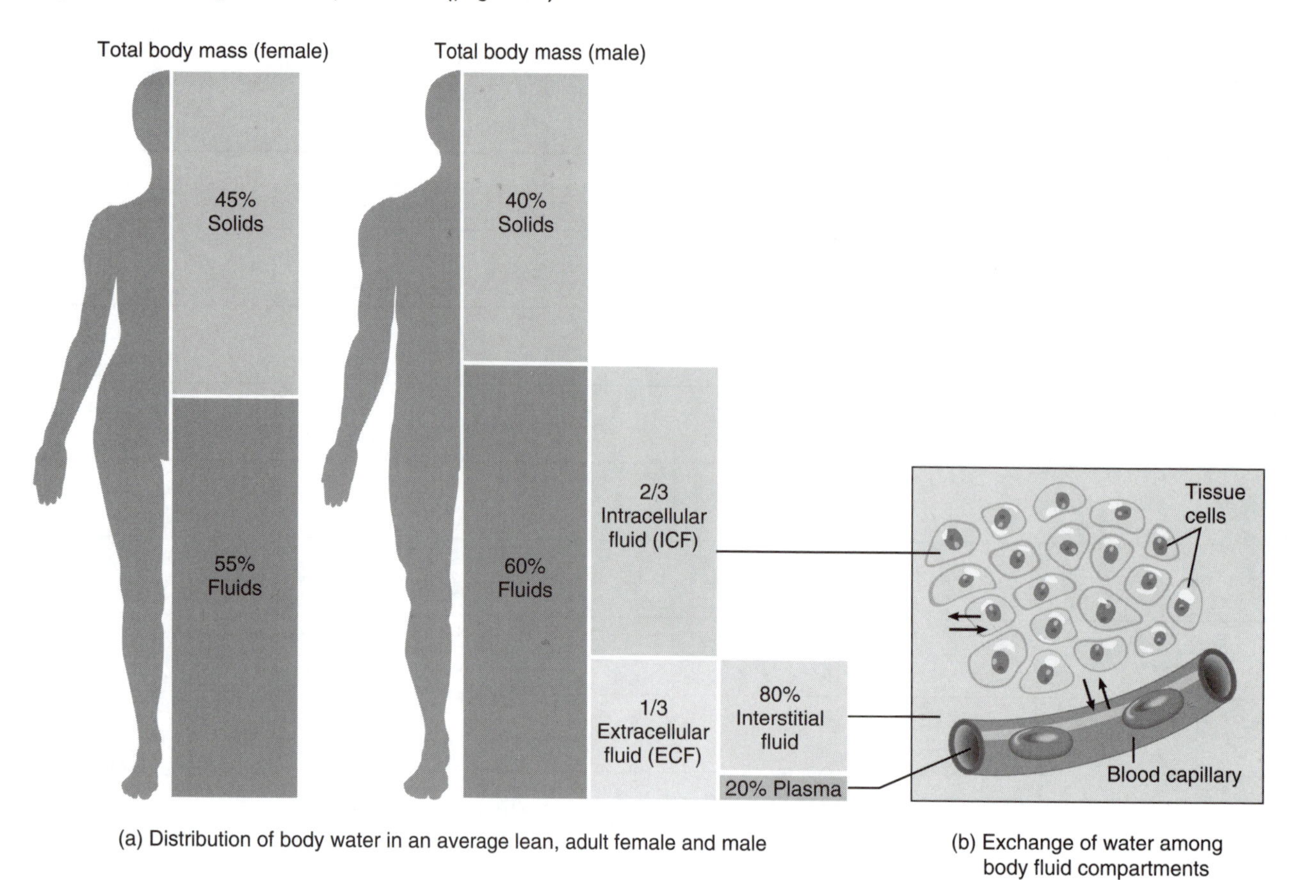
Total body mass (female)
Total body mass (male)
45% Solids
55% Fluids
40% Solids
60% Fluids
2/3 Intracellular fluid (ICF)
1/3 Extracellular fluid (ECF)
80% Interstitial fluid
20% Plasma
Tissue cells
Blood capillary

(a) Distribution of body water in an average lean, adult female and male

(b) Exchange of water among body fluid compartments

Figure 27.2 Sources of daily water gain and loss under normal conditions (page 958).

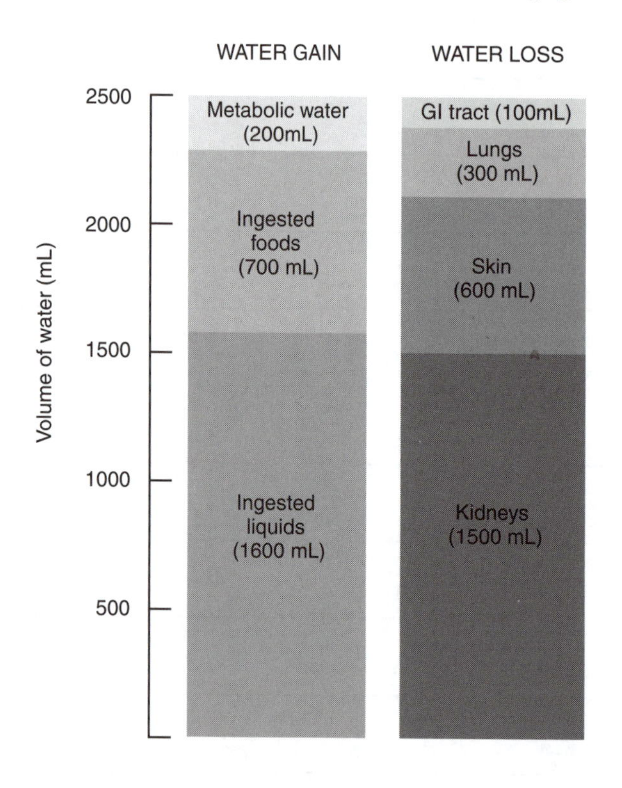
WATER GAIN
WATER LOSS
Volume of water (mL)
2500
2000
1500
1000
500
Metabolic water (200mL)
Ingested foods (700 mL)
Ingested liquids (1600 mL)
GI tract (100mL)
Lungs (300 mL)
Skin (600 mL)
Kidneys (1500 mL)

NOTES

27

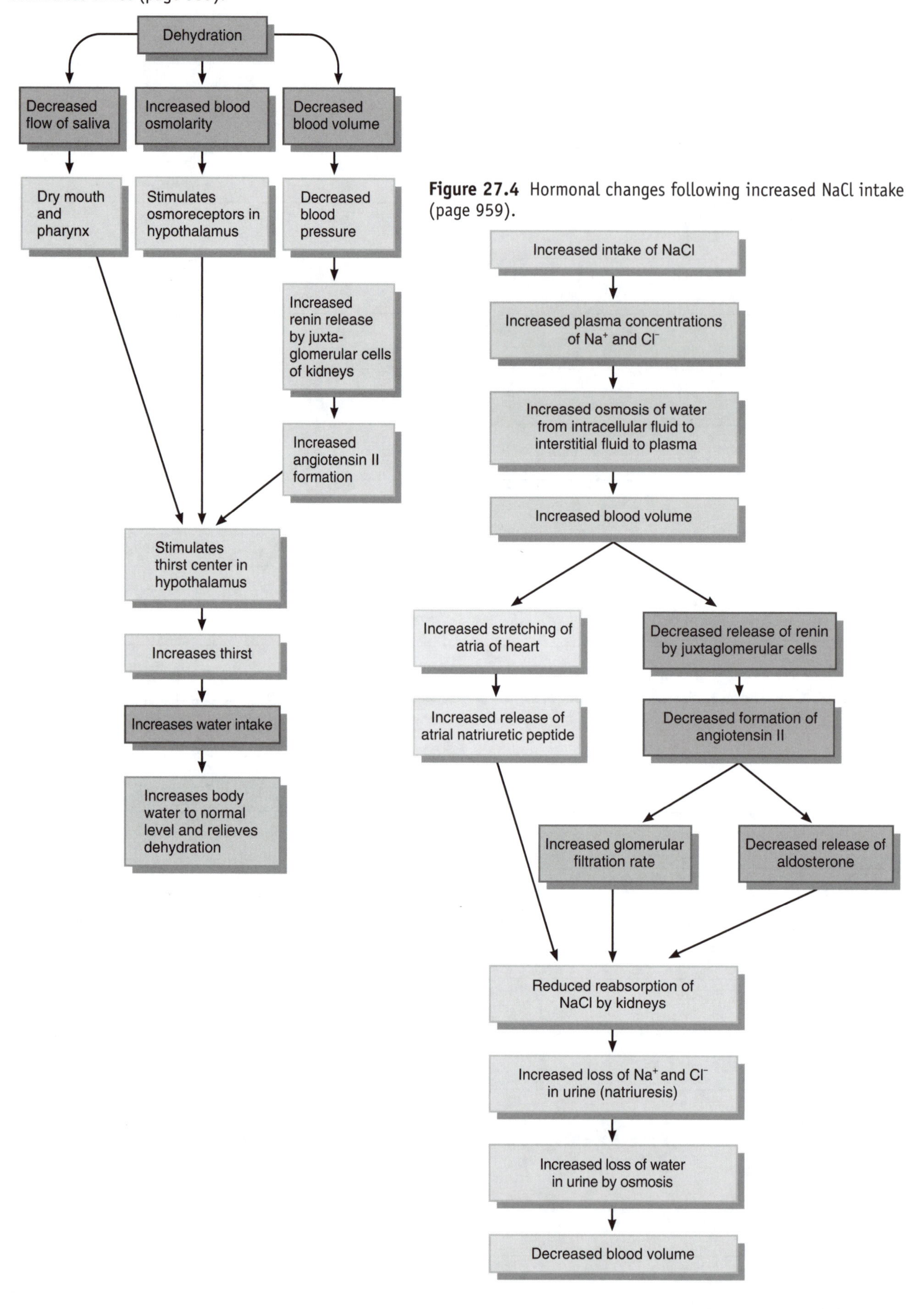

Dehydration
Decreased flow of saliva
Increased blood osmolarity
Decreased blood volume
Dry mouth and pharynx
Stimulates osmoreceptors in hypothalamus
Decreased blood pressure
Increased renin release by juxta-glomerular cells of kidneys
Increased angiotensin II formation
Stimulates thirst center in hypothalamus
Increases thirst
Increases water intake
Increases body water to normal level and relieves dehydration
Increased intake of NaCl
Increased plasma concentrations of Na+ and Cl−
Increased osmosis of water from intracellular fluid to interstitial fluid to plasma
Increased blood volume
Increased stretching of atria of heart
Decreased release of renin by juxtaglomerular cells
Increased release of atrial natriuretic peptide
Decreased formation of angiotensin II
Increased glomerular filtration rate
Decreased release of aldosterone
Reduced reabsorption of NaCl by kidneys
Increased loss of Na+ and Cl− in urine (natriuresis)
Increased loss of water in urine by osmosis
Decreased blood volume

Figure 27.3 Pathways through which dehydration stimulates thirst (page 959).

Figure 27.4 Hormonal changes following increased NaCl intake (page 959).

NOTES

27

Figure 27.5 Series of events in water intoxication (page 960).

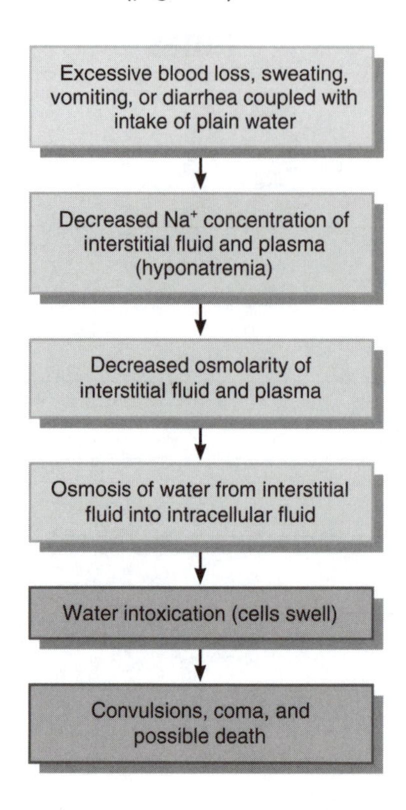
Excessive blood loss, sweating, vomiting, or diarrhea coupled with intake of plain water
Decreased Na^+ concentration of interstitial fluid and plasma (hyponatremia)
Decreased osmolarity of interstitial fluid and plasma
Osmosis of water from interstitial fluid into intracellular fluid
Water intoxication (cells swell)
Convulsions, coma, and possible death

Figure 27.6 Electrolyte and protein anion concentrations in plasma, interstitial fluid, and intracellular fluid (page 962).

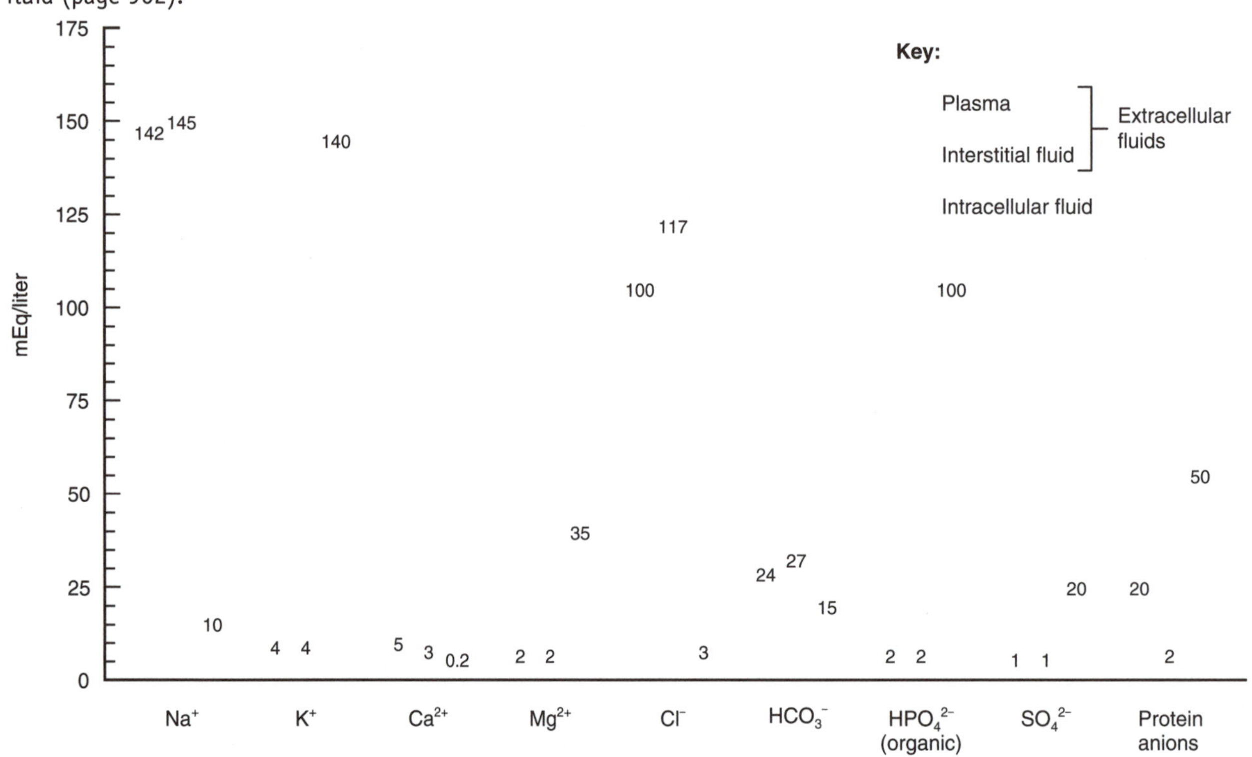
175
150
125
100
75
50
25
0
mEq/liter
Key:
Plasma
Interstitial fluid
Intracellular fluid
Extracellular fluids
142 145
10
140
4 4
5 3 0.2
2 2
35
100 117
3
24 27
15
2 2
100
1 1
20
20
50
2
Na^+
K^+
Ca^{2+}
Mg^{2+}
Cl^-
HCO_3^-
HPO_4^{2-} (organic)
SO_4^{2-}
Protein anions

NOTES

27

Figure 27.7 Negative feedback regulation of blood pH by the respiratory system (page 967).

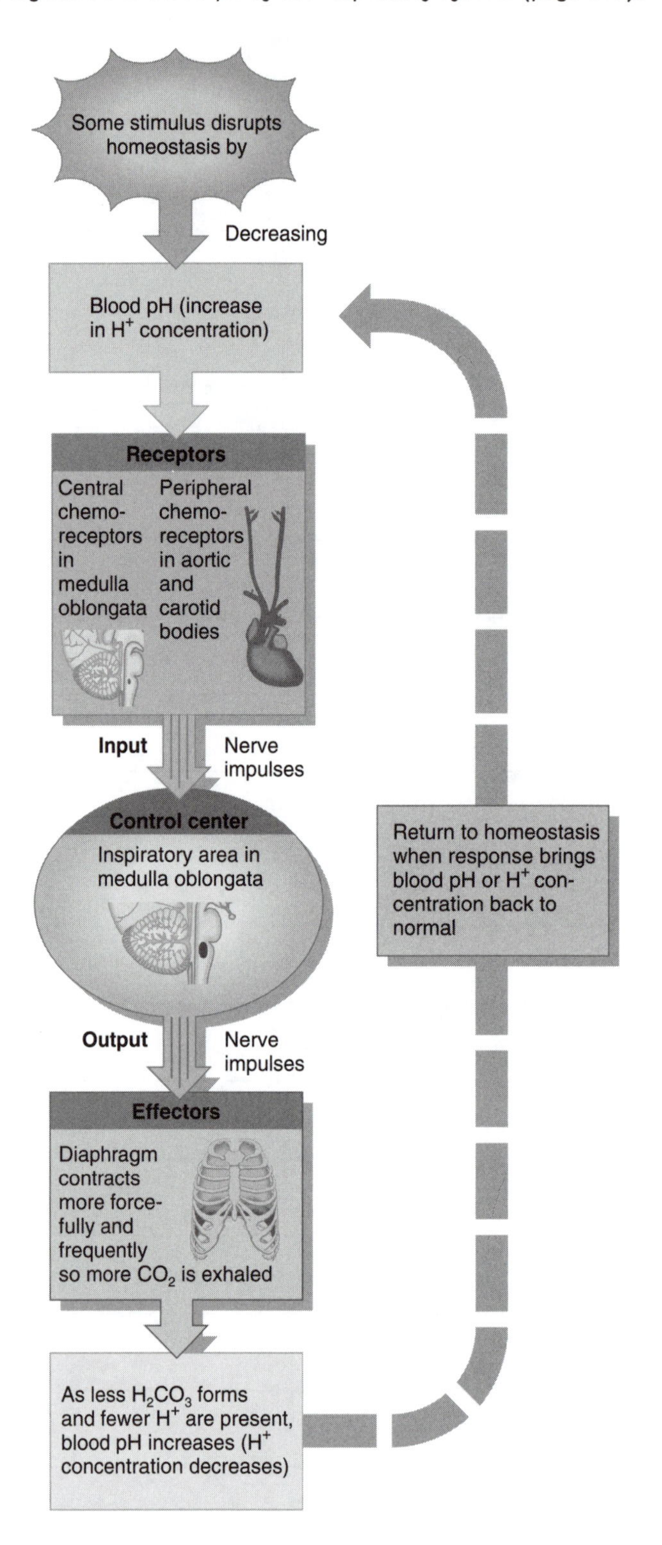

Some stimulus disrupts homeostasis by
Decreasing
Blood pH (increase in H^+ concentration)
Receptors
Central chemo- receptors in medulla oblongata
Peripheral chemo- receptors in aortic and carotid bodies
Input
Nerve impulses
Control center
Inspiratory area in medulla oblongata
Output
Nerve impulses
Effectors
Diaphragm contracts more force- fully and frequently so more CO_2 is exhaled
As less H_2CO_3 forms and fewer H^+ are present, blood pH increases (H^+ concentration decreases)
Return to homeostasis when response brings blood pH or H^+ con- centration back to normal

NOTES

27

Figure 28.1 Meiosis, reproductive cell division (page 975).

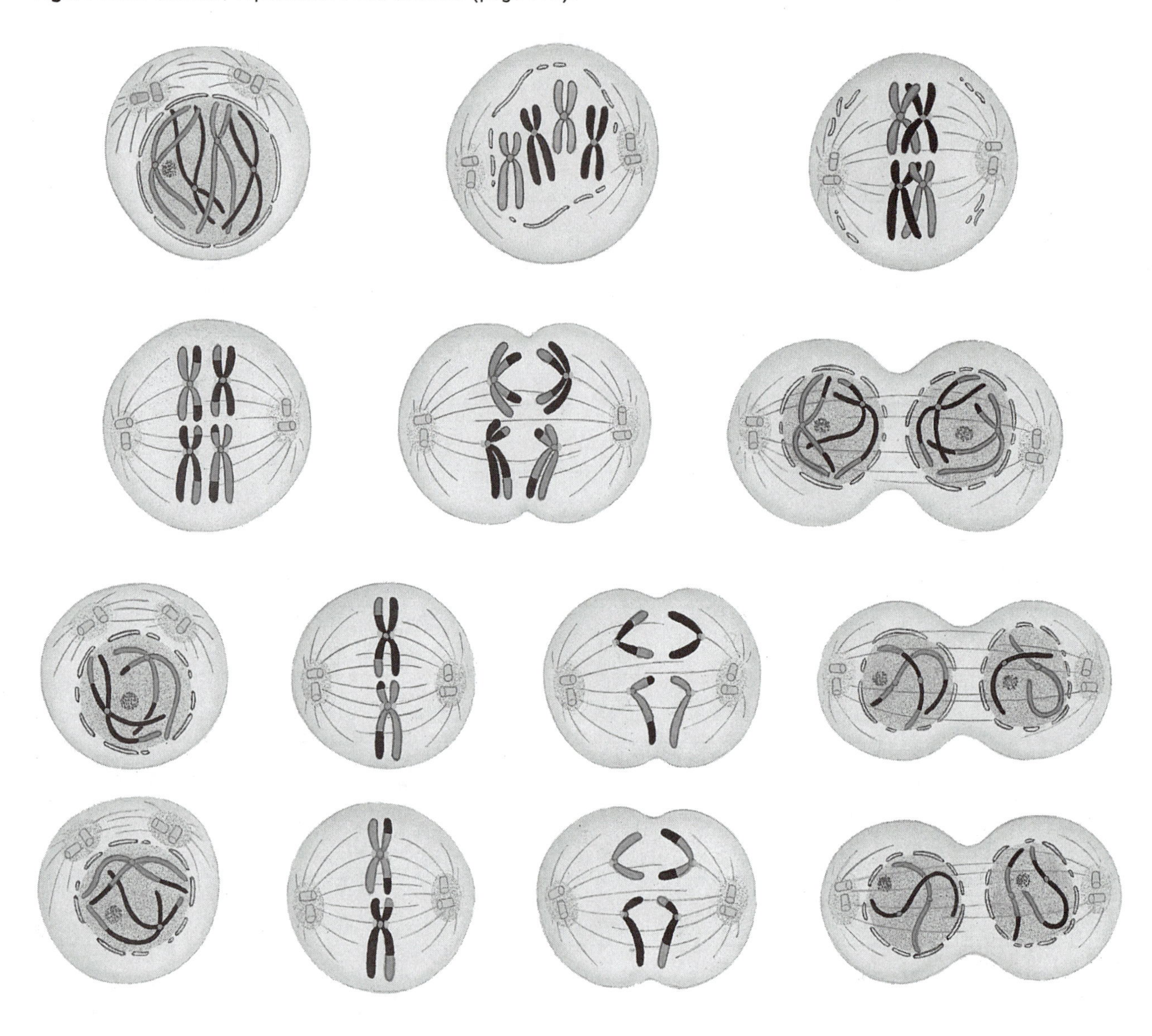

Figure 28.2 Crossing over within a tetrad during prophase I of meiosis (page 976).

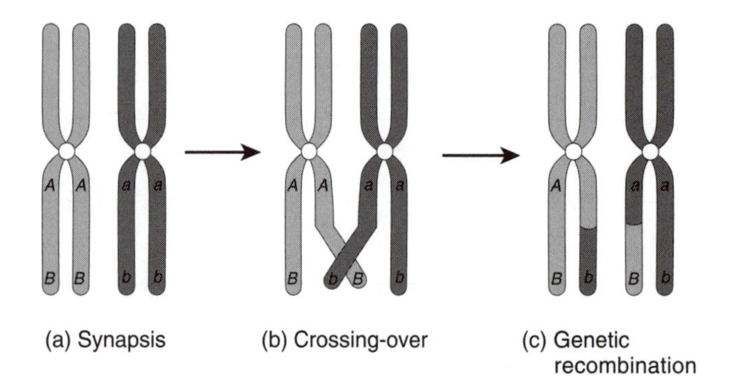
A A a a
B B b b
A A a a
B b B b
A a a
B b B b
(a) Synapsis
(b) Crossing-over
(c) Genetic
recombination

NOTES

28

Figure 28.3 Male organs of reproduction and surrounding structures (see page 977).

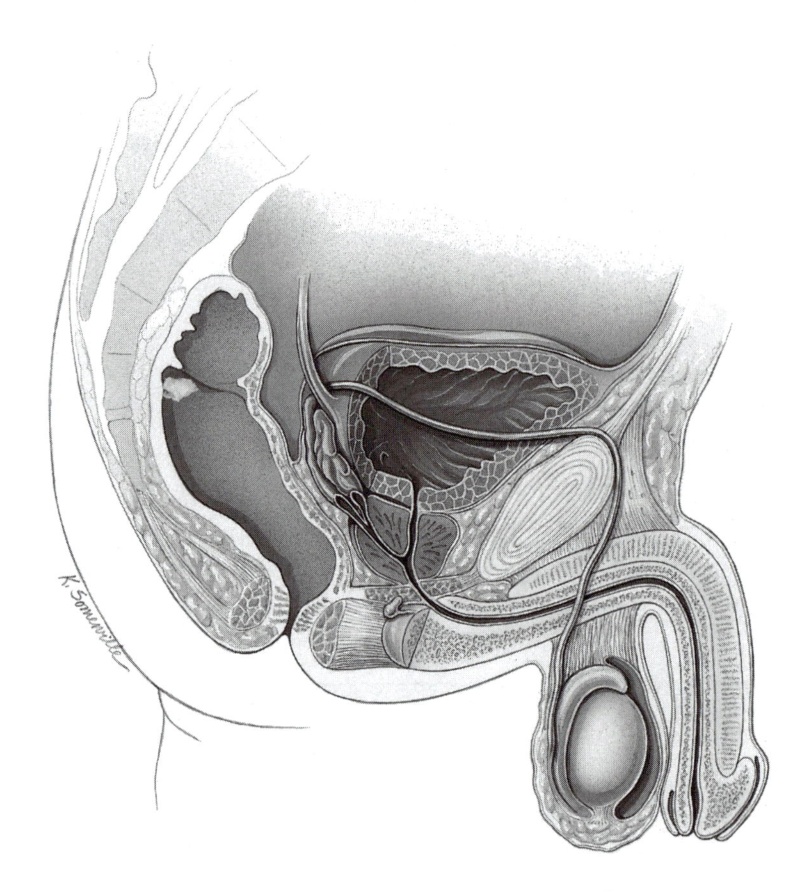
K. Somerville

Figure 28.4 The scrotum, the supporting structure for the testes (page 979).

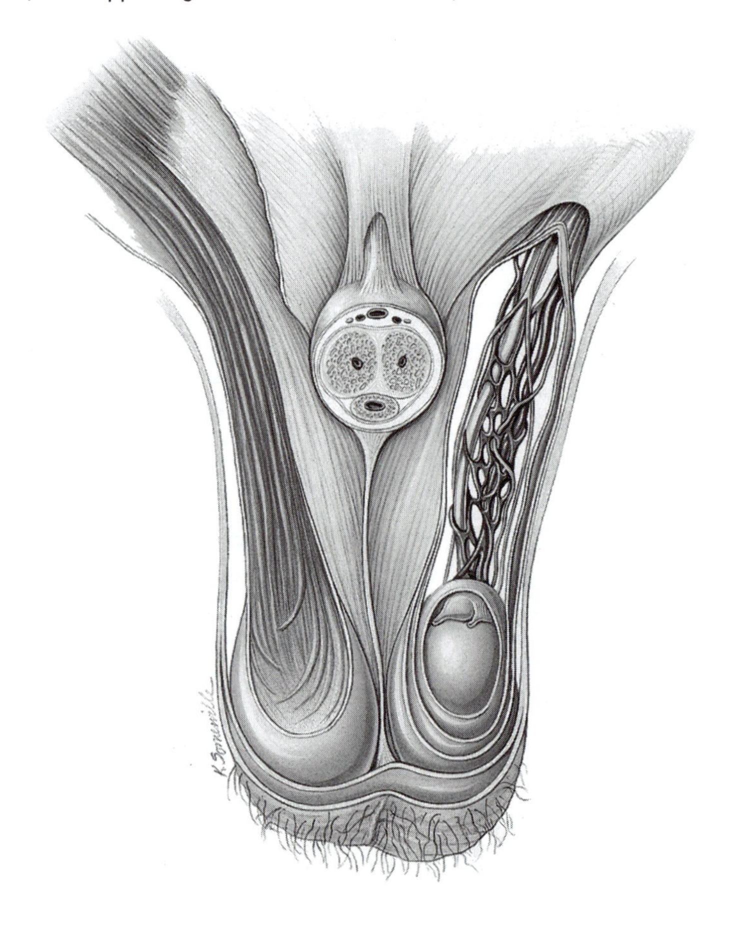
K. Somerville

NOTES

28

Figure 28.5 Internal and external anatomy of a testis (page 980).

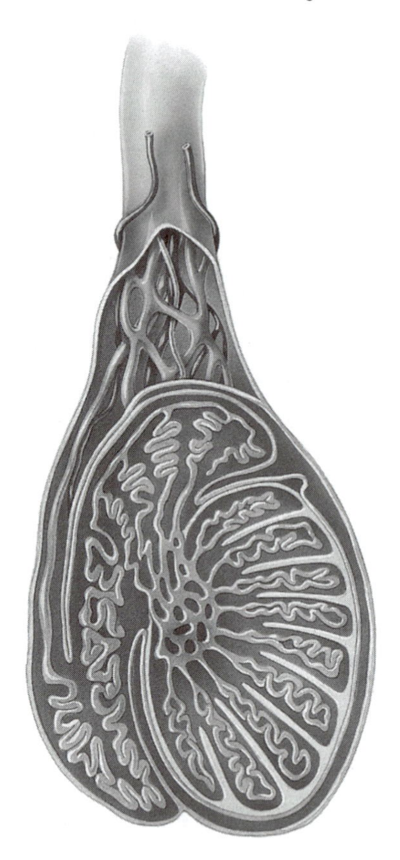

Figure 28.6b Microscopic anatomy of the seminiferous tubules and stages of sperm production (spermatogenesis) (page 981).

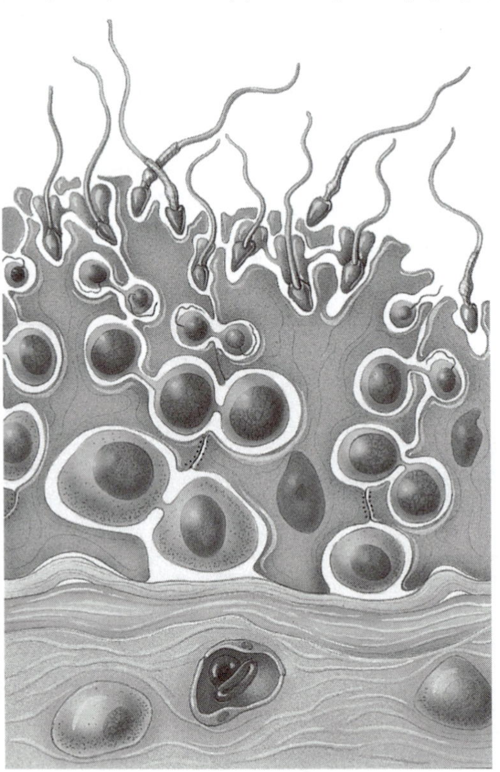

NOTES

28

Figure 28.7 Events in spermatogenesis (page 982).

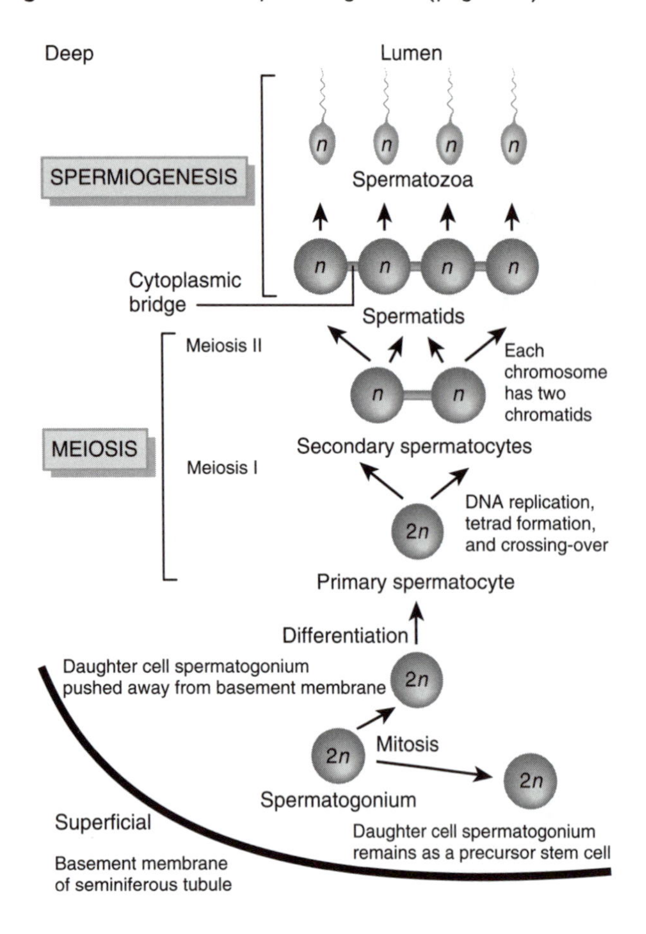
Deep
Lumen
SPERMIOGENESIS
Spermatozoa
Cytoplasmic bridge
Spermatids
Meiosis II
Each chromosome has two chromatids
MEIOSIS
Meiosis I
Secondary spermatocytes
DNA replication, tetrad formation, and crossing-over
Primary spermatocyte
Differentiation
Daughter cell spermatogonium pushed away from basement membrane
Mitosis
Spermatogonium
Superficial
Daughter cell spermatogonium remains as a precursor stem cell
Basement membrane of seminiferous tubule

Figure 28.8 A sperm cell (page 982).

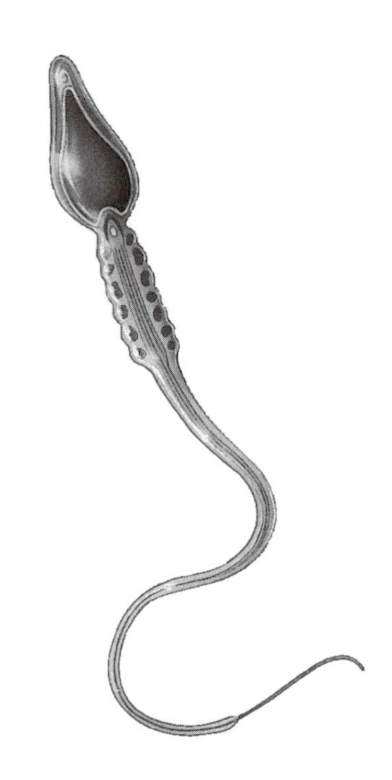

NOTES

28

Figure 28.9 Hormonal control of testicular functions (page 983).

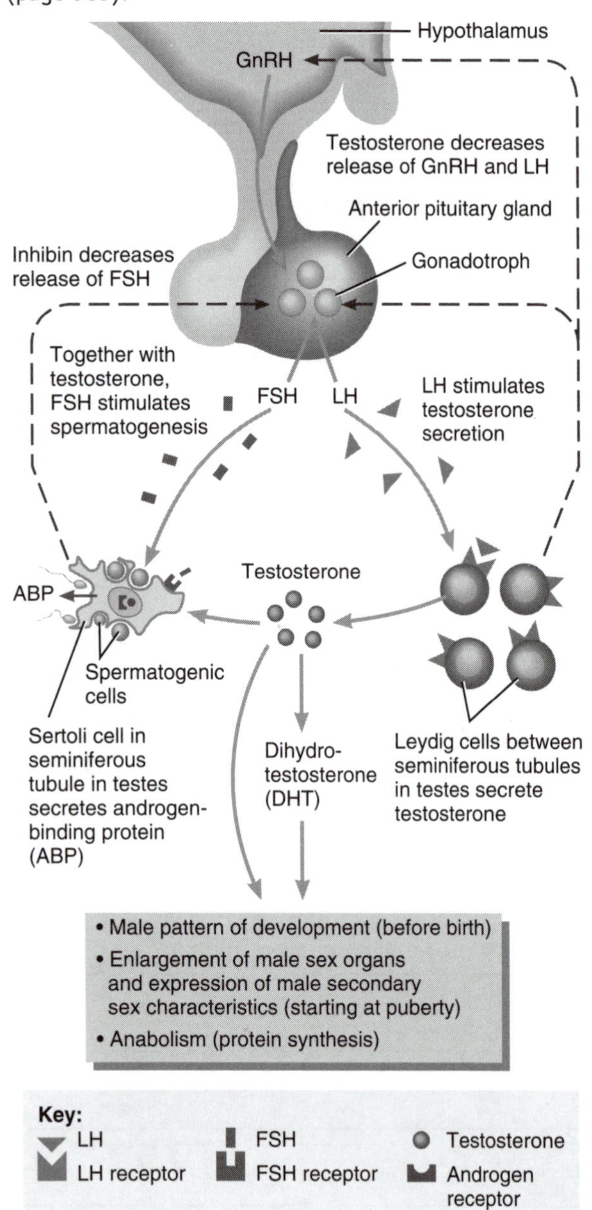

Hypothalamus
GnRH
Testosterone decreases release of GnRH and LH
Anterior pituitary gland
Gonadotroph
Inhibin decreases release of FSH
Together with testosterone, FSH stimulates spermatogenesis
FSH
LH
LH stimulates testosterone secretion
ABP
Testosterone
Spermatogenic cells
Sertoli cell in seminiferous tubule in testes secretes androgen-binding protein (ABP)
Dihydro-testosterone (DHT)
Leydig cells between seminiferous tubules in testes secrete testosterone
• Male pattern of development (before birth)
• Enlargement of male sex organs and expression of male secondary sex characteristics (starting at puberty)
• Anabolism (protein synthesis)
Key:
LH
LH receptor
FSH
FSH receptor
Testosterone
Androgen receptor

Figure 28.10 Negative feedback control of blood level of testosterone (page 984).

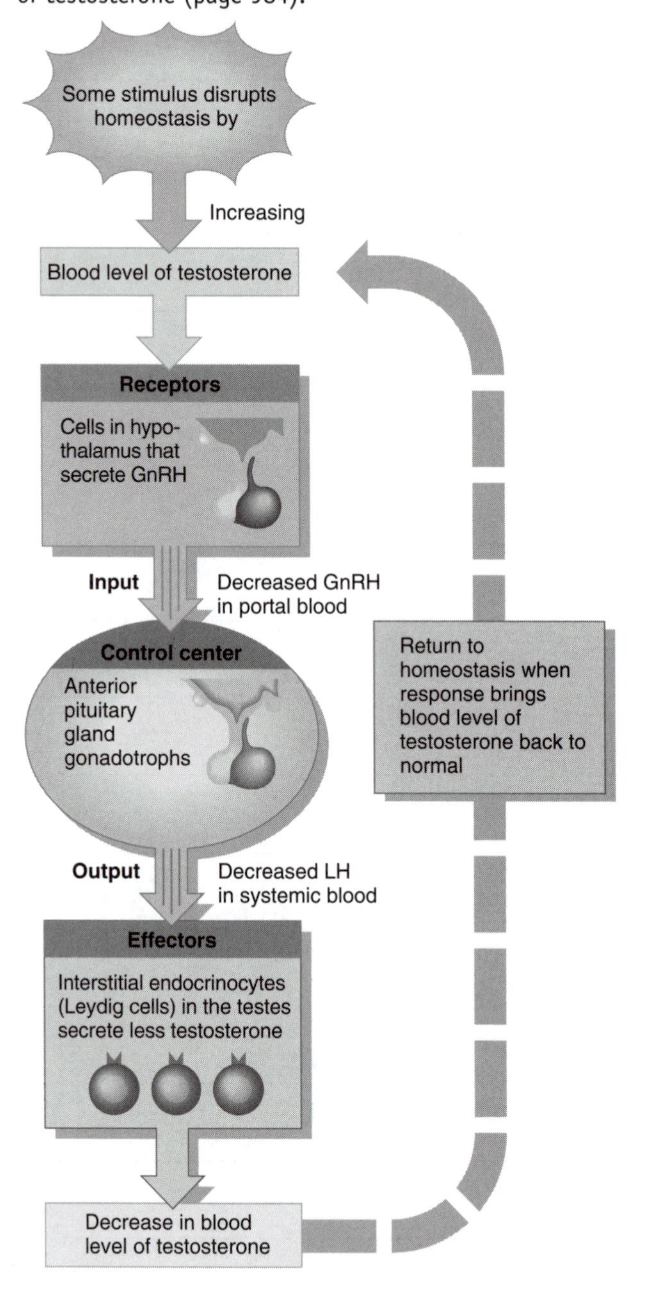

Some stimulus disrupts homeostasis by
Increasing
Blood level of testosterone
Receptors
Cells in hypo-thalamus that secrete GnRH
Input
Decreased GnRH in portal blood
Control center
Anterior pituitary gland gonadotrophs
Output
Decreased LH in systemic blood
Effectors
Interstitial endocrinocytes (Leydig cells) in the testes secrete less testosterone
Decrease in blood level of testosterone
Return to homeostasis when response brings blood level of testosterone back to normal

NOTES

28

Figure 28.11 Locations of several accessory reproductive organs in males (page 986).

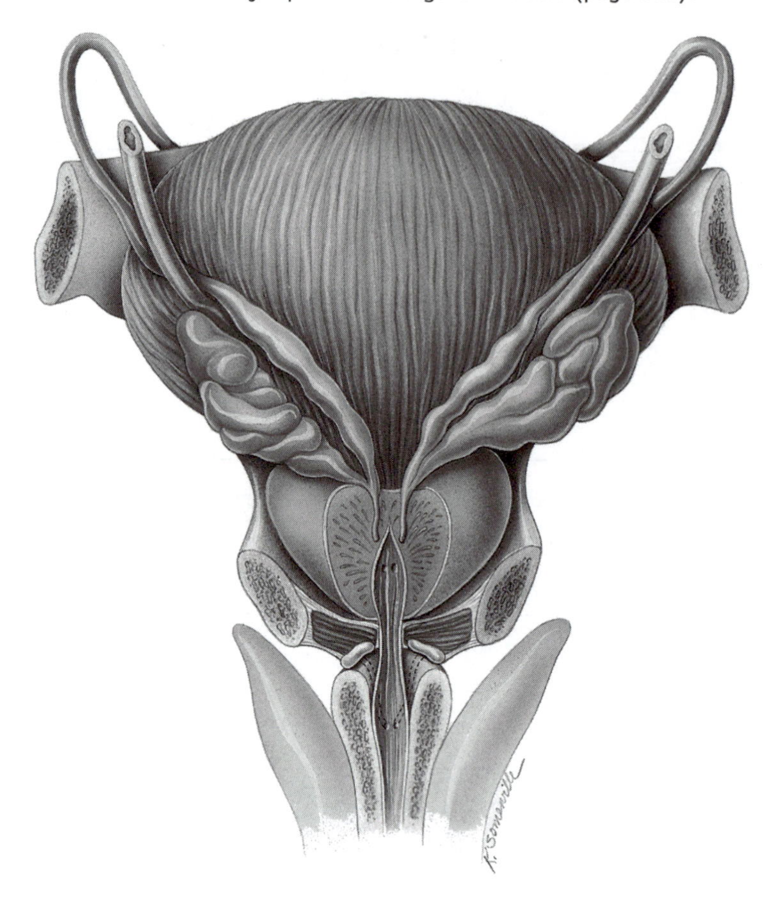

Figure 28.12 Internal structure of the penis (page 988).

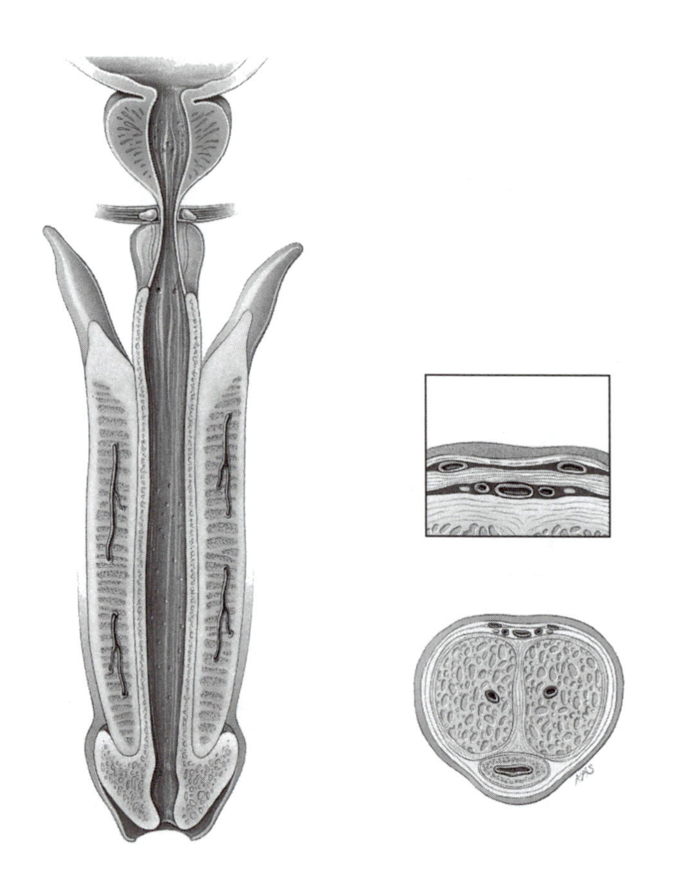

NOTES

28

Figure 28.13 Organs of reproduction and surrounding structures in females (page 990).

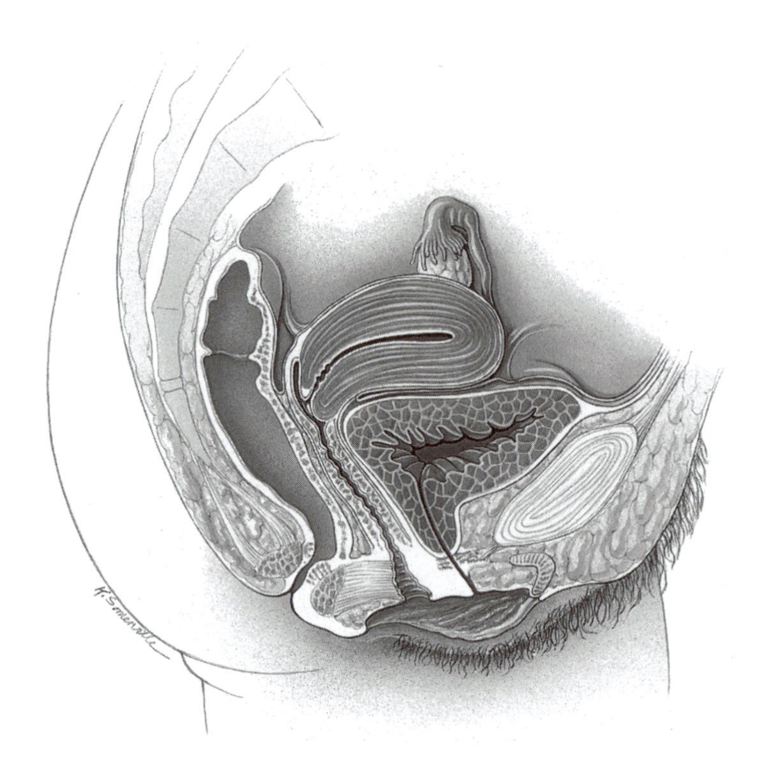

Figure 28.14 Relative positions of the ovaries, the uterus, and the ligaments that support them (page 992).

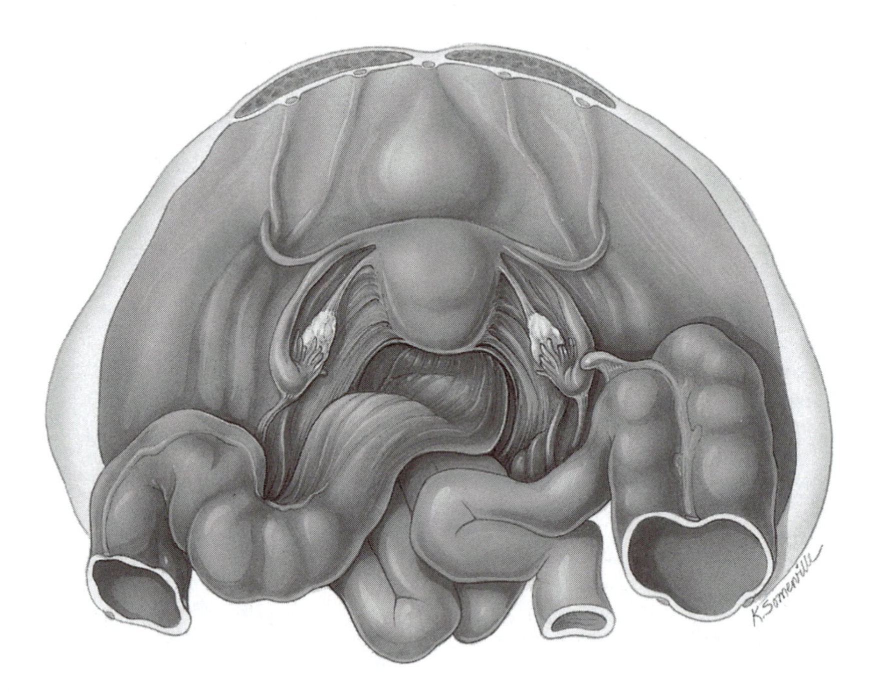

NOTES

28

Figure 28.15 Histology of the ovary (page 993).

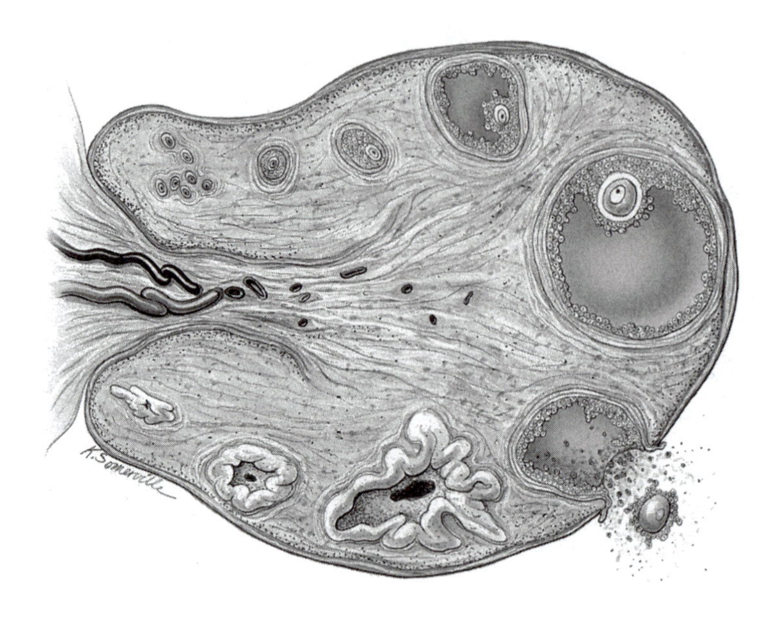

Figure 28.17 Oogenesis (page 994).

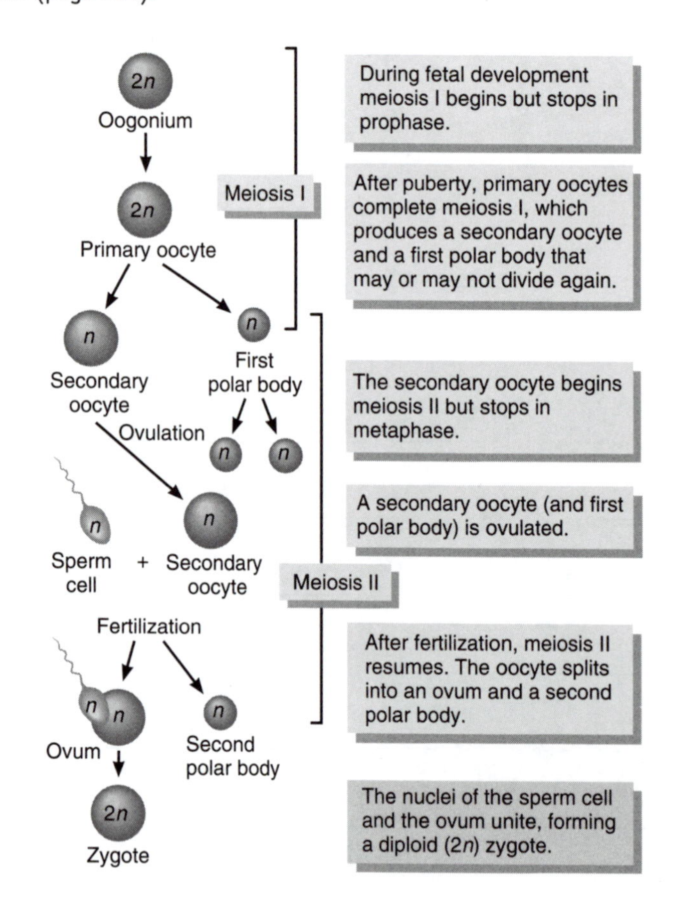
2n
Oogonium
2n
Primary oocyte
Meiosis I
n
Secondary oocyte
n
First polar body
Ovulation
n
n
n
Sperm cell
+
n
Secondary oocyte
Meiosis II
Fertilization
n
n
Ovum
n
Second polar body
2n
Zygote
During fetal development meiosis I begins but stops in prophase.
After puberty, primary oocytes complete meiosis I, which produces a secondary oocyte and a first polar body that may or may not divide again.
The secondary oocyte begins meiosis II but stops in metaphase.
A secondary oocyte (and first polar body) is ovulated.
After fertilization, meiosis II resumes. The oocyte splits into an ovum and a second polar body.
The nuclei of the sperm cell and the ovum unite, forming a diploid (2n) zygote.

NOTES

28

Figure 28.18 Relationship of the uterine (Fallopian) tubes to the ovaries, uterus, and associated structures (page 995).

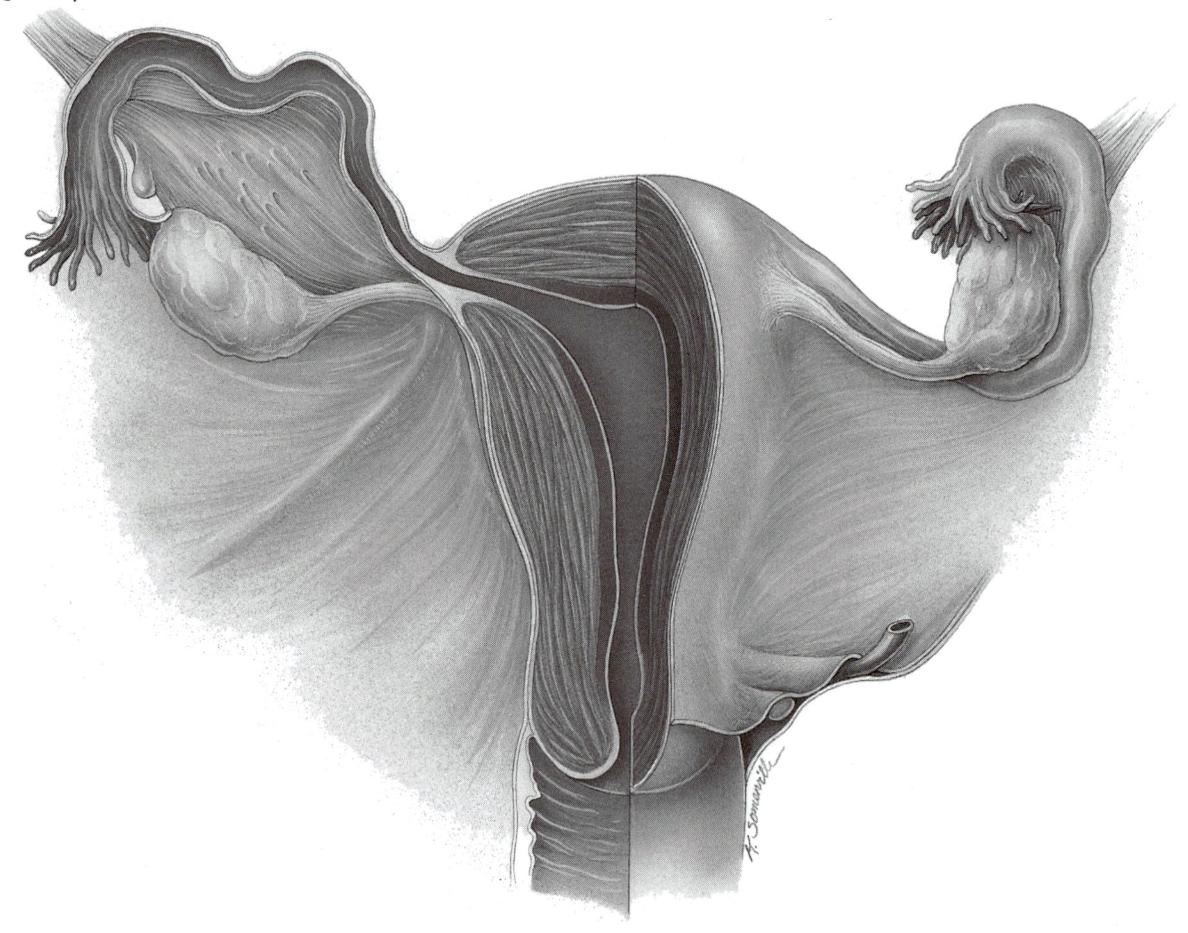

Figure 28.21 Blood supply of the uterus (page 997).

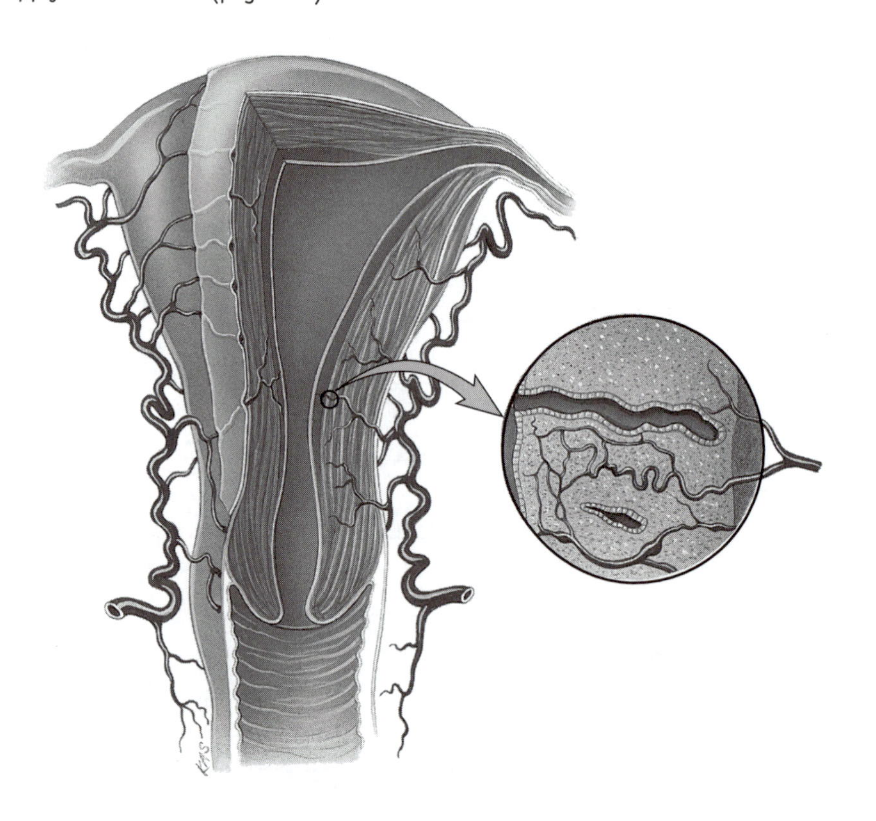

NOTES

28

Figure 28.22 Components of the vulva (page 999).

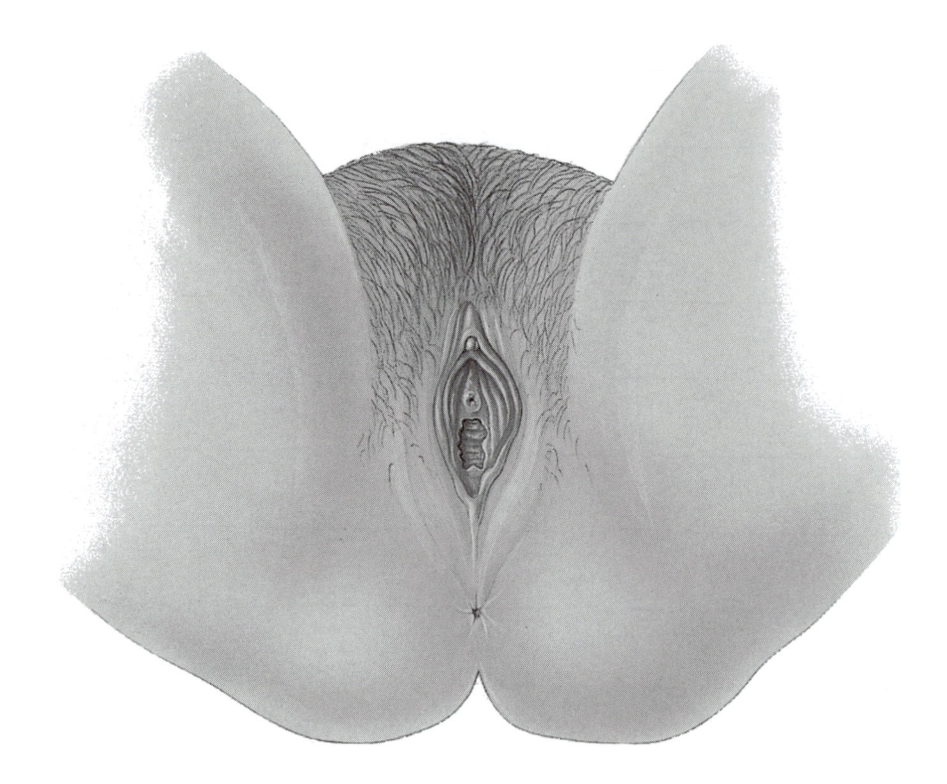

Figure 28.23 Perineum of a female (page 1000).

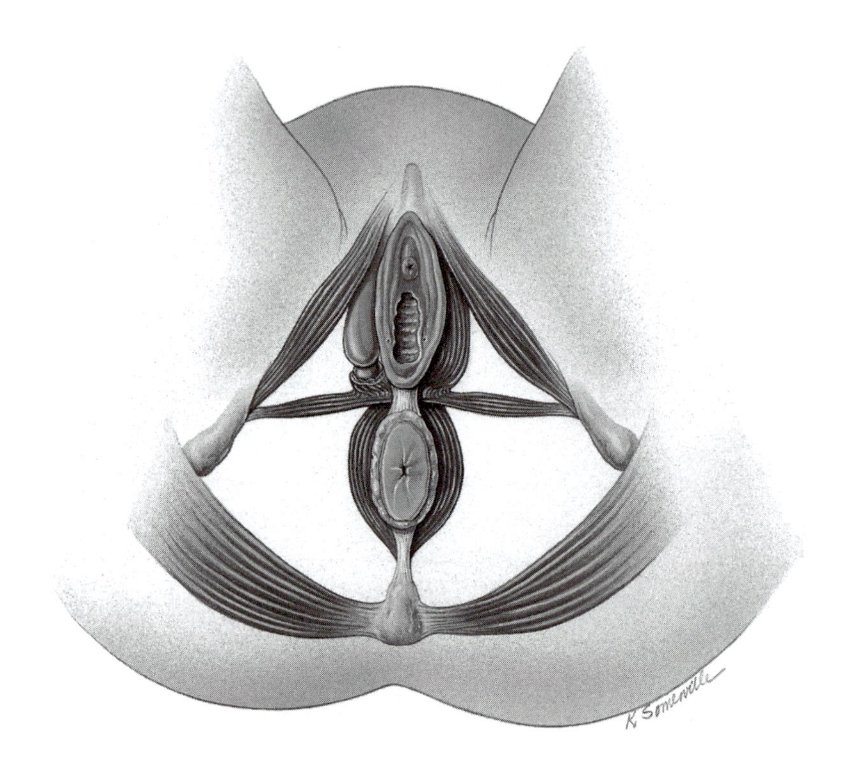

NOTES

28

Figure 28.24 Mammary glands (page 1001).

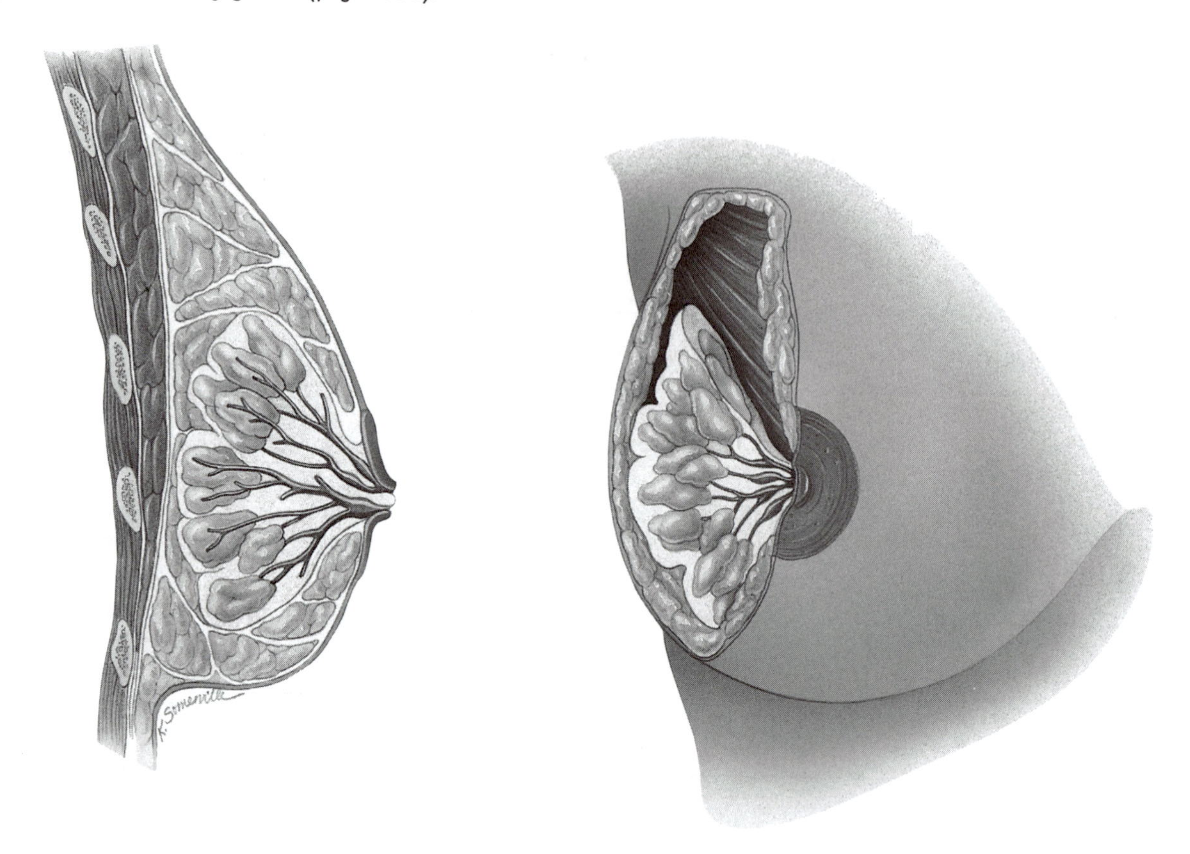

Figure 28.25 Secretion and physiological effects of estrogens, progesterone, relaxin, and inhibin (page 1002).

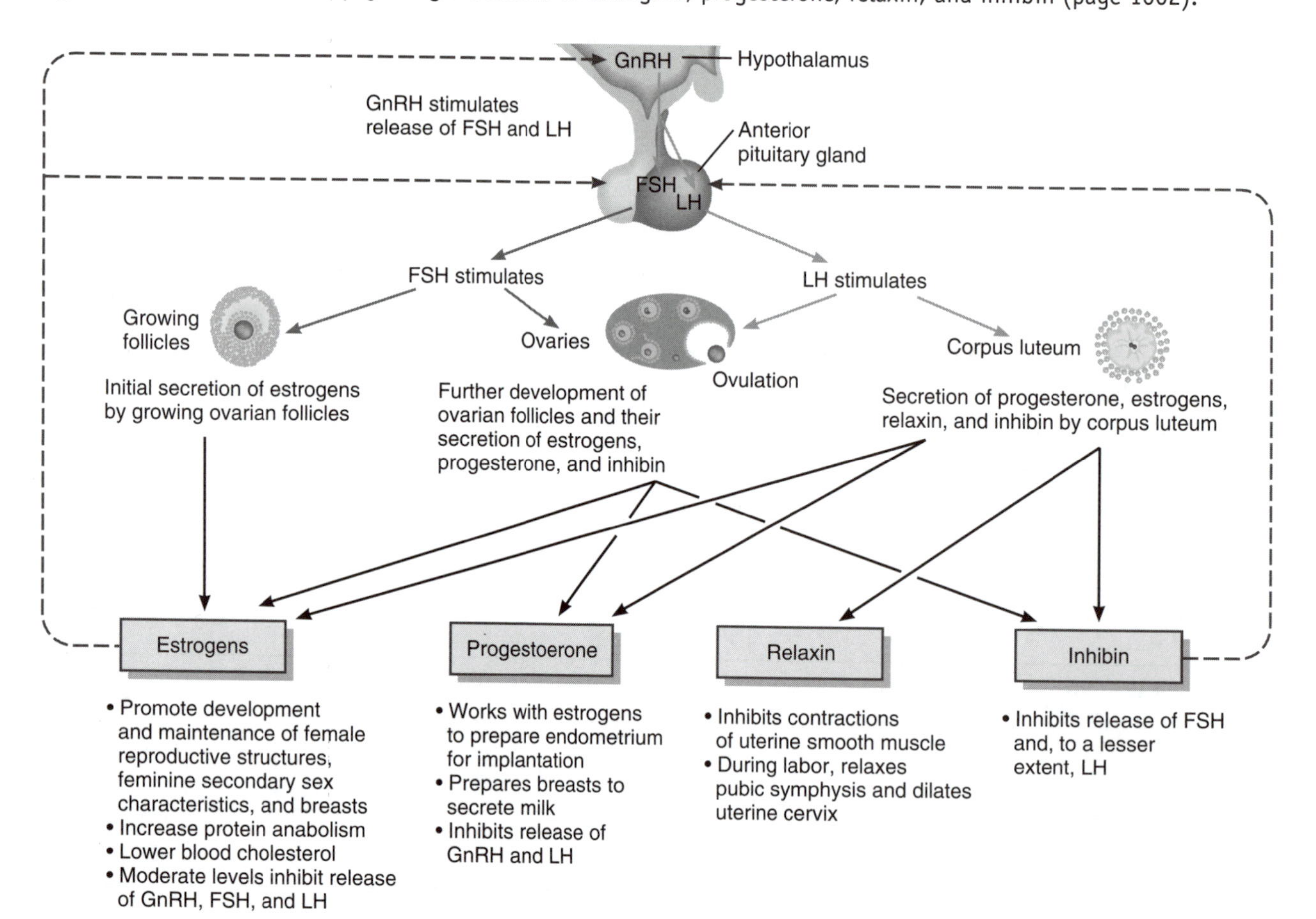

GnRH
Hypothalamus
GnRH stimulates release of FSH and LH
Anterior pituitary gland
FSH
LH
FSH stimulates
LH stimulates
Growing follicles
Ovaries
Ovulation
Corpus luteum
Initial secretion of estrogens by growing ovarian follicles
Further development of ovarian follicles and their secretion of estrogens, progesterone, and inhibin
Secretion of progesterone, estrogens, relaxin, and inhibin by corpus luteum
Estrogens
Progestoerone
Relaxin
Inhibin
• Promote development and maintenance of female reproductive structures, feminine secondary sex characteristics, and breasts
• Increase protein anabolism
• Lower blood cholesterol
• Moderate levels inhibit release of GnRH, FSH, and LH
• Works with estrogens to prepare endometrium for implantation
• Prepares breasts to secrete milk
• Inhibits release of GnRH and LH
• Inhibits contractions of uterine smooth muscle
• During labor, relaxes pubic symphysis and dilates uterine cervix
• Inhibits release of FSH and, to a lesser extent, LH

NOTES

Figure 28.26 The female reproductive cycle (page 1003).

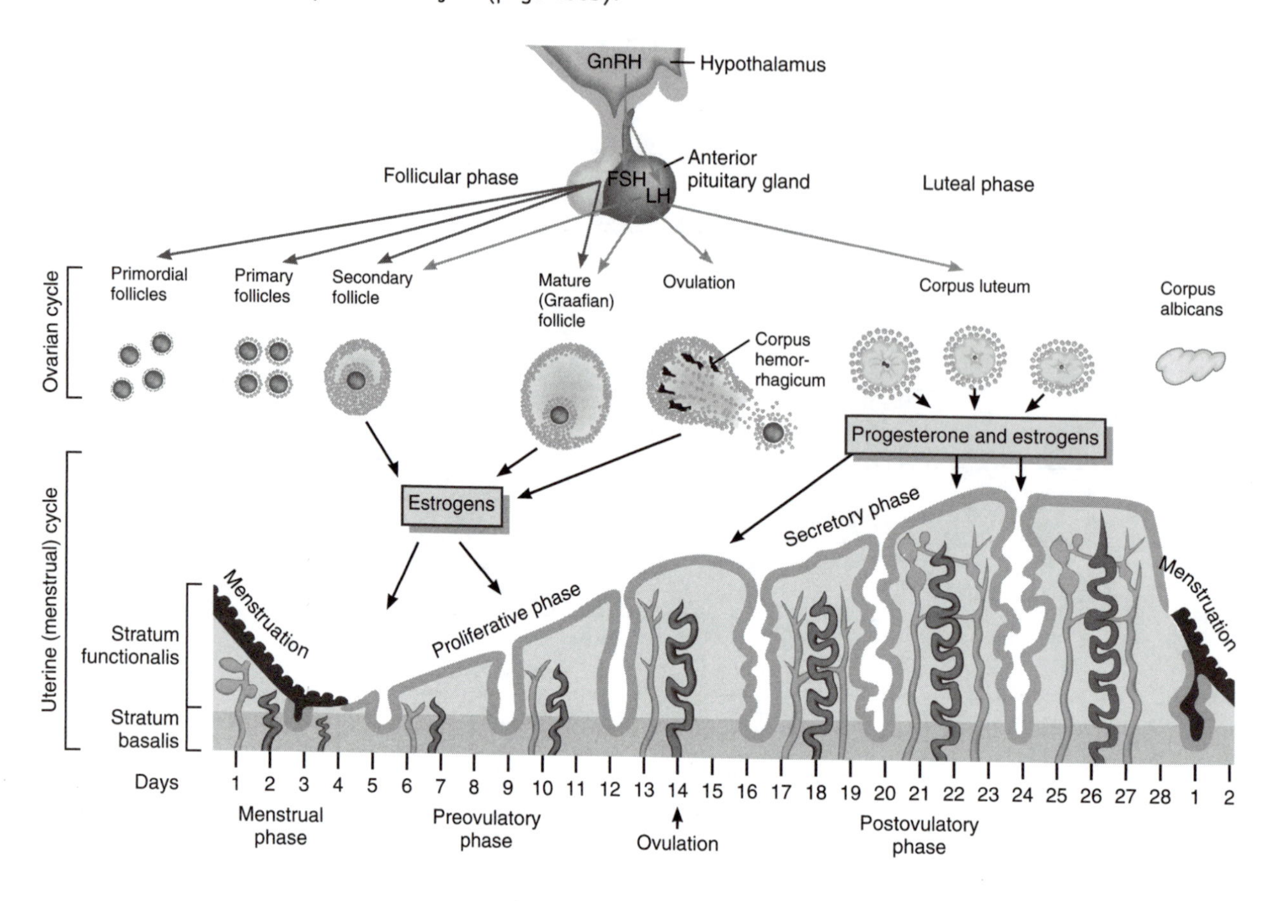

GnRH
Hypothalamus
Follicular phase
FSH
LH
Anterior pituitary gland
Luteal phase
Ovarian cycle
Primordial follicles
Primary follicles
Secondary follicle
Mature (Graafian) follicle
Ovulation
Corpus luteum
Corpus albicans
Corpus hemor-rhagicum
Progesterone and estrogens
Estrogens
Uterine (menstrual) cycle
Secretory phase
Menstruation
Proliferative phase
Stratum functionalis
Stratum basalis
Days
1 2 3 4 5 6 7 8 9 10 11 12 13 14 15 16 17 18 19 20 21 22 23 24 25 26 27 28 1 2
Menstrual phase
Preovulatory phase
Ovulation
Postovulatory phase

Figure 28.27 Relative concentrations of anterior pituitary gland hormones and ovarian hormones during a normal female reproductive cycle (page 1004).

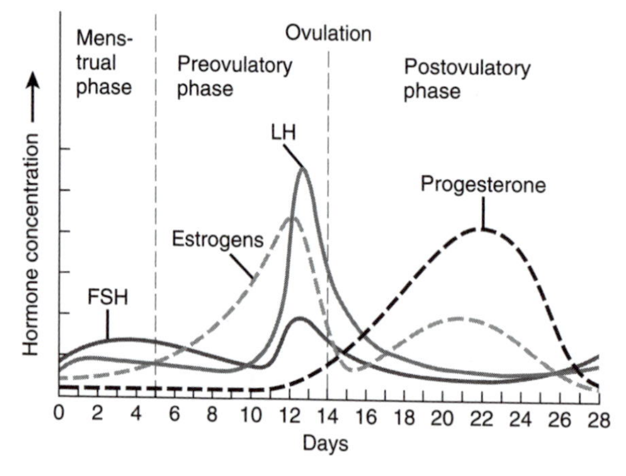

Mens-trual phase
Ovulation
Preovulatory phase
Postovulatory phase
Hormone concentration
LH
Progesterone
Estrogens
FSH
0 2 4 6 8 10 12 14 16 18 20 22 24 26 28
Days

NOTES

Figure 28.28 Role of high levels of estrogens in increasing secretion of GnRH and LH and initiating ovulation (page 1005).

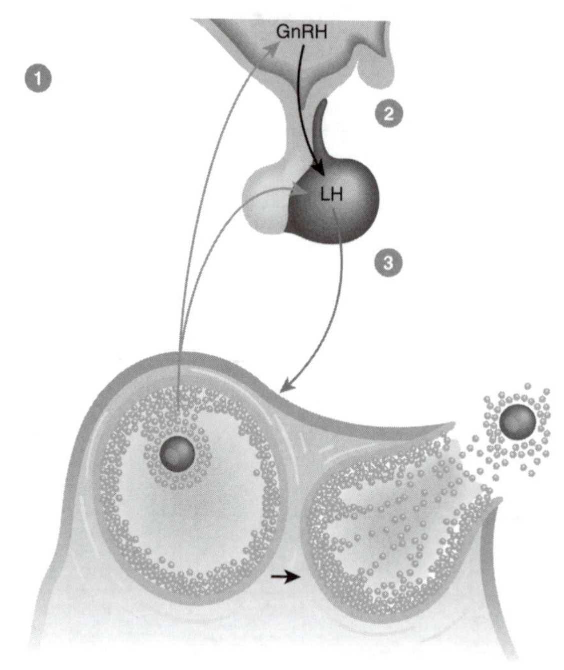
GnRH
1
2
LH
3

Figure 28.29 Summary of hormonal interactions in the ovarian and uterine cycles (page 1006).

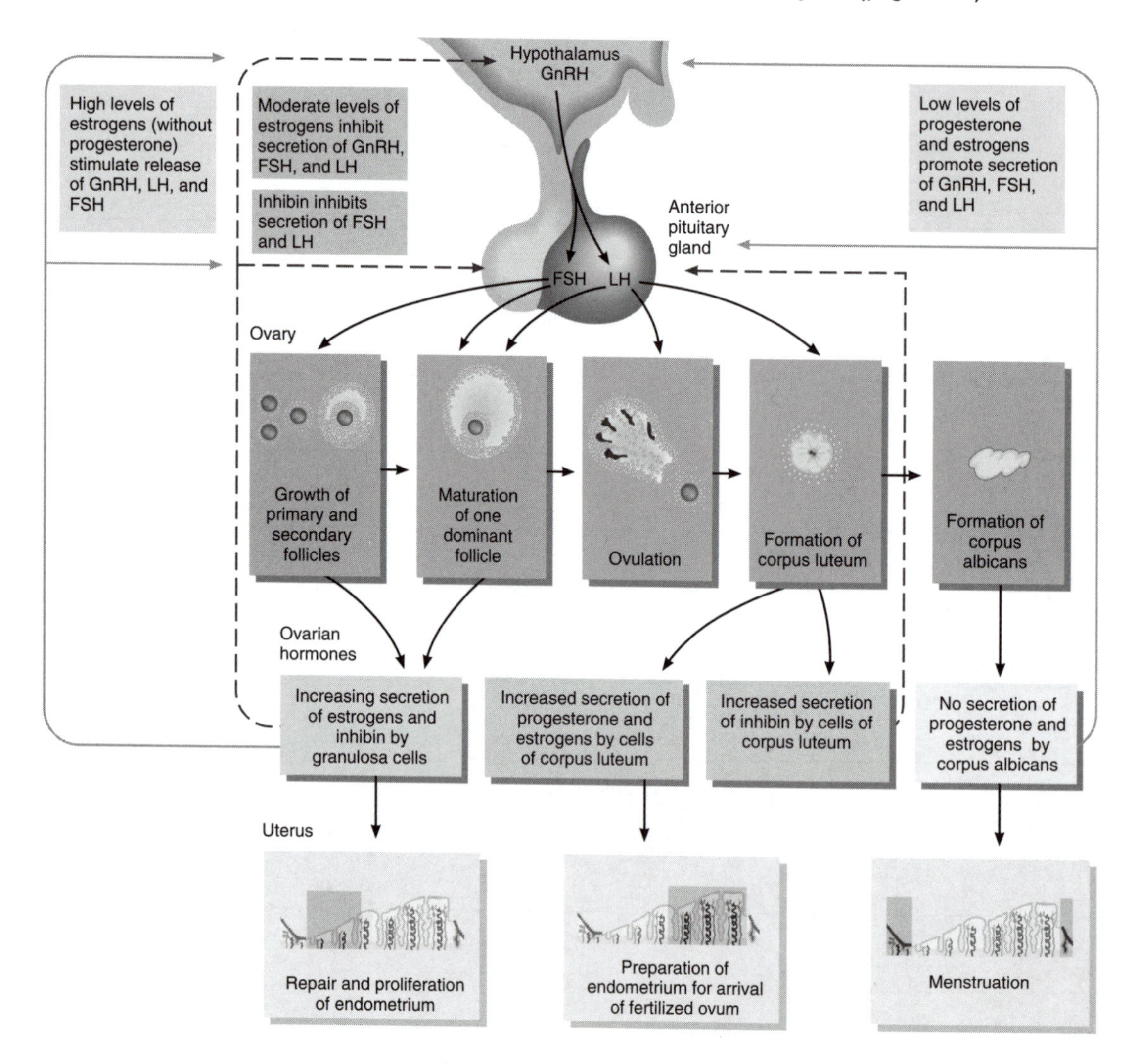
Hypothalamus
GnRH
High levels of estrogens (without progesterone) stimulate release of GnRH, LH, and FSH
Moderate levels of estrogens inhibit secretion of GnRH, FSH, and LH
Inhibin inhibits secretion of FSH and LH
Low levels of progesterone and estrogens promote secretion of GnRH, FSH, and LH
Anterior pituitary gland
FSH
LH
Ovary
Growth of primary and secondary follicles
Maturation of one dominant follicle
Ovulation
Formation of corpus luteum
Formation of corpus albicans
Ovarian hormones
Increasing secretion of estrogens and inhibin by granulosa cells
Increased secretion of progesterone and estrogens by cells of corpus luteum
Increased secretion of inhibin by cells of corpus luteum
No secretion of progesterone and estrogens by corpus albicans
Uterus
Repair and proliferation of endometrium
Preparation of endometrium for arrival of fertilized ovum
Menstruation

NOTES

28

Figure 28.30 Development of the internal reproductive systems (page 1011).

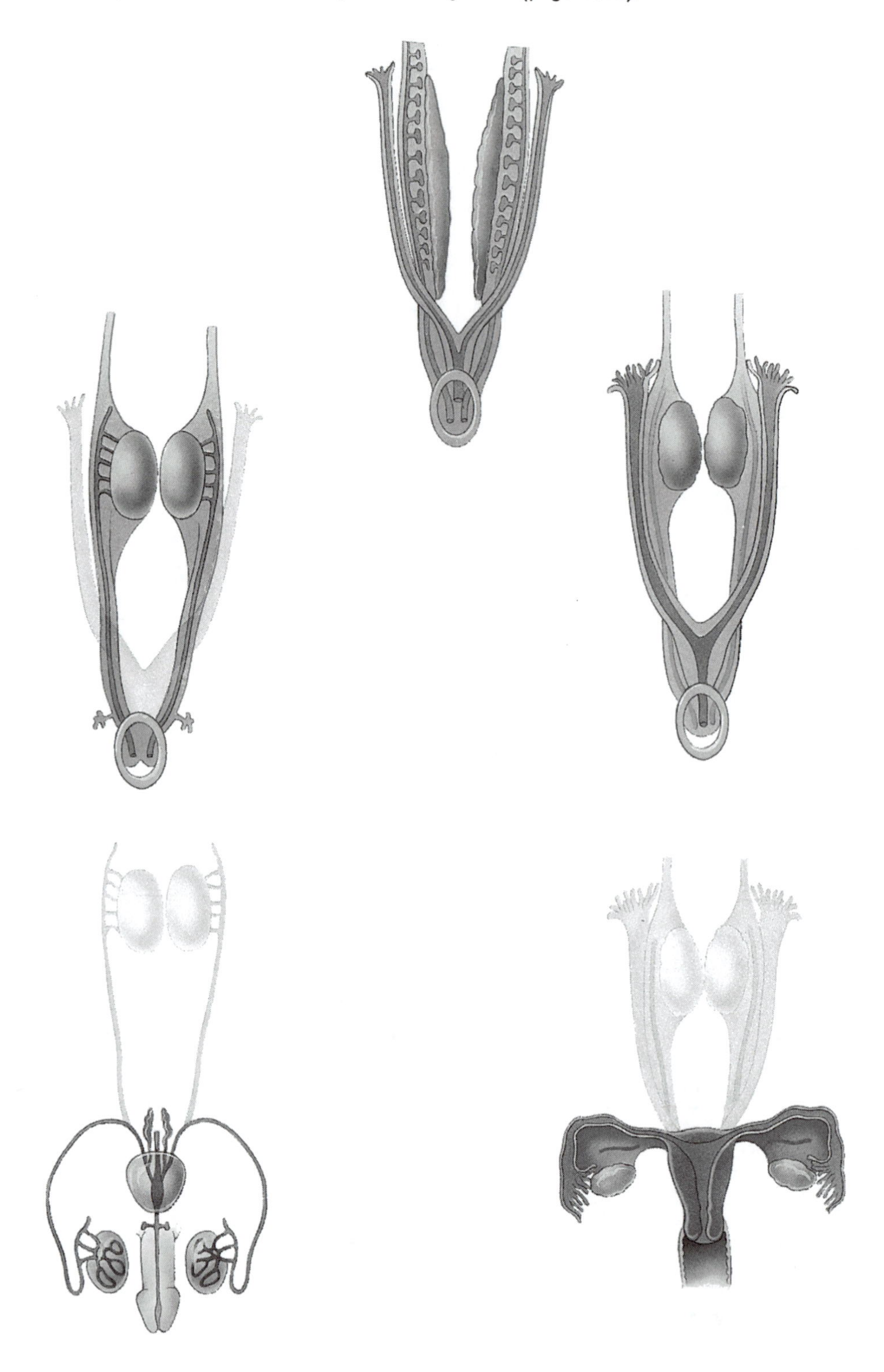

NOTES

28

Figure 28.31 Development of the external genitals (page 1013).

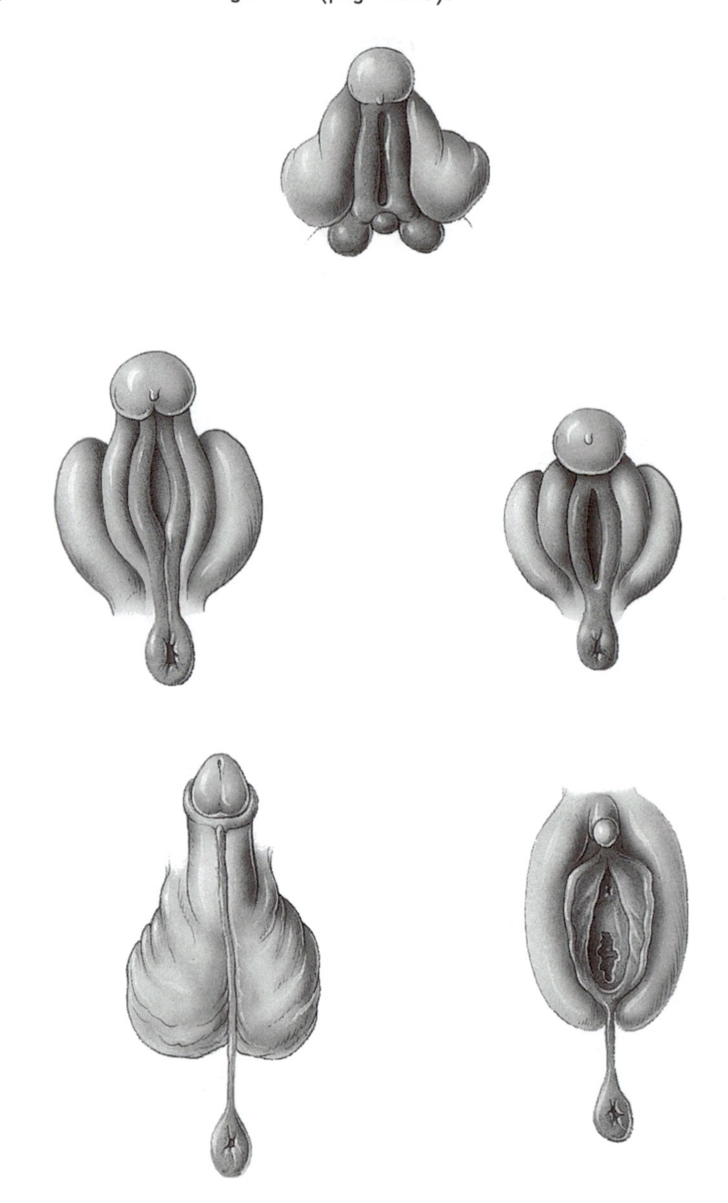

NOTES

28

Figure 29.1 Selected structures and events in fertilization (page 1023).

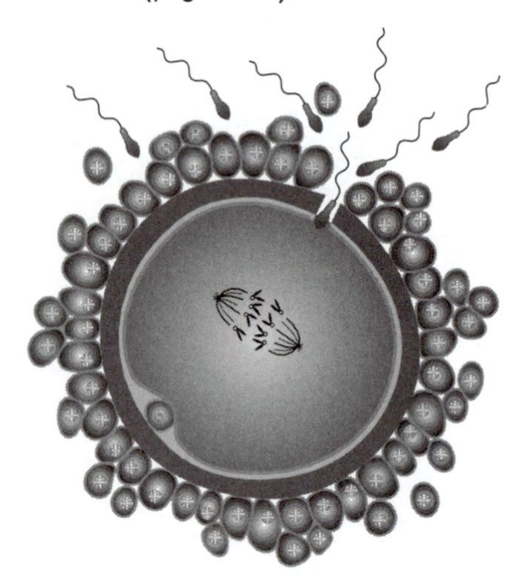

Figure 29.2 Cleavage and the formation of the morula and blastocyst (page 1024).

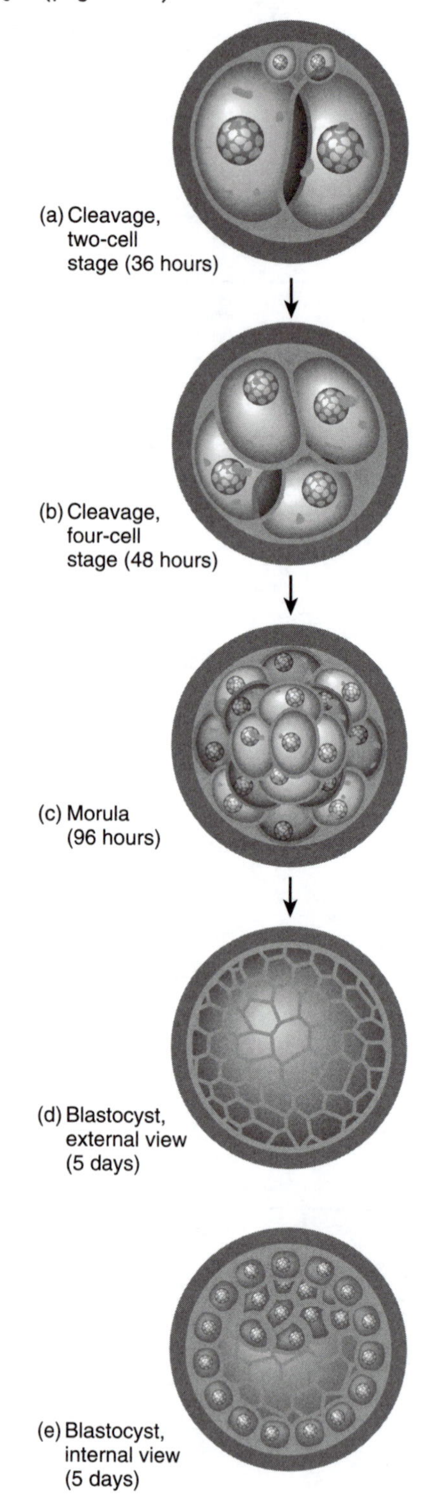
(a) Cleavage, two-cell stage (36 hours)
(b) Cleavage, four-cell stage (48 hours)
(c) Morula (96 hours)
(d) Blastocyst, external view (5 days)
(e) Blastocyst, internal view (5 days)

NOTES

Figure 29.3 Implantation of a blastocyst to the endometruim of the uterus (page 1025).

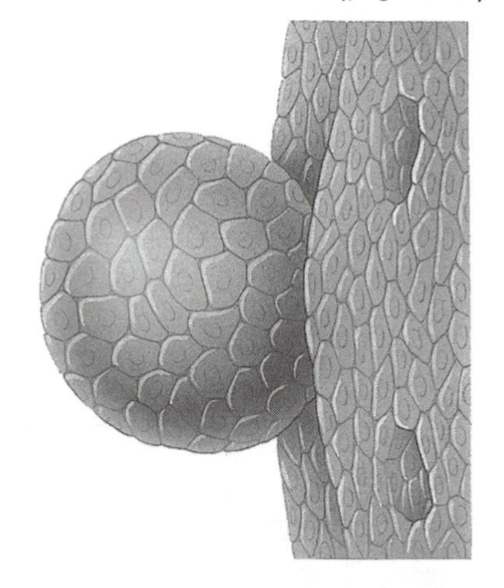

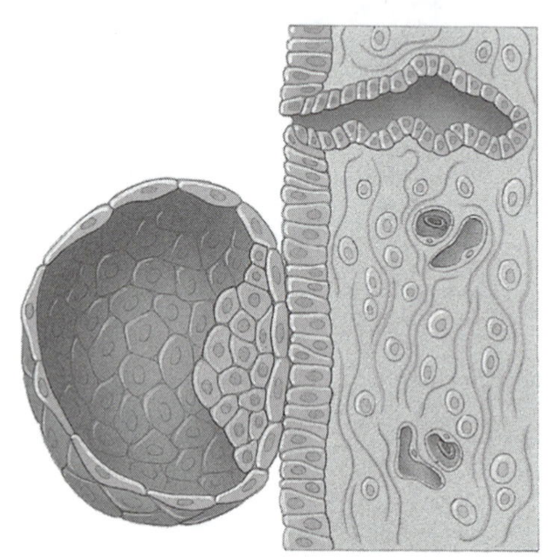

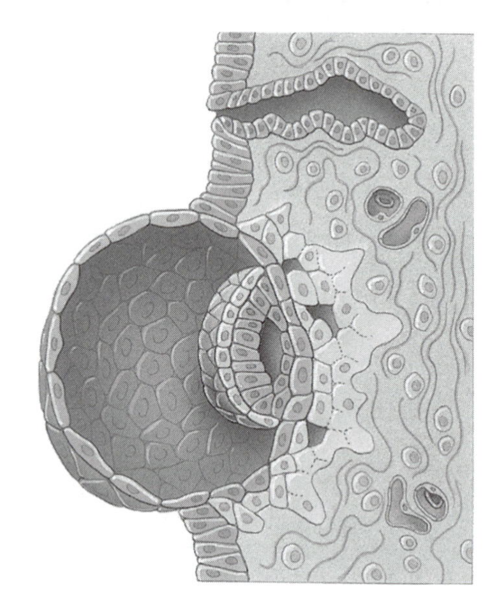

Figure 29.4 Summary of the events associated with fertilization and implantation (page 1026).

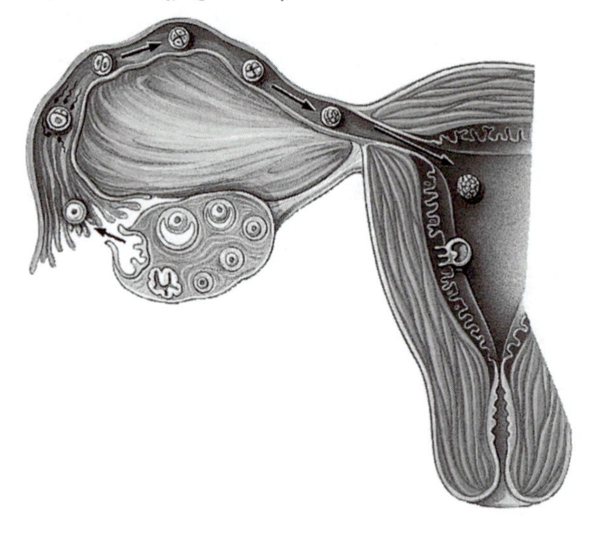

NOTES

Figure 29.5a–b Formation of the primary germ layers and associated structures (page 1028).

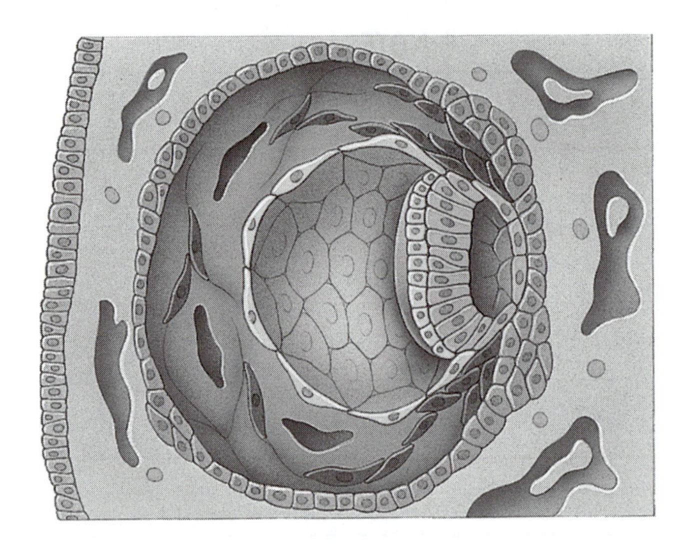

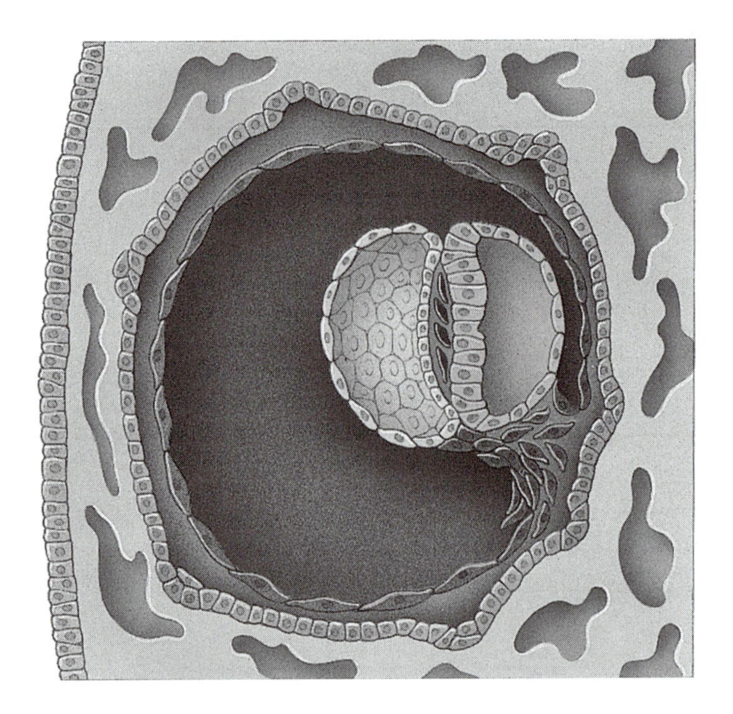

NOTES

Figure 29.5c Formation of the primary germ layers and associated structures (page 1029).

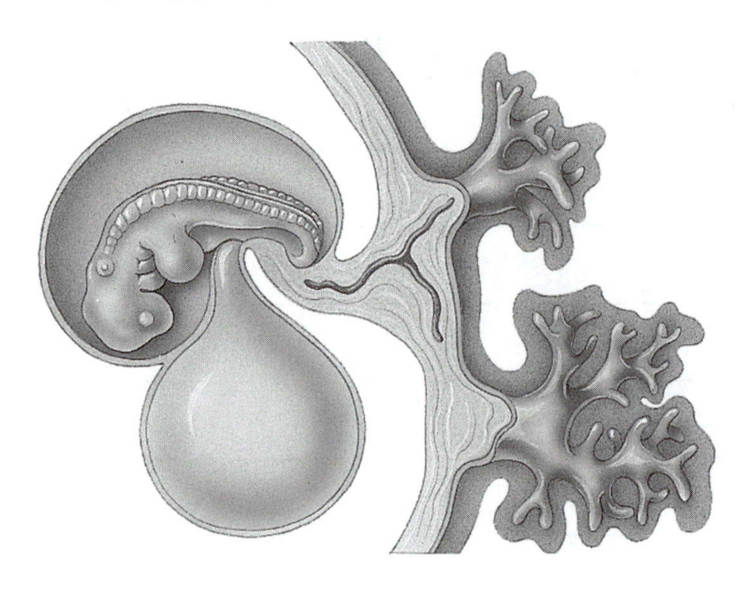

Figure 29.6a Embryonic membranes (page 1030).

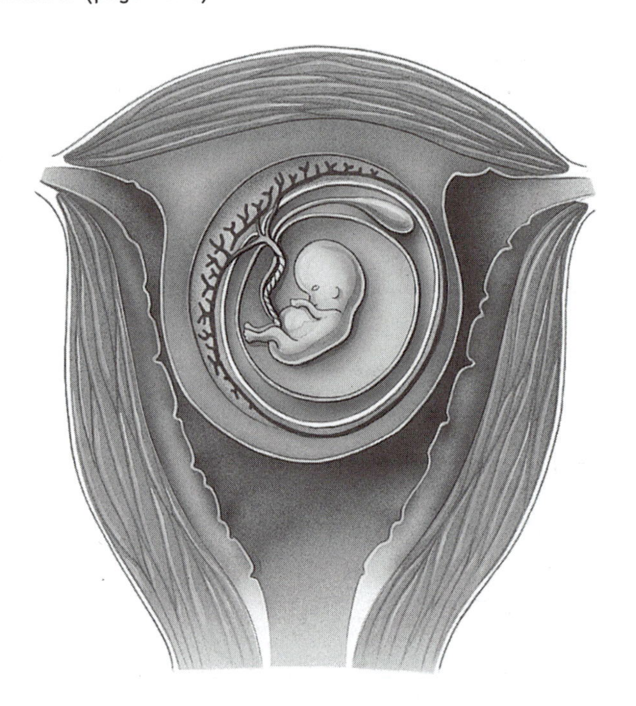

NOTES

Figure 29.7 Regions of the decidua (page 1031).

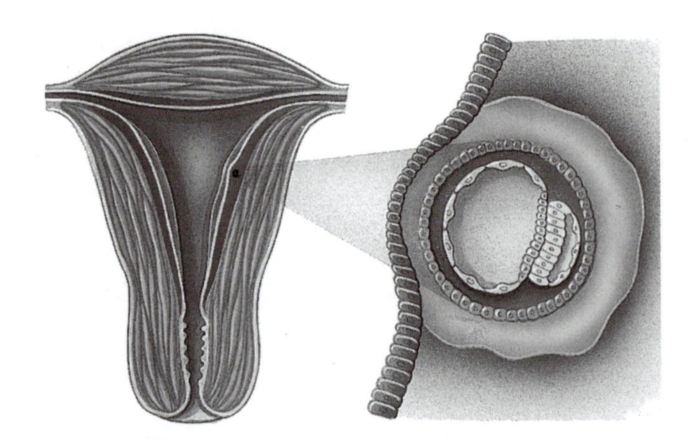

Figure 29.8a Placenta and umbilical cord (page 1032).

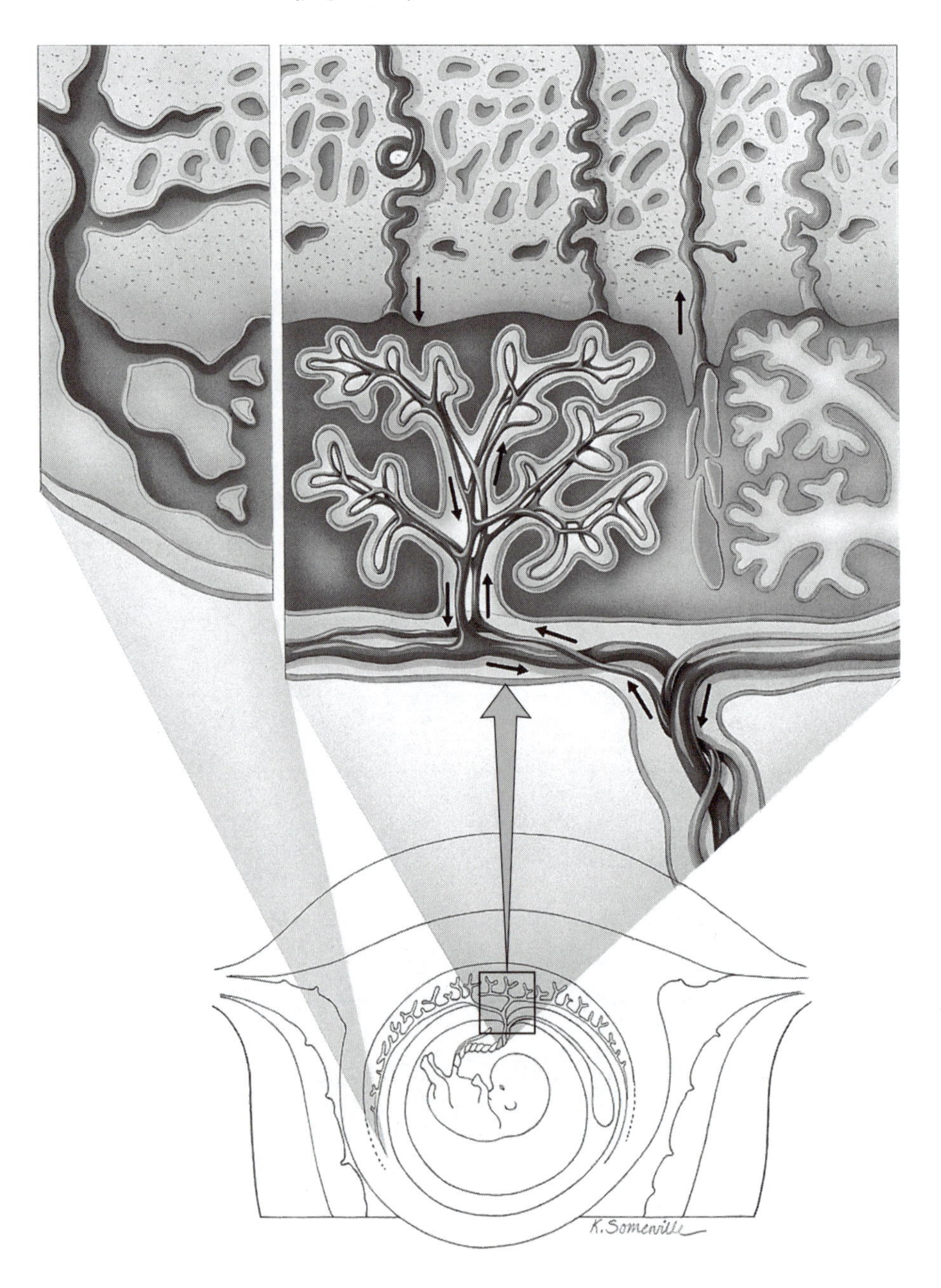
K. Somerville

NOTES

Figure 29.9 Amniocentesis and chorionic villi sampling (page 1035).

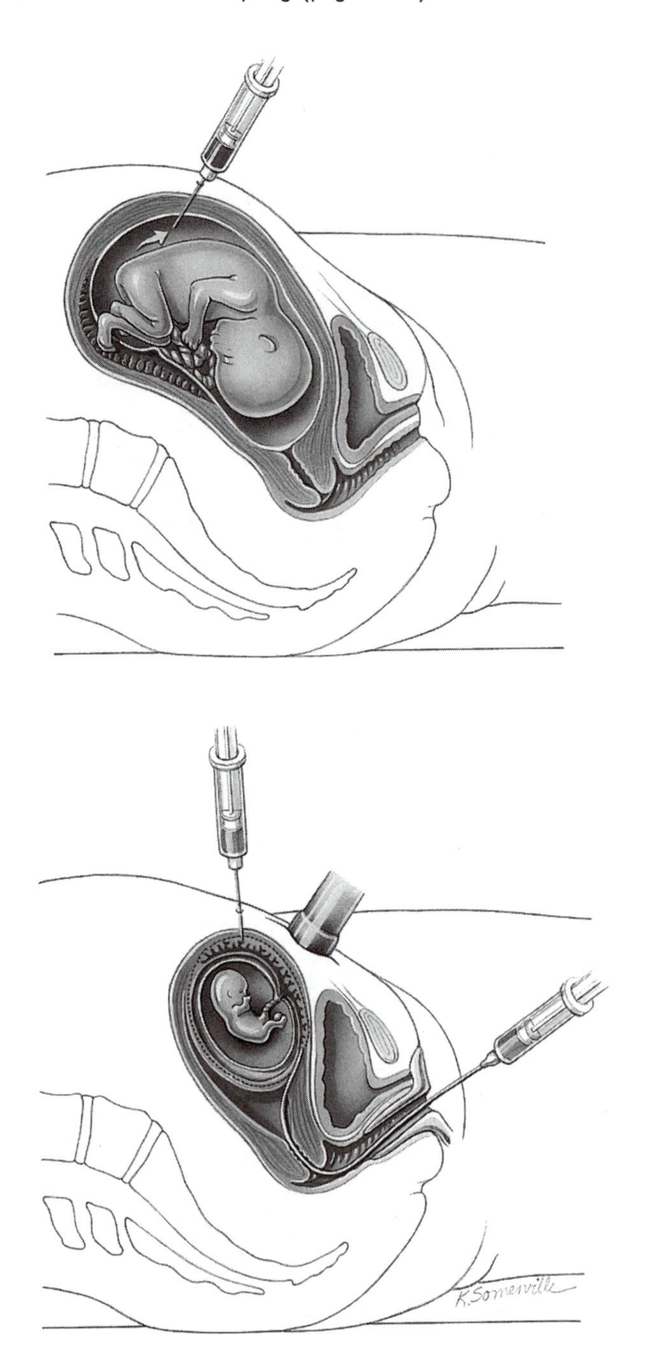
K.Somerville

NOTES

Figure 29.10 Hormones during pregnancy (page 1036).

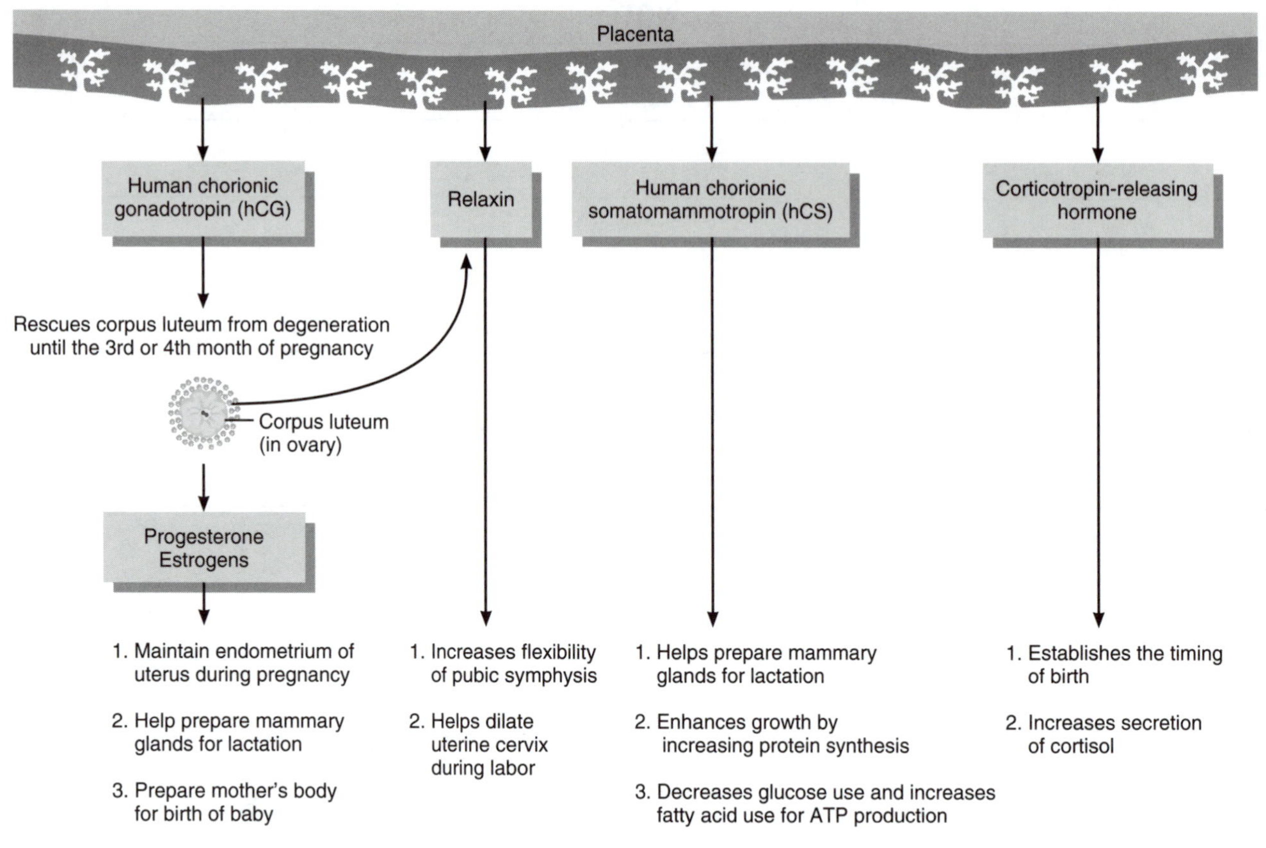

Placenta
Human chorionic gonadotropin (hCG)
Relaxin
Human chorionic somatomammotropin (hCS)
Corticotropin-releasing hormone
Rescues corpus luteum from degeneration until the 3rd or 4th month of pregnancy
Corpus luteum (in ovary)
Progesterone Estrogens
1. Maintain endometrium of uterus during pregnancy
2. Help prepare mammary glands for lactation
3. Prepare mother's body for birth of baby
1. Increases flexibility of pubic symphysis
2. Helps dilate uterine cervix during labor
1. Helps prepare mammary glands for lactation
2. Enhances growth by increasing protein synthesis
3. Decreases glucose use and increases fatty acid use for ATP production
1. Establishes the timing of birth
2. Increases secretion of cortisol

(a) Sources and functions of hormones

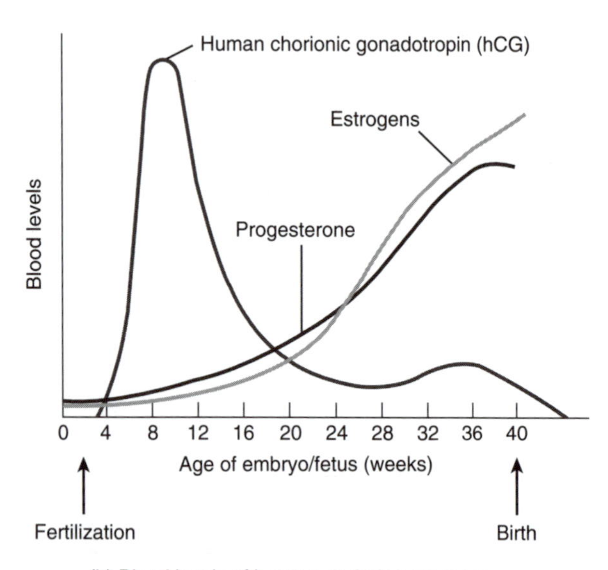

Human chorionic gonadotropin (hCG)
Estrogens
Progesterone
Blood levels
0 4 8 12 16 20 24 28 32 36 40
Age of embryo/fetus (weeks)
Fertilization
Birth

(b) Blood levels of hormones during pregnancy

NOTES

Figure 29.11 Normal fetal location and position at the end of a full-term pregnancy (page 1038).

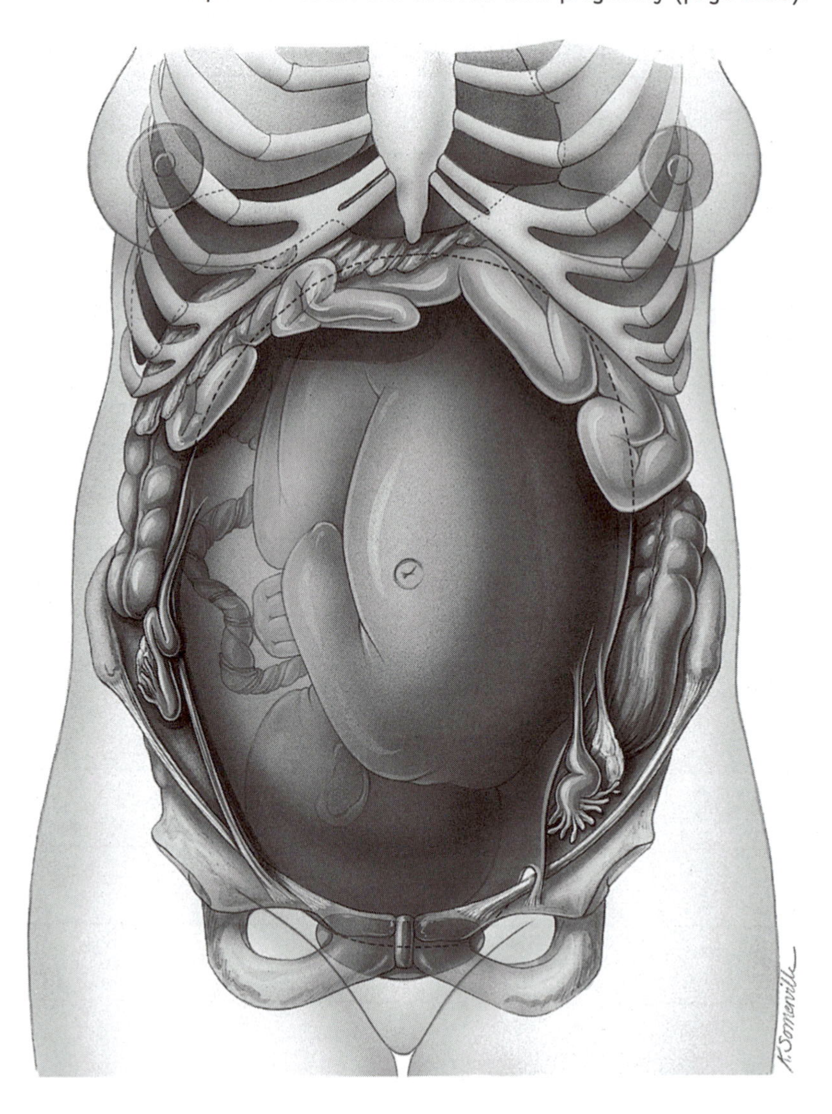
K. Somerville

NOTES

Figure 29.12 Stages of true labor (page 1040).

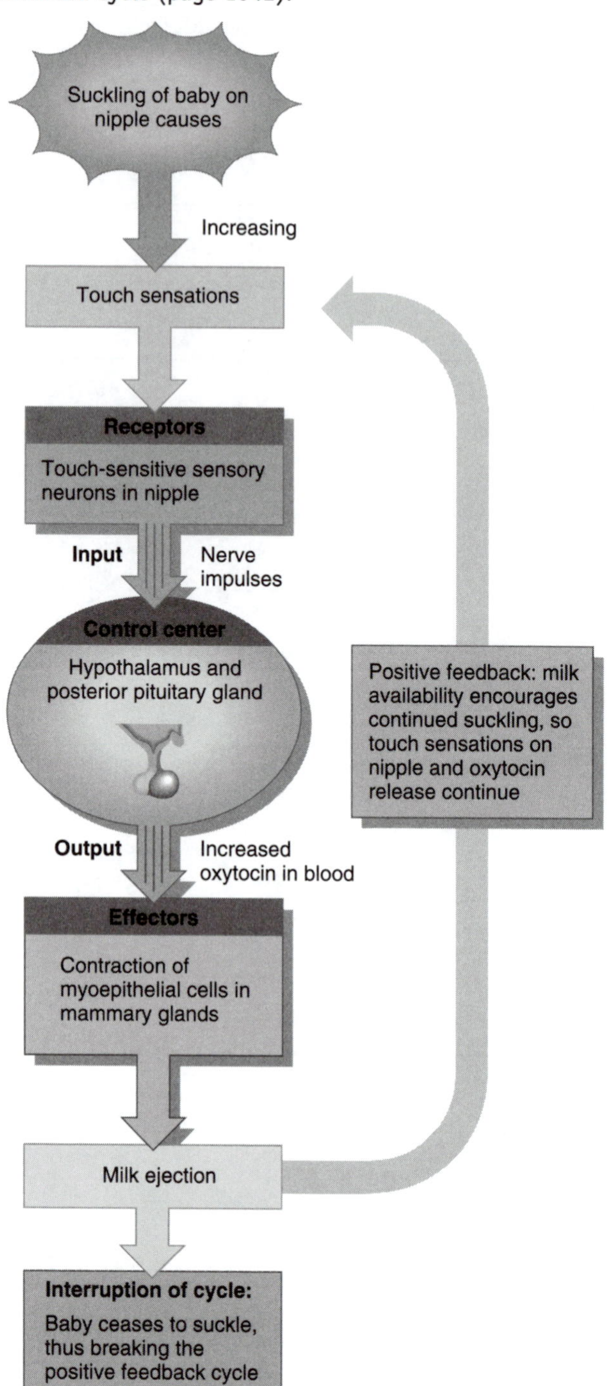

Suckling of baby on nipple causes
Increasing
Touch sensations
Receptors
Touch-sensitive sensory neurons in nipple
Input
Nerve impulses
Control center
Hypothalamus and posterior pituitary gland
Output
Increased oxytocin in blood
Effectors
Contraction of myoepithelial cells in mammary glands
Milk ejection
Positive feedback: milk availability encourages continued suckling, so touch sensations on nipple and oxytocin release continue
Interruption of cycle:
Baby ceases to suckle, thus breaking the positive feedback cycle

Figure 29.13 The milk ejection reflex, a positive feedback cycle (page 1042).

NOTES

Figure 29.14 Inheritance of phenylketonuria (page 1044).

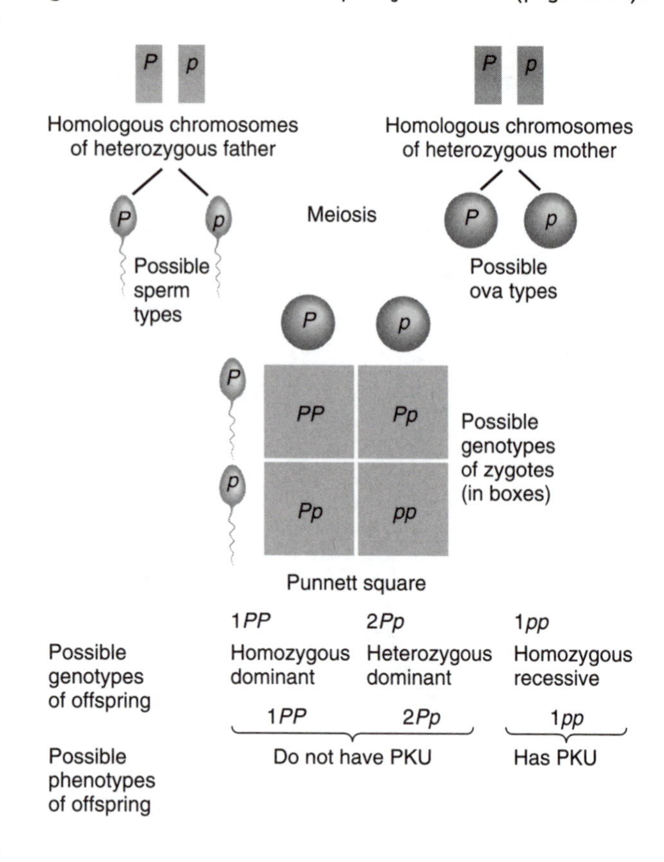

P p
Homologous chromosomes of heterozygous father
P p
Homologous chromosomes of heterozygous mother
Meiosis
Possible sperm types
Possible ova types
PP
Pp
Pp
pp
Possible genotypes of zygotes (in boxes)
Punnett square
1PP
2Pp
1pp
Possible genotypes of offspring
Homozygous dominant
Heterozygous dominant
Homozygous recessive
Possible phenotypes of offspring
Do not have PKU
Has PKU

Figure 29.15 Inheritance of sickle-cell disease (page 1046).

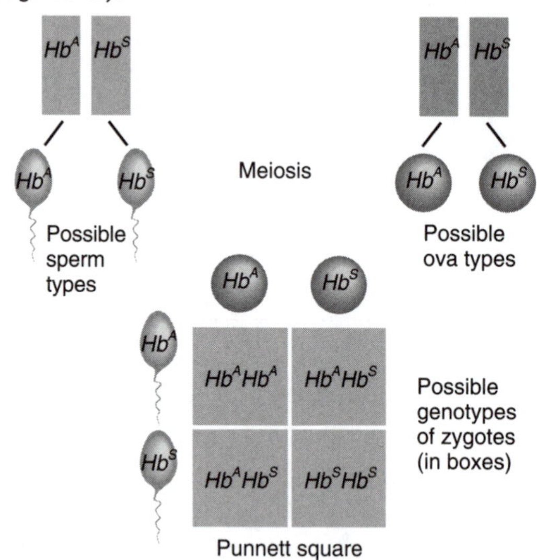

Hb^A Hb^S
Meiosis
Possible sperm types
Possible ova types
Hb^AHb^A
Hb^AHb^S
Hb^AHb^S
Hb^SHb^S
Possible genotypes of zygotes (in boxes)
Punnett square

Hb^AHb^A = normal

Hb^AHb^S = carrier of sickle-cell disease

Hb^SHb^S = has sickle-cell disease

NOTES

Figure 29.16 Ten possible combinations of parental ABO blood types (page 1046).

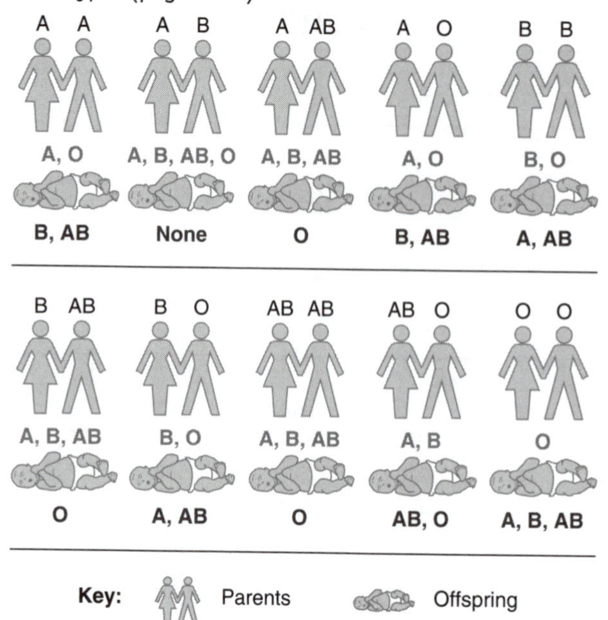

A A
A B
A AB
A O
B B
A, O
A, B, AB, O
A, B, AB
A, O
B, O
B, AB
None
O
B, AB
A, AB
B AB
B O
AB AB
AB O
O O
A, B, AB
B, O
A, B, AB
A, B
O
O
A, AB
O
AB, O
A, B, AB
Key:
Parents
Offspring

Figure 29.17 Polygenic inheritance of skin color (page 1047).

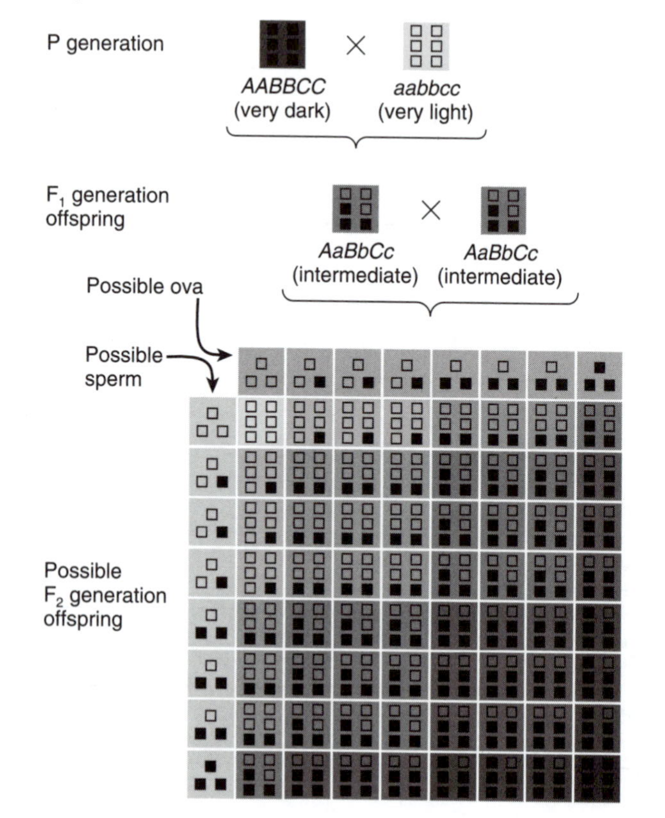

P generation
AABBCC
(very dark)
aabbcc
(very light)
F1 generation offspring
AaBbCc
(intermediate)
AaBbCc
(intermediate)
Possible ova
Possible sperm
Possible F2 generation offspring

Figure 29.18 Autosomes and sex chromosomes (page 1047).

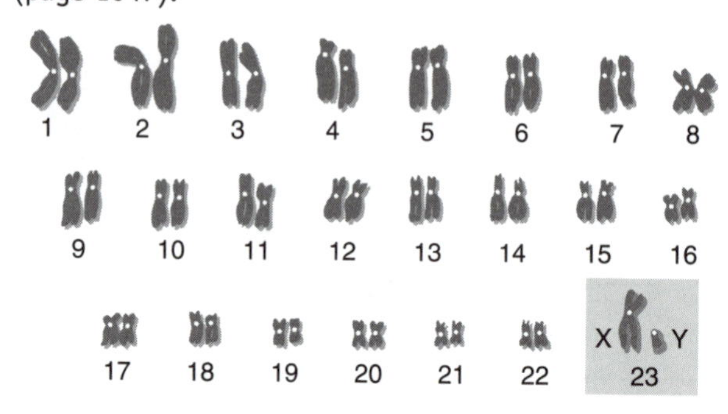

1
2
3
4
5
6
7
8
9
10
11
12
13
14
15
16
17
18
19
20
21
22
X
Y
23

NOTES

Figure 29.19 Sex determination (page 1048).

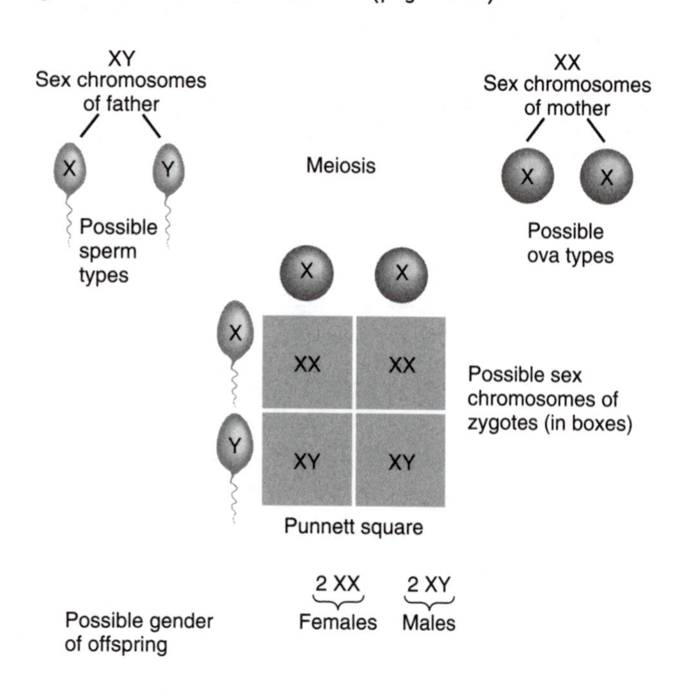

XY
Sex chromosomes of father
XX
Sex chromosomes of mother
X
Y
Meiosis
X
X
Possible sperm types
Possible ova types
X
X
X
XX
XX
Y
XY
XY
Possible sex chromosomes of zygotes (in boxes)
Punnett square
2 XX
Females
2 XY
Males
Possible gender of offspring

Figure 29.20 An example of the inheritance of red-green color blindness (page 1049).

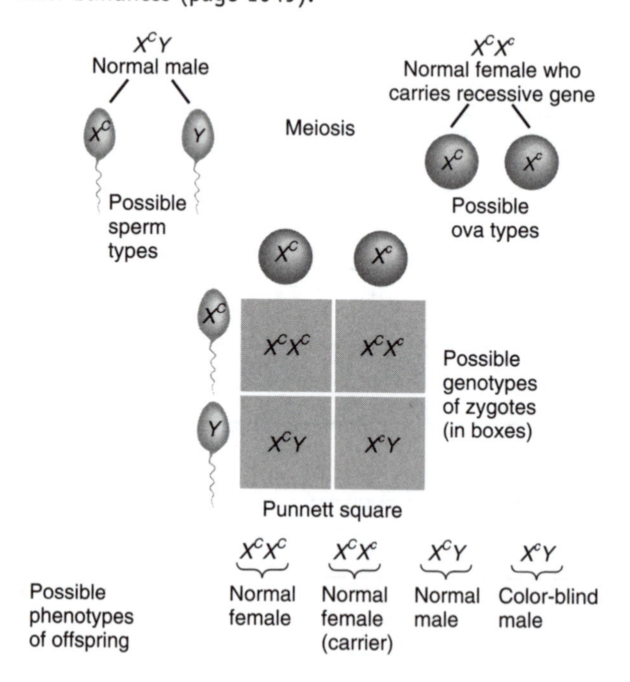

X^CY
Normal male
X^CX^c
Normal female who carries recessive gene
X^C
Y
Meiosis
X^C
X^c
Possible sperm types
Possible ova types
X^C
X^c
X^C
X^CX^C
X^CX^c
Y
X^CY
X^cY
Possible genotypes of zygotes (in boxes)
Punnett square
X^CX^C
Normal female
X^CX^c
Normal female (carrier)
X^CY
Normal male
X^cY
Color-blind male
Possible phenotypes of offspring

NOTES

CREDITS

Illustration Credits

Chapter 1: 1.1: Tomo Narashima. 1.2: Jared Schneidman Design. 1.3: Jared Schneidman Design. 1.4: Jared Schneidman Design. 1.5: Kevin Somerville. 1.6: Lynn O'Kelley. 1.8: Kevin Somerville. 1.9: Kevin Somerville. 1.10: Kevin Somerville. 1.11: Kevin Somerville. 1.12: Kevin Somerville.

Chapter 2: 2.1: Imagineering. 2.2: Jared Schneidman Design. 2.3: Jared Schneidman Design. 2.4: Imagineering. 2.5: Jared Schneidman Design. 2.6: Imagineering. 2.7: Imagineering. 2.8: Imagineering. 2.9: Imagineering. 2.10: Imagineering. 2.11: Imagineering. 2.12: Jared Schneidman Design. 2.13: Adapted from Karen Timberlake, Chemistry 6e, F8.5, p255 (Menlo Park, CA; Addison Wesley Longman, 1999). ©1999 Addison Wesley Longman, Inc. 2.14: Jared Schneidman Design. 2.15: Jared Schneidman Design. 2.16: Jared Schneidman Design. 2.17: Jared Schneidman Design. 2.18: Imagineering. 2.19: Jared Schneidman Design. 2.20: Jared Schneidman Design. 2.21: Jared Schneidman Design. 2.22: Adapted from Neil Campbell, Jane Reece, and Larry Mitchell, Biology 5e, F5.24, p75 (Menlo Park, CA; Addison Wesley Longman, 1999). ©1999 Addison Wesley Longman, Inc. 2.23: Adapted from Neil Campbell, Jane Reece, and Larry Mitchell, Biology 5e, F5.22, p74 (Menlo Park, CA; Addison Wesley Longman, 1999). ©1999 Addison Wesley Longman, Inc. 2.24: Jared Schneidman Design. 2.25: Imagineering. 2.26: Jared Schneidman Design.

Chapter 3: 3.1: Tomo Narashima. 3.2: Tomo Narashima. 3.3: Imagineering. 3.4: Imagineering. 3.5: Adapted from Bruce Alberts et al., Essential Cell Biology, F12.5, p375 and F12.12, p380 (New York: Garland Publishing Inc., 1998). ©1998 Garland Publishing Inc. 3.7: Adapted from Fundamentals of Anatomy and Physiology 4e, by Martini, Frederic H., ©1998 F3.7, p75. Reprinted by permission of Prentice-Hall, Inc., Upper Saddle River, NJ. 3.8: Jared Schneidman Design. 3.9: Imagineering. 3.10: Imagineering. 3.11: Imagineering. 3.12: Imagineering. 3.13: Imagineering. 3.14: Imagineering. 3.15: Imagineering. 3.16: Imagineering. 3.17: Tomo Narashima, Imagineering. 3.18: Tomo Narashima, Imagineering. 3.19: Tomo Narashima, Imagineering. 3.20: Tomo Narashima, Imagineering. 3.21: Tomo Narashima, Imagineering. 3.22: Tomo Narashima, Imagineering. 3.23: Tomo Narashima. 3.24: Tomo Narashima, Imagineering. 3.25: Tomo Narashima, Imagineering. 3.26: Imagineering. 3.27: Imagineering. 3.28: Imagineering. 3.29: Imagineering. 3.30: Imagineering. 3.31: Imagineering. 3.32: Imagineering. 3.33: Lauren Keswick. 3.34: Hilda Muinos.

Chapter 4: Table 4.1: Kevin Somerville, Nadine Sokol, Imagineering. Table 4.2: Kevin Somerville, Nadine Sokol. Table 4.3: Kevin Somerville, Nadine Sokol, Imagineering, Leonard Dank. Table 4.4: Kevin Somerville, Nadine Sokol. Table 4.5: Nadine Sokol. 4.1: Adapted from Lewis Kleinsmith and Valerie Kish, Principles of Cell and Molecular Biology 2e, F6.44, p237; F6.46, p238; F6.47, p239; F6.50, p241 (New York: HarperCollins, 1995). ©1995 HarperCollins College Publishers. By permission of Addison Wesley Longman. 4.2: Nadine Sokol. 4.3: Adapted from Leslie P. Gartner and James L. Hiatt, Color Textbook of Histology, F5.22, p89 (Philadelphia, PA: Saunders, 1997). ©1997 Saunders College Publishing. 4.4: Imagineering. 4.5: Imagineering.

Chapter 5: 5.1: Kevin Somerville. 5.2: Adapted from Ira Telford and Charles Bridgman, Introduction to Functional Histology 2e, p84, p261, p262 (New York: HarperCollins, 1995). ©1995 HarperCollins College Publishers. By permission of Addison Wesley Longman. 5.3: Imagineering. 5.4: Kevin Somerville. 5.5: Kevin Somerville. 5.6: Imagineering. 5.7: Lauren Keswick.

Chapter 6: 6.1: Leonard Dank. 6.2: Lauren Keswick. 6.3: Leonard Dank. 6.4: Kevin Somerville. 6.5: Kevin Somerville. 6.6: Leonard Dank. 6.8: Kevin Somerville. 6.9: Leonard Dank. 6.10: From Priscilla LeMone and Karen M. Burke, Medical-Surgical Nursing, p1560 (Menlo Park, CA: Benjamin/Cummings, 1996). ©1996 The Benjamin/Cummings Publishing Company. 6.11: Jared Schneidman Design. 6.12: Leonard Dank.

Chapter 7: 7.1: Leonard Dank. 7.2: Leonard Dank. 7.3: Leonard Dank. 7.4: Leonard Dank. 7.5: Leonard Dank. 7.6: Leonard Dank. 7.7: Leonard Dank. 7.8: Leonard Dank. 7.9: Leonard Dank. 7.10: Leonard Dank. 7.11: Leonard Dank. 7.12: Leonard Dank. 7.13: Leonard Dank. 7.14: Leonard Dank. 7.15: Leonard Dank. 7.16: Leonard Dank. 7.17: Leonard Dank. 7.18: Leonard Dank. 7.19: Leonard Dank. 7.20: Leonard Dank. 7.21: Leonard Dank. 7.22: Leonard Dank. 7.23: Leonard Dank. 7.24: Leonard Dank.

Chapter 8: 8.1: Leonard Dank. 8.2: Leonard Dank. 8.3: Leonard Dank. 8.4: Leonard Dank. 8.5: Leonard Dank. 8.6: Leonard Dank. 8.7: Leonard Dank. 8.8: Leonard Dank. 8.9: Leonard Dank. 8.10: Leonard Dank. 8.11a Leonard Dank. 8.12: Leonard Dank. 8.13: Leonard Dank. 8.14: Leonard Dank. 8.15: Leonard Dank. 8.16: Leonard Dank. 8.17: Leonard Dank.

Chapter 9: 9.1: Leonard Dank. 9.2: Leonard Dank. 9.3: Leonard Dank. 9.4: Leonard Dank. 9.11: Leonard Dank. 9.12: Leonard Dank. 9.13: Leonard Dank. 9.14: Leonard Dank.

Chapter 10: 10.1: Kevin Somerville. 10.3: Kevin Somerville. Adapted from Martini, Frederic H., Fundamentals of Anatomy and Physiology 4e, F10.2, p280 (Upper Saddle River, NJ: Prentice-Hall/Pearson Education, 1998). ©1998 Prentice-Hall. 10.4: Imagineering. 10.6: Imagineering. 10.7: Hilda Muinos. 10.8: Imagineering. 10.9: Imagineering. 10.10: Imagineering. 10.11: Hilda Muinos. 10.12: Imagineering. 10.13: Jared Schneidman Design. 10.14: Jared Schneidman Design. 10.15: Jared Schneidman Design. 10.16: Imagineering. 10.18: Imagineering. 10.19: Beth Willert. 10.20: Beth Willert.

Chapter 11: 11.1: Leonard Dank. 11.2: Leonard Dank. 11.3: Leonard Dank. 11.4: Leonard Dank. 11.5: Leonard Dank. 11.6: Leonard Dank. 11.7: Leonard Dank. 11.8: Leonard Dank. 11.9: Leonard Dank. 11.10: Leonard Dank. 11.11: Leonard Dank. 11.12: Leonard Dank. 11.13: Leonard Dank. 11.14: Leonard Dank. 11.15: Leonard Dank. 11.16: Leonard Dank. 11.17: Leonard Dank. 11.18: Leonard Dank. 11.19: Leonard Dank. 11.20: Leonard Dank. 11.21: Leonard Dank. 11.22: Leonard Dank. 11.23: Leonard Dank.

Chapter 12: 12.1: Kevin Somerville, Imagineering. 12.2: Imagineering. 12.3: Sharon Ellis. 12.4: Nadine Sokol. 12.5: Nadine Sokol. 12.7: Sharon Ellis. 12.8: Imagineering. 12.9: Imagineering. 12.10: Imagineering. 12.11: Imagineering. 12.12: Imagineering. Adapted from Becker et al., The World of the Cell 3e, F22.18, p732 (Menlo Park, CA: Benjamin/Cummings, 1996) ©1996 The Benjamin/Cummings Publishing Company. 12.13: Jared Schneidman Design. 12.14: From Becker et al., The World of the Cell 3e, F22.28, p741 (Menlo Park, CA: Benjamin/Cummings, 1996) ©1996 The Benjamin/Cummings Publishing Company. 12.15: Jared Schneidman Design. 12.16: Nadine Sokol. 12.17: Nadine Sokol.

Chapter 13: 13.1: Sharon Ellis. 13.2: Sharon Ellis. 13.3: Sharon Ellis. 13.4: Sharon Ellis. 13.5: Sharon Ellis. 13.6: Leonard Dank. 13.7: Leonard Dank. 13.8: Leonard Dank. 13.9: Leonard Dank. 13.10: Sharon Ellis. 13.11: Sharon Ellis. 13.12: Imagineering. 13.13: Precision Graphics. 13.14: Precision Graphics. 13.15: Precision Graphics, Imagineering. 13.16: Imagineering. 13.17: Imagineering.

Chapter 14: 14.1: Sharon Ellis. 14.2: Sharon Ellis. 14.3: Sharon Ellis. 14.4: Sharon Ellis, Imagineering. 14.5: Sharon Ellis. 14.6: Sharon Ellis. 14.7: Sharon Ellis. 14.8: Sharon Ellis. 14.9: Sharon Ellis. 14.10: Sharon Ellis. 14.11: Sharon Ellis. 14.13: Sharon Ellis. 14.14: Sharon Ellis. 14.15: Sharon Ellis. 14.16: Hilda Muinos. 14.18: Kevin Somerville. 14.19: Kevin Somerville.

Chapter 15: 15.1: Imagineering. 15.2: Kevin Somerville. 15.3: Kevin Somerville. 15.4: Beth Willert. 15.5: Sharon Ellis. 15.6: Lynn O'Kelley. 15.7: Sharon Ellis. 15.8: Jared Schneidman Design. 15.9: Sharon Ellis. 15.10: Sharon Ellis. 15.11: Adapted from Purves et al., Neuroscience 2e, F26.1 and F26.2, p498 (Sunderland, MA: Sinauer Associates, 1997). ©1997 Sinauer Associates.

Chapter 16: 16.1: Tomo Narashima. 16.2: Lynn O'Kelley. 16.4: Sharon Ellis. 16.5: Tomo Narashima. 16.8: Lynn O'Kelley. 16.9: Tomo Narashima. 16.10: Jared Schneidman Design. 16.11: Nadine Sokol, Imagineering. 16.12: Lynn O'Kelley. 16.13: Jared Schneidman Design. 16.14: Lynn O'Kelley. 16.15: Adapted from Seeley et al., Anatomy and Physiology 4e, F15.22, p480 (New York: WCB McGraw-Hill, 1998) ©1998 The McGraw-Hill Companies. 16.16: Tomo Narashima. 16.17: Tomo Narashima. 16.18: Tomo Narashima. 16.19: Tomo Narashima. 16.20: Tomo Narashima. 16.21: Tomo Narashima, Sharon Ellis. 16.22: Tomo Narashima, Sharon Ellis.

Chapter 17: 17.1: Jared Schneidman Design. 17.2: Hilda Muinos. 17.3: Kevin Somerville. 17.4: Sharon Ellis. 17.5: Imagineering.

Chapter 18: 18.1: Lynn O'Kelley. 18.2: Jared Schneidman Design. 18.3: Jared Schneidman Design. 18.4: Jared Schneidman Design. 18.5: Lynn O'Kelley. 18.6: Imagineering. 18.7: Jared Schneidman Design. 18.8: Lynn O'Kelley. 18.9: Jared Schneidman Design. 18.10: Lynn O'Kelley. 18.11: Jared Schneidman Design. 18.12: Jared Schneidman Design. 18.13: Lynn O'Kelley. 18.14: Jared Schneidman Design. 18.15: Lynn O'Kelley. 18.16: Nadine Sokol. 18.17: Jared Schneidman Design. 18.18: Lynn O'Kelley. 18.19: Jared Schneidman Design. 18.20: Hilda Muinos. 18.21: Nadine Sokol. 18.22: Lynn O'Kelley.

Chapter 19: 19.1: Hilda Muinos. 19.3: Nadine Sokol. 19.4: Nadine Sokol. 19.5: Jared Schneidman Design. 19.6: Jared Schneidman Design. 19.8: Jared Schneidman Design. 19.9: Nadine Sokol. 19.11: Imagineering. 19.12: Jean Jackson. 19.13: Nadine Sokol.

Chapter 20: 20.1: Kevin Somerville. 20.2: Kevin Somerville. 20.3a, c: Hilda Muinos. 20.4: Hilda Muinos, Kevin Somerville. 20.5: Nadine Sokol. 20.6: Kevin Somerville. 20.7: Nadine Sokol, Hilda Muinos. 20.8: Hilda Muinos. 20.9: Kevin Somerville. 20.10: Kevin Somerville. 20.11: Burmar Technical Corp. 20.12: Burmar Technical Corp. 20.13: Hilda Muinos. 20.14: Kevin Somerville. 20.15: Hilda Muinos. 20.17: Hilda Muinos. 20.19: Hilda Muinos.

Chapter 21: Exhibit 21.1: Keith Ciociola. Exhibit 21.2: Keith Ciociola. Exhibit 21.3: Keith Ciociola. Exhibit 21.4: Keith Ciociola. Exhibit 21.5: Keith Ciociola. Exhibit 21.6: Keith Ciociola. Exhibit 21.7: Keith Ciociola. Exhibit 21.8: Keith Ciociola. Exhibit 21.9: Keith Ciociola. Exhibit 21.10: Keith Ciociola. Table 21.2: Imagineering. 21.1: Hilda Muinos. 21.2: Hilda Muinos. 21.3: Nadine Sokol, Imagineering. 21.4: Hilda Muinos. 21.6: Jared Schneidman Design. 21.7: Jared Schneidman Design. 21.8: Jared Schneidman Design. 21.9: Imagineering. 21.10: Imagineering. 21.11: Jared Schneidman Design. 21.12: Jared Schneidman Design. 21.13: Imagineering. 21.14: Jared Schneidman Design. 21.15: Jared Schneidman Design. 21.16: Jared Schneidman Design. 21.17: Hilda Muinos. 21.18: Hilda Muinos. 21.19: Hilda Muinos. 21.20: Kevin Somerville, Hilda Muinos. 21.21: Kevin Somerville. 21.22: Hilda Muinos. 21.23: Hilda Muinos. 21.24: Kevin Somerville. 21.25: Kevin Somerville. 21.26: Kevin Somerville. 21.27: Kevin Somerville. 21.28: Kevin Somerville. 21.29: Kevin Somerville, Nadine Sokol. 21.30: Hilda Muinos. 21.31: Hilda Muinos, Keith Ciociola. 21.32: Hilda Muinos.

Chapter 22: 22.1: Sharon Ellis. 22.2: Sharon Ellis. 22.3: Sharon Ellis. 22.4: Nadine Sokol. 22.5: Steve Oh. 22.6: Sharon Ellis. 22.7: Steve Oh. 22.8: Sharon Ellis. 22.9: Nadine Sokol. 22.10: Nadine Sokol. 22.11: Jared Schneidman

Design. 22.12: Jared Schneidman Design. 22.13: Jared Schneidman Design. 22.14: Jared Schneidman Design. 22.15: Jared Schneidman Design. 22.16: Jared Schneidman Design. 22.17: Jared Schneidman Design. 22.18: Jared Schneidman Design. 22.19: Jared Schneidman Design. 22.20: Jared Schneidman Design. 22.21: Nadine Sokol, Imagineering.

Chapter 23: 23.1: Lynn O'Kelley. 23.2: Kevin Somerville, Lynn O'Kelley. 23.4: Lynn O'Kelley. 23.5: Lynn O'Kelley. 23.6: Steve Oh. 23.8: Lynn O'Kelley. 23.10: Lynn O'Kelley. 23.11: Kevin Somerville. 23.12: Kevin Somerville. 23.13: Jared Schneidman Design. 23.14: Kevin Somerville. 23.15: Nadine Sokol. 23.16: Jared Schneidman Design. 23.17: Jared Schneidman Design. 23.18: Jared Schneidman Design. 23.19: Jared Schneidman Design. 23.20: Jared Schneidman Design. 23.21: Jared Schneidman Design. 23.22: Jared Schneidman Design. 23.23: Jared Schneidman Design. 23.24: Jared Schneidman Design. 23.25: Hilda Muinos. 23.26: Jared Schneidman Design. 23.27: Jared Schneidman Design. 23.28: Jared Schneidman Design. 23.29: Jared Schneidman Design.

Chapter 24: 24.1: Nadine Sokol. 24.2: Steve Oh. 24.3: Nadine Sokol. 24.4: Nadine Sokol. 24.5: Nadine Sokol. 24.6: Nadine Sokol. 24.7: Nadine Sokol. 24.8: Nadine Sokol. 24.10: Nadine Sokol. 24.11: Nadine Sokol. 24.12: Hilda Muinos. 24.13: Jared Schneidman Design. 24.14: Jared Schneidman Design. 24.15: Jared Schneidman Design. 24.16: Nadine Sokol. 24.17: Jared Schneidman Design. 24.18: Nadine Sokol. 24.19: Jared Schneidman Design. 24.20: Jared Schneidman Design. 24.21: Kevin Somerville. 24.22: Hilda Muinos. 24.24: Jared Schneidman Design. 24.25: Jared Schneidman Design. 24.26: Nadine Sokol. 24.27: Hilda Muinos. 24.28: Nadine Sokol.

Chapter 25: 25.1: Imagineering. 25.2: Imagineering. 25.3: Imagineering. 25.4: Imagineering. 25.5: Imagineering. 25.6: Imagineering. 25.7: Imagineering. 25.8: Imagineering. 25.9: Imagineering. 25.10: Imagineering. 25.11: Imagineering. 25.12: Imagineering. 25.13: Imagineering. 25.14: Imagineering. 25.15: Imagineering. 25.16: Imagineering. 25.17: Jared Schneidman Design. 25.18: Imagineering.

Chapter 26: 26.1: Kevin Somerville. 26.2: Kevin Somerville. 26.3: Steve Oh. 26.4: Nadine Sokol. 26.5: Imagineering. 26.6: Kevin Somerville. 26.7: Nadine Sokol. 26.8: Kevin Somerville. 26.9: Nadine Sokol. 26.10: Jared Schneidman Design. 26.11: Imagineering. 26.12: Jared Schneidman Design. 26.13: Jared Schneidman Design. 26.14: Imagineering. 26.15: Jared Schneidman Design. 26.16: Imagineering. 26.17: Imagineering. 26.18: Jared Schneidman Design. 26.19: Jared Schneidman Design. 26.20: Jared Schneidman Design. 26.21: Jared Schneidman Design. 26.22: Nadine Sokol. 26.23: Nadine Sokol.

Chapter 27: 27.1: Jared Schneidman Design. 27.2: Jared Schneidman Design. 27.3: Jared Schneidman Design. 27.4: Imagineering. 27.5: Imagineering. 27.6: Imagineering. 27.7: Jared Schneidman Design.

Chapter 28: 28.1: Lauren Keswick. 28.2: Imagineering. 28.3: Kevin Somerville. 28.4: Kevin Somerville. 28.5: Kevin Somerville. 28.6: Kevin Somerville. 28.7: Jared Schneidman Design. 28.8: Kevin Somerville. 28.9: Jared Schneidman Design. 28.10: Jared Schneidman Design. 28.11: Kevin Somerville. 28.12: Kevin Somerville. 28.13: Kevin Somerville. 28.14: Kevin Somerville. 28.15: Kevin Somerville. 28.17: Jared Schneidman Design. 28.18: Kevin Somerville. 28.21: Kevin Somerville. 28.22: Kevin Somerville. 28.23: Kevin Somerville. 28.24: Kevin Somerville. 28.25: Jared Schneidman Design. 28.26: Jared Schneidman Design. 28.27: Jared Schneidman Design. 28.28: Jared Schneidman Design. 28.29: Jared Schneidman Design. 28.30: Kevin Somerville. 28.31: Kevin Somerville.

Chapter 29: 29.1: Nadine Sokol. 29.2: Jared Schneidman Design. 29.3: Kevin Somerville. 29.4: Kevin Somerville. 29.5: Kevin Somerville. 29.06: Kevin Somerville. 29.07: Kevin Somerville. 29.08: Kevin Somerville. 29.09: Kevin Somerville. 29.10: Jared Schneidman Design. 29.11: Kevin Somerville. 29.12: Kevin Somerville. 29.13: Imagineering. 29.14: Jared Schneidman Design. 29.15: Jared Schneidman Design. 29.16: Jared Schneidman Design. 29.17: Jared Schneidman Design. 29.18: Jared Schneidman Design. 29.19: Jared Schneidman Design. 29.20: Jared Schneidman Design.